Calmodulin and Signal Transduction

Calmodulin and Signal Transduction

Edited by

Linda J. Van Eldik
Northwestern University Medical School, Chicago, Illinois

and

D. Martin Watterson
Northwestern University Medical School, Chicago, Illinois

ACADEMIC PRESS
San Diego London Boston New York Sydney Tokyo Toronto

Front cover photograph: Three-dimensional structure of a calcium-saturated calmodulin (CaM) complexed with the CaM recognition peptide from smooth muscle myosin light chain kinase. The N-terminal helix of CaM is shown in blue, the C-terminal helix is red, the peptide is magenta, and the Ca^{2+} ions are gray balls. Courtesy of S. Weigand, L. Shuvalova, T. J. Lukas, S. Mirzoeva, D. M. Watterson, and W. F. Anderson, Northwestern University Medical School, Chicago, Illinois.

This book is printed on acid-free paper. ∞

Copyright © 1998 by ACADEMIC PRESS

All Rights Reserved.
No part of this publication may be reproduced or transmitted in any form pr by any means, electronic or mechanical, including photocopy, recording, or any information storage and retrieval system, without permission in writing from the publisher.

Academic Press
15 East 26th St. 15th Floor, New York, New York 10010, USA
http://www.academicpress.com

Academic Press Limited
24-28 Oval Road, London NW1 7DX, UK
http://www.hbuk.co.uk/ap/

Library of Congress Cataloging-in-Publication Data

Calmodulin and signal transduction / edited by Linda J. Van Eldik and
 D. Martin Watterson.
   p. cm.
   Includes bibliographical references and index.
   ISBN 0-12-713860-9 (hc. : alk. paper)
   1. Calmodulin. 2. Cellular signal transduction. 3. Calcium-
-Physiological effect. I. Van Eldik, Linda J. II. Watterson, D.
Martin
   QP552.C28C344   1998
   572'.516—DC21   98-4732
                                                                      CIP

PRINTED IN THE UNITED STATES OF AMERICA
98 99 00 01 02 03 MM 9 8 7 6 5 4 3 2 1

Contents

Contributors vii
Preface ix

1 Calmodulin and Calcium Signal Transduction: An Introduction
Linda J. Van Eldik and D. Martin Watterson

I. Background 1
II. Phylogenetic Conservation of Calmodulin 3
III. Molecular Mechanisms of Calmodulin-Mediated Ca^{2+} Signal Transduction 4
References 14

2 Calmodulin as a Calcium Sensor
Melanie R. Nelson and Walter J. Chazin

I. Introduction 17
II. Ca^{2+}-Binding Affinity, Cooperativity, and Selectivity 21
III. Structural Basis for Calmodulin as a Calcium Sensor 27
IV. Transduction of the Calcium Signal 42
References 58

3 Calmodulin-Regulated Protein Kinases
Thomas J. Lukas, Salida Mirzoeva, and D. Martin Watterson

I. Overview 66
II. Myosin Light-Chain Kinase 78
III. Calmodulin-Dependent Protein Kinase II 101
IV. Phosphorylase Kinase 116
V. Calmodulin-Dependent Protein Kinase I 124
VI. Calmodulin-Dependent Protein Kinase III 132
VII. Calmodulin-Dependent Protein Kinase IV 136
VIII. Other Calmodulin-Dependent Protein Kinases 140

IX.	Summary	145
	References	146

4 Biochemistry and Pharmacology of Calmodulin-Regulated Phosphatase Calcineurin

Brian A. Perrino and Thomas R. Soderling

I.	Introduction	170
II.	Subunit Composition and Functional Domain Organization	171
III.	Catalytic Properties	183
IV.	Physiological Roles of Calmodulin-Regulated Serine Threonine Phosphatase	192
	References	221

5 Cyclic Nucleotide Regulation by Calmodulin

William K. Sonnenburg, Gary A. Wayman, Daniel R. Storm, and Joseph A. Beavo

I.	Introduction	237
II.	Calmodulin-Regulated Adenylyl Cyclase	238
III.	Calmodulin-Stimulated Phosphodiesterase Isoforms: Genetic Diversity	251
IV.	Concluding Remarks	277
	References	278

6 Regulation of Nitric Oxide Synthase by Calmodulin

Jingru Hu and Linda J. Van Eldik

I.	Introduction	288
II.	Nitric Oxide Synthase	288
III.	Calmodulin and Nitric Oxide Synthase Activation	300
IV.	Regulation of Nitric Oxide Synthase Activity by Phosphorylation/Dephosphorylation	319
V.	Physiological and Pathophysiological Functions of Nitric Oxide	325
VI.	Conclusion	333
	References	334

7 Calmodulin-Binding Proteins of the Cytoskeleton

Nathalie M. Bonafé and James R. Sellers

I.	Introduction	348
II.	Calmodulin-Binding Motor Proteins	350
III.	Calmodulin-Binding Muscle Proteins	366
IV.	Actin-Binding Proteins of the Membrane Cytoskeleton	375
V.	Other Cytoplasmic Calmodulin-Binding Proteins	378

Contents

VI.	Calmodulin-Binding Microtubule-Binding Proteins	380
VII.	Summary	382
	References	383

8 Calmodulin and Ion Flux Regulation
Paul C. Brandt and Thomas C. Vanaman

I.	Introduction: The Nature and Physiological Importance of Ion Translocation	397
II.	Mechanisms of Cellular Calcium Regulation	398
III.	Plasma Membrane Ca^{2+}-ATPases (PMCAs): A Complex Family of Related Enzymes Whose Activities Are Differentially Regulated by Multiple Signaling Pathways in Animal Cells	403
IV.	Other Ion Channels Directly Regulated by Calmodulin	457
	References	458

Index 473

Contributors

Numbers in parentheses indicate the pages on which the authors' contributions begin.

Joseph A. Beavo (237), Department of Pharmacology, University of Washington, Seattle, Washington 98195

Nathalie M. Bonafé (347), Laboratory of Molecular Cardiology, National Heart, Lung and Blood Institute, National Institutes of Health, Bethesda, Maryland 20892

Paul C. Brandt (397), Department of Biochemistry, University of Kentucky Medical Center, Lexington, Kentucky 40536

Walter J. Chazin (17), Department of Molecular Biology, The Scripps Research Institute, La Jolla, California 92037

Jingru Hu (287), Department of Cell and Molecular Biology, Northwestern University Medical School, Chicago, Illinois 60611

Thomas J. Lukas (65), Northwestern Drug Discovery Program and Department of Molecular Pharmacology and Biological Chemistry, Northwestern University School of Medicine, Chicago, Illinois 60611

Salida Mirzoeva (65), Northwestern Drug Discovery Program and Department of Molecular Pharmacology and Biological Chemistry, Northwestern University School of Medicine, Chicago, Illinois 60611

Melanie R. Nelson (17), Department of Molecular Biology, The Scripps Research Institute, La Jolla, California 92037

Brian A. Perrino (169), Department of Physiology and Cell Biology, University of Nevada School of Medicine, Reno, Nevada 89557

James R. Sellers (347), Laboratory of Molecular Cardiology, National Heart, Lung and Blood Institute, National Institutes of Health, Bethesda, Maryland 20892

Thomas R. Soderling (169), Vollum Institute for Advanced Biomedical Research, Oregon Health Sciences University, Portland, Oregon 97201

William K. Sonnenburg (237), Department of Pharmacology, University of Washington, Seattle, Washington 98195

Daniel R. Storm (237), Department of Pharmacology, University of Washington, Seattle, Washington 98195

Thomas C. Vanaman (397), Department of Biochemistry, University of Kentucky Medical Center, Lexington, Kentucky 40536

Linda J. Van Eldik (1, 287), Northwestern Drug Discovery Program and Department of Cell and Molecular Biology, Northwestern University Institute for Neuroscience, Northwestern University Medical School, Chicago, Illinois 60611

D. Martin Watterson (1, 65), Drug Discovery Program and Department of Molecular Pharmacology and Biological Chemistry, Northwestern University School of Medicine, Chicago, Illinois 60611

Gary A. Wayman (237), Department of Pharmacology, University of Washington, Seattle, Washington 98195

Preface

The focus of this book is on emerging themes in molecular mechanisms of Ca^{2+} signal transduction through calmodulin (CaM)-regulated pathways. It attempts to provide the reader with selected examples and experimental precedents that underlie current models of cell regulation through CaM-regulated pathways and their linkage with other regulatory pathways. CaM is now known to be involved in a large number of different Ca^{2+}-dependent signal transduction pathways modulating processes as diverse as cellular motility, metabolism, transport, secretion, fertilization, proliferation, structural integrity, and intercellular communication. It is little wonder, then, that this remarkable multifunctional protein has been the center of attention of so many scientific investigations.

This book presents selected case studies for which extensive data are available on the molecular mechanisms of CaM-mediated signal transduction. Clearly, based on the large body of literature on CaM over the past twenty-five years and the continuing identification of new CaM-regulated proteins, there are many aspects of CaM research that cannot be covered adequately nor can every CaM binding protein be discussed. Earlier comprehensive reviews are available, and this book builds on these previous contributions. A number of minireviews or short opinion articles on CaM research have appeared in the past several years. The intent of this volume is *not* to comprehensively review all aspects of CaM research, but to provide a more in-depth analysis of selected aspects of CaM signal transduction pathways that serve as relevant precedents or demonstrate mechanistic themes that have emerged over the past decade.

The emphasis of each chapter is to review the fundamental knowledge and basic principles of specific CaM-mediated calcium signal transduction pathways and to place the current state of knowledge in a sound factual context with a perspective that looks to the next decade of research. Reflecting the state of the field, many of the examples covered are those for which insight into molecular mechanisms is rapidly emerging, and are generally from vertebrate organisms. For most of the critical CaM regulatory pathways, the more experimentally tractable organisms such as yeast, *Dictyostelium,* and *Drosophila* are becoming increasingly useful as their inherent power as research systems is applied to specific scientific questions. In a similar manner, the study of CaM regulation in plants, algae, and *Paramecium* have provided important precedents, especially about new pathways and *in vivo* functions. However, the full complement of CaM-regulated pathways and their integration appear to be limited to vertebrate organisms due to the late phylogenetic emergence of some essential CaM pathways and the specialized nature of the Ca^{2+} signaling pathways in the phylogenetically earlier organisms. For this reason, the chapters in this book have sought to bring general themes elucidated in these organisms into the context of a discussion centered on regulation of pathways in vertebrates. It is hoped that the molecular emphasis on defined and well-accepted CaM target proteins will provide the field with a CaM reference book of both depth and breadth and that it will provide a firm foundation for new investigators to enter the field and rapidly make significant contributions.

A brief introductory chapter by Van Eldik and Watterson provides an overview and discusses, in a historical context, the directions of current research on CaM. The chapter by Nelson and Chazin introduces how the CaM molecule can act as a Ca^{2+} sensor. Information from three-dimensional structures of CaM in the apo form, in the calcium-saturated form, and in the calcium-saturated form complexed with a CaM-binding peptide provides insights into the diverse conformations that CaM can assume. The authors present models of how these structures and those of related Ca^{2+}-binding proteins can potentially allow high affinity recognition of Ca^{2+} in the presence of other ions such as Mg^{2+}. Higher resolution structures of CaM in supramolecular complexes will continue to enhance our understanding of how CaM's structure is related to its ability to bind Ca^{2+} with high affinity, selectivity, and reversibility, as well as how the conformation of the CaM-binding structure can modulate CaM's Ca^{2+} affinity.

The chapter by Lukas, Mirzoeva, and Watterson reviews the CaM-regulated protein kinases, one of the better characterized group of enzymes regulated by CaM. The first physiological CaM recognition peptides were those isolated from the myosin light chain kinases (MLCKs), two of the more extensively characterized CaM-regulated enzymes available. Based on the paradigms originally described for the MLCKs, similar themes were identified in the other CaM-regulated protein kinases. The segmental organization of the protein kinases is a theme that has been found with other CaM-regulated enzymes, and the mechanistic principles in CaM recognition and activation by a relief of autoinhibition are also seen in other CaM-regulated enzymes. The emphasis in this chapter is on the molecular mechanisms by which CaM interacts with and activates its target kinases to transduce calcium signals into the amplification step of protein phosphorylation.

In addition to phosphorylation, CaM also regulates dephosphorylation of proteins through its modulation of protein phosphatases. Perrino and Soderling discuss the biochemistry and pharmacology of the most extensively characterized CaM-regulated phosphatase, calcineurin. The chapter focuses on recent advances using site-directed mutagenesis, cell-permeable selective inhibitors, and high resolution structural analysis that have led to an increased understanding of the structure, biochemical regulation, and physiological functions of calcineurin as a CaM target. An emerging theme is that targeting mechanisms, such as association with membranes, presence of anchoring proteins, and nuclear cotransport with substrates, may localize calcineurin to distinct subcellular sites to provide additional mechanisms for regulating substrate specificity of calcineurin isoforms.

Since its discovery as an activator of cyclic nucleotide phosphodiesterase (PDE), CaM has been known to mediate the interaction of calcium signal transduction pathways with cyclic nucleotide signaling. CaM regulates degradation of cyclic nucleotides by activating CaM-stimulated PDE and regulates synthesis of cyclic nucleotides by activating adenylyl cyclase (AC). The chapter by Sonnenburg, Wayman, Storm, and Beavo reviews CaM regulation of PDE and AC. The focus is on the molecular mechanisms by which CaM regulates the enzymes, the distribution of these enzymes in different cells and tissues, and the physiological context in which different isoforms of these CaM-regulated enzymes are important for normal signal transduction and functioning.

Hu and Van Eldik discuss CaM regulation of the nitric oxide synthase (NOS) family of proteins, enzymes responsible for the synthesis of nitric oxide (NO), an important signaling molecule in most cell types. The chapter emphasizes the mechanisms by which CaM interacts with NOS isoforms, the significance of this interaction for NOS function, the multiple modes of regulation of the enzymes, and the consequences of NOS activation to physiology and pathology.

In addition to regulation of intracellular enzymes by CaM, the structure and functional properties of the cytoskeleton are also tightly linked to CaM. The chapter by Bonafé and Sellers covers a large group of cytoskeletal proteins that interact with CaM, including motor proteins (unconventional myosins, kinesin), muscle proteins (caldesmon, calponin, titin, nebulin, dystrophin), actin-binding proteins (spectrin, adducin, synapsin, MARCKS), and microtubule-binding proteins (MAPs, Mip-90 STOPs). The interaction of the proteins with CaM and the consequences of that interaction are emphasized. For example, the interaction of some cytoskeletal proteins with CaM can be a regulatory switch for an enzymatic reaction. In other cases, the interaction of CaM with a cytoskeletal protein might be a nonenzymatic modulatory switch controlling protein–protein interaction or might provide structural stability to the cytoskeletal protein. The interaction between CaM and the cytoskeletal target can be Ca^{2+}-independent or strengthened or weakened by Ca^{2+}.

Another important aspect of CaM regulation of calcium signal transduction is its ability to regulate ion flux systems, particularly those involved in modulating the intracellular concentration of free calcium responsible for the calcium signal itself. The maintenance of intracellular ion concentrations is accomplished through both passive and active transport systems present in cell membranes. The chapter by Brandt and Vanaman discusses ion flux systems that appear to be regulated in a calcium-dependent fashion by direct and/or indirect actions of CaM. The focus is on the most well-characterized of these CaM-regulated ion flux systems, the plasma membrane Ca^{2+}-ATPases (PMCAs). The structural features and genetic regulation of the PMCA family, the molecular mechanism of enzyme activation by CaM, other regulatory mechanisms, and the physiological roles of the PMCAs are reviewed. The chapter concludes with a brief discussion of CaM-regulated ion channels in *Paramecium* and *Drosophila*.

We would like to thank the contributors for their cooperation and readiness to write the chapters. This nontrivial and often unappreciated task is important to the stimulation of scientific progress. Finally, we thank the staff of Academic Press for their patience and assistance in the completion of this work.

Linda J. Van Eldik
D. Martin Watterson

1

Calmodulin and Calcium Signal Transduction: An Introduction

LINDA J. VAN ELDIK[*,†] **AND D. MARTIN WATTERSON**[*,‡]

[*] Northwestern Drug Discovery Program, and
Departments of [†]Cell and Molecular Biology and
[‡]Molecular Pharmacology and Biological Chemistry
Northwestern University Medical School
Chicago, Illinois 60611

I. Background. 1
II. Phylogenetic Conservation of Calmodulin 3
III. Molecular Mechanisms of Calmodulin-Mediated Ca^{2+} Signal Transduction. 4
    References 14

I. BACKGROUND

Calcium has diverse and broad roles in nature. This book is concerned with only a small part of the world of calcium—its functioning as a critical component of a dynamic system in eukaryotic cells that uses the protein calmodulin (CaM) as a calcium sensor and transducer. Calmodulin has the broadest phylogenetic and ontogenetic distribution in its class of eukaryotic proteins and is the most extensively studied. Calmodulin, therefore, has been called a prototype for eukaryotic cell Ca^{2+} signal transduction and homeostasis. The mechanism by which CaM is able to transduce Ca^{2+} signals into biological responses is through its modulation of the structure and function of other proteins that usually lack the ability to bind Ca^{2+} with the affinity and kinetics required for the biological response (see chapter by Nelson and Chazin). Most of these CaM-modulated proteins are enzymes (see chapters by Lukas et al.; Perrino and Soderling; Sonnenburg et al.; Hu and Van Eldik; Brandt and Vanaman). This provides the potential for CaM to amplify, through the regulation of enzyme-based catalysis, a stoichiometric bioinorganic equilibrium into a series of reiterative chemical reactions. Because CaM is the calcium sensor for multiple enzymes, there is amplification of diverse chemical reactions in response to a given Ca^{2+} signal. In addition to enzyme-based catalysis, the cytoskeleton is an area of CaM regulation that is receiving increasing attention and provides a different perspective on how CaM can be involved in cellular homeostasis (see chapter by Bonafé and Sellers). Because many cytoskeletal proteins serve as both structural proteins and regulatory systems, they offer another level, or mode, of regulation. If CaM is involved in the regulation of nodes in an extended array of biopolymers, then there is the opportunity to transduce a signal through a physical transducer system, rather than through an enzymatic amplification system. All in all, CaM itself appears to be an integrator through its presence as a calcium sensor in a diverse array of regulatory pathways.

Calmodulin activity was discovered in the early 1970s when the late W. Y. (George) Cheung and S. Kakiuchi independently described (Cheung, 1970; Kakiuchi and Yamazaki, 1970) an activating factor for cyclic nucleotide phosphodiesterase (PDE). A few years later Wang and co-workers (Teo and Wang, 1973) showed that the PDE-activating factor required calcium for activity. An early example of diverse fields of research converging at the subject of CaM research was with the first reports of the purification of CaM to chemical homogeneity (Stevens et al., 1976; Watterson et al., 1976). Stevens et al. (1976) isolated the protein as part of attempts to purify and characterize the PDE activator protein and noted that it had some similarities to troponin C from vertebrate skeletal muscle. Watterson et al. (1976) purified the CaM to homogeneity in their search for nonmuscle proteins that had physical, chemical, and calcium-binding properties similar to those of skeletal muscle troponin C. The search was not to find a protein that would function as a troponin C, but to find a new class of proteins that would have the general calcium-binding properties required to serve as a calcium signal transducer. Troponin C was the best characterized calcium signal transducer protein at the time and served as the standard of comparison. Since that time, CaM has become the standard of comparison, with troponin Cs being viewed as tissue-specific CaM-like proteins.

With the availability of a homogeneous protein, there was further convergence of diverse fields as several investigators proceeded to show that the calcium-binding subunit, or calcium-activating factor, of other enzymes was indistinguishable from this new standard of comparison, which had not yet been named CaM. By the late 1970s, it was apparent that there were several enzymes that were probable physiological targets for CaM. In addition to PDE (see chapter by Sonnenburg *et al.*), these included adenylyl cyclase (see chapter by Sonnenburg *et al.*), the ATPase pump (see chapter by Brant and Vanaman), myosin light chain kinase (see chapter by Lukas *et al.*), and phosphorylase kinase (see chapter by Lukas *et al.*). The name calmodulin, derived from calcium modulator protein, was coined by Cheung in 1978 (Cheung *et al.*, 1978) to reflect the multifunctional nature of CaM as a calcium regulatory protein. It was at this time that the dogma of CaM being present in a great molar excess over its targets became established. This was based on the observation that the concentration of CaM in tissues such as vertebrate brain far exceeded the estimated concentrations of the limited number of enzymes known at the time. However, the number of CaM targets proteins was soon to be expanded.

II. PHYLOGENETIC CONSERVATION OF CALMODULIN

The first complete amino acid sequence of CaM (Watterson *et al.*, 1980) was reported in 1980. The 1980s witnessed the expansion of the field to phylogenetics with the purification and characterization of CaMs from diverse species. For example, the first amino acid sequences of CaMs from higher plants (Lukas *et al.*, 1984), a slime mold (Marshak *et al.*, 1984), a unicellular alga (Schleicher *et al.*, 1984; Zimmer *et al.*, 1988), a ciliated protozoan (Schaefer *et al.*, 1987), the fruit fly *Drosophila* (Smith *et al.*, 1987; Yamanaka *et al.*, 1987), and yeast (Davis *et al.*, 1986) were reported. It was the phylogenetic conservation of residues among diverse CaMs with retention of *in vitro* activity that provided the basis of the first site-directed mutagenesis and chimeric CaM analyses (Roberts *et al.*, 1985; Craig *et al.*, 1987; Putkey *et al.*, 1988). As discussed in the following chapters in this volume, the conclusions from these results were consistent with the structures of CaM:peptide complexes reported several years later, and the two types of approaches are proving to be complementary in the dissection of the detailed relationships among structures and functions (see chapter by Lukas *et al.*).

Based on the large number of biochemical characterizations of CaMs from multiple species and analysis of the amino acid variability allowed in the CaM sequence for isofunctionality, the current state of knowledge about CaM sequences can be summarized in Fig. 1. A generic CaM sequence and the vertebrate CaM sequence are shown. Calmodulin is one of the most highly conserved eukaryotic proteins known. The positions of selected conserved residues are landmarked in the three-dimensional structure of CaM shown in Fig. 2. In contrast to the high conservation of the amino acid sequence throughout species, the CaM gene varies among eukaryotes in both the number of genes and the structural organization. Calmodulin can be encoded by a single gene in an organism or by multiple genes, the number and position of introns differ among CaM genes, and the length of 5' and 3' flanking regions varies. Although the number and structural organization of CaM genes vary among species, these genes encode identical amino acid sequences in most cases. The highly conserved primary structure of CaM may reflect the necessity for a highly conserved structure for highly conserved functions or may reflect the importance of CaM as a multifunctional transducer of calcium signals through target protein regulation and the need to conserve a diversity of recognition interfaces in a single, relatively small protein.

III. MOLECULAR MECHANISMS OF CALMODULIN-MEDIATED Ca^{2+} SIGNAL TRANSDUCTION

Concurrent with the study of CaM biochemistry and genetics, progress was being made in the probing of the initial steps in the molecular mechanism of CaM action. The development of CaM–Sepharose chromatography (Watterson and Vanaman, 1976) in the 1970s provided a facile and gentle step late in purifications of CaM-regulated proteins, resulting in an increased number of purified CaM-binding proteins and the analysis of the next step in the mechanism. Although efficient use of the calcium-dependent chromatography step required the prior depletion of the endogenous CaM, or CaM-like protein, by differential centrifugation or ion-exchange chromatography, there was a tendency to equate the biochemical properties with the physiological state of the calcium signal transduction complex. From the 1980s and well into the 1990s,

```
→
XXXLXXζQXXζφδEAXXLFDδDXXXIXXXELXXVφRSφXXXPXXEφXD
↓
ADQLTEEQIAEFKEAFSLFDKDGDGTITTKELGTVMRSLGQNPTEAELQD
        10        20        30        40        50

XXXEXDXXXXXXXIζFXEφLXφMXδXφXXDXEXEφXEAFXXFDδXXXGφX
MINEVDADGNGTIDFPEFLTMMARKMKDTDSEEEIREAFRVFDKDGNGYI
        60        70        80        90        100

                                    →
XXXζφXHφφXXXφGEδXXXXζVXXMφδζXDXXXXXXXXXXFXXφφXXδ
SAAELRHVMTNLGEKLTDEEVDEMIREADIDGDGQVNYEEFVQMMTAK
        110       120       130       140
```

Fig. 1. Consensus sequence in naturally occurring CaMs. A consensus sequence (upper rows) and a vertebrate CaM sequence (lower rows) are shown. For the vertebrate CaM sequence, standard single-letter amino acid codes are used. To evaluate the CaM consensus sequence, sequences of naturally occurring CaMs obtained from the Swiss-Prot or GenBank databases were aligned: φ, Hydrophobic amino acids (I,L,V,M,F,Y); ζ, negatively charged amino acids (D,E); δ, positively charged amino acids (K,R); and X, any amino acid. Positions where amino acid insertions are found in some CaMs are indicated by arrows. Amino acid numbering is that from vertebrate CaM.

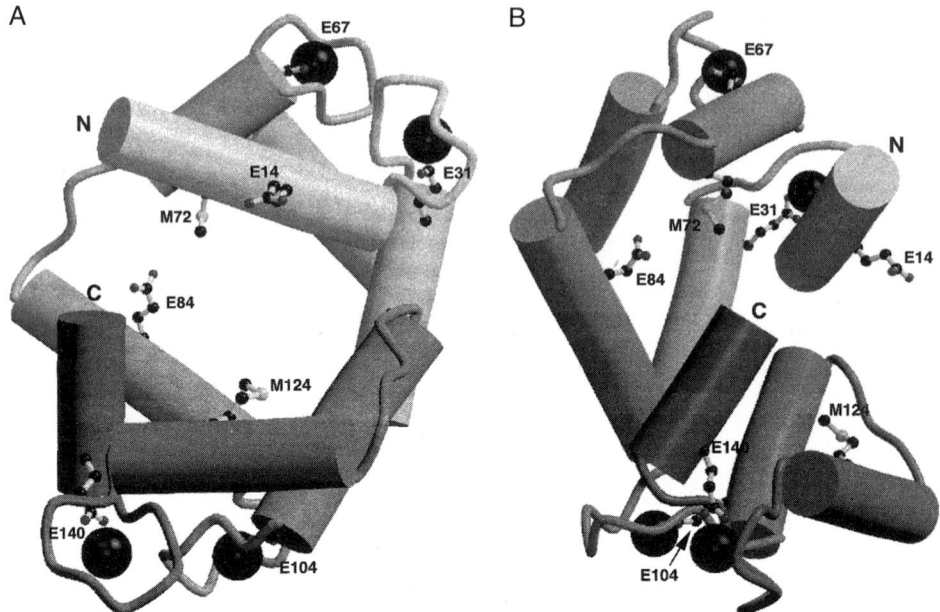

Fig. 2. Phylogenetically conserved amino acid residues in the three-dimensional structure of CaM. These ribbon drawings are based on coordinates in the Brookhaven Protein Data Bank (PDB) entry 1CDL and represent one of the multiple conformations that CaM can assume. Two different views of CaM are shown, with side chains of selected amino acids that are phylogenetically conserved and noted in Fig. 1. Ca^{2+} ions are solid spheres. The amino (N) and carboxy (C) terminal ends of CaM are indicated.

new CaM-binding proteins were discovered as described in the following chapters in this volume. The focus on characterizing the components of Ca^{2+} signal transduction pathways continued to bring investigators in diverse fields together as they found a CaM-regulated protein as a component of their pathway of interest. The concept of CaM being the calcium-binding subunit of a diversity of CaM-regulated enzymes and a subunit of abundant regulatory proteins, such as the myosins, in the same tissue or cell is now generally accepted. This requires that the question of molar stoichiometry be revisited, especially if the field is to move to the next level of research detail in which pathways are modeled as a basis for forecasting the response of cells to a given stimulus. In this regard, the field of CaM research is poised to take advantage of the emerging discipline of bioinformatics and its application to drug discovery and the analysis of signal transduction homeostasis.

Based on the increasing database of knowledge about CaM and its targets, there are a number of general principles and emerging themes concerning the molecular mechanisms of action of CaM as a transducer of biological Ca^{2+} signals. Certain structural and mechanistic themes are now being recognized with a variety of CaM-regulated enzymes, in addition to other regulatory mechanisms that may be specific to an individual CaM-regulated enzyme. By utilizing both common and unique regulatory mechanisms in the biological response to Ca^{2+} signals, the organism has both a level of redundancy and discrimination to allow fine tuning and modulation of the response. The elucidation of the structural basis of selective Ca^{2+} recognition in a sea of magnesium and other ions is required for a full understanding of how biological systems decode a Ca^{2+} signal. The emerging themes in Ca^{2+} recognition (see chapter by Nelson and Chazin) provide a firm foundation for such future investigations.

Probing further into the molecular mechanisms of Ca^{2+} signal transduction through CaM-regulated pathways, the 1980s and 1990s saw the surprising discovery of a common theme among CaM-regulated enzymes: the CaM recognition activity of many large enzymes could be mimicked by relatively short segments of the enzyme (see chapters by Lukas *et al.*; Perrino and Soderling; Hu and Van Eldik). The first identification of CaM recognition sequences was with CaM-regulated protein kinases (Blumenthal *et al.*, 1985; Lukas *et al.*, 1986). Since those initial studies, a large body of work with synthetic peptide analogs of the CaM recognition regions of enzyme targets, enzymes isolated from tissue sources, and site-directed mutagenesis approaches have yielded insight into the molecular mechanisms of CaM recognition and enzyme regulation. Many of the peptide analogs of the CaM-binding regions of the enzymes have a common basic amphiphilic α-helix motif. The original hypothesis (Lukas *et al.*, 1986) that the potential secondary structure and the presence of appropriately spaced positively charged and hydrophobic amino acid side chains, not primary sequence similarity alone, might be a common characteristic feature among divergent CaM recognition sequences has been verified. However, the presence of related sequence motifs in a protein does not mean that the region is a physiological CaM recognition sequence. More extensive analyses, especially with site-directed mutagenesis and enzymology of the purified mutant enzymes (Shoemaker *et al.*, 1990), are required to establish the short CaM-binding segments as physiologically relevant regions. In a related manner, the absence of such sequence motifs in a protein does not imply necessarily that a protein is not a CaM-binding protein (Bosc *et al.*, 1996; Wes *et al.*, 1996).

The correlation of structures with earlier functional studies of CaMs and CaM recognition peptides from enzymes (see chapters by Lukas *et al.*; Perrino and Soderling; Sonnenburg *et al.*; Hu and Van Eldik) has revealed another theme found among many CaM-regulated enzymes: that each CaM recognition peptide uses a unique set of contact points. If one makes a list of amino acids in CaM that are within approximately 4 Å of the amino acids in the CaM-binding peptides in structures of CaM:peptide complexes (see chapter by Lukas *et al.*), it is found that a number of the same amino acids in CaM are potential contact residues, based on this criteria, across multiple enzymes. There are also several individual points for each complex that are specific to a given complex, making the full set unique for a given enzyme. Even at amino acids of CaM that are potential contact residues among multiple complexes, the details of side chain interactions are often distinct for each complex. Thus, each enzyme might have functionally critical contacts, even though the same amino acid in CaM may be used across a variety of enzymes. This high level of discrimination provides a means by which CaM-binding regions of similar secondary structure (e.g., amphiphilic α helix with net basic charge) from multiple enzymes can be appropriately interpreted by the CaM molecule. It is interesting to speculate that these multiple CaM recognition features across diverse CaM-regulated enzymes might contribute to the phylogenetic conservation of the CaM amino acid sequence.

How is Ca^{2+} binding by CaM linked to enzyme regulation? The increasing availability of structural information for CaMs in different conformational states, including CaM complexed with Ca^{2+} and with CaM-binding peptides, has provided insight (see chapters by Nelson and Chazin; Lukas *et al.*). It is clear from a number of studies of mutant CaMs and CaM-binding peptides that CaM can exist in a variety of conformational states. The nuclear magnetic resonance and X-ray crystallographic structures of CaM are snapshots of only a few of the conformational states in which CaM can likely exist. Data to date are consistent with the conformational restriction hypothesis, which assumes that CaM can exist in a variety of conformational states, with the distribution being increasingly restricted as an increasing number of Ca^{2+} ions are bound and the complexes with CaM-binding structures are formed (see chapters by Nelson and Chazin; Lukas *et al.*). A correlate is that more than one CaM conformation can be active, with the profile of active conformations being characteristic for each given CaM:enzyme complex. This emphasizes the need for additional high-resolution structures of CaM in various conformational states with Ca^{2+} or CaM-binding structures in order to determine which conformations and interactions are truly correlated with function.

Another linkage between Ca^{2+} binding and enzyme recognition is that the binding of Ca^{2+} by CaM influences its ability to regulate enzymes, and the enzymes can, in turn, influence the characteristics of Ca^{2+} binding by CaM. This also provides an example of discrimination in the response of a cell to transient Ca^{2+} fluxes through the differential Ca^{2+} sensitivity of the CaM:enzyme supramolecular complex. The situation is complicated by the fact that CaM has four Ca^{2+}-binding sites and uses an ordered and cooperative binding mechanism. The Ca^{2+} sensitivity of a particular enzyme activation, therefore, represents a nontrivial relationship among multiple equilibria (see chapter by Lukas *et al.*). Further, for most CaM-regulated enzymes, it is not known for certain how the calcium occupancy state of CaM is related to enzyme activation.

Although much insight has been gained through the use of small peptides to model a CaM-binding site on an enzyme, this is not sufficient to understand recognition by the holoenzyme complex. Although small peptide fragments have much of the CaM-binding affinity and selective CaM recognition features of the enzyme, there clearly are additional potential contacts that must be discovered. The more extreme cases of this point are the phosphorylase kinase and inducible form of nitric oxide synthase (NOS), which appear to require multiple regions of the enzyme linear sequence for full CaM-binding activity (see chapters by Lukas *et al.*; Hu and Van Eldik).

A fundamental question in Ca^{2+}-mediated signal transduction is how activation of multiple enzymes by changes in a diffusible second messenger like calcium leads to temporally and spatially restricted effects. In addition to control at the level of enzyme and substrate turnover and the differential calcium sensitivity of CaM:enzyme complexes, another general principle of mechanistic control of Ca^{2+} signal transduction pathways is the concept of specific targeting of supramolecular association complexes within the cell. In contrast to the dominant theme of other signaling systems (such as cAMP and the cAMP-dependent protein kinase, which use separate targeting proteins as well as separate catalytic and regulatory subunits), CaM-regulated enzymes often have their catalytic, CaM recognition, and supramolecular recognition, or subcellular targeting, sequences within the same polypeptide chain. This is a feature that distinguishes many of the CaM-regulated enzymes from other signal transduction complexes. Differential localization control mechanisms can allow the activity of CaM-regulated enzymes at sites where their particular substrates are required to produce the desired physiological responses.

Several CaM-regulated enzymes exist as multimeric assemblies composed of individual subunits that may be single or multiple isoforms. The composition of the holoenzyme may affect both the catalytic properties and the localization of the enzyme. For example, vertebrate CaM-dependent protein kinase II (CaMPKII) is encoded by multiple genes, each of which encodes a distinct isoform that can have multiple splice variants (see chapter by Lukas *et al.*). One of the isoform variants has a consensus nuclear localization signal that targets the enzyme to the nucleus, thus providing a mechanism for catalytic activity in a specific subcellular location. Another example is in protein phosphatase I, which has a glycogen-binding subunit that allows association with glycogen in striated muscle (see chapter by Perrino and Soderling). Targeting motifs within a CaM-binding protein may influence subcellular localization of the protein, as in the case of motifs in myosin light-chain kinase that lead to association with the actin cytoskeleton (see chapter by Lukas *et al.*). In addition, the subcellular location of CaM-regulated enzyme complexes may provide mechanisms for appropriate physiological regulation, as in the case of phosphorylase kinase at the glycogen particle (see chapter by Lukas *et al.*), ATPases at the plasma membrane (see chapter by Brandt and Vanaman), or cytoskeletal CaM-binding proteins (see chapter by Bonafé and Sellers).

Related to mechanisms for restricting CaM-binding proteins to appropriate subcellular locations, CaM itself is localized within the cell. As discussed earlier and in light of the accumulating evidence, the old concept of CaM free-floating in the cytoplasm, waiting to bind a CaM target when the Ca^{2+} rises, needs to be subjected to experimental testing. As discussed in the following chapters for specific enzymes, CaM binding to an enzyme is not equivalent to CaM activation of catalytic activity. Binding is required but not necessarily sufficient. Experimental data are strong for some CaM-regulated enzymes that CaM is an integral subunit of the holoenzyme under physiological conditions. Calmodulin can be thought of as a common Ca^{2+}-sensitizing subunit among the various enzymes. It is likely that CaM is associated *in vivo* with diverse targets, but activation of the activity of the target protein is modulated by the Ca^{2+} signaling dynamics. An emerging theme is that various subcellular targeting mechanisms localize CaM through its target protein to distinct subcellular locations, providing additional selectivity and specificity in the temporal and spatial aspects of cellular responses.

Another emerging theme is coordinate regulation of CaM and specific target proteins. For example, an increase in CaM levels in oncogene-transformed cells is accompanied by a concomitant change in the levels of selected CaM-binding proteins. A complementary example is in phosphorylase kinase deficiencies. In the I strain of mice, which have an enzyme deficiency caused by a frameshift mutation in one of the phosphorylase kinase subunit genes, there is also a 40% decrease in the skeletal muscle CaM concentration compared to normal mice (see chapter by Lukas *et al.*). Thus, perturbation of the biosynthetic levels of CaM may be reflected in changes in the levels of CaM-regulated enzymes, and vice versa. Clearly, most studies to date suggest that CaM is critical in eukaryotic cell life. When present, multiple genes still lead to a conserved protein, possibly providing redundancy in critical functions. Although the absence of CaM can be lethal, mutations in CaM are allowed. The regulation of CaM and CaM-regulated enzyme biosynthesis is an area with much left to be explored.

Analysis of the molecular mechanisms of CaM-regulated signal transduction is revealing an intricate web of interacting events in response to Ca^{2+} signals and the extensive cross-talk among diverse signaling pathways (Fig. 3). This interaction and cross-talk occur at the level of both CaM-dependent and CaM-independent pathways. For example, there appear to be CaM kinase cascades initiated in response to Ca^{2+} transients. Upstream CaM-dependent kinase kinases (CaMKK) can phosphorylate and activate other kinases that are also CaM regulated (e.g., CaMPKI and CaMPKIV), which in turn can phosphorylate downstream substrates. In this example, the activity of CaM-regulated kinases CaMPKI and CaMPKIV can be regulated by both Ca^{2+}/CaM at the CaM regulatory domain and through phosphorylation by a distinct CaM-dependent kinase kinase (see chapter by Lukas *et al.*). Another example of cross-talk among CaM-dependent pathways is the observation that CaM-dependent protein kinases can phosphorylate and modulate the activity of other CaM-regulated enzymes. CaMPKIV, for example, can phosphorylate a CaM-regulated adenylyl cyclase (adenylate cyclase) and inhibit its Ca^{2+}-stimulated activity (see chapter by Sonnenburg *et al.*). CaMPKII can phosphorylate a variety of CaM-regulated enzymes *in vitro* (including NOS, PDE, kinases) and modulate their catalytic activity. How many of these CaM-dependent phosphorylations occur *in vivo* is

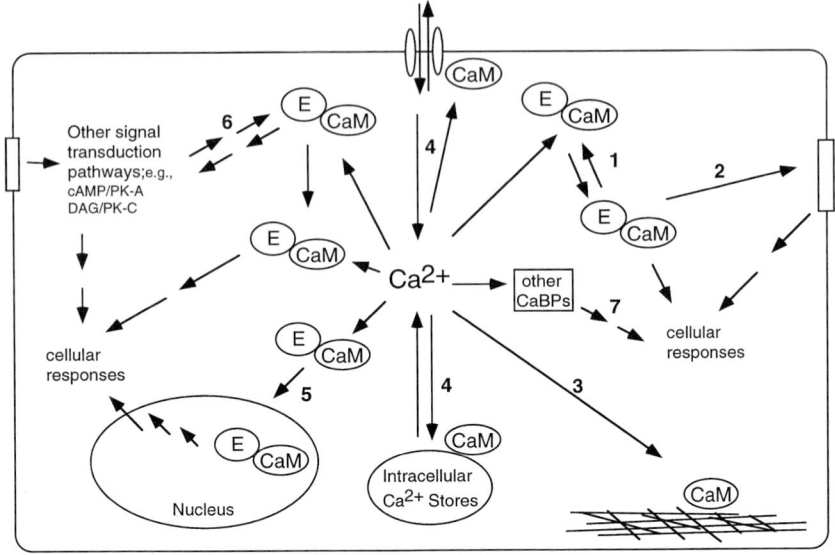

Fig. 3. Multiple CaM-regulated signal transduction pathways and cross-talk among intracellular signals. In response to a variety of physiological, pharmacological, or pathological stimuli, cytoplasmic free Ca^{2+} levels rise through influx of extracellular calcium or release of Ca^{2+} from intracellular stores. This calcium signal is transduced into cellular responses through interaction with calcium-binding proteins such as CaM. Analysis of the molecular mechanisms of CaM-regulated signal transduction reveals an intricate web of interacting events in response to Ca^{2+} signals and extensive cross-talk among diverse signaling pathways. Selected examples of the interconnections are depicted in this diagram for CaM-regulated enzymes (E) that are discussed in the chapters in this book. Example 1: CaM-regulated phosphatases can dephosphorylate another CaM-regulated enzyme or CaM-regulated kinases can phosphorylate another CaM-regulated enzyme. Example 2: A CaM-regulated enzyme can directly interact with membrane-associated, extracellular receptor-linked signal transduction complexes to affect cellular responses. Example 3: CaM can affect cytoskeletal integrity and interactions. Example 4: CaM can interact with transporter systems such as the Ca^{2+} pumps and ion channels. Example 5: CaM-regulated enzymes can be targeted to specific subcellular locations, such as the nucleus, to exert their biological effects, such as regulation of gene transcription. Example 6: One of the more intricate and elaborate interdigitations is the cross-talk and convergence with other signal transduction pathways, such as cAMP-regulated pathways and diacylglycerol (DAG)-regulated pathways. In the case of cAMP-regulated pathways, certain isoforms of phosphodiesterase or adenylate cyclase are CaM-regulated enzymes, and protein kinase A (PK-A) can phosphorylate CaM-regulated enzymes. In the case of DAG-regulated pathways, protein kinase C (PK-C) can phosphorylate CaM-regulated enzymes, and CaM-regulated kinases and PK-C can converge at the level of a common substrate protein. Example 7: Other calcium-binding proteins (CaBP) have the calcium-binding features of CaM and have the potential to influence cellular responses to Ca^{2+} signals.

not known, but the observations *in vitro* provide a framework for how one CaM-regulated enzyme activity can be modulated by another CaM-regulated enzyme. Further, multiple phosphorylation sites are often present on individual CaM-regulated enzymes or substrates, and the site of phosphorylation may have distinct effects on activity or expression of the target. For example, phosphorylation of a particular transcription factor (CREB) at one site by CaMPKII leads to a decrease in gene expression, whereas phosphorylation by CaMPKIV at a different site leads to upregulated gene expression (see chapter by Lukas *et al.*).

In addition to a high degree of interaction among CaM-regulated pathways, there is also extensive cross-talk among diverse signal transduction pathways (e.g., cyclic nucleotide and calcium signaling). For example, the cyclic nucleotide-dependent kinase, PK-A, can phosphorylate, and in many cases modulate, the activity of a variety of CaM-regulated enzymes, including CaM-dependent kinases, PDE, ATPases, and NOS. Activation of the lipid-dependent kinase, PK-C, can also affect selective CaM-regulated enzymes such as adenylyl cyclases and NOS. In some cases, such as the plasma membrane Ca^{2+}-ATPases, there is also a direct activation of enzyme activity by lipids.

Overall, the signal transduction pathways modulated by Ca^{2+}–CaM in response to an individual stimulus are interdigitated with other signaling pathways, allowing for discrimination, rapid responses, and fine tuning of the responses of an organism to stimuli. Disease states may result from disruption at specific points in these interconnected pathways. In addition, small disruptions or aberrant regulation at early points in the signal transduction cascades could have devastating consequences for the final physiological responses. For example, a small change in CaM activity or concentration can lead to severalfold changes in more downstream responses because of the amplification nature of the signaling pathways and the fact that CaM works early in these amplification pathways. How CaM-regulated enzymes maintain cellular homeostasis and use the intricate linked webs of signal transduction pathways to respond to external stimuli is an active frontier in the field.

ACKNOWLEDGMENTS

The authors acknowledge the financial support of National Institutes of Health Grants GM 30861 and AG13939. We thank Drs. Salida Mirzoeva and Longyin Chen for assistance in the preparation of the figures.

REFERENCES

Blumenthal, D. K., Takio, K., Edelman, A. M., Charbonneau, H., Titani, K., Walsh, K. A., and Krebs, E. G. (1985). Identification of the calmodulin-binding domain of skeletal muscle myosin light chain kinase. *Proc. Natl. Acad. Sci. U.S.A.* **82,** 3187–3191.

Bosc, C., Cronk, J. D., Pirollet, F., Watterson, D. M., Haiech, J., Job, D., and Margolis, R. L. (1996). Cloning, expression, and properties of the microtubule-stabilizing protein STOP. *Proc. Natl. Acad. Sci. U.S.A.* **93,** 2125–2130.

Cheung, W. Y. (1970). Cyclic 3',5'-nucleotide phosphodiesterase. Demonstration of an activator. *Biochem. Biophys. Res. Commun.* **38,** 533–538.

Cheung, W. Y., Lynch, T. J., and Wallace, R. W. (1978). An endogenous Ca^{2+}-dependent activator protein of brain adenylate cyclase and cyclic nucleotide phosphodiesterase. *Adv. Cyclic Nucleotide Res.* **9,** 233–251.

Craig, T. A., Watterson, D. M., Prendergast, F. G., Haiech, J., and Roberts, D. M. (1987). Site-specific mutagenesis of the α-helices of calmodulin. Effects of altering a charge cluster in the helix that links the two halves of calmodulin. *J. Biol. Chem.* **262,** 3278–3284.

Davis, T. N., Urdea, M. S., Masiarz, F. R., and Thorner, J. (1986). Isolation of the yeast calmodulin gene: Calmodulin is an essential protein. *Cell (Cambridge, Mass.)* **47,** 423–431.

Kakiuchi, S., and Yamazaki, R. (1970). Calcium dependent phosphodiesterase activity and its activating factor (PAF) from brain. *Biochem. Biophys. Res. Commun.* **41,** 1104–1110.

Lukas, T. J., Iverson, D. B., Schleicher, M., and Watterson, D. M. (1984). Structural characterization of a higher plant calmodulin. *Plant Physiol.* **75,** 788–795.

Lukas, T. J., Burgess, W. H., Prendergast, F. G., Lau, W., and Watterson, D. M. (1986). Calmodulin binding domains: Characterization of a phosphorylation and calmodulin binding site from myosin light chain kinase. *Biochemistry* **25,** 1458–1464.

Marshak, D. R., Clarke, M., Roberts, D. M., and Watterson, D. M. (1984). Structural and functional properties of calmodulin from the eukaryotic microorganism *Dictyostelium discoideum. Biochemistry* **23,** 2891–2899.

Putkey, J. A., Ono, T., VanBerkum, M. F. A., and Means, A. R. (1988). Functional significance of the central helix in calmodulin. *J. Biol. Chem.* **263,** 11242–11249.

Roberts, D. M., Crea, R., Malecha, M., Alvarado-Urbina, G., Chiarello, R. H., and Watterson, D. M. (1985). Chemical synthesis and expression of a calmodulin gene designed for site-specific mutagenesis. *Biochemistry* **24,** 5090–5098.

Schaefer, W. H., Lukas, T. J., Blair, I. A., Schultz, J. E., and Watterson, D. M. (1987). Amino acid sequence of a novel calmodulin from *Paramecium tetraurelia* that contains dimethyllysine in the first domain. *J. Biol. Chem.* **262,** 1025–1029.

Schleicher, M., Lukas, T. J., and Watterson, D. M. (1984). Isolation and characterization of calmodulin from the motile green alga *Chlamydomonas reinhardtii. Arch. Biochem. Biophys.* **229,** 33–42.

Shoemaker, M. O., Lau, W., Shattuck, R. L., Kwiatkowski, A. P., Matrisian, P. E., Guerra-Santos, L., Wilson, E., Lukas, T. J., Van Eldik, L. J., and Watterson, D. M. (1990). Use of DNA sequence and mutant analyses and antisense oligodeoxynucleotides to examine the molecular basis of nonmuscle myosin light chain kinase autoinhibition, calmodulin recognition, and activity. *J. Cell Biol.* **111,** 1107–1125.

Smith, V. L., Doyle, K. E., Maune, J. F., Munjaal, R. P., and Beckingham, K. (1987). Structure and sequence of the *Drosophila melanogaster* calmodulin gene. *J. Mol. Biol.* **196,** 471–485.

Stevens, F. C., Walsh, M., Ho, H. C., Teo, T. S., and Wang, J. H. (1976). Comparison of calcium-binding proteins. Bovine heart and brain protein activators of cyclic nucleotide phosphodiesterase and rabbit skeletal muscle troponin C. *J. Biol. Chem.* **251,** 4495–4500.

Teo, T. S., and Wang, J. H. (1973). Mechanism of activation of a cyclic adenosine 3':5'-monophosphate phosphodiesterase from bovine heart by calcium ions. *J. Biol. Chem.* **248,** 5950–5955.

Watterson, D. M., and Vanaman, T. C. (1976). Affinity chromatography purification of a cyclic nucleotide phosphodiesterase using immobilized modulator protein, a troponin C-like protein from brain. *Biochem. Biophys. Res. Commun.* **73,** 40–46.

Watterson, D. M., Harrelson, W. G., Jr., Keller, P. M., Sharief, F., and Vanaman, T. C. (1976). Structural similarities between the Ca^{2+}-dependent regulatory proteins of 3':5'-cyclic nucleotide phosphodiesterase and actomyosin ATPase. *J. Biol. Chem.* **251,** 4501–4513.

Watterson, D. M., Sharief, F., and Vanaman, T. C. (1980). The complete amino acid sequence of the Ca^{2+}-dependent modulator protein (calmodulin) of bovine brain. *J. Biol. Chem.* **255,** 962–975.

Wes, P. D., Yu, M., and Montell, C. (1996). RIC, a calmodulin-binding Ras-like GTPase. *EMBO J.* **15,** 5839–5848.

Yamanaka, M. K., Saugstad, J. A., Hanson-Painton, O., McCarthy, B. J., and Tobin, S. L. (1987). Structure and expression of the *Drosophila* calmodulin gene. *Nucleic Acids Res.* **15,** 3335–3348.

Zimmer, W. E., Schloss, J. A., Silflow, C. D., Youngblom, J., and Watterson, D. M. (1988). Structural organization, DNA sequence, and expression of the calmodulin gene. *J. Biol. Chem.* **263,** 19370–19383.

2

Calmodulin as a Calcium Sensor

MELANIE R. NELSON AND WALTER J. CHAZIN

Department of Molecular Biology
The Scripps Research Institute
La Jolla, California 92037

I. Introduction 17
II. Ca^{2+}-Binding Affinity, Cooperativity, and Selectivity 21
III. Structural Basis for Calmodulin as a Calcium Sensor 27
    A. Three-Dimensional Structure of Ca^{2+}-Loaded Calmodulin 27
    B. The HMJ Model 30
    C. Structural Analysis of the Calcium Trigger 31
    D. Consideration of Energetics 37
    E. Role of Methionines 39
IV. Transduction of the Calcium Signal 42
    A. Binding of Peptides Derived from Calmodulin Targets 42
    B. Interactions with Other Ligands 50
    C. Binding to Targets in the Absence of Ca^{2+} 53
    References 58

I. INTRODUCTION

Calcium signaling systems must be finely tuned for rapid and effective response to transient variations in Ca^{2+} concentration. The central role in the transduction of calcium signals is played by members of a family of highly homologous Ca^{2+}-binding proteins (CaBPs). These proteins function by undergoing conformational changes on binding of Ca^{2+}, thereby translating the transient influx of Ca^{2+} into metabolic or mechanical responses. Calmodulin is the most well studied of these Ca^{2+}-signaling proteins, commonly known as Ca^{2+} sensors (da Silva and Reinach, 1991).

The ability to bind Ca^{2+} selectively, yet respond to the rather narrow range of change in ion concentration, demonstrates an exquisite level of fine tuning of the binding properties of the CaBPs. The special relationship between these proteins and calcium extends far beyond signal transduction activities, as other members of this protein family are found to have critical roles in the uptake, transport, and numerous other aspects of homeostasis of calcium (Dieter and Marmé, 1988; Kawasaki and Kretsinger, 1994, 1995). The wide diversity in the biological activities of these proteins, all centered on calcium, suggests an extensive evolutionary optimization of the fit between the EF-hand CaBP fold and the ion. In fact, the extent of fine tuning of binding sites extends even further. The optimized fit must be sufficiently dynamic to allow the very substantial differences in binding affinities, kinetics, and conformational response to Ca^{2+} binding that are required for the regulatory and various accessory functions in the signaling system. Clearly, the biological roles of calmodulin and other CaBPs are intrinsically connected to their biophysical properties.

The calmodulin superfamily of CaBPs is characterized by a common helix–loop–helix (HLH) structural motif in their Ca^{2+}-binding sites. The term EF-hand was introduced for these motifs by Kretsinger on examination of the three-dimensional structure of parvalbumin, the first member of the family whose structure was solved by X-ray crystallography (Kretsinger and Nockolds, 1973). A great deal of insight into the EF-hand CaBPs has since been obtained from X-ray diffraction and, more recently, by nuclear magnetic resonance (NMR) (reviews in Strynadka and James, 1989; Kawasaki and Kretsinger, 1995; Ikura, 1996).

The EF-hand CaBPs can be recognized via distinct homologies in their Ca^{2+}-binding loops. They bind Ca^{2+} with seven oxygen ligands in a pentagonal bipyramidal arrangement, with six of the ligands from the protein and one oxygen atom supplied by a water molecule (Fig. 1). The consensus sequence consists of a 12-residue loop with coordination provided by residues at positions 1, 3, 5, 7, 9, and 12. The amino acids at positions 1, 6, and 12 are the most highly conserved. Residues 1, 3, and 5 provide monodentate ligands via side-chain carboxylates, whereas residue 7 directly coordinates the Ca^{2+} via its main-chain oxygen. An unusual backbone conformation with a positive ϕ angle is required at position 6 to enable main-chain ligation at position 7 while maintaining side-chain ligation at position 5. Positive ϕ angles are unfavorable for

2. Calmodulin as a Calcium Sensor

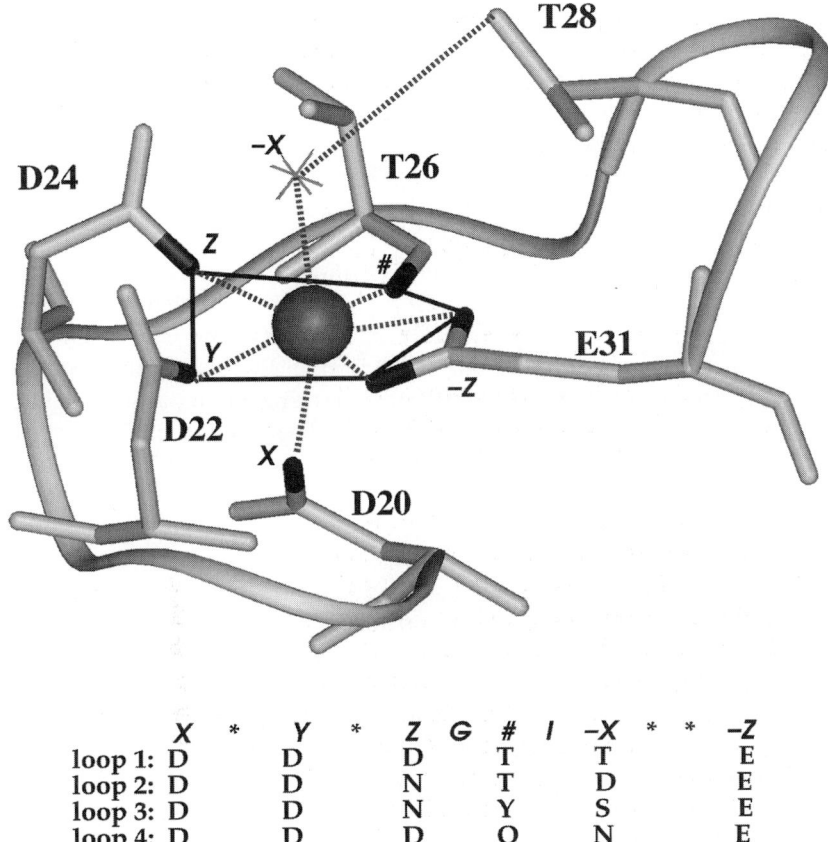

	X	*	Y	*	Z	G	#	I	-X	*	*	-Z
loop 1:	D		D		D		T		T			E
loop 2:	D		D		N		T		D			E
loop 3:	D		D		N		Y		S			E
loop 4:	D		D		D		Q		N			E

Fig. 1. Calcium coordination in calmodulin. The first EF-hand in calmodulin is used to illustrate the pentagonal bipyramidal ligation of the Ca^{2+} ion. Oxygen atoms ligating Ca^{2+} are colored black. The water ligand is shown as a star. The five ligands that form the base of the pentagonal bipyramid are connected with black lines. The identities of the protein ligands in all four Ca^{2+}-binding loops of calmodulin are indicated and labeled on the molecular structure with the traditional ligand nomenclature. X, Y, Z, and $-Z$ ligate Ca^{2+} with side-chain carboxylates, $-X$ is water mediated, and # indicates backbone carbonyl ligation. The molecular structure was rendered in Insight (MSI, San Diego) using PDB coordinates 1CLL.

all amino acids except glycine, which explains why glycine is highly conserved in this position. The coordination of Ca^{2+} by residue 9 is actually indirect, mediated by a water molecule that is hydrogen bonded to its side chain (and often to the side chain of residue 3 as well). Residue 12 is a strictly conserved glutamic acid bidentate ligand that coordinates the Ca^{2+} ion via both oxygen atoms of the side-chain carboxylate. A definitive analysis of the binding loops was reported by Strynadka and James (1989).

The basic structural/functional unit of the EF-hand CaBPs are pairs of EF-hands rather than single binding sites, a property that is presumed to stabilize the protein conformation and increase the Ca^{2+} affinity of each site over that of isolated sites (Seamon and Kretsinger, 1983). The paired EF-hands are invariably packed face to face in a parallel manner and form a stable globular domain (Fig. 2). The integral pairing of sites provides a ready means for cooperativity in the binding of Ca^{2+}. A variety of experimental evidence indicates extensive interactions between the paired Ca^{2+}-binding sites within each domain, in most cases resulting in cooperative Ca^{2+}-binding phenomena.

Calmodulin has four EF-hands organized into two distinct globular domains, usually referred to as N(-terminal) and C(-terminal) domains. One interesting feature of the protein is that the portion of the polypeptide chain that links the two globular domains is very sensitive to mild proteolytic digestion, enabling the isolation of the two domains. This strategy has been used extensively for both biophysical and biological

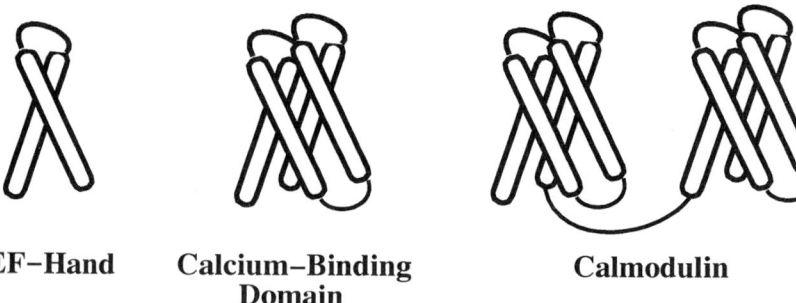

EF–Hand   **Calcium–Binding Domain**   **Calmodulin**

Fig. 2. Hierarchical structure of calmodulin. The fundamental Ca^{2+}-binding motif is the helix–loop–helix EF-hand. Two EF-hand units comprise a globular calcium-binding domain. Calmodulin is composed of two calcium-binding domains.

2. Calmodulin as a Calcium Sensor

studies in an effort to assign specific functions to one of the two domains. The domains bind Ca^{2+} and undergo Ca^{2+}-dependent conformational changes independently, but usually act in concert to activate signaling pathways.

Calmodulin has been studied by a variety of biophysical techniques (reviewed in Forsén et al., 1986), with an accentuation on defining the changes in the protein induced on binding of Ca^{2+}. The apoprotein exhibits some unique properties. For example, microcalorimetric measurements on apo calmodulin do not show a sigmoidal "two-state" melting curve characteristic of many globular proteins (Tsalkova and Privalov, 1985), and NMR reveals conformational heterogeneity in the C-terminal domain (Kuboniwa et al., 1995). The binding of Ca^{2+} greatly stabilizes the protein, resulting in a significantly lower sensitivity to general proteolysis and a substantial increase in the thermal denaturation midpoint (Brzeska et al., 1983). Although the ultraviolet spectra of the two states of the protein are very similar, circular dichroism experiments indicated a higher helical content for the Ca^{2+}-loaded protein relative to the apo state (Martin and Bayley, 1986). However, as discussed in detail later, subsequent NMR analysis revealed that this effect is not due to changes in secondary structure. Fluorescence anisotropy (on calmodulin cross-linked to fluorescent dyes) and small-angle X-ray scattering have been used to monitor the global shape of the protein, and only a small increase in the average length of the protein occurs as the protein binds Ca^{2+} (Small and Anderson, 1988; Yoshino et al., 1989; Matsushima et al., 1989). Although the ensemble of this evidence provided invaluable insights into the effects of Ca^{2+} binding on the protein, these results could only be fully rationalized once the three-dimensional structures of the apo and Ca^{2+}-loaded states became available.

II. Ca^{2+}-BINDING AFFINITY, COOPERATIVITY, AND SELECTIVITY

The four Ca^{2+}-binding sites in calmodulin each have a reasonably high affinity ($K_a \sim 10^{-6}\ M^{-1}$) that is optimally tuned to read out the calcium signal generated by a transient increase in Ca^{2+} concentration. Stimulation and a subsequent influx of calcium through the calcium channels raise the cellular Ca^{2+} concentration from the basal level of $\sim 10^{-7}\ M$ to $\sim 10^{-5}\ M$. As a consequence of the precise tuning of the Ca^{2+} affinities, only a minute number of calmodulin molecules will have Ca^{2+} bound in the resting state of the cell, but on release of the calcium signal all calmodulin molecules will be activated. Cooperative binding is particularly important in this context, as it allows for a tightly controlled "all or nothing" response to the changes in Ca^{2+} concentration and, consequently, very clean separation of the "off" and "on" states. The effect of cooperativity on the binding curve is to effectively increase the slope in the region of the binding event and flatten the curve before and after (Fig. 3). If binding events were spread over a large range, then the system would be leaky, i.e., calmodulin would be partially activated over a large range of Ca^{2+} concentrations, and the protein would not be effective in its biological role as a calcium sensor.

The measurement of the binding of Ca^{2+} to calmodulin has been carried out by a variety of techniques over a number of years (reviewed in Forsén et al., 1986). However, due to the complexities associated with a system containing four similar binding sites dispersed over two structural domains, detailed mechanisms for the binding events remain a topic of considerable debate. Techniques used to measure the Ca^{2+}-binding constants can be grouped into two distinct categories. The first

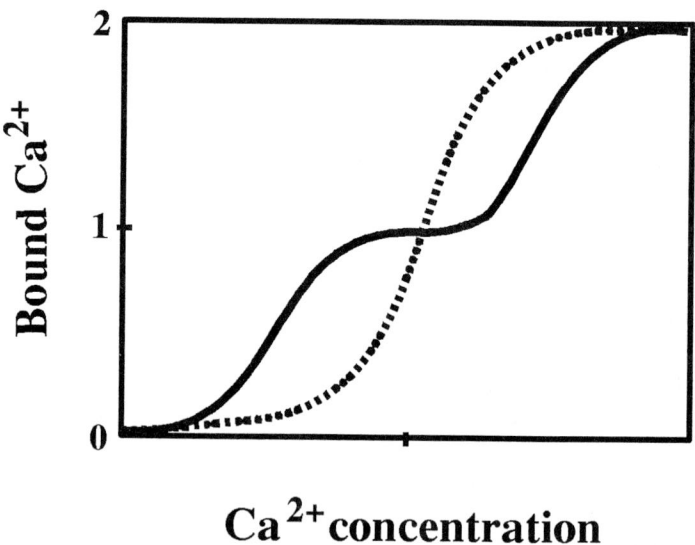

Fig. 3. Schematic diagram of the effect of cooperativity on Ca^{2+} binding. The solid curve is for noncooperative binding and the dashed curve for cooperative binding.

consists of measurements of the residual Ca^{2+} concentration in solutions of protein plus calcium, such as equilibrium dialysis. The second consists of measurements of changes in some property of the protein, such as a spectroscopic signal, as increasing amounts of Ca^{2+} are added to a protein solution. Regardless of what technique is used to make the measurement, the primary problem with almost all methods is the interpretation of the response of a four-site system when only a single macroscopic observable is available. Theoretically, NMR spectroscopy offers an opportunity to monitor the titration at a large number of nuclear sites and to overcome the problems of lack of specificity in the measurements. This approach has been demonstrated for troponin C (Li *et al.*, 1995) and should be readily applicable to calmodulin.

The interpretation of the Ca^{2+}-binding curves for calmodulin has varied from the existence of four independent sites, to two independent pairs of interacting sites, to four mutually interacting sites. Consequently, depending on the model used to explain the data, the sites are purported to have either identical or different Ca^{2+} affinities. The different methods of analyzing primary data have been discussed (Forsén *et al.*, 1986; Kilhoffer *et al.*, 1992), as well as the difficulties in distinguishing the binding curves for systems with four cooperative sites, two slightly different pairs of cooperative sites, or four very similar but noncooperative sites.

One portion of the debate over the Ca^{2+}-binding constants of calmodulin arises from differences in the solution conditions under which the measurements have been made. In particular, the pH, ionic strength, and temperature all have significant effects on Ca^{2+} affinity. Studies have also demonstrated that the concentration of the protein must be taken into consideration when measuring binding affinities with many techniques due to electrostatic screening, especially for a highly charged protein such as calmodulin (Linse *et al.*, 1995). The Ca^{2+}-binding constants measured for calmodulin under a variety of experimental conditions range over two to three orders of magnitude. The strongest effect is due to differences in the ionic strength, which varies the degree of electrostatic screening by the solvent. Perhaps the most important point arising from the ensemble of these results is that to make accurate comparisons, it is absolutely imperative to carry out the measurements under identical conditions.

The selectivity in the binding of Ca^{2+} is another important factor in the effective transduction of the calcium signal. In particular, calmodulin must be able to bind Ca^{2+} selectively in the background of $\sim 10^4$ greater concentration of Mg^{2+} and even greater amounts of monovalent cations. The affinity for Mg^{2+} in calmodulin and other EF-hand CaBPs can be sufficiently high that Mg^{2+} serves as an effective Ca^{2+} antagonist. In fact, Tsai et al. (1987) report Mg^{2+} association constants of $2 \times 10^3\ M^{-1}$ for the N-terminal domain of calmodulin. This implies that sites I and II may be occupied by Mg^{2+} in resting cells. However, when the Ca^{2+} concentration is increased on release of the Ca^{2+} signal, Mg^{2+} will be displaced from the binding sites.

Ion selectivity appears to be achieved by a combination of optimization of the identity of the ligands and the corresponding binding loop geometry, which is adjusted to provide a specific site size and charge distribution. For example, the ability to discriminate Ca^{2+} versus Mg^{2+} can be attributed in part to the reduction in the size of the site that would be required for optimal binding of the smaller Mg^{2+} ion (Falke et al., 1994). A collapsed binding site would cause repulsion of large negatively charged side chain ligands, thereby destabilizing the Mg^{2+}–protein complex relative to the Ca^{2+} complex.

Calcium affinities and functions cannot be expected to be completely attributable to the local loop structure. Possible correlations between Ca^{2+} affinities and structural features of the Ca^{2+}-loaded states of EF-hand CaBPs have been examined (Linse and Forsén, 1995). It was concluded that a variety of ways exist to achieve high-affinity binding in CaBPs and, consequently, that the conformation of the protein is not the sole determinant of Ca^{2+} affinity. Extensive treatises reviewing the current level of understanding of the factors governing the Ca^{2+}-binding properties of EF-hand CaBPs have been published (Linse and Forsén, 1995; Falke et al., 1994).

How is Ca^{2+} affinity fine-tuned to meet the requirements of protein function? The most direct approach to modulating the affinity for Ca^{2+} is via the residues in the binding loops. Mutations of binding loop residues in calmodulin have been used for a variety of purposes, ranging from selective deactivation of single sites in the protein to probing the central helix and critical elements of cooperative ion binding (e.g., Maune et al., 1992; Craig et al., 1987; Persechini and Kretsinger, 1988; Putkey et al., 1988; Starovasnik et al., 1992). Mutagenesis experiments on calmodulin and other EF-hand CaBPs such as troponin C (e.g., Putkey et al., 1989) and calbindin D_{9k} (e.g., Carlström and Chazin, 1993) have revealed that removing one side-chain ligand oxygen is sufficient to significantly reduce the Ca^{2+} affinity. The Ca^{2+} affinity in a site can also be altered by variation in the effective electrostatic potential in the binding loop region. Indeed, in calbindin D_{9k}, charge-neutralizing mutations of nonligating side chains reduce the Ca^{2+} affinity (e.g., Linse et al., 1991). A similar effect is expected in calmodulin and other EF-hand proteins due to the high homology of the binding sites.

The deactivation of binding sites by site-directed mutagenesis is commonly used in biochemical assays to examine the specific functional roles of a domain or of one of the two binding sites within a domain. However, great care must be taken in analyzing the results from these experiments, particularly in the absence of biophysical characterization of the mutant protein. It is important to understand the structural consequences of mutations in advance of analyzing biological or biochemical data, as these alterations can perturb the protein in a manner that affects more than just the local site.

For example, consider the very typical mutation of the critical bidentate glutamic acid at position 12 of the Ca^{2+}-binding loop. Several investigators have shown that even very minimal mutations, such as Glu → Gln, produce a drastic reduction in binding affinity (Haiech et al., 1991; Beckingham, 1991; Maune et al., 1992). Careful examination of the three-dimensional structure of calmodulin provides insights into why and strongly suggests that even this minimal mutation may have far greater implications than might be anticipated initially. At an elementary level, a mutation of the side chain from carboxylate to carboxylamide simply results in the removal of one of the oxygen atom Ca^{2+} ligands. However, the amide group is larger than the oxygen atom, a possible steric factor, and some hydrogen-bonding potential is lost. It is important to also consider the potential loss of negative charge (which helps attract the dication to the site), and especially the perturbation of the intricate network of hydrogen bonds in the binding loop. Figure 4 shows the very central role of the bidentate Glu-31 carboxylate ligand in site I of the N-terminal domain of calmodulin. Interactions from the carboxylate to three backbone residues are present, which in turn are hydrogen bonded to other residues in the binding loop. This network of interactions is critical to the stabilization of the conformation of the binding loop in the Ca^{2+}-loaded state (McPhalen et al., 1991).

Nuclear magnetic resonance studies on a series of calmodulins mutated at the critical bidentate glutamic acid show that in some instances, the progression of the changes in conformation on binding of Ca^{2+} can be different from that seen in wild-type protein (Starovasnik et al., 1992). Even differences in site–site interactions within each of the two globular domains have been noted. Nuclear magnetic resonance studies on the

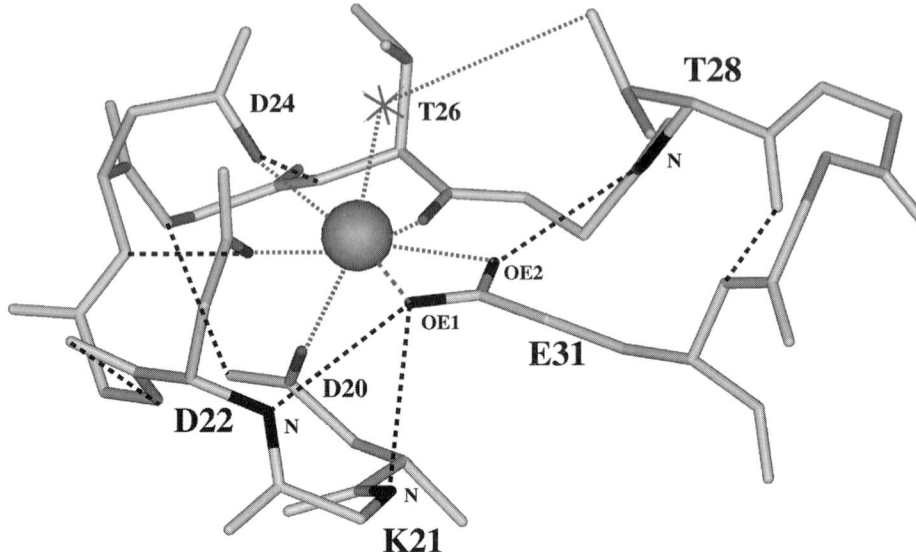

Fig. 4. The integral network of hydrogen bonds in calcium-binding loops. The Ca^{2+}-binding loop of the first EF-hand of calmodulin is used to illustrate the interconnected network of hydrogen bonds that stabilizes the calcium-binding loop. Calcium ligands are colored dark gray. Atoms involved in hydrogen bonds with the side-chain oxygens of the bidentate ligand Glu-31 are colored black. Calcium ligation is indicated with gray dotted lines. Interresidue hydrogen bonds are indicated with black dotted lines. The molecular structure was rendered as for Fig. 1.

corresponding mutation in the consensus type EF-hand of calbindin D_{9k} have shown that the simple Glu → Gln substitution is sufficient to prevent the domain from being able to attain the conformation observed for the Ca^{2+}-loaded state of the wild-type protein, even in the presence of large excesses of Ca^{2+} (Carlström and Chazin, 1993; Wimberly et al., 1995). Thus, alteration of the bidentate glutamic acid ligand is seen to severely upset the delicate balance of forces in the Ca^{2+}-binding loop.

A complete understanding of the biochemical implications of mutational experiments requires knowledge of the effects on the conformation of the protein. Although it is ultimately the ability to transduce the Ca^{2+} signal (i.e., binding to target proteins) that is most biologically relevent, perturbations of the conformational response to the binding of Ca^{2+} may be substantial. This can complicate the interpretation of the results of functional assays. In fact, there are several examples of native or modified EF-hand CaBPs, including native yeast CaM, that function even though one of the two sites within a domain does not bind Ca^{2+} (Putkey et al., 1989; Shaw et al., 1991; Starovasnik et al., 1993). At the molecular level, as Ca^{2+} affinity is known to be increased on the binding of receptor peptides to wild-type calmodulin (see later), the ability to attain the conformation of the Ca^{2+}-loaded state in domains with mutated or naturally nonfunctional sites could conceivably be driven or recovered by initial interactions with a receptor. Clearly, the molecular mechanisms associated with the biochemical outcomes of specific mutation experiments are not obvious in the absence of biophysical characterization. Structural studies are required to establish what effects are associated with loss of Ca^{2+} affinity in a binding site.

III. STRUCTURAL BASIS FOR CALMODULIN AS A CALCIUM SENSOR

A. Three-Dimensional Structure of Ca^{2+}-Loaded Calmodulin

The three-dimensional X-ray crystal structure of calmodulin in the Ca^{2+}-loaded state was first reported in 1985 (Babu et al., 1985) and has been refined to high resolution (1.7 Å) (Chattopadhyaya et al., 1992). The structure revealed a very surprising dumbbell shape (Fig. 5) containing two structurally homologous globular domains with dimensions of approximately 25 × 25 × 20 Å and an overall length of ~65 Å along the long axis. The domains are each composed of a pair of EF-hands that are connected to each other by a short linker polypeptide as well as by an antiparall β-type interaction between the two 12-residue Ca^{2+}-binding loops (Fig. 6). The helices are labeled I–IV or A–D in the N-terminal domain, and V–VIII or E–H in the C-terminal domain. The binding loops actually extend a few residues into the two adjacent helices, which, combined with the hydrogen bonds between the two binding loops, appear to provide a molecular basis for the site–site interactions that lead to cooperative calcium binding.

The two globular domains are separated by a long (28 residue) "central helix" that incorporates the fourth (D) helix in the N-terminal domain, an additional seven residues (Met-76 to Glu-82), and the first (E) helix

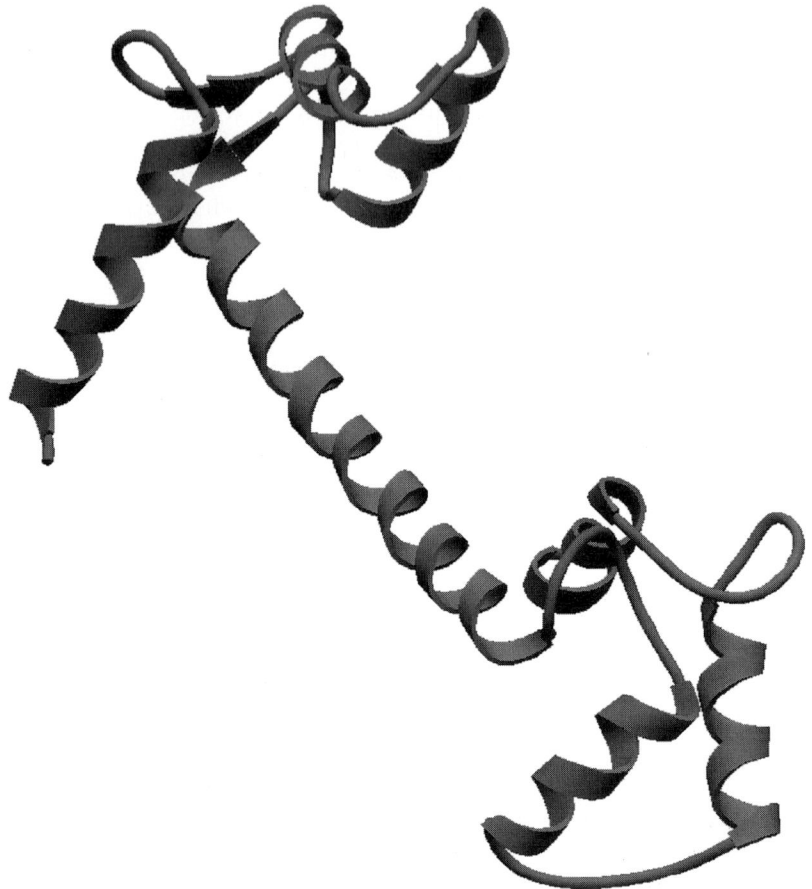

Fig. 5. Crystal structure of calcium-loaded calmodulin. The ribbon diagram of the 1.7-Å X-ray crystal structure of calcium-loaded calmodulin was rendered using the "PDB Ribbons" module (Mike Carson and Alex Shah) written for AVS (Upson et al., 1989) and PDB coordinates 1CLL.

in the C-terminal domain. The conformation of the long central helix was a topic of great debate for many years. The controversy arose because the all-helical conformation observed in the high-resolution crystal structures was not fully consistent with the extensive biochemical data on these proteins indicating that the two domains can interact. Furthermore, such a highly solvent-exposed helix is not expected to be stable.

2. Calmodulin as a Calcium Sensor

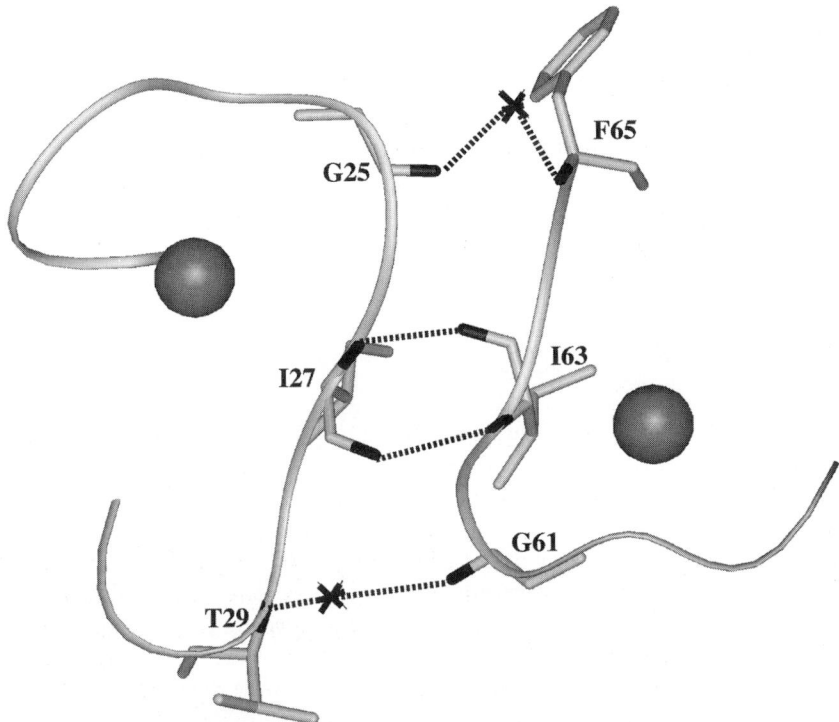

Fig. 6. Hydrogen bonding interactions between two Ca^{2+}-binding loops of the N-terminal domain of calmodulin. Hydrogen bonds are indicated by black dotted lines. Two crystallographic waters are shown as stars. The two calcium ions are shown as spheres. The molecular structure was rendered as for Fig. 1.

Several lines of evidence, including somewhat elevated B-factors in the crystal structures, indicated that the central helix was not, in fact, a well-formed helix throughout its entire length. Fluorescence anisotropy and small-angle X-ray scattering studies (e.g., Heidorn and Trewhella, 1988), as well as numerous site-directed mutagenesis experiments (e.g., Persechini and Kretsinger, 1988; Raghunathan *et al.*, 1993), provided experimental evidence that this linker serves more as a "flexible tether" as opposed to a rigid spacer between the two domains. Definitive proof of the flexible tether concept was eventually obtained by NMR. ^{15}N backbone relaxation and amide proton exchange experiments showed that the seven residue polypeptide segment linking helix D to helix E is flexible in both apo and Ca^{2+}-loaded states (Spera *et al.*, 1991; Barbato *et al.*, 1992; Tjandra *et al.*, 1995).

The final affirmation of the flexible tether concept came from the structures of calmodulin in complex with peptide fragments of its intracellular receptors: myosin light-chain kinase and calmodulin-dependent kinase II (see Section IV). These structures showed that the linker serves as a pivot, allowing the two domains of calmodulin to swing around and completely enclose the target peptide.

B. The HMJ Model

The mode of action of calmodulin and other regulatory EF-hand CaBPs is based on conformational changes induced by the binding of Ca^{2+}. The long-standing absence of knowledge of the details of these structural changes was a major barrier to an understanding of this critical aspect of calcium signaling at the molecular level (Strynadka and James, 1991). For many years, the only view of the conformational changes induced by Ca^{2+} binding in calmodulin was derived from comparative modeling of the N-terminal and C-terminal domains of the highly homologous troponin C (the HMJ model, Herzberg *et al.*, 1986). The modeling was possible because a difference in the ion occupancy of the two domains occurred as a result of the high ionic strength and low pH ($\sim$5) of the solution from which the troponin C crystals were obtained. In the crystal structure, the N-terminal domain containing the lower affinity regulatory sites is Ca^{2+} free, whereas the C-terminal domain containing the higher affinity sites is Ca^{2+} loaded. Comparison of the two domains after best-fit superposition revealed very large differences in the relative orientations of the four helices; the C-terminal domain adopts a much more open conformation than the N-terminal domain, with a substantially greater solvent-exposed surface.

James and co-workers reasoned that since the sequences of the two domains of troponin C are so similar to each other and to the sequences of the two domains of calmodulin, all four domains are likely to undergo very similar changes in conformation on ion binding. Thus, they utilized the apo N-terminal domain structure to homology model the apo "off" state of the troponin C C-terminal domain and, conversely, used the Ca^{2+}-filled C-terminal domain structure to homology model the activated state of the N-terminal domain of troponin C (Herzberg *et al.*, 1986).

2. Calmodulin as a Calcium Sensor

Subsequently, these same concepts were used to model the changes induced in calmodulin (Strynadka and James, 1988). Models derived in this manner were consistent with numerous biochemical and biophysical experiments carried out on calmodulin and troponin C and have been tested by site-directed mutagenesis experiments from several laboratories (e.g., da Silva and Reinach, 1991).

C. Structural Analysis of the Calcium Trigger

After years of study by NMR, the solution structure of apo calmodulin was finally solved in 1995. In fact, three separate three-dimensional structures were reported for the Ca^{2+}-free state, including two independent studies of intact calmodulin (Zhang et al., 1995; Kuboniwa et al., 1995) and a third study of the isolated C-terminal domain (Finn et al., 1995). In addition, complementary information was obtained at the same time on the apo states of three related systems: troponin C (Gagné et al., 1995), recoverin (Tanaka et al., 1995), and an EF-hand domain from α-spectrin (Travé et al., 1995).

It is remarkable that the critical barrier to understanding the structural consequences of Ca^{2+} binding was overcome using the NMR approach to structure determination in solution. This happened despite the rich history of X-ray crystallographic analysis of EF-hand CaBPs that laid much of the groundwork in our general understanding of the biophysical properties of these proteins (Strynadka and James, 1989; McPhalen et al., 1991). However, nearly all EF-hand CaBP structures determined to date have ions bound and no X-ray crystal structures are available for a Ca^{2+}-binding site crystalized in both Ca^{2+}-free and Ca^{2+}-bound states. The bottleneck for crystallographic analysis has been the inability to obtain suitable crystals in the Ca^{2+}-free state, presumably because the protein is more flexible in the absence of bound ions. Only the N-terminal domain of troponin C and three of the four sites in recoverin have been crystallized in the Ca^{2+}-free state (Herzberg and James, 1985; Flaherty et al., 1993), but as noted none of these could also be crystallized with bound ions. The NMR method is uniquely suited to overcome this problem, as the studies are carried out in solution where experimental parameters such as the absence or presence of a ligand such as Ca^{2+} can be freely adjusted, and regions of high flexibility do not prohibit obtaining the structure of the molecule. Furthermore, EF-hand CaBPs are of a small enough size to be readily amenable to NMR spectroscopic analysis.

The critical importance of the apo calmodulin, troponin C, and recoverin structures is that they permit a detailed analysis of the conformational changes induced in these regulatory proteins by the binding of Ca^{2+}. (Although the Ca^{2+}-free and Ca^{2+}-bound NMR structures of the EF-hand domain from α-spectrin were determined at the same time, a regulatory function has not yet been assigned due to its low, nonphysiological Ca^{2+} affinity *in vitro*.) Previous to 1995, the only EF-hand CaBP whose structure had been determined in the presence and the absence of Ca^{2+} was calbindin D_{9k} (Skelton *et al.*, 1994). However, this protein is involved in some form of calcium transport or buffering activity as opposed to regulation of one or more target proteins and does not share several key calcium-dependent properties characteristic of regulatory CaBPs.

When the NMR structures of apo calmodulin became available, it was evident that the essential elements of the HMJ model were correct. Overall, the secondary structure of calmodulin in the Ca^{2+}-free and Ca^{2+}-bound states is very similar, but the tertiary structure within a given domain differs. Thus, the conformational response to the binding of Ca^{2+} is dominated by substantial reorganization of the packing of the helices within the individual domains. At the local level, changes are much more limited, e.g., reorganization of the backbone at the termini of the helices near the binding loops. One striking observation in these structures is a well-characterized conformational heterogeneity in the C-terminal domain of the intact protein (Kuboniwa *et al.*, 1995), yet no evidence for a similar phenomenon was reported in the study of the isolated domain (Finn *et al.*, 1995). This heterogeneity in the intact protein results in lower effective resolution in the C-terminal domain.

In the absence of Ca^{2+}, calmodulin adopts a closed conformation. Each of its two domains is a twisted bundle of four nearly antiparallel helices with interhelical angles of approximately 180° (Table I). When Ca^{2+} binds to calmodulin, it triggers a pronounced conformational change, and the protein adopts an open conformation with near-perpendicular interhelical angles. This rearrangement exposes a hydrophobic surface of approximately 10×12.5 Å in each domain (Fig. 7) (Zhang *et al.*, 1995; Babu *et al.*, 1988), which permits the interaction with various cellular targets. Although early evidence from circular dichroism spectroscopy implied that there is an increase in helical content associated with binding of Ca^{2+} (Martin and Bayley, 1986), NMR assignment analy-

2. Calmodulin as a Calcium Sensor

TABLE I

Interhelical Angles in Calmodulin[a]

Domain and state	Helical pairs					
N-terminal	A/B	A/C	A/D	B/C	B/D	C/D
Apo	133 ± 1	95 ± 2	124 ± 1	124 ± 3	52 ± 2	129 ± 2
Ca^{2+}	88	162	105	109	40	87
Difference	45	−67	19	15	12	42
C-terminal	E/F	E/G	E/H	F/G	F/H	G/H
Apo	131 ± 2	51 ± 3	150 ± 7	154 ± 3	45 ± 3	150 ± 4
Ca^{2+}	105	125	111	129	25	121
Difference	26	−74	39	25	20	29

[a] Angles measured using CALC_ALL (K. Yap and M. Ikura, unpublished, 1995). Values for the apo state are averages over the ensemble of structures of Kuboniwa et al. (1995; PDB accession number 1CFC). Values for the calcium-bound state are from the 1.7-Å crystal structure (Chattopadhyaya et al., 1992; PDB accession number 1CLL).

sis showed that the overall content and distribution of secondary structure was retained (Ikura et al., 1991). In the case of troponin C, these changes in the CD spectrum have been interpreted as the result of the rearrangement of existing secondary structural elements (Gagné et al., 1995).

Detailed insights into the conformational response to Ca^{2+} binding can be obtained by comparing the closed apo state conformation and the open calcium-loaded state conformation. In the following, the N-terminal domain of calmodulin is used as a representative case. The discussion here focuses on the global fold of the protein as opposed to a precise atomic description, which is beyond the scope of this chapter. In fact, elucidation of the atomic level mechanisms for the changes in conformation in terms of specific side-chain interactions is an active area of research.

Difference distance matrix (DDM) and contact map analyses are chosen for in-depth comparisons over the more typical graphical comparisons after best-fit superposition, as these methods do not suffer from the bias intrinsic to the process of selecting atoms to perform the superposition (see discussion in Akke et al., 1995). A difference distance (DD) can be defined for the same pair of atoms (i, j) in two structures by taking the distance (d) between them in one structure and subtracting

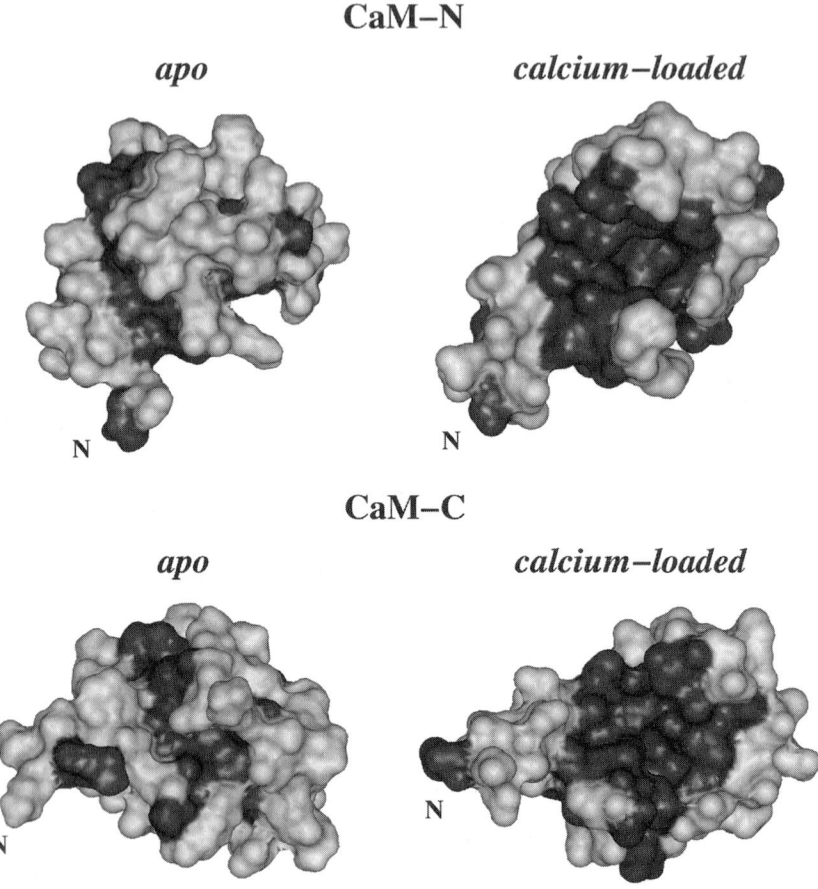

Fig. 7. Exposure of hydrophobic surface on binding of Ca^{2+} to calmodulin. Connolly surfaces of both the apo and the Ca^{2+}-loaded states of the two domains of calmodulin are shown with the same orientation in both states. The surfaces of hydrophobic residues (Ala, Val, Leu, Ile, Met, Phe, Tyr) are shaded dark gray. Surfaces were rendered in Insight (MSI, San Diego, CA) using PDB coordinates 1CFC (apo state) and 1CLL (Ca^{2+}-loaded state).

the corresponding distance in the other structure. For example, a DD for the apo and Ca^{2+}-loaded states of calmodulin is defined as $DD_{ij} = d_{ij}(\text{apo}) - d_{ij}(Ca^{2+})$. The DDM is a matrix containing the values of these

2. Calmodulin as a Calcium Sensor

differences, for all or any defined subset of atoms in the structures. As defined earlier, a negative entry in this matrix means that the two atoms are closer together in the apo state than they are in the Ca^{2+}-loaded state, and a positive entry means the two atoms are closer together in the Ca^{2+}-loaded state. For simplicity, the discussion here focuses only on the DDM of α-carbons.

Whereas the DDM provides information about each pair of residues in the structure, contact maps are more selective, only providing information about two residues close enough to be "in contact" on the basis of a user-defined cutoff. Hence, the contact map is calculated for a single structure only. Here, contact maps have been constructed by evaluating the distance between the centers of geometry of each amino acid side chain in the structure. Two residues are considered to be in contact if they come within 8 Å of each other. The cutoff was derived from the distance between the centers of geometry of two phenylalanine residues positioned to just barely contact each other. Taken together, the DDM and contact map analyses provide a detailed picture of the differences between the two conformations adopted by calmodulin.

Both the DDM and contact map analyses show that the opening of the calmodulin N-terminal domain on binding of Ca^{2+} occurs almost exclusively within the EF-hands, not between them (Nelson and Chazin, 1998). The paired helices within the EF-hands (A/B and C/D) move closer together near the binding loop and further apart at the opposite ends (Fig. 8). This is reflected in negative entries in the DDM between the C terminus of helix A and the N terminus of helix B, and positive entries at the opposite ends of these two helices. A similar trend is seen for helices C and D. Correspondingly, the contact maps indicate that on binding of Ca^{2+}, additional contacts are made between the C terminus of helix A and the N terminus of helix B, whereas contacts are lost at the opposite ends (Table II).

The concerted changes within the two EF-hands on Ca^{2+} binding are also accompanied by helices B and D moving further away from each other, such that the contacts between these two helices observed in the apo state are all absent in the Ca^{2+}-loaded state. The interface between helices A and D remains fairly stable, as does the interface between helices B and C. Helices A and C have no contacts in either state.

Although this analysis was performed on the N-terminal domain of calmodulin, it is believed that similar changes are likely to occur in all calcium-binding domains that undergo the closed to open conformational switch, including the C-terminal domain of calmodulin. Contact map analyses and DDM have also been made comparing the two calmodulin

Helix A/B Interface

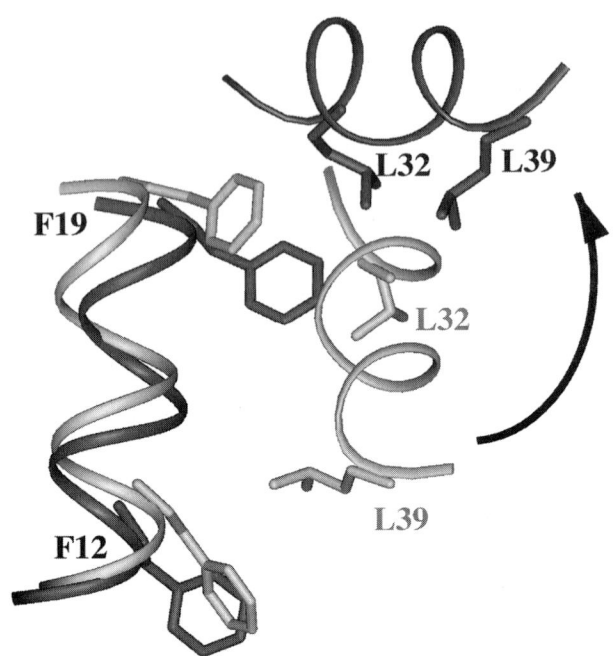

Helix C/D Interface

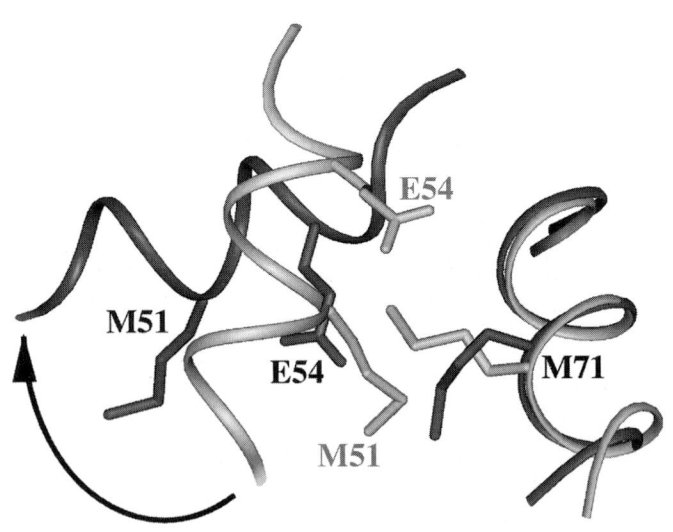

domains in both their apo and calcium-loaded states. These show that the two domains are very similar in both conformations, as will be discussed later in more detail in the context of calmodulin–receptor interactions.

D. Consideration of Energetics

Having characterized the changes in the structure of the protein, it is important to address how these different conformational states relate to the energetics of calcium signaling, with a goal of understanding the specific driving forces for Ca^{2+} binding and the corresponding conformational change. Calmodulin and other regulatory CaBPs provide an interesting puzzle regarding the relationship between structure and energetics. The exposure of hydrophobic residues to solvent is expected to be enthalpically and entropically unfavorable, yet on binding to calcium, calmodulin exposes a large hydrophobic surface. Somehow this open conformation is energetically more favorable than the well-packed apo state (it must be or else calcium would not bind). How is the cost of exposing the hydrophobic surface overcome?

Clearly, the affinity for calcium must be viewed in the context of both states of the protein, not just the free energy of a fixed ligand–Ca^{2+} complex. The binding free energy is dependent on the difference in free energies of the complex versus that of the separated solvated ion and solvated apoprotein. The total energy of the system includes the internal energy of the protein (arising from the folded conformation of the pro-

---

Fig. 8. Opening of the EF-hands on calcium binding to calmodulin. Illustration of the calcium-induced opening determined from DDMs and contact maps of the helix A/B interface and helix C/D interface in the N-terminal domain of calmodulin. The apo structure (shown in light gray) is from PDB coordinates 1CFC. The Ca^{2+}-loaded structure (dark gray) is from PDB coordinates 1CLL. The apo and Ca^{2+}-loaded structures were superimposed on the relatively stable helix A/D interface and rendered using Insight (MSI, San Diego, CA). Examples of contacts made in only one of the two states are highlighted. In the A/B interface, Phe-19 and Leu-32, which are near the Ca^{2+}-binding loop, contact only in the Ca^{2+}-loaded state, whereas Phe-12 and Leu-39, found at the other end of the helices, contact only in the apo state. In the C/D interface, Met-71 is different in the two states, contacting Glu-54 only in the Ca^{2+}-loaded state and Met-51 only in the apo state.

TABLE II

Unique Interhelical Contacts in N-Terminal Domain of Apo and Ca^{2+}-Loaded Calmodulin

Helix interface	State observed	Contact residue	Distance (Å)
A/B	Apo	E11–L39	6.5
		F12–L39	5.8
		A15–S38	4.6
		A15–L39	4.5
		F16–V35	6.1
		L18–S38	6.4
		F19–E31	5.5
		F19–T34	5.2
		F19–S38	6.4
	Ca^{2+}	F19–L32	6.4
		D20–E31	6.1
A/D	Apo	K13–L69	7.7
		S17–F65	7.8
	Ca^{2+}	F12–A73	6.3
		F19–F68	6.3
B/C	Apo	L32–V55	7.5
	Ca^{2+}	R37–L48	7.4
B/D	Apo	L32–F68	6.8
		L32–M71	6.0
		V35–F68	6.0
		M36–F68	6.6
		M36–M71	6.6
		M36–M72	6.4
		L39–F68	7.8
		L39–M72	6.3
C/D	Apo	M51–M71	5.5
		E54–M71	7.3
		D56–E67	5.9
		A57–E67	4.5
		D58–E67	5.7

tein), the interactions of solvent with the protein, and the effects of Ca^{2+} ions on the protein and solvent. Calcium–solvent interactions are dominated by the large gain in solvent entropy on binding of the calcium to the protein due to the release of the cage of water molecules solvating the ion. This effect is thought to be fairly constant per calcium ion bound and actually dominates the binding free energy (Linse and Forsén, 1995).

2. Calmodulin as a Calcium Sensor

The energetic cost of exposing the large hydrophobic surface therefore seems to be paid for by a reduction in calcium affinity with respect to a theoretical maximum that could be attained if the unfavorable effects of the protein conformation did not counteract some of the favorable effects of releasing the solvating waters. It is precisely this "lowered" binding affinity that enables the protein to sense variations in Ca^{2+} concentration in the range appropriate to intracellular signals. A portion of the theoretical maximum in Ca^{2+} affinity is recovered on binding to peptides. This correlates well with the decrease in the exposure of the hydrophobic surface of calmodulin in the complex. Here again we see the interrelatedness of the biological function and biophysical properties of calmodulin.

However, the relationship between the structure and energetics of calmodulin is still not entirely clear. For example, the fact that apo calmodulin occupies the closed conformation strongly implies that the internal energy of the protein and the interaction of the protein with solvent is more favorable in the closed conformation than in the open conformation. When Ca^{2+} binds, the open conformation is occupied, and so it must now be more favorable than the closed conformation. The specific reasons for this switch in relative stabilities of the two conformations on calcium binding are not known. One hypothesis is that the closed conformation is somehow destabilized by calcium binding, perhaps because the coordination geometry required for binding the Ca^{2+} ion imposes local geometrical constraints on the protein, creating conformational strain in the closed conformation that is relieved in the open conformation. It is also possible that the energetic cost of exposing the hydrophobic surface is not as high as intuitively imagined. For instance, a possible role for methionine in making the open conformation more energetically accessible is discussed in the next section.

E. Role of Methionines

Analysis of the amino acid sequences of the EF-hand CaBPs shows that calmodulin has an unusual number of methionines: 5.3% in the N-terminal domain and 6.8% in the C-terminal domain. This contrasts sharply with an average of only 1.5% in a standard protein database (Rose *et al.*, 1985). A higher than usual percentage of methionine is seen in other regulatory EF-hand CaBPs as well. In an alignment of 38 EF-hand calcium-binding domains (Nelson and Chazin, 1996), 29 are found to have above average (greater than 2%) methionine content. Within the subfamily of proteins most closely related to calmodulin, the average methionine content is 5.8%. Furthermore, in these closely related proteins, methionines are overwhelmingly found in four conserved positions.

Every EF-hand Ca^{2+}-binding domain known from structural studies to undergo a conformational change (both domains of calmodulin and troponin C, as well as α-spectrin) contains more than 5% methionine. Furthermore, calbindin D_{9k}, a CaBP known not to undergo a calmodulin-like conformational change, contains no methionines. Even within a given protein, there are noticeable differences in methionine content in functional regulatory domains versus ancestral domains containing nonfunctional sites, with ancestral domains having fewer methionines. For example, calbindin D_{28k} has nonfunctional sites in domains 1 and 3, both of which are approximately 1% methionine. Domain 2, which contains only functional calcium-binding sites, is 4.2% methionine. Hippocalcin is an extreme example, with only 1% methionine in an ancestral domain and 7% methionine in its putative regulatory domain.

The reason for the inordinately large number of methionines in calmodulin appears to be associated with the unique chemical properties of this residue, which are critically important to the biochemical functions of the protein. The open conformation of calmodulin exposes a large hydrophobic patch to solvent, which, as discussed in the previous section, intuitively seems both enthalpically and entropically unfavorable. A surprisingly large 46% of this exposed surface consists of methionine side chains (Fig. 9) (O'Neil and DeGrado, 1990). Methionine is normally classified as a hydrophobic amino acid, yet it is unlike any other hydrophobic residue. The side chain is very flexible, due to the combination of the unbranched chain and the long C—S bond. This longer bond allows essentially free rotation about the χ_3 torsion angle: there are no preferred rotamers as seen for most other side chains (Gellman, 1991). Another important property is that sulfur is more polarizable than car-

Fig. 9. Disposition of methionine residues in the calcium-loaded state of calmodulin. The space-filling model of Ca^{2+}-loaded calmodulin with methionine residues shaded in dark gray was rendered in Insight (MSI, San Diego, CA) using PDB coordinates 1CLL. Two orientations are provided to visualize all of the methionines in the N-terminal and C-terminal domains.

2. Calmodulin as a Calcium Sensor

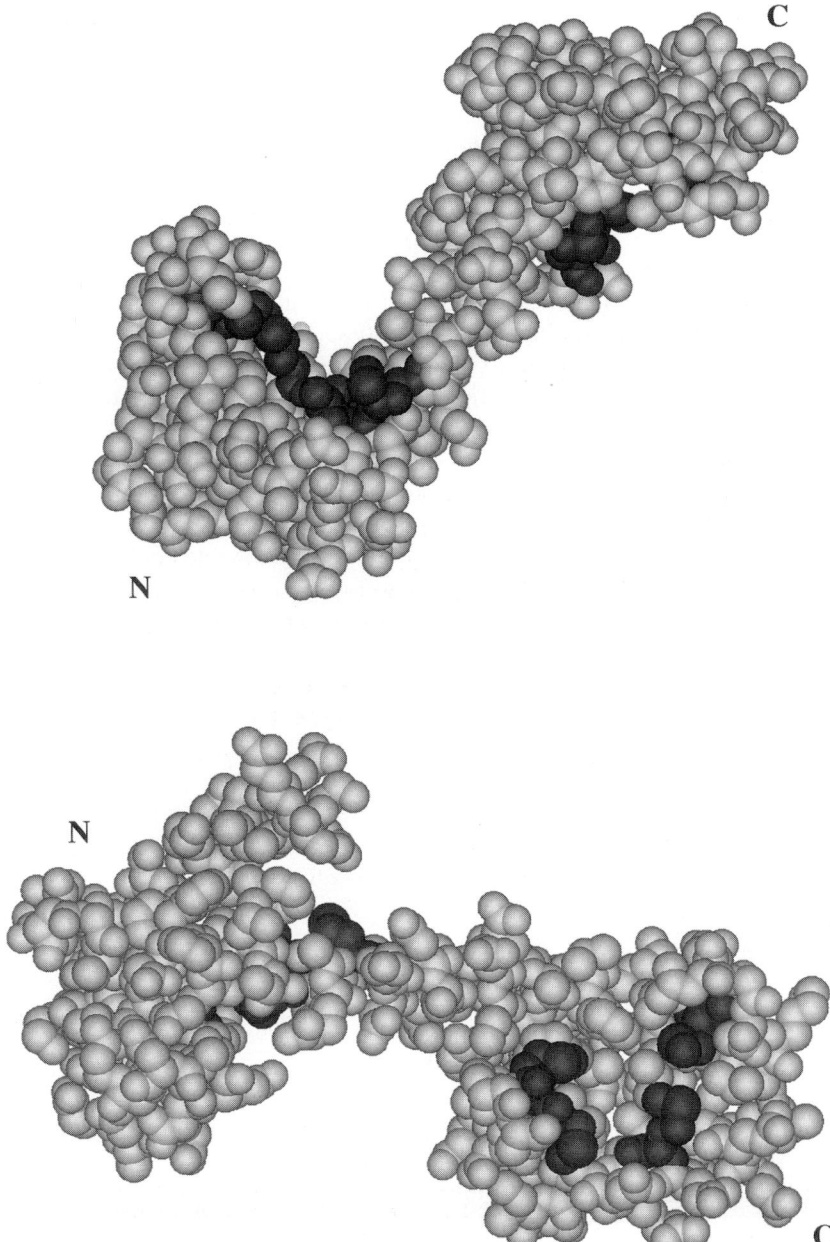

bon. This leads to stronger van der Waals contacts and also gives methionine a more plastic nature (Gellman, 1991), presumably allowing it to adapt to both sequestered and solvent accessible environments. In fact, experimental measurements of solvation energies using the octanol to water partitioning of host–guest peptides find that methionine is easier to solvate than most other hydrophobic amino acids (Wimley et al., 1996).

The arguments presented earlier all support the concept that methionine residues are important for making the open conformation more energetically accessible. This does not preclude the generally accepted role for methionines in allowing calmodulin to interact with a wide range of intracellular targets, which will be discussed in the next section. Several studies have shown that the oxidation of methionine residues decreases the ability of calmodulin to activate its targets (e.g., Walsh and Stevens, 1978; Gopalakrishna and Anderson, 1985; Guerini et al., 1987; O'Neil et al., 1989). Studies in which the methionine residues have been mutated to leucine (Zhang et al., 1994) or glutamine (Chin and Means, 1996) further demonstrate the importance of these residues in calmodulin–target interactions. It has been suggested that the flexible and polarizable methionine side chains allow calmodulin to bind to a variety of targets (O'Neil and DeGrado, 1990; Vogel and Zhang, 1995). However, troponin C also has a high percentage of methionines, yet it has only one target. This implies an additional role for the methionine residues in EF-hand Ca^{2+}-binding proteins, which we propose is the stabilization of the open conformation (Nelson and Chazin, 1998).

IV. TRANSDUCTION OF THE CALCIUM SIGNAL

A. Binding of Peptides Derived from Calmodulin Targets

Calcium-loaded calmodulin binds to and activates a wide range of target proteins. Perhaps the most well studied of these are protein kinases and phosphatases. These enzymes are activated by the release of an autoinhibitory domain that is bound to the active site in the resting state. Ca^{2+}-loaded calmodulin activates these proteins by binding to a site near to or overlapping with the autoinhibitory domain, causing the autoinhibitory domain to dissociate from the active site. Proteins regulated in this manner can be made constitutively active (i.e., Ca^{2+} and calmodulin independent) by either cleaving the entire regulatory domain (which includes the autoinhibitory domain and the calmodulin-binding site) or removing only the autoinhibitory domain (Crivici and Ikura, 1995). Examples of such proteins include myosin light-chain kinase (MLCK) (Blumenthal and Krebs, 1987; Lukas et al., 1986; Persechini and Kretsinger, 1988) and calmodulin-dependent protein kinase II (CaMKII) (Colbran et al., 1989, Colbran and Soderling, 1990; Hanson and Schulman, 1992; Payne et al., 1988).

Calmodulin also regulates various proteins involved in the generation of second messengers, including calmodulin-dependent cyclic nucleotide phosphodiesterase (PDE) (Charbonneau, 1990; Novack et al., 1991) and nitric oxide synthase (Lowenstein and Snyder, 1992; Vorherr et al., 1993). The precise mechanism of regulation of this class of proteins is not fully understood, but it is believed to be similar to the regulatory mechanism used by the kinases and phosphatases (Crivici and Ikura, 1995).

Even less is known about the role of calmodulin in the regulation of cytoskeletal proteins and their regulatory proteins. Examples of such proteins with which calmodulin has been reported to interact include spectrin (Steiner et al., 1989; Stromqvist et al., 1988) and brush-border myosin I (Collins et al., 1990; Swanljung-Collins and Collins, 1991). The functional relevance of calmodulin interaction with these proteins is not clear.

Because of the multidomain nature and large size of the target proteins of calmodulin, there is currently no three-dimensional structure determined for a complex between Ca^{2+}-loaded calmodulin and any of its protein targets. However, critical structural knowledge about effector–receptor interactions has been obtained from three structures of calmodulin complexed with small peptides derived from the calmodulin-binding domains of target proteins. The first structure to appear was an NMR solution structure of calmodulin complexed with a 26-residue peptide corresponding to the calmodulin-binding domain of skeletal MLCK (sMLCK) (Fig. 10; Ikura et al., 1992). This was followed by X-ray crystal structures of calmodulin bound to a 20-residue peptide representing the calmodulin-binding domain of smooth muscle MLCK (smMLCK) (Meador et al., 1992) and bound to a 24-residue peptide corresponding to the calmodulin-binding domain of CaMKII (Meador et al., 1993).

There are many similarities among these three structures. The overall structure of calmodulin in all three complexes is globular and ellipsoid. The central helix connecting the two globular domains of calmodulin is disrupted in the bound structures, allowing the N-terminal and

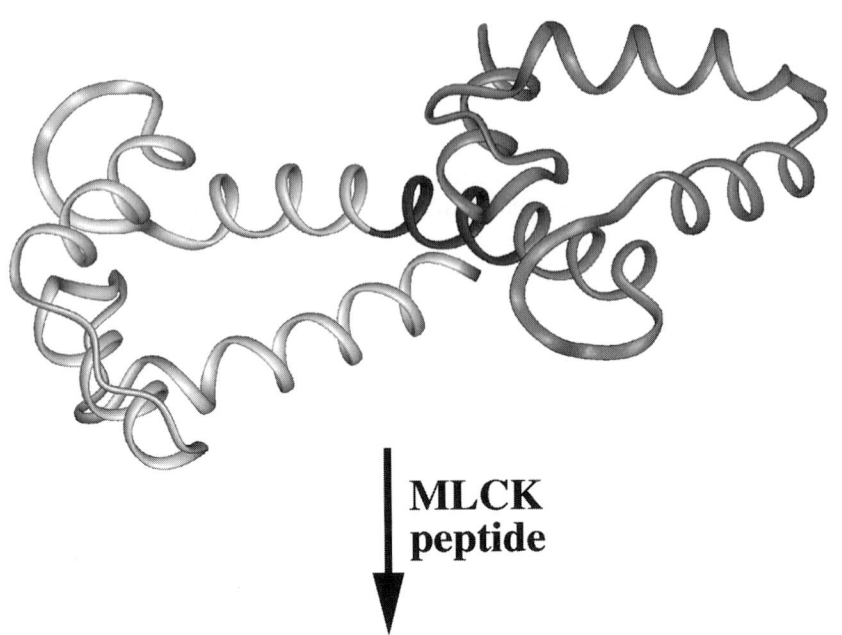

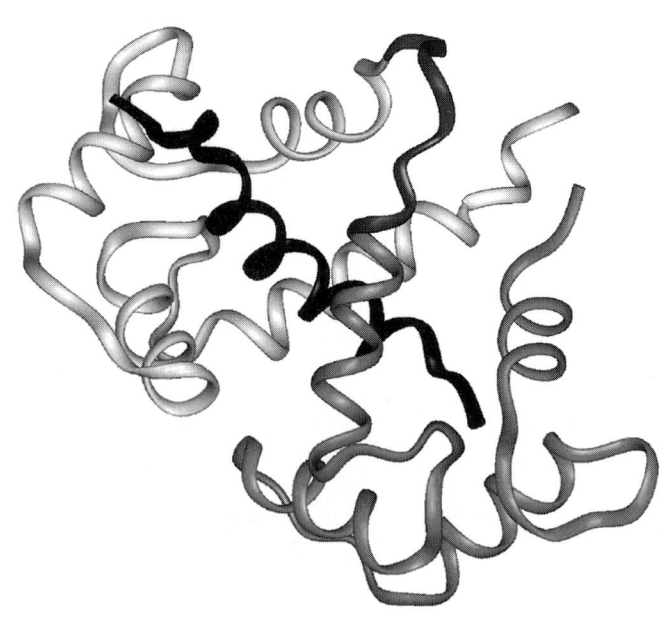

2. Calmodulin as a Calcium Sensor

C-terminal domains, which are quite far apart in the uncomplexed Ca^{2+}-loaded calmodulin structure, to come into close proximity with each other and even form contacts (Fig. 10). Although the orientation of the two domains with respect to each other is drastically different in the uncomplexed versus the complexed structure, the domains themselves are largely unaffected by peptide binding. The preexisting hydrophobic surfaces of the two domains exposed on Ca^{2+} binding form a channel that runs diagonal to the long axis of the ellipsoid.

The general mode of binding of the peptide is also conserved. In all three structures, the peptide is engulfed in the interdomain hydrophobic channel, anchored at both ends by large hydrophobic residues from the peptide (Fig. 11). Thus, all three complexes are stabilized by extensive hydrophobic interactions. Electrostatic interactions are also seen between the basic residues found at the N terminus of the peptide and nearby acidic residues on calmodulin.

Whereas the two structures of calmodulin–MLCK peptide complexes are extremely similar, there are some significant differences between them and the calmodulin–CaMKII peptide structure (Fig. 12). The key difference is in the number of residues between the two anchoring hydrophobic residues on either terminus of the peptides. In the two MLCK structures, there are 12 residues between the anchors, whereas in the CaMKII structure, there are only 8. The overall mode of binding remains very similar, however. In order to accommodate the shorter binding site on the CaMKII peptide, the two domains of calmodulin are closer together in the calmodulin–CaMKII structure than in the two calmodulin–MLCK structures (Meador *et al.*, 1993). This is accomplished via a further unraveling of the central helix, clearly confirming its role as a flexible "expansion joint" or "tether." The additional unraveling occurs at the C-terminal end of the central helix. In all three structures, the tether begins at Ala-73. In MLCK–calmodulin structures, the tether

---

Fig. 10. Change in quaternary structure on binding of the myosin light-chain kinase peptide to Ca^{2+}-loaded calmodulin. The calmodulin N-terminal domain (residues 1–75) is colored light gray, the calmodulin C-terminal domain (residues 84–148) is colored medium gray, the linker between the two domains is colored dark gray, and the MLCK peptide is colored black. The molecular structures were rendered in Insight (MSI, San Diego, CA), using PDB coordinates 1CLL (Ca^{2+}-loaded calmodulin) and 2BBM (calmodulin–MLCK peptide complex).

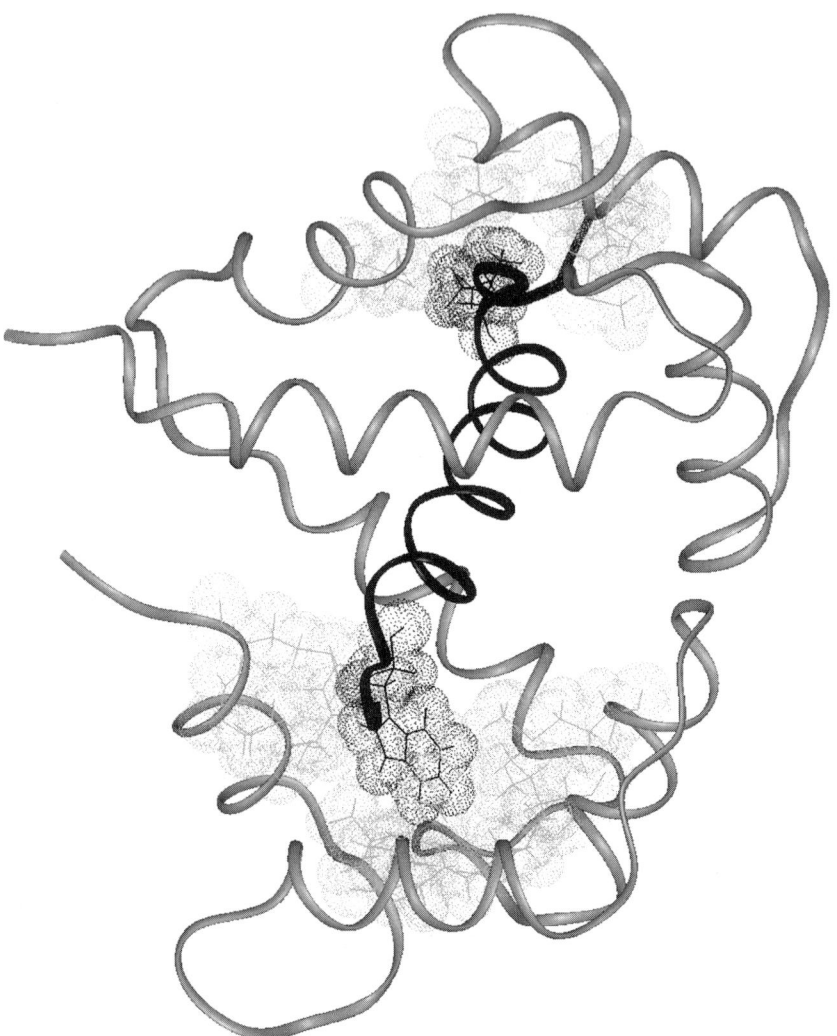

Fig. 11. Hydrophobic interactions between calmodulin and the myosin light-chain kinase peptide. Calmodulin is colored gray and the MLCK peptide is colored black, The van der Waals surfaces of the hydrophobic anchors on either end of the peptide (Trp-800 and Leu-813) are shown as black dots. The van der Waals surfaces of the hydrophobic residues in calmodulin, which interact with these anchors, are shown as gray dots. The molecular structure and surfaces were rendered in Insight (MSI, San Diego, CA) using PDB coordinates 2BBM.

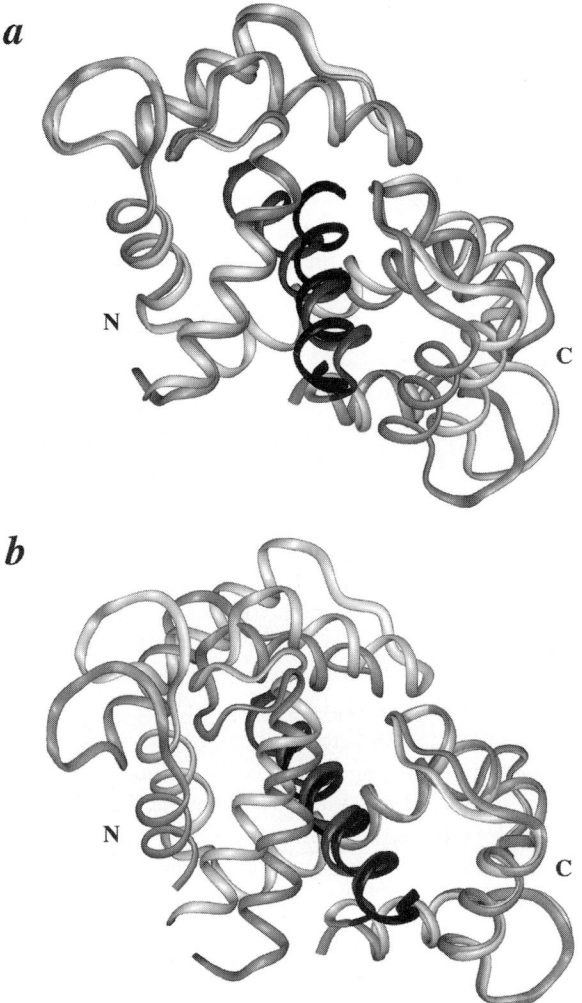

Fig. 12. Difference in the domain orientation of calmodulin in complexes with the myosin light-chain kinase peptide and the calmodulin kinase II peptide. The structure of the calmodulin complex with the 8-residue CaMKII peptide is superimposed on the structure of the complex with the 12-residue MLCK peptide. The backbone atoms of the N-terminal domain of calmodulin in the two complexes were superimposed in (a) and the backbone atoms of the C-terminal domain were superimposed in (b). The two domains of calmodulin are closer together in the CaMKII complex than in the MLCK complex. Calmodulin is shown in light gray in the complex with the MLCK peptide and in medium gray in the complex with the CaMKII peptide. The MLCK peptide is dark gray, and the CaMKII peptide is black. The structures were superimposed and rendered in Insight (MSI, San Diego, CA) using PDB coordinates 1CDL (MLCK complex) and 1CDM (CaMKII complex).

terminates at residue Lys-77, whereas in the CaMKII complex it extends to residue Glu-83. As the tether in the CaMKII complex encompasses the entire length of the linker between the end of calmodulin helix D and the beginning of calmodulin helix E, it is unlikely that calmodulin can accommodate a peptide with fewer than 8 residues between the two anchoring hydrophobic residues when using this mode of binding.

Despite the similarities seen in the three structures just described, calmodulin target peptides do not have highly similar sequences. However, the peptides can usually be mapped as amphiphilic helices and tend to have two conserved bulky hydrophobic residues at either end, often spaced 12 residues apart. They also tend to have an arginine residue adjacent to the second hydrophobic anchor in the N-terminal direction. In the calmodulin–MLCK crystal structure, this arginine is involved in extensive contacts with the region of the bend in the central helix, as shown in Fig. 13. It is the only peptide residue involved in such contacts, and it has been postulated that this may help maintain or even initiate the bent structure (Meador et al., 1993).

The binding of calmodulin to targets has been shown to have a significant effect on its affinity for Ca^{2+} (see chapter by Hu and Van Eldik). Because the open conformation of the individual domains of Ca^{2+}-loaded calmodulin is not affected greatly by binding target peptides, thermodynamics requires that the binding of the peptide will increase the Ca^{2+} affinity of the protein (and conversely, the binding of Ca^{2+} increases the affinity for the peptide). This is indeed what is seen experimentally (Linse and Forsén, 1995). The phenomenon can be rationalized in molecular terms by considering the Ca^{2+} affinity of calmodulin to be composed of a portion due to effects directly involving the Ca^{2+} ion and a portion due to the effect of the protein conformation. The effects of the protein conformation include the energetic cost of exposing the large hydrophobic surface, which, as discussed earlier, is presumably paid for in part by a reduction in the Ca^{2+} affinity. In the protein–peptide complex, because the hydrophobic surface is sequestered by contacts to the peptide instead of being exposed to solvent, the penalty is reduced, and consequently the calcium affinity should increase (Linse and Forsén, 1995).

Calmodulin binds to its targets extremely tightly: K_d values range from 10^{-7} to 10^{-11} (Crivici and Ikura, 1995). Calmodulin is able to bind so tightly to such a wide array of targets due to its own plasticity. As described earlier, the central helix functions as a flexible tether, allowing calmodulin to bind to peptides with different numbers of residues between the two hydrophobic anchors. Furthermore, the contacts between

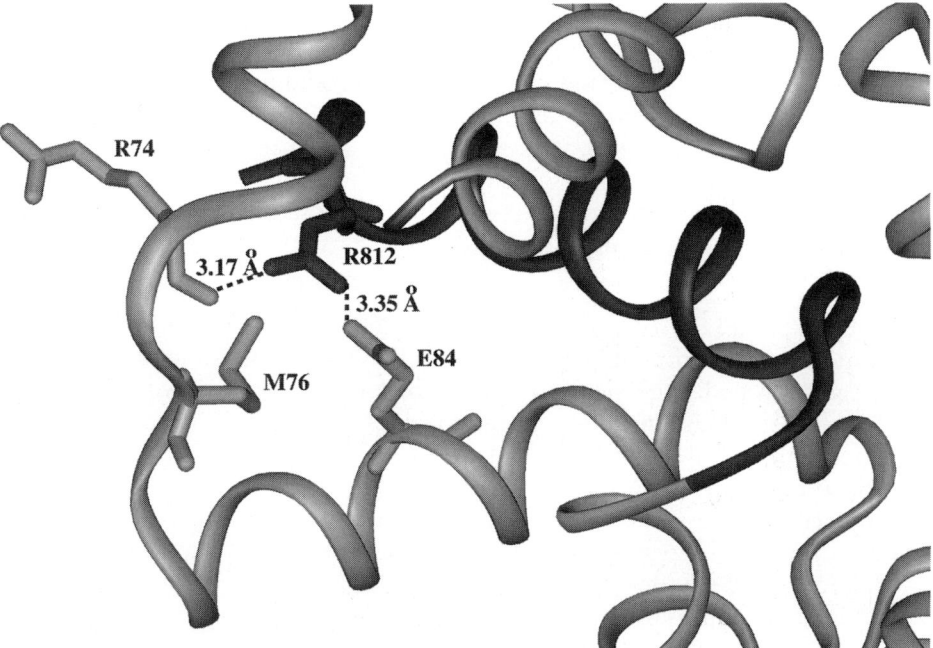

Fig. 13. Interactions between the conserved arginine from the target peptide and the calmodulin linker region. Arginine-812 from the MLCK peptide hydrogen bonds with two residues of calmodulin at the beginning and at the end of the flexible tether between the N-terminal and the C-terminal domains. This arginine, found one residue before the C-terminal hydrophobic anchor of the peptide, is well conserved in calmodulin target peptides. Calmodulin is light gray and the MLCK peptide is dark gray. The structure was rendered in Insight (MSI, San Diego, CA) using PDB coordinates 1CDL.

calmodulin and the peptides are mostly van der Waals interactions, which can be rather nonspecific, as opposed to more specific hydrogen bonds. For example, in the calmodulin–smMLCK crystal structure, 80% of all calmodulin–peptide contacts are van der Waals interactions, and only 15% involve hydrogen bonding (Meador *et al.*, 1992). This nonspecificity may be further aided by the amino acid composition of the hydrophobic surface of calmodulin. The methionines, which constitute such a substantial portion of this hydrophobic interface, appear to have an important role to play, as all nine are found to interact with the peptide. Here again, the special features of the methionine side chain (the flexibility and increased polarizability of the sulfur atom relative to carbon) make it an ideal residue for the strong, but nonspecific van der Waals interactions required for binding to the multitude of calmodulin receptors (O'Neil and DeGrado, 1990; Vogel and Zhang, 1995).

B. Interactions with Other Ligands

Even before the calmodulin–peptide structures were solved, an important role for hydrophobic interactions in the binding of calmodulin to its targets was deduced from the existence of several classes of hydrophobic antagonist drugs, including phenothiazines (Levin and Weiss, 1978), naphthalene sulfonamides (Hidaka et al., 1981), imidazoles (Gietzen et al., 1981), and dihydropyridines (Boström et al., 1981). All of these molecules were shown to bind to calmodulin and inhibit its ability to activate target enzymes. Several fluorescent probes also bind to calmodulin, including 9-anthroylcholine (9AC), 8-anilino-1-napthalenesulfonate (ANS), and 2-p-toluidinyl-napthalene-6-sulfonate (TNS) (LaPorte et al., 1980; Tanaka and Hidaka, 1980). On binding to calmodulin, the fluorescence optima of these probes shift to shorter wavelengths, indicating that they bind to a hydrophobic site (Forsén et al., 1986). Furthermore, the Ca^{2+}-dependent interaction of calmodulin with hydrophobic supports such as phenyl–Sepharose has long been used as a method of purification (Gopalakrishna and Anderson, 1982; Vogel et al., 1983).

Complexes of calmodulin with various drugs have been studied extensively. Because these drugs antagonize calmodulin function, it is likely that they bind in the same site on calmodulin as its target peptides do (Forsén et al., 1986). Nuclear magnetic resonance studies also indicate that the drugs bind in the same hydrophobic pockets used by the target peptides (Dalgarno et al., 1984; Craven et al., 1996). However, NMR studies on J-8 (a naphthalene sulfonamide) and trifluoperazine (a phenothiazine derivative) indicate that, unlike the peptides with which calmodulin interacts, the drugs are in fast exchange between bound and free states (Craven et al., 1996). This is consistent with the micromolar affinity of calmodulin for these drugs seen in many biochemical studies (Levin and Weiss, 1978; Jackson and Puett, 1986; Massom et al., 1990, 1991). Thus, the affinity of calmodulin for these drugs is typically 10^3–10^5 weaker than for target peptides.

2. Calmodulin as a Calcium Sensor

The stoichiometry of calmodulin–drug interactions remains unclear. For example, equilibrium methods have found a stoichiometry of between two and seven equivalents of trifluoperazine (TFP) bound per calmodulin molecule (Levin and Weiss, 1978; Jackson and Puett, 1986; Massom et al., 1990, 1991). Furthermore, although some studies find no cooperativity in the binding of TFP molecules (Massom et al., 1990), others find positive cooperativity (Levin and Weiss, 1978). Some of this confusion may be due to differences in ionic strength in the various binding assays (Massom et al., 1990). It does seem certain that calmodulin binds at least two equivalents of TFP using the same hydrophobic surface that binds peptides. Mellitin (a natural venom peptide which, as discussed later, binds calmodulin with nanomolar affinity) competes with two equivalents of TFP. These two equivalents are presumably bound to the peptide-binding hydrophobic surface (Massom et al., 1990). Furthermore, NMR methods find two major binding sites for TFP on calmodulin, one in each domain, and localize these sites to the same region as used in peptide binding. Secondary binding sites were also seen (Craven et al., 1996).

Even the two crystal structures of the calmodulin–TFP complexes do not clear up the confusion over the number of binding sites. One structure has only one TFP molecule bound per molecule of calmodulin (Cook et al., 1994), whereas the other has four TFP molecules bound to one molecule of calmodulin (Vandonselaar et al., 1994). Both structures show a globular conformation of calmodulin very similar to that seen in calmodulin–peptide complexes. Both also have a TFP molecule bound in the hydrophobic pocket of the C-terminal domain, although the phenothiazine ring in one structure is rotated 180° with respect to its position in the other structure. This causes the trifluoromethyl group and the piperazine ring to point in opposite directions in the two structures. This recent structural information is consistent with earlier reports that the C-terminal domain of calmodulin has a higher affinity for TFP than the N-terminal domain (Thulin et al., 1984). Not only does the single TFP molecule in the Cook et al. (1994) structures bind to the C-terminal domain, but in the structure with four TFP molecules bound, the TFP molecule in the C-terminal pocket has the best density of the four molecules. Interestingly, the TFP molecule with the poorest density is bound in the hydrophobic pocket of the N-terminal domain. The remaining two TFP molecules are bound to other parts of the hydrophobic "tunnel" created by the hydrophobic pockets of the two domains. In both structures, the majority of the binding interactions between the TFP molecule(s) and calmodulin are hydrophobic.

The important role of hydrophobic, nonspecific van der Waals interactions is further supported by a wide array of studies on binding of various natural and synthetic peptides to calmodulin. Calmodulin has been shown to bind neurotransmitters such as substance P (Yoshino *et al.*, 1993), hormones such as glucagon and adrenocorticotropin (ACTH) (Malencik and Anderson, 1983), and venoms such as mellitin and mastoporan (Kataoka *et al.*, 1989; Matsushima *et al.*, 1989) in a Ca^{2+}-dependent manner. Because these peptides are either exogenous or secreted from the cell, these interactions are of questionable physiological relevance and may simply be due to the general "stickiness of Ca^{2+}-loaded calmodulin. However, solution X-ray scattering studies indicate that these calmodulin complexes adopt a globular conformation similar to that seen in the calmodulin–target peptide complexes, and not a dumbell conformation as seen in free Ca^{2+}-loaded calmodulin (Yoshino *et al.*, 1989, 1993). Furthermore, NMR studies indicate that these natural peptides bind to calmodulin in the same general region as peptides derived from protein targets (Ohki *et al.*, 1991).

The putative role for nonspecific interactions in calmodulin target binding has been tested using synthetic peptides. DeGrado and co-workers hypothesized that the structural feature calmodulin recognizes in its targets is that of a basic amphiphilic α-helix (O'Neil and DeGrado, 1990). They designed a series of peptides composed only of leucine, lysine, and tryptophan, unrelated to any known calmodulin-binding sequence. These peptides bound calmodulin with K_d values of 10^{-9} to 10^{-10} M, which is in the range of the K_d values seen for calmodulin target peptides. The conformation of one of these synthetic peptides has been analyzed using NMR techniques (Prêcheur *et al.*, 1992). When not bound to calmodulin, the peptide exchanges rapidly between the unfolded state and a helical conformation, which is populated a significant percentage of the time. When the peptide binds to calmodulin, it adopts a stable, helical conformation.

The important role of nonspecific interactions is further indicated by the fact that even peptides synthesized from all D-amino acids will bind to calmodulin (Fisher *et al.*, 1994). Presumably, these D analogs can make nonspecific van der Waals interactions with calmodulin, but will be unable to make the same specific hydrogen bonding and electrostatic interactions. The D analogs of mellitin and a peptide derived from the smMLCK calmodulin-binding site were synthesized and shown to bind to calmodulin. However, although NMR showed the calmodulin–L-smMLCK complex to be in slow exchange with its dissociated components ($k_{off} \sim 1$ sec^{-1}), the calmodulin–D-smMLCK complex was in fast exchange ($k_{off} \sim 10^2$ sec^{-1}), indicative of weaker binding. Furthermore, although both D and L peptides seem to interact with the same region of calmodulin, the differences in chemical shift seen in comparing Ca^{2+}-loaded calmodulin to the D-smMLCK complex were smaller in magnitude than those seen between Ca^{2+}-loaded calmodulin and the L-smMLCK complex (Fisher et al., 1994).

The differences between the binding of the D peptide and the L peptide may be due to a difference between the formation of a complex based only on nonspecific van der Waals interactions and a more specific complex also involving the key hydrogen-bonding and electrostatic interactions seen in the three calmodulin–peptide structures. This interpretation is supported by kinetic data on the binding of sMLCK peptide to calmodulin. The binding event was shown to involve two steps: a fast association step and a slower isomerization step (Torok and Trentham, 1994). Presumably the fast association step only involves nonspecific van der Waals interactions. After this initial step the peptide may not be aligned optimally in the binding pocket of calmodulin. The slower isomerization would then occur when the specific interactions between calmodulin and the target peptide are formed. The studies of Wand and co-workers (Ehrhardt et al., 1995) showed that the MLCK peptide has little helical content until binding to CaM. The initial binding and subsequent folding of the peptide could also explain some of the observed kinetic phenomena.

Even tight, specific binding is not equivalent to activation. There are numerous studies in which a mutant form of calmodulin only partially activates target proteins. In fact, the activation of MLCK can be dissociated from the binding event. For example, George and co-workers (1990) constructed a chimera protein composed of the first EF-hand of cardiac troponin C and the remainder of calmodulin. The troponin C binding loop, normally nonfunctional, was restored via mutagenesis. Although this chimera protein is unable to activate MLCK, it does serve as a potent competitive inhibitor ($K_i = 66$ nM), implying that the chimera is able to bind tightly to MLCK, but cannot activate it.

C. Binding to Targets in the Absence of Ca^{2+}

A new mode of calmodulin regulation has begun to be elucidated. Neuromodulin and some unconventional myosins have been shown to bind to calmodulin with higher affinity in the absence of Ca^{2+} than in the presence of Ca^{2+} (Alexander et al., 1987, 1988; Houdusse and Cohen, 1995). It has been suggested that perhaps neuromodulin, which is plasma membrane associated, functions as a trap for calmodulin, keeping it sequestered near the plasma membrane in the absence of the activating calcium signal (Liu and Storm, 1990).

This interaction with calmodulin is presumed to be mediated by an IQ motif on neuromodulin. The IQ motif (IQXXXRGXXXR) was first identified in myosin (Xie et al., 1994), where the C-terminal domains of the essential and regulatory light chains (ELC and RLC) bind to the heavy chain via an IQ motif on the heavy chain. This binding occurs to a "semi-open" conformation of the light-chain C-terminal domain, which is distinct from the previously described closed and open conformations of calmodulin and troponin C (Houdusse and Cohen, 1995, 1996; Houdusse et al., 1996). Because myosin ELC and calmodulin are highly similar in both sequence and structure and have analogous receptor-binding motifs, the conformation of the C-terminal domain of calmodulin in the absence of Ca^{2+} is of interest. In fact, Houdusse et al. (1996) have built a model of calmodulin complexed with an unconventional myosin based on their structure of the scallop myosin regulatory domain. The C-terminal domain of calmodulin was modeled in the semi-open conformation. It has also been proposed that the apo state of this domain occupies a semi-open conformation (Swindells and Ikura, 1996) or has a predisposition to form the semi-open conformation (M. B. Swindells and M. Ikura, personal communication, 1997).

The semi-open conformation can be defined and distinguished from the open and closed conformations by comparing the orientations and interactions of the helices of the EF-hands in the two domains of myosin ELC (Nelson and Chazin, 1998). As in calmodulin, the helices in the myosin ELC N-terminal domain are labeled A–D and the helices in the C-terminal domain are labeled E–H. A DDM and contact map-based comparison of the closed ELC N-terminal domain and the semi-open ELC C-terminal domain reveals that the disposition of helices G and H in the C-terminal domain of ELC is quite distinct from that of the analogous helices C and D in the N-terminal domain (Tables III and IV). The two helices do not move apart in the semi-open conformation of the myosin ELC. Rather, they are pivoted like scissors, with the pivot point near the middle of the two helices. The interhelical angle is nearly 90° in the semi-open C-terminal domain compared to 145° for the corresponding angle in the N-terminal domain. Compared to the disposition

2. Calmodulin as a Calcium Sensor

TABLE III

Range of DDM Entries for Apo and Ca^{2+}-Loaded Calmodulin (CaM), and Myosin Essential Light-Chain N-Terminal and C-Terminal Domains[a]

Helical interface	Ca^{2+}-loaded CaM N vs C	Apo CaM N vs C	ELC N vs C
First/second	0–1.81	−2.77–2.19	−9.18–0
First/third	*	−1.47–2.9	−5.68–3.71
First/fourth	*	−1.67–2.89	−7.38–3.15
Second/third	*	2.5–2.15	−9.82–5.47
Second/fourth	0–1.72	−1.99–3.19	−14.28–0
Third/fourth	0–1.71	−1.85–3.87	−7.40–7.08

[a] Distance difference matrices (*DDM*) were calculated using DISCOM (Gippert, 1995). Entries with an absolute value less than 1 are not considered significant, and an asterisk indicates no significant entries. Coordinates: ELC X-ray crystal structure (Houdusse and Cohen, 1996; PDB accession number 1WDC), apo (Kuboniwa *et al.*, 1995; PDB accession number 1CFC), and calcium-loaded calmodulin (Chattopadhyaya *et al.*, 1992, PDB accession number 1CLL).

of helices C and D in the N-terminal domain, the N terminus of helix G is further away from the C terminus of helix H and the C terminus of helix G is further away from the N terminus of helix H. This difference

TABLE IV

Contact Analysis of Apocalmodulin and Myosin Essential Light Chain[a]

		Helical interface					
Protein	Type of contact	First/second	First/third	First/fourth	Second/third	Second/fourth	Third/fourth
CaM	Unique to N domain	1	0	4	2	3	0
	Unique to C domain	6	0	5	3	2	7
	Both	10	0	6	5	4	3
ELC	Unique to N domain	10	0	3	8	8	8
	Unique to C domain	3	0	5	6	0	7
	Both	3	0	6	2	2	0

[a] Contact maps were calculated using CHARMM (Brooks *et al.*, 1982). Two residues are considered to be in contact if their centers of geometry are within 8 Å of each other. Coordinates: apocalmodulin (Kuboniwa *et al.*, 1995; PDB accession number 1CFC) and ELC (Houdusse and Cohen 1996; PDB accession number 1WDC).

in the interface of the third and fourth helices of the domain uniquely defines the semi-open conformation. The conformational differences seen between the closed and the semi-open conformations do not appear to be intermediate to the differences between the closed and the open conformations of calmodulin, but rather seem to be of a different type altogether.

To search for the semi-open conformation in calmodulin, fitting-independent DDM and contact map analysis has been applied to the N-terminal and C-terminal domains of calmodulin and the results compared to the myosin ELC analysis (Nelson and Chazin, 1998). The differences between calmodulin N-terminal and C-terminal domains, first noted by Swindells and Ikura (1996), are found to be of a smaller magnitude and a different type than those seen between ELC N-terminal and C-terminal domains (Tables III and IV). The most critical structural element is the interface between the third and the fourth helix within each globular domain. In this interface, differences between the N-terminal and C-terminal domains of calmodulin are completely unlike those between the corresponding helices in the myosin ELC domains (Table IV, boldface numbers). In the C-terminal domain of the ELC, helices G and H scissor apart, disrupting most of the interface. In calmodulin, helices G and H in the C-terminal domain are actually closer together than corresponding helices C and D in the N-terminal domain. Although the two domains of calmodulin do differ, the C-terminal domain does not occupy the archetypal semiopen conformation of the C-terminal domain of myosin ELC (Fig. 14).

This does not imply, however, that the C-terminal domain of calmodulin does not adopt the semi-open conformation when it is bound to neuromodulin. In fact, based on homology to the myosin case, it seems likely that the calmodulin C-terminal domain occupies a semi-open conformation in the complex. Preliminary NMR studies using a peptide derived from neuromodulin indicate that neuromodulin binds to the C-terminal domain of calmodulin, analogous to the myosin case (Urbauer *et al.*, 1995). However, proof that neuromodulin interacts with calmodulin in a manner similar to the interaction of myosin heavy and light chains will require a structure of the complex.

The structures determined for calmodulin in the presence and absence of ligands represent major steps toward developing an understanding of its role as a calcium sensor. Further structural work is crucial for determining if calmodulin binds to all of its targets in the same manner as it binds to the protein kinase targets studied structurally so far. A

2. Calmodulin as a Calcium Sensor

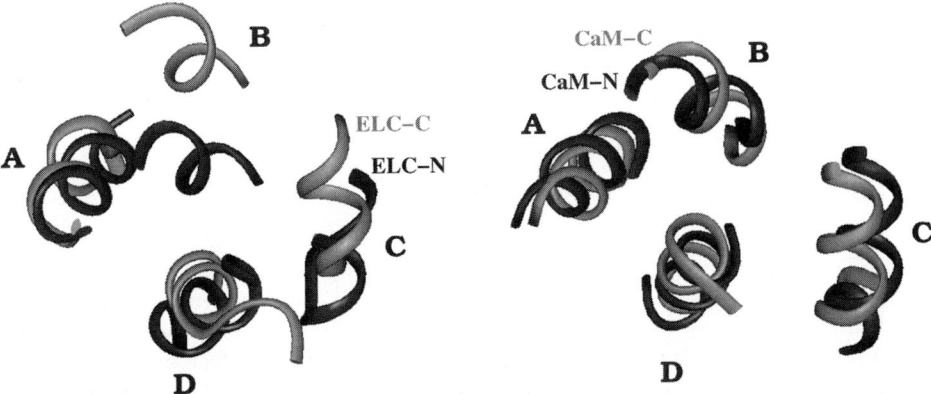

Fig. 14. Comparison of the N-terminal and C-terminal domains of calmodulin and the myosin essential light chain. (Left) Superposition of the backbone atoms in the first and fourth helices of the N-terminal (A,D) and C-terminal (E,H) domains of myosin ELC. Contact maps and the DDM show that these two helices have a fairly similar interface in the two domains. The N-terminal domain (darky gray) is in a closed conformation. The C-terminal domain (light gray) is in a semi-open conformation. (Right) Superposition of the backbone atoms of the first and fourth helices in the N-terminal (A,D) and C-terminal (E,H) domains of apo calmodulin. Contact maps and the DDM show that these two helices have a fairly similar interface in the two domains. The N-terminal domain (dark gray) is in a closed conformation and the C-terminal domain (light gray) is in a closed conformation. All structures were rendered in Insight (MSI, San Diego, CA) using PDB coordinates 1WDC (myosin ELC) and 1CFC (apo calmodulin).

calmodulin–PDE peptide structure would be of particular interest, as the activation of PDE by calmodulin has been studied extensively and shows some different characteristics relative to the activation of MLCK and CaMKII. A structure of a complete target protein complexed with calmodulin would also answer many fundamental questions about how calmodulin activates its targets, providing a deeper understanding of this important event in calcium signal transduction.

ACKNOWLEDGMENTS

We thank Dr. Mitsu Ikura for helpful discussions and suggestions and Brian Sheehan for assistance in preparing figures. Research on Ca^{2+}-binding proteins in this laboratory is supported by operating grants from the National Institutes of Health (GM 40120 and GM 48495), an NSF predoctoral fellowship to MRN, and a Faculty Research Award to WJC from the American Cancer Society.

REFERENCES

Akke, M., Forsén, S., and Chazin, W. (1995). Solution structure of $(Cd^{2+})_1$-calbindin D_{9k} reveals details of the stepwise structural changes along the apo $\rightarrow (Ca^{2+})^{II}_1 \rightarrow (Ca^{2+})^{I,II}_2$ binding pathway. *J. Mol. Biol.* **252**, 102–121.

Alexander, K. A., Cimler, B. M., Meier, K. E., and Storm, D. R. (1987). Regulation of calmodulin binding to P-57. *J. Biol. Chem.* **262**, 6108–6113.

Alexander, K. A., Wakim, B. T., Doyle, G. S., Walsh, K. A., and Storm, D. R. (1988). Identification and characterization of the calmodulin-binding domain of neuromodulin, a neurospecific calmodulin-binding protein. *J. Biol. Chem.* **263**, 7544–7549.

Babu, Y. S., Sack, J. S., Greenhough, T. J., Bugg, C. E., Means, A. R., and Cook, W. J. (1985). Three-dimensional structure of calmodulin. *Nature (London)* **315**, 37–40.

Babu, Y. S., Bugg, C. E., and Cook, W. J. (1988). Structure of calmodulin refined at 2.2 Å resolution. *J. Mol. Biol.* **204**, 191–204.

Barbato, G., Ikura, M., Kay, L. E., Pastor, R. W., and Bax, A. (1992). Backbone dynamics of calmodulin studied by ^{15}N relaxation using inverse detected two-dimensional NMR spectroscopy: The central helix is flexible. *Biochemistry* **31**, 5269–5278.

Beckingham, K. (1991). Use of site-directed mutations in the individual Ca^{2+}-binding sites of calmodulin to examine Ca^{2+}-induced conformational changes. *J. Biol. Chem.* **266**, 6027–6030.

Blumenthal, D. K., and Krebs, E. G. (1987). Preparation and properties of the calmodulin-binding domain of skeletal muscle myosin light chain kinase. *Methods Enzymol.* **139**, 115–126.

Boström, S.-L., Ljung, B., Mardh, S., Forsén, S., and Thulin, E. (1981). Interaction of the antihypertensive drug felodipine with calmodulin. *Nature (London)* **292**, 777–778.

Brooks, B. R., Bruccoleri, R. E., Olafson, B. D., States, D. J., Swaminathan, S., and Karplus, M. (1983). CHARMM: A program for macromolecular energy minimization and dynamics calculations. *J Comput. Chem.* **4**, 187–217.

Brzeska, H., Venyaminov, S., Grabarek, Z., and Drabikowski, W. (1983). Comparative studies on thermostability of calmodulin, skeletal muscle troponin C and their tryptic fragments. *FEBS Lett.* **153**, 169–173.

Carlström, G., and Chazin, W. J. (1993). Two-dimensional ^1H nuclear magnetic resonance studies of the half-saturated $(Ca^{2+})_1$ state of calbindin D_{9k}: Further implications for the molecular basis of cooperative Ca^{2+} binding. *J. Mol. Biol.* **231**, 415–430.

Charbonneau, H. (1990). Structure–function relationships among cyclic nucleotide phosphodiesterases. *In* "Cyclic Nucleotide Phosphodiesterases: Structure, Regulation and Drug Action" (J. Beavo and M. D. Houslay, eds.), pp. 267–296. Wiley, New York.

Chattopadhyaya, R., Meador, W., Means, A., and Quiocho, F. (1992). Calmodulin structure refined at 1.7 Å resolution. *J. Mol. Biol.* **228**, 1177–1192.

2. Calmodulin as a Calcium Sensor

Chin, D., and Means, A. R. (1996). Methionine to glutamine substitutions in the C-terminal domain of calmodulin impair the activation of three protein kinases. *J. Biol. Chem.* **271,** 30465–30471.

Colbran, R. J., and Soderling, T. R. (1990). Calcium/calmodulin dependent protein kinase II. *Curr. Top. Cell. Regul.* **31,** 181–221.

Colbran, R. J., Schworer, C. M., Hashimoto, Y., Fong, Y.-L., Rich, D. P., Smith, M. K., and Soderling, T. R. (1989). Calcium/calmodulin-dependent protein kinase II. *Biochem. J.* **258,** 313–325.

Collins, K., Sellers, J. R., and Matsudaira, P. (1990). Calmodulin dissociation regulates brush border myosin I (110-kD-calmodulin) mechanochemical activity in vitro. *J. Cell Biol.* **110,** 1137–1147.

Cook, W. J., Walter, L. J., and Walter, M. R. (1994). Drug binding by calmodulin: Crystal structure of a calmodulin–trifluoperazine complex. *Biochemistry* **33,** 15259–15265.

Craig, T. A., Watterson, D. M., Prendergast, F. G., Haiech, J., and Roberts, D. M. (1987). Site-specific mutagenesis of the α-helices of calmodulin. *J. Biol. Chem.* **262,** 3278–3284.

Craven, C. J., Whitehead, B., Jones, S. K. A., Thulin, E., Blackburn, G. M., and Waltho, J. P. (1996). Complexes formed between calmodulin and the antagonists J-8 and TFP in solution. *Biochemistry* **35,** 10287–10299.

Crivici, A., and Ikura, M. (1995). Molecular and structural basis of target recognition by calmodulin. *Annu. Rev. Biophys. Biomol. Struct.* **24,** 85–116.

da Silva, A. C. R., and Reinach, F. C. (1991). Calcium binding induces conformational changes in muscle regulatory proteins. *Trends Biochem. Sci.* **16,** 53–57.

Dalgarno, D. C., Klevit, R. E., Levine, B. A., Scott, G. M. M., Williams, R. J. P., Gergely, J., Grabarek, Z., Leavis, P. C., Grand, R. J. A., and Drabikowski, W. (1984). The nature of the trifluoperazine binding sites on calmodulin and troponin C. *Biochim. Biophys. Acta* **791,** 164–172.

Dieter, P., and Marmé, D. (1988). The history of calcium-binding proteins. *In* "Calcium-Binding Proteins" (M. P. Thompson, ed.), Vol. 1, pp. 1–9. CRC Press, Boca Raton, Florida.

Ehrhardt, M. R., Urbauer, J. L. and Wand, A. J. (1995). The energetics and dynamics of molecular recognition by calmodulin. *Biochemistry* **265,** 2731–2738.

Falke, J. J., Drake, S. K., Hazard, A. L., and Peersen, O. B. (1994). Molecular tuning of ion binding to calcium signaling proteins. *Q. Rev. Biophys.* **27,** 219–290.

Finn, B. E., Evenas, J., Crakenberg, T., Waltho, J. P., Thulin, E., and Forsén, S. (1995). Calcium-induced structural changes and domain autonomy in calmodulin. *Nat. Struct. Biol.* **2,** 777–783.

Fisher, P. J., Prendergast, F. G., Ehrhardt, M. R., Urbauer, J. L., Wand, A. J., Sedarous, S. S., McCormick, D. J., and Buckley, P. J. (1994). Calmodulin interacts with amphiphillic peptides composed of all D-amino acids. *Nature (London)* **368,** 651–653.

Flaherty, K. M., Zozulya, S., Stryer, L., and McKay, D. B. (1993). Three-dimensional structure of recoverin, a calcium sensor in vision. *Cell (Cambridge, Mass.)* **75,** 709–716.

Forsén, S., Vogel, H. J., and Drakenberg, T. (1986). Biophysical studies of calmodulin. *In* "Calcium and Cell Function" (E. Cheung, ed.), Vol. 6, pp. 113–157. Academic Press, San Diego.

Gagné, S., Tsuda, S., Li, M., Smillie, L., and Sykes, B. (1995). Structures of the troponin C regulatory domains in the apo and calcium-saturated states. *Nat. Struct. Biol.* **2,** 784–789.

Gellman, S. (1991). On the role of methionine residues in the sequence-independent recognition of nonpolar protein surfaces. *Biochemistry* **30,** 6633–6636.

George, S. E., VanBerkum, M. F. A., Ono, T., Cook, R., Hanley, R. M., Putkey, J. A., and Means, A. R. (1990). Chimeric calmodulin-cardiac troponin C proteins differentially activate calmodulin target enzymes. *J. Biol. Chem.* **265,** 9228–9235.

Gietzen, K., Wuthrich, A., and Bader, H. (1981). R 24571: A new powerful inhibitor of red blood cell Ca^{2+}-transport ATPase and of calmodulin-related functions. *Biochem. Biophys. Res. Commun.* **101,** 418–425.

Gippert, G. (1995). New computational methods for 3D NMR data analysis and protein structure determination in high-dimensional internal coordinate space. Ph.D. Thesis, The Scripps Research Institute, La Jolla, California.

Gopalakrishna, R., and Anderson, W. B. (1982). Ca^{2+}-induced hydrophobic site on calmodulin: Application for purification by phenyl-sepharose chromatography. *Biochem. Biophys. Res. Commun.* **104,** 830–836.

Gopalakrishna, R., and Anderson, W. B. (1985). The effects of chemical modification of calmodulin Ca^{2+}-induced exposure of a hydrophobic region. Separation of active and inactive forms of calmodulin. *Biochim. Biophys. Acta* **844,** 265–269.

Guerini, D., Krebs, J., and Carafoli, E. (1987). Stimulation of the erythrocyte Ca^{2+}-ATPase and of bovine brain cyclic nucleotide phosphodiesterase by chemically modified calmodulin. *Eur. J. Biochem.* **170,** 35–42.

Haiech, J., Kilhoffer, M.-C., Lukas, T. J., Craig, T. A., Roberts, D. M., and Watterson, D. M. (1991). Restoration of calcium binding activity of mutant calmodulins toward normal by presence of a calmodulin binding structure. *J. Biol. Chem.* **266,** 3427–3431.

Hanson, P. I., and Schulman, H. (1992). Neuronal Ca^{2+}/calmodulin-dependent protein kinases. *Annu. Rev. Biochem.* **61,** 559–601.

Heidorn, D. B., and Trewhalla, J. (1988). Comparison of the crystal and solution structures of calmodulin and troponin C. *Biochemistry* **27,** 909–915.

Herzberg, O., and James, M. N. G. (1985). Structure of the calcium regulatory muscle protein troponin-C at 2.8 Å resolution. *Nature (London)* **313,** 653–659.

Herzberg, O., Moult, J., and James, M. N. G. (1986). A model for the Ca^{2+}-induced conformational transition of troponin C. *J. Biol. Chem.* **261,** 2638–2644.

Hidaka, H., Asano, M., and Tanaka, T. (1981). Activity–structure relationship of calmodulin antagonists: Naphthalene derivatives. *Mol. Pharmacol.* **20,** 571–578.

Houdusse, A., and Cohen, C. (1995). Target sequence recognition by the calmodulin superfamily: Implications from light chain binding to the regulatory domain of scallop myosin. *Proc. Natl. Acad. Sci. U.S.A.* **92,** 10644–10647.

Houdusse, A., and Cohen, C. (1996). Structure of the regulatory domain of scallop myosin at 2 Å resolution: Implications for regulation. *Structure* **4,** 21–32.

Houdusse, A., Silver, M., and Cohen, C. (1996). A model of Ca^{2+}-free calmodulin binding to unconventional myosin reveals how calmodulin acts as a regulatory switch. *Structure* **4,** 1475–1490.

Ikura, M. (1996). Calcium binding and conformational response in EF-hand proteins. *Trends Biochem. Sci.* **21,** 14–17.

Ikura, M., Spera, S., Barbato, G., Kay, L. E., Krinks, M., and Bax, A. (1991). Secondary structure and side-chain ^1H and ^{13}C resonance assignments of calmodulin in solution by heteronuclear multidimensional NMR spectroscopy. *Biochemistry* **30,** 9216–9228.

Ikura, M., Clore, G. M., Gronenborn, A. M., Zhu, G., Klee, C. B., and Bax, A. (1992). Solution structure of a calmodulin-target peptide complex by multidimensional NMR. *Science* **256,** 632–638.

2. Calmodulin as a Calcium Sensor

Jackson, A. E., and Puett, D. (1986). Binding of trifluoperazine and fluorene-containing compounds to calmodulin and adducts. *Biochem. Pharmacol.* **25,** 4395–4400.

Kataoka, M., Head, J. F., Seaton, B. A., and Engelman, D. M. (19890. Melittin binding causes a large calcium-dependent conformational change in calmodulin. *Proc. Natl. Acad. Sci. U.S.A.* **86,** 6944–6948.

Kawasaki, H., and Kretsinger, R. H. (1994). Calcium-binding proteins P:1. *Protein Profile* **1,** 343–517.

Kawasaki, H., and Kretsinger, R. H. (1995). Calcium-binding proteins:1. *Protein Profile* **2,** 297–490.

Kilhoffer, M.-C., Kubina, M., Travers, F., and Haiech, J. (1992). Use of engineered proteins with internal tryptophan reporter groups and perturbation techniques to probe the mechanism of ligand–protein interactions: Investigation of the mechanism of calcium-binding to calmodulin. *Biochemistry* **31,** 8098–8106.

Kretsinger, R. H., and Nockolds, C. E. (1973). Carp muscle calcium binding protein: Structure determination and general description. *J. Biol. Chem.* **248,** 3313–3326.

Kuboniwa, H., Tjandra, N., Grzesiek, S., Ren, H., Klee, C. B., and Bax, A. (1995). Solution structure of calcium-free calmodulin. *Nat. Struct. Biol.* **2,** 768–776.

LaPorte, D. C., Wierman, B. M., and Storm, D. R. (1980). Calcium-induced exposure of a hydrophobic surface on calmodulin. *Biochemistry* **19,** 3814–3819.

Levin, R. M., and Weiss, B. (1978). Specificity of the binding of trifluoperazine to the calcium-dependent activator of phosphodiesterase and to a series of other calcium-binding proteins. *Biochim. Biophys. Acta* **540,** 197–204.

Li, M. X., Gagné, S. M., Tsuda, S., Kay, C. M., Smillie, L. B., and Sykes, B. D. (1995). Calcium binding to the regulatory N-domain of skeletal muscle troponin C occurs in a stepwise manner. *Biochemistry* **34,** 8330–8340.

Linse, S., and Forsén, S. (1995). Determinants that govern high-affinity calcium binding. *Adv. Second Messenger Phosphoroprotein Res.* **30,** 89–151.

Linse, S., Johansson, C., Brodin, P., Grundstrom, T., Drakenberg, T., and Forsén, S. (1991). Electrostatic contributions to the binding of Ca^{2+} in calbindin D_{9k}. *Biochemistry* **30,** 154–183.

Linse, S., Jonsson, B., and Chazin, W. J. (1995). The effect of protein concentration on ion binding. *Proc. Natl. Acad. Sci. U.S.A.* **92,** 4748–4752.

Liu, Y., and Storm, D. R. (1990). Regulation of free calmodulin levels by neuromodulin: Neuron growth and regeneration. *Trends Pharmacol. Sci.* **11,** 107–111.

Lowenstein, C. J., and Snyder, S. H. (1992). Nitric oxide, a novel biologic messenger. *Cell (Cambridge, Mass.)* **70,** 705–707.

Lukas, T. J., Burgess, W. H., Prendergast, F. G., Lau, W., and Watterson, D. M. (1986). Calmodulin binding domains: Characterization of a phosphorylation and calmodulin binding site from myosin light chain kinase. *Biochemistry* **25,** 1458–1464.

McPhalen, C. A., Strynadka, N. C. J., and James, M. N. G. (1991). Calcium-binding sites in proteins: A structural perspective. *Adv. Protein Chem.* **42,** 77–144.

Malencik, D. A., and Anderson, S. R. (1983). Binding of hormones and neuropeptides by calmodulin. *Biochemistry* **22,** 1995–2001.

Martin, S. R., and Bayley, P. M. (1986). The effects of Ca^{2+} and Cd^{2+} on the secondary and tertiary structure of bovine testis calmodulin. *Biochem. J.* **238,** 485–490.

Massom, L., Lee, H., and Jarrett, H. W. (1990). Trifluoperazine binding to porcine brain calmodulin and skeletal muscle troponin C. *Biochemistry* **29,** 671–681.

Massom, L. R., Lukas, T. J., Persechini, A., Kretsinger, R. H., Watterson, D. M., and Jarrett, H. W. (1991). Trifluoperazine binding to mutant calmodulins. *Biochemistry* **30**, 663–667.

Matsushima, N., Izumi, Y., Matsuo, T., Yoshino, H., Ueki, T., and Miyake, Y. (1989). Binding of Ca^{2+} and mastoparan to calmodulin induces a large change in the tertiary structure. *J. Biochem. (Tokyo)* **105**, 883–887.

Maune, J. F., Klee, C. B., and Beckingham, K. (1992). Ca^{2+}-binding and conformational change in two series of point mutations to the individual Ca^{2+}-binding sites of calmodulin. *J. Biol. Chem.* **267**, 5286–5295.

Meador, W. E., Means, A. R., and Quiocho, F. A. (1992). Target enzyme recognition by calmodulin: 2.4 Å structure of a calmodulin–peptide complex. *Science* **257**, 1251–1255.

Meador, W. E., Means, A. R., and Quiocho, F. (1993). Modulation of calmodulin plasticity in molecular recognition on the basis of x-ray structures. *Science* **262**, 1718–1721.

Nelson, M. R., and Chazin, W. J. (1996). The EF-hand calcium-binding proteins data library. http://chazin.scripps.edu/cabp_database/seq/align.html

Nelson, M. R., and Chazin, W. J. (1998). An interaction-based analysis of calcium-induced conformational changes in Ca^{2+} sensor proteins. *Protein Sci.* **7**, 270–282.

Novack, J. P., Charbonneau, H., Bentley, J. K., Walsh, K. A., and Beavo, J. A. (1991). Sequence comparison of the 63-, 61-, and 59-kDa calmodulin-dependent cyclic nucleotide phosphodiesterases. *Biochemistry* **30**, 7940–7947.

O'Neil, K. T., and DeGrado, W. F. (1990). How calmodulin binds its targets: Sequence independent recognition of amphiphilic α-helices. *Trends Biochem. Sci.* **15**, 59–64.

O'Neil, K. T., Erickson-Viitanen, S., and DeGrado, W. F. (1989). Photolabeling of calmodulin with basic, amphiphilic α-helical peptides containing *p*-benzoylphenylalanine. *J. Biol. Chem.* **264**, 14571–14578.

Ohki, S., Tsuda, S., Joko, S., Yazawa, M., Yagi, K., and Hikichi, K. (1991). ^{1}H NMR study of amide proton exchange of calmodulin-mastoparan complex. *J. Biochem. (Tokyo)* **109**, 234–237.

Payne, M. E., Fong, Y.-L., Ono, T., Colbran, R. J., Kemp, B. E., Soderling, T. R., and Means, A. R. (1988). Calcium/calmodulin-dependent protein kinase II: Characterization of distinct calmodulin binding and inhibitory domains. *J. Biol. Chem.* **263**, 7190–7195.

Persechini, A., and Kretsinger, R. (1988). The central helix of calmodulin functions as a flexible tether. *J. Biol. Chem.* **263**, 12175–12178.

Prêcheur, B., Munier, H., Mispelter, J., Bârzu, O., and Craescu, C. (1992). ^{1}H and ^{15}N characterization of free and bound states if an amphiphilic peptide interacting with calmodulin. *Biochemistry* **31**, 229–236.

Putkey, J. A., Ono, T., VanBerkum, M. F. A., and Means, A. R. (1988). Functional significance of the central helix in calmodulin. *J. Biol. Chem.* **263**, 11242–11249.

Putkey, J. A., Sweeney, H. L., and Campbell, S. T. (1989). Site-directed mutation of the trigger calcium-binding sites in cardiac troponin C. *J. Biol. Chem.* **264**, 12370–12378.

Raghunathan, S., Chandross, R. J., Cheng, B.-P., Persechini, A., Sobottka, S. E., and Kretsinger, R. H. (1993). The linker of des-Glu^{84}-calmodulin is bent. *Proc. Natl. Acad. Sci. U.S.A.* **90**, 6869–6873.

Rose, G. D., Geselowitz, A. R., Lesser, G. J., Lee, R. H., and Zehfus, M. H. (1985). Hydrophobicity of amino acid residues in globular proteins. *Science* **229**, 834–838.

Seamon, K. B., and Kretsinger, R. H. (1983). Calcium-modulated proteins. *Met. Ions Biol.* **6**, 1–51.

2. Calmodulin as a Calcium Sensor

Shaw, G. S., Golden, L. F., Hodges, R. S., Sykes, B. D. (1991). Interactions between paired calcium-binding sites in proteins: NMR determination of the stoichiometry of calcium binding to a synthetic troponin-C peptide. *J. Am. Chem. Soc.* **113,** 5557–5563.

Skelton, N. J., Kördel, J., Akke, M., Forsén, S., and Chazin, W. J. (1994). Signal transduction versus buffering activity in Ca^{2+}-binding proteins. *Nat. Struct. Biol.* **1,** 239–245.

Small, E. W., and Anderson, S. R. (1988). Fluorescence anisotropy decay demonstrates calcium-dependent shape changes in photo-cross-linked calmodulin. *Biochemistry* **27,** 419–428.

Spera, S., Mitsuhiko, I., and Bax, A. (1991). Measurement of the exchange rates of rapidly exchanging amide protons: Application to the study of calmodulin and its complex with a myosin light chain kinase fragment. *J. Biomol. NMR* **1,** 155–165.

Starovasnik, M. A., Su, D.-R., Beckingham, K., and Klevit, R. E. (1992). A series of point mutations reveal interactions between calcium-binding sites of calmodulin. *Protein Sci.* **1,** 245–253.

Starovasnik, M. A., Davis, T. N., and Klevit, R. E. (1993). Similarities between yeast and vertebrate calmodulin: An examination of the calcium-binding and structural properties of calmodulin from the yeast *Saccharomyces cerevisiae*. *Biochemistry* **32,** 3261–3270.

Steiner, J. P., Walke, H. T., Jr., and Bennet, V. (1989). Calcium/calmodulin inhibits direct binding of spectrin to synaptosomal membranes. *J. Biol. Chem.* **264,** 2783–2791.

Stromqvist, M., Berglund, A., Shanbhag, V. P., and Backman, L. (1988). Influence of calmodulin on the human red cell membrane skeleton. *Biochemistry* **27,** 1104–1110.

Strynadka, N. C. J., and James, M. N. G. (1988). Two trifluoperazine-binding sites on calmodulin predicted from comparative modeling with troponin-C. *Proteins* **3,** 1–17.

Strynadka, N. C. J., and James, M. N. G. (1989). Crystal structures of the helix-loop-helix calcium-binding proteins. *Annu. Rev. Biochem.* **58,** 951–998.

Strynadka, N. C. J., and James, M. N. G. (1991). Towards an understanding of the effects of calcium on protein structure and function. *Curr. Opin. Struct. Biol.* **1,** 905–914.

Swanljung-Collins, H., and Collins, J. H. (1991). Ca^{2+} stimulates the Mg^{2+}-ATPase activity of brush border myosin I with three or four calmodulin light chains but inhibits with less than two bound. *J. Biol. Chem.* **266,** 1312–1319.

Swindells, M. B., and Ikura, M. (1996). Pre-formation of the semi-open conformation by the apo-calmodulin C-terminal domain and implications for binding IQ motifs. *Nat. Struct. Biol.* **3,** 501–504.

Tanaka, T., and Hidaka, H. (1980). Hydrophobic regions function in calmodulin–enzyme(s) interactions. *J. Biol. Chem.* **255,** 11078–11080.

Tanaka, T., Ames, J. B., Harvey, T. S., Stryer, L., and Ikura, M. (1995). Sequestration of the membrane-targeting myristoyl group of recoverin in the calcium-free state. *Nature (London)* **376,** 444–446.

Thulin, E., Andersson, A., Drakenberg, T., Forsén, S., and Vogel, H. J. (1984). Metal ion and drug binding to proteolytic fragments of calmodulin: Proteolytic, cadmium-113, and proton nuclear magnetic resonance studies. *Biochemistry* **23,** 1862–1870.

Tjandra, N., Kuboniwa, H., Ren, H., and Bax, A. (1995). Rotational dynamics of calcium-free calmodulin studied by ^{15}N-NMR relaxation measurements. *Eur. J. Biochem.* **230,** 1014–1024.

Torok, K., and Trentham, D. R. (1994). Mechanism of 2-chloro-(e-amino-lys(75))-[6-[4-(N,N-diethylamino)phenyl]-1,3,5-triazin-4-y1]calmodulin interactions with smooth muscle myosin light chain kinase and derived peptides. *Biochemistry* **33,** 12807–12820.

Travé, G., Lacombe, P.-J., Pfuhl, M., Saraste, M., and Pastore, A. (1995). Molecular mechanism of the calcium-induced conformational change in the spectrin EF-hands. *EMBO J.* **14,** 4922–4931.

Tsai, M.-D., Drakenberg, T., Thulin, E., and Forsén, S. (1987). Is the binding of magnesium(II) to calmodulin significant? An investigation by magnesium-25 nuclear magnetic resonance. *Biochemistry* **26,** 3635–3643.

Tsalkova, T. N., and Privalov, P. L. (1985). Thermodynamic study of domain organization in troponin C and calmodulin. *J. Mol. Biol.* **181,** 533–544.

Urbauer, J. L., Short, J. H., Dow, L. K., and Wand, A. J. (1995). Structural analysis of a novel interaction by calmodulin: High-affinity binding of a peptide in the absence of calcium. *Biochemistry* **34,** 8099–8109.

Upson, C., Faulheber, T., Jr., Kamins, D., Laidloaw, D., Shlegel, D., Vroom, J., Gurwitz, R., and van Dam, A. (1989). The application visualization system: A computational environment for scientific visualization. *Comput. Graphics App.* **9,** 30–42.

Vandonselaar, M., Hickie, R. A., Quail, J. W., and Delbaere, L. T. J. (1994). Trifluoperazine-induced conformational change in Ca^{2+}-calmodulin. *Nat. Struct. Biol.* **1,** 795–801.

Vogel, H. J., and Zhang, M. (1995). Protein engineering and NMR studies of calmodulin. *Mol. Cell. Biochem.* **149/150,** 3–15.

Vogel, H. J., Lindahl, L., and Thulin, E. (1983). Calcium-dependent hydrophobic interaction chromatography of calmodulin, troponin C and their proteolytic fragments. *FEBS Lett.* **157,** 241–246.

Vorherr, T., Knopfel, L., Hofmann, F., Mollner, S., Pfeuffer, T., and Carofoli, E. (1993). The calmodulin binding domain of nitric oxide synthase and adenylyl cyclase. *Biochemistry* **32,** 6081–6088.

Walsh, M., and Stevens, F. C. (1978). Chemical modification studies on the Ca^{2+}-dependent protein modulator: The role of methionine residues in the activation of cyclic nucleotide phosphodiesterase. *Biochemistry* **17,** 3924–3930.

Wimberly, B., Thulin, E., and Chazin, W. (1995). Characterization of the N-terminal half-saturated state of calbindin D_{9k}: NMR studies of the N56A mutant. *Protein Sci.* **4,** 1045–1055.

Wimbley, W., Creamer, T., and White, S. (1996). Solvation energies of amino acid side chains and backbone in a family of host–guest pentapeptides. *Biochemistry* **35,** 5109–5124.

Xie, X., Harrison, D. H., Schlichting, I., Sweet, R. M., Kalabokis, V. N., and Szent-Gyorgyi, A. G. (1994). Structure of the regulatory domain of scallop myosin at 2.8 Å resolution. *Nature (London)* **368,** 306–312.

Yoshino, H., Minari, O., Matsushima, N., Ueki, T., Miyake, Y., Matsuo, T., and Izumi, Y. (1989). Calcium-induced shape change of calmodulin with mastoparan studied by solution X-ray scattering. *J. Biol. Chem.* **264,** 19706–19709.

Yoshino, H., Wakita, M., and Izumi, Y. (1993). Calcium-dependent changes in structure of calmodulin with substance P. *J. Biol. Chem.* **268,** 12123–12128.

Zhang, M., Li, M., Wang, J. H., and Vogel, H. J. (1994). The effect of met → leu mutations on calmodulin's ability to activate cyclic nucleotide phosphodiesterase. *J. Biol. Chem.* **269,** 15546–15552.

Zhang, M., Tanaka, T., and Ikura, M. (1995). Calcium-induced conformational transition revealed by the solution structure of apo calmodulin. *Nat. Struct. Biol.* **2,** 758–767.

3

Calmodulin-Regulated Protein Kinases

**THOMAS J. LUKAS, SALIDA MIRZOEVA, AND
D. MARTIN WATTERSON**

Drug Discovery Program and
Department of Molecular Pharmacology and Biological Chemistry
Northwestern University School of Medicine
Chicago, Illinois 60611-3008

I. Overview	66
A. Introduction	66
B. Protein Kinases	68
C. Protein Kinase Segmental Organization and Calmodulin Regulation	69
D. Calmodulin Recognition	72
E. Linkage of Calcium and Enzyme Recognition	73
II. Myosin Light-Chain Kinase	78
A. Myosin Light-Chain Kinase Function and Cellular Ca^{2+} Signal Transduction	78
B. Myosin Light-Chain Kinase as a Calmodulin-Regulated Enzyme	81
C. Molecular Genetics of Myosin Light-Chain Kinase and Related Proteins	97
III. Calmodulin-Dependent Protein Kinase II	101
A. Calmodulin-Dependent Protein Kinase II Function and Cellular Ca^{2+} Signal Transduction	101
B. Calmodulin-Dependent Protein Kinase II as a Calmodulin-Regulated Enzyme	105
C. Molecular Genetics of Calmodulin-Dependent Protein Kinase II and Related Proteins	114
IV. Phosphorylase Kinase	116
A. Phosphorylase Kinase Function and Ca^{2+} Signal Transduction	116
B. Phosphorylase Kinase as a Calmodulin-Regulated Enzyme	117
C. Molecular Genetics of Phosphorylase Kinase	123
V. Calmodulin-Dependent Protein Kinase I	124
A. Calmodulin-Dependent Protein Kinase I Function and Cellular Ca^{2+} Signal Transduction	124
B. Calmodulin-Dependent Protein Kinase I as a Calmodulin-Regulated Enzyme	126

VI. Calmodulin-Dependent Protein Kinase III 132
   A. Calmodulin-Dependent Protein Kinase III Function and Cellular Ca^{2+} Signal Transduction... 132
   B. Calmodulin-Dependent Protein Kinase III as a Calmodulin-Regulated Enzyme 133
VII. Calmodulin-Dependent Protein Kinase IV 136
   A. Calmodulin-Dependent Protein Kinase IV Function and Calcium Signal Transduction.. 136
   B. Calmodulin-Dependent Protein Kinase IV as a Calmodulin-Regulated Enzyme 138
   C. Molecular Genetics of Calmodulin-Dependent Protein Kinase IV and Related Proteins.. 139
VIII. Other Calmodulin-Dependent Protein Kinases 140
   A. Calmodulin-Dependent Protein Kinase Kinases..................... 140
   B. Death-Associated Protein Kinase 143
   C. Caki 144
   D. Possible Roles of Calmodulin-Regulated Protein Kinases in G-Protein Signaling and MAP Kinase Pathways............................... 144
IX. Summary 145
   References.. 146

I. OVERVIEW

A. Introduction

Calmodulin (CaM)-regulated protein kinases have provided insight into the regulatory mechanisms that underlie Ca^{2+} signal transduction through calmodulin. Specifically, structural and mechanistic themes discovered through the analysis of CaM-regulated protein kinases are now being recognized in various forms with other CaM-regulated enzymes, with details specific to each enzyme. The control of biological homeostasis through CaM-regulated protein kinases involves more than one regulatory mechanism. The common mechanistic steps among all of the protein kinases include the selective recognition of calcium and initial decoding of the calcium signal by CaM, followed by enzyme activation through a relief-of-autoinhibition mechanism. However, other regulatory steps in the mechanism vary among the protein kinases, providing a level of discrimination in the biological response to the calcium signal and potential safeguards for the organism.

The activation of the kinase phosphotransferase activity occurs through the functional coupling of the CaM recognition and autoinhibitory segments through a relief-of-autoinhibition mechanism. The various CaM-regulated protein kinases begin to add variations to a common theme after the activation step. An example of a second control element is through autophosphorylation of the autoinhibitory segment of the kinase, CaM-regulated protein kinase II (CaMPKII) (Schworer et al., 1986; Miller and Kennedy, 1986; Lai et al., 1986). The autophosphorylation allows the kinase activity to persist for a period after calcium dissociation from CaM. This can be viewed as a fast on, slow decay process. For other CaM kinases (e.g., CaM-regulated protein kinase I, CaMPKI), the catalytic activity is further modulated by exogenous phosphorylation through a kinase kinase mechanism that works in concert with CaM activation (Mochizuki et al., 1993a; Lee and Edelman, 1994, 1995) and has similarity to other protein kinases with activation loops in their catalytic segments. Another type of second control mechanism can involve desensitization through phosphorylation of the CaM recognition segment by exogenous protein kinases, as found with vertebrate smooth muscle/nonmuscle myosin light-chain kinase, or sm/nmMLCK (Adelstein et al., 1978; Lukas et al., 1986). For large multisubunit kinases (e.g., phosphorylase kinase, PhosK), Ca^{2+}/CaM activation is only part of a larger control scheme that includes additional subunits that are substrates for other protein kinases (for a review see Heilmeyer, 1991).

The first identification of physiological CaM recognition sequences was with protein kinases (Blumenthal et al., 1985; Lukas et al., 1986). The availability of synthetic peptide analogs of CaM recognition regions (Blumenthal et al., 1985; Lukas et al., 1986) provided an experimentally tractable system that, when combined with results with enzymes isolated from tissue sources (Dabrowska et al., 1978; Adelstein et al., 1978; Lukas et al., 1986; Blumenthal et al., 1985; Thiel et al., 1988) and site-directed mutagenesis studies (Shoemaker et al., 1990; Hanson et al., 1989; Waxham et al., 1990), yielded insight into the molecular mechanisms of CaM recognition and enzyme regulation. The increasing availability of information for CaMs in different conformational states, including CaM complexed with calcium and CaM recognition peptides from enzymes, has added structural correlates (Zhang et al., 1995; Kuboniwa et al., 1995; Babu et al., 1988; Meador et al., 1992, 1993; Ikura et al., 1992a) to the proposed mechanisms. The conformational restriction hypothesis (Lukas et al., 1994; Chabbert et al., 1991), for example, assumes that CaM can exist in a variety of conformational states, with the distribution being increasingly restricted as an increasing number of calcium ions are bound and as complexes with CaM-binding structures are formed. A correlate is that more than one CaM conformation can be active, with the profile of active conformations being characteristic for each given enzyme complex. Much of the experimental detail that forms the foundation of this hypothesis comes from the use of mutant CaMs and CaM regulatory site peptides from protein kinases.

This chapter deals with the general principles that are emerging from the study of CaM-regulated protein kinases and the potential molecular mechanisms by which they transduce calcium signals into biological responses, with some historical perspective to assist the nonspecialist in understanding the evolution of certain concepts and terms. Many details of each enzyme cannot be covered adequately here, and every reference cannot be cited. Therefore, the interested reader is referred to more specialized reviews or key references that are given at various points in this chapter. There has been a major change in the field from a decade ago (Watterson and Vincenzi, 1980; Klee and Vanaman, 1982; Cohen and Klee, 1988) when more simplified models were dominant. This introduction and overview section (Section I) provides general background and a glimpse at some of the emerging themes, with case studies from specific classes of CaM-regulated protein kinases presented in the succeeding sections.

B. Protein Kinases

Protein kinases catalyze the transfer of a phosphoryl group from an ATP substrate to a protein substrate, a stoichiometric covalent modification that can be an important allosteric regulator of biological activity (Krebs, 1993). Kinases are usually divided into two broad classes, tyrosine kinases or serine/threonine kinases, on the basis of the hydroxyamino acid phosphorylated in their protein substrates. Among the serine/threonine protein kinases, several CaM-regulated enzymes exist. The Ca^{2+}/CaM-regulated protein kinase family currently consists of at least 10 enzyme groups: skeletal muscle phosphorylase kinase, nonmuscle phosphorylase kinase, skeletal muscle myosin light chain kinase (skMLCK), smooth muscle/nonmuscle myosin light chain kinase (sm/nmMLCK), CaM-dependent protein kinase I (CaMPKI), CaM-dependent protein kinase II (CaMPKII), CaM-dependent protein kinase III (CaMPKIII),

CaM-dependent protein kinase IV (CaMPKIV), CaM kinase kinases, and other CaM-regulated kinases not yet characterized as extensively as the preceding groups. The most extensively characterized group as far as CaM regulation is concerned are the MLCKs, especially the sm/nmMLCK. The discussion in this chapter of the principles involved in the molecular mechanisms of calcium signal transduction through CaM-regulated protein kinases is divided into subsections based on the just-described groups, with emphasis given to MLCK because of the more detailed knowledge about CaM regulation for this kinase.

Calmodulin-regulated protein kinases, like other serine/threonine kinases, prefer Mg^{2+}-ATP as a phosphotransfer substrate. The protein substrate specificity for CaM-regulated protein kinases ranges from kinases with narrow protein substrate specificity, such as the phosphorylase kinases and MLCKs, to those with comparatively broad substrate recognition properties, such as CaMPKII. The substrate specificity of protein kinases has been reviewed (Pinna and Ruzzene, 1996) and will not be expanded on here.

Protein kinases have been implicated in a broad array *in vivo* functions, especially those where the signals of second and third messengers are amplified (Krebs, 1993). Where the intracellular signals are Ca^{2+} signals, these are transduced into biological responses through CaM-regulated phosphorylation mechanisms. The Ca^{2+}/CaM-regulated protein kinases are only one component of the interdigitated pathways that bring about cellular and organismal homeostasis, and are involved in a diverse array of critical cellular functions, such as metabolism, motility, gene transcription, and cell division.

C. Protein Kinase Segmental Organization and Calmodulin Regulation

Calmodulin-regulated protein kinases are characterized by a segmental organization of their primary structures (Lukas *et al.,* 1988). A diagrammatic summary is presented in Fig. 1. There is a core composed of a catalytic segment containing serine/threonine protein kinase motifs and a CaM regulatory segment on the carboxyl side of the catalytic segment. Flanking sequences on the amino or carboxyl terminus of this catalytic/CaM regulatory core vary with the particular protein kinase. Flanking sequences appear to be responsible for supramolecular association so

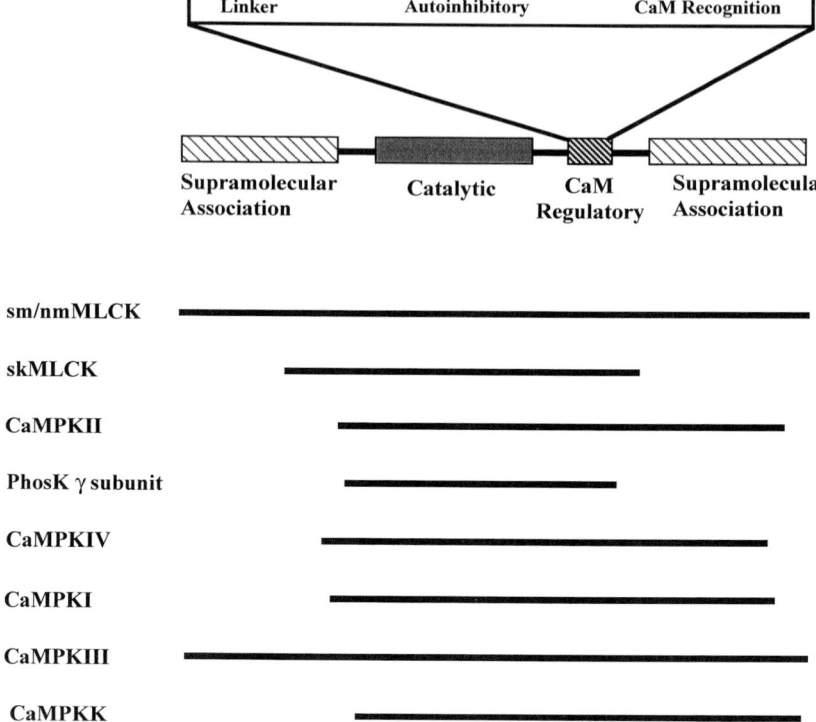

Fig. 1. Segmental organization of CaM-regulated protein kinases. A generalized layout of a CaM-regulated kinase is shown. All CaM-regulated protein kinases contain catalytic and CaM regulatory segments. Most of them also have supramolecular association segments that flank the core catalytic and CaM regulatory segments. The supramolecular association segments have been implicated in subcellular targeting or multimeric protein assembly.

that the kinase is more than a CaM-regulated catalytic activity. Supramolecular association segments may be responsible for intracellular enzyme targeting or the assembly of a large multimeric complex. For example, flanking segments may be critical for the proper targeting of MLCK to the contractile apparatus of a cell (Shirinsky *et al.*, 1993) and are key to the assembly of CaMPKII into a large multimeric holoenzyme and to its nuclear localization (Srinivasan *et al.*, 1994) for the regulation of transcription factors.

3. Calmodulin-Regulated Protein Kinases

The CaM regulatory segment can be further divided into three regions based on experimental results (Shoemaker et al., 1990): the functional center of autoinhibitory activity, the functional center of CaM recognition, and a third segment that links autoinhibitory and CaM recognition domains with the catalytic domain (Fig. 1). As suggested by the use of the term autoinhibitory, the mechanism of CaM activation of protein kinases is one of relief of autoinhibition. The center of autoinhibitory activity in CaM-regulated protein kinases has been defined by the use of the site-directed mutagenesis. Usually, mutations in this region render the particular protein kinase constitutively active with retention of CaM-binding activity by the neighboring CaM recognition segment. The CaM recognition segment is usually defined by a combination of approaches. The two main approaches used are fragmentation of the kinase, followed by purification and characterization of short segments that have CaM-binding activity that mimics the activity of the intact enzyme, or site-directed mutagenesis of the enzyme with the identification of regions that alter CaM-binding activity. Clearly, a consensus in the results from more than one approach is preferable, and this consensus has been important in correctly identifying key areas and amino acid residues for CaM recognition.

The CaM regulatory segment, as well as other segments of CaM-regulated protein kinases, can also contain phosphorylation sites. These sites may be phosphorylated by exogenous kinases or may be autophosphorylation sites, but the functional effect is dependent on where the phosphorylation occurs. For example, autophosphorylation of CaMPKII in its autoinhibitory segment (Miller and Kennedy, 1986; Schworer et al., 1986; Lai et al., 1986) renders the enzyme relatively CaM independent with the retention of CaM recognition activity (i.e., there is an uncoupling of CaM recognition from the ability of CaM to modulate the enzyme activity through the relief-of-autoinhibition mechanism). Phosphorylation by an exogenous kinase of sm/nmMLCK (Adelstein et al., 1978; Hashimoto and Soderling, 1990; Lukas et al., 1986) decreases its Ca^{2+}/CaM sensitivity and alters the interaction between CaM and the CaM recognition site of MLCK, with retention of the autoinhibitory state of the MLCK. The theme of exogenous phosphorylation affecting the function of a CaM-dependent protein kinase, first described for the CaM recognition segment of sm/nmMLCK (Adelstein et al., 1978; Lukas et al., 1986), is observed with regions outside of the CaM regulatory segment as well. For example, phosphorylation of CaMPKI (Mochizuki et al., 1993a; Lee and Edelman, 1994, 1995) and CaMPKIV (Okuno and Fujisawa, 1993; Selbert *et al.*, 1995) at a specific residue within the activation loop in the catalytic segment is essential for their maximal catalytic activity. Potential phosphorylation sites in the flanking sequences of some CaM-regulated protein kinases, such as the tail and kinase-related protein (KRP) segments of sm/nmMLCK (Shoemaker *et al.*, 1990), have been identified based on similarity to consensus substrate recognition sites for other protein kinases, but their functional significance has yet to be established. In general, the trend is one in which CaM-regulated protein kinases can serve as physiological protein kinase substrates, thereby allowing modulation of the CaM-regulated enzyme activity and providing a mechanism for cross-talk across diverse signal transduction pathways (e.g., lipid, cyclic nucleotide, and calcium signaling) at the level of CaM-regulated kinases.

D. Calmodulin Recognition

Examples of some CaM recognition sequences are given in Fig. 2. In general, the potential secondary structure and the presence of appropriately spaced positively charged and hydrophobic amino acid side chains, not primary sequence similarity alone, have been the characteristic feature common to the CaM recognition sequences found in protein kinases. This potential theme was first noted (Lukas *et al.*, 1986; O'Neil *et al.*,

Kinase	CaM Recognition Sequence
sm/nmMLCK	ARRKWQKTGHAVRAIGRLSS
skMLCK	KRRWKKNFIAVSAANRFKKISSSGALM
CaMPKII	FNARRKLKGAILTTMLATR
CaMPKI	AKSKWKQAFNATAVVRHM
CaMPKIV	ARRKLKAAVKAVVASSRLG

Fig. 2. Alignment of proposed CaM-binding sequences. The actual and proposed CaM-binding amino acid sequences of five protein kinases are aligned. Extensive experimental data support the sequences shown for MLCKs and CaMPKII. The remaining sequences are proposed based on sequence of alignment and initial experimental data.

1987) during studies of the sm/nmMLCK and synthetic peptide analogs of its CaM recognition region. Subsequent investigations with other protein kinases and with synthetic peptide analogs of their CaM recognition regions confirmed this early hypothesis. More recently, the structures of calcium-saturated CaM alone (Babu et al., 1988) or complexed with the CaM recognition peptides from CaM-regulated protein kinases (Meador et al., 1992, 1993; Ikura et al., 1992a) show that CaM can assume a broad spectrum of conformations, including a more compact structure when the calcium-saturated form is complexed with a CaM recognition peptide from a protein kinase. Two examples are used to illustrate this point in Fig. 3.

Interestingly, each protein kinase CaM recognition peptide uses a unique set of contact points based on crystal and nuclear magnetic resonance structures of CaMs complexed with a CaM recognition peptide (Meador et al., 1992, 1993; Ikura et al., 1992a). A summary of the relative position of these potential contact points in the CaM sequence is shown in Fig. 4. Although a number of the CaM:peptide contact points are the same across multiple enzymes, there are several individual points for each enzyme that make the full set unique for a given enzyme. Early site-directed mutagenesis studies of CaM (Craig et al., 1987; Lukas et al., 1988; Weber et al., 1989) showed that particular residues in CaM had differential functional importance that varied with the CaM-regulated enzyme. This suggested that each enzyme might have a unique set of functionally critical contacts, although some individual residues would be common across a variety of enzymes. Potential contact points derived from structures of the CaM:peptide complexes (Meador et al., 1992, 1993; Ikura et al., 1992a), with peptides from different CaM regulated enzymes, are strikingly consistent with the paradigms derived from earlier site-directed mutagenesis studies (Craig et al., 1987; Lukas et al., 1988; Weber et al., 1989) of CaM, coupled with enzyme activation analysis. This congruence of results with the CaM-regulated enzyme and its corresponding CaM recognition peptide has allowed more detailed analyses through the use of the more experimentally tractable CaM:peptide complexes.

E. Linkage of Calcium and Enzyme Recognition

The binding of Ca^{2+} by CaM influences its ability to interact with and regulate protein kinases, and the kinases can, in turn, influence the

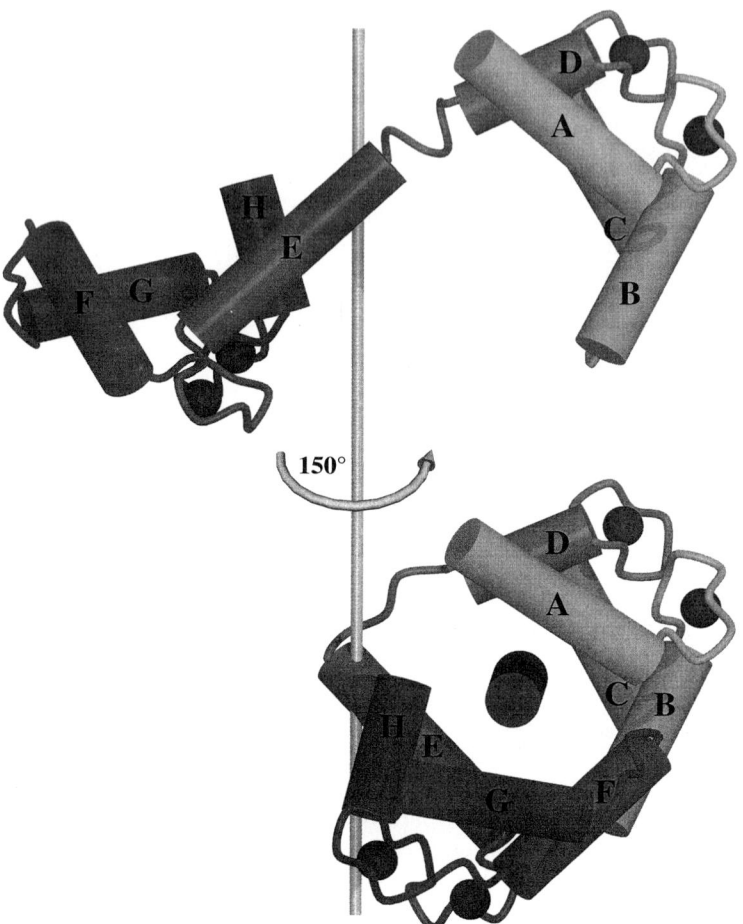

Fig. 3. Ribbon diagrams of three-dimensional structures for CaM and a CaM:peptide complex. These drawings are based on the coordinates deposited with the Brookhaven Protein Data Bank (PDB) and represent two of the multiple conformations that CaM can assume. The upper structure (entry 1CLL) is that for CaM with four calcium ions (solid spheres), whereas the lower structure (entry 1CDL) is that of the complex between calcium-saturated CaM and the CaM recognition peptide from sm/nmMLCK. CaM has eight helices (cylinders labeled A–H from the amino terminus). The difference between CaM alone and the CaM:peptide complex can be described by a 150° rotation of the domain containing the E–H helices. Helices in both the amino and the carboxyl halves of CaM come sufficiently close in three-dimensional space to allow interaction with the helical CaM-binding peptide (unlabeled cylinder).

3. Calmodulin-Regulated Protein Kinases

characteristics of Ca^{2+} binding by CaM. This dynamic equilibrium among a series of linked equilibria is summarized by the simplified scheme (Scheme I)

$$\begin{array}{ccccccccc}
& K_1 & & K_2 & & K_3 & & K_4 & \\
\text{CaM} & \rightleftharpoons & \text{CaM-Ca}_1 & \rightleftharpoons & \text{CaM-Ca}_2 & \rightleftharpoons & \text{CaM-Ca}_3 & \rightleftharpoons & \text{CaM-Ca}_4 \\
K_a \updownarrow & & K_b \updownarrow & & K_c \updownarrow & & K_d \updownarrow & & K_e \updownarrow \\
\text{E-CaM} & \rightleftharpoons & \text{E-CaM-Ca}_1 & \rightleftharpoons & \text{E-CaM-Ca}_2 & \rightleftharpoons & \text{E-CaM-Ca}_3 & \rightleftharpoons & \text{E-CaM-Ca}_4 \\
& K_{1'} & & K_{2'} & & K_{3'} & & K_{4'} &
\end{array}$$

where K_{1-4} are association constants for Ca^{2+} binding by CaM. K_{a-e} are the association constants for the interaction of E (= enzyme or CaM-binding peptide) with CaM in the presence of 0–4 Ca^{2+} ions. $K_{1'-4'}$ are corresponding association constants for Ca^{2+} binding by a CaM–E complex.

The relationship between Ca^{2+} binding by CaM, CaM binding to the enzyme, and enzyme activation is complex. However, several simplifications can be applied to most CaM-regulated protein kinases. For example, phosphorylase kinase is a CaM-regulated enzyme in which it is difficult to study the dissociated enzyme (Heilmeyer, 1991). As isolated from vertebrate tissues, dissociation of the CaM and catalytic subunits usually requires both a calcium chelator and chaotropic salts. Therefore, much of the data for the calcium regulation of phosphorylase kinase is that along the pathways described by $K_{1'}$ through $K_{4'}$. In systems such as MLCK, where one can readily dissociate and reconstitute a heterodimer complex and where CaM recognition site peptides are available (Lukas et al., 1988; Kilhoffer et al., 1992), it is possible to explore more readily the coupling between Ca^{2+} recognition and enzyme recognition implied by Scheme I.

Calmodulin has four Ca^{2+}-binding sites and uses an ordered and cooperative mechanism that is maintained in the presence of CaM-regulated enzymes or their respective CaM recognition site peptides (Haiech et al., 1991). The Ca^{2+}-binding process is usually described by the multisite binding relationship:

$$y = (ax + 2bx^2 + 3cx^3 + 4dx^4) / (1 + ax + bx^2 + cx^3 + dx^4),$$

where $a = K_1$; $b = K_1 K_2$; $c = K_1 K_2 K_3$; $d = K_1 K_2 K_3 K_4$; K_1, K_2, K_3, and K_4 = macroscopic association constants; x is the concentration of

	*		✝✝	✝✝	✝✝	Δ			Δ		✝	Δ✝		✝✝	x					
4	L T E E Q I A E F K E A F S L F D K D G D G T I T T K E L G T V M R S L G Q N P T E A E L Q D	Vertebrate
4	L T D E Q I A E F K E A F S L F D K D G D G T I T T K E L G T V M R S L G Q N P T E A E L Q D	Spinach
7	L T E E Q I A E F K E A F A L F D K D G D G C I T T K E L G T V M R S L G Q N P T E A E L Q D	Chlamydomonas
4	L T E E Q I A E F K K E A F S L F D K D G D G T I T T K E L G T V M R S L G Q N P T E A E L Q D	Paramecium
4	L T E E Q I A E F K E A F S L F D K D G D G T I T T K E L G T V M R S L G Q N P T E A E L Q D	Drosophila
6	L L T E E Q I A E F K E A F S L F D K D G D G S I T T K E L G T V M R S L G Q N P T E A E L Q D	Dictyostelium
	10					20				30				40				50

	✝Δ				Δ		✝✝	✝✝	+	+	✝✝	Δ		Δ✝✝	✝✝	Δ
51	M I N E V D A D G N G T I H I D F P E F L T M M A R K M K D T D S E E E I R E A F R V F D K D G N G Y I	Vertebrate
51	M I N E V D A D G N G T I T T K F P E F L N L M A R K M K D T D S E E E L K E A F R V F D K D G N G F I	Spinach
54	M I N S V D A D G N G T I D F K E F L T M M A R K M K D T D H E D E L I E A F K V F D R D G N G L I	Chlamydomonas
51	M I N E V D A D G N G T I D F P E F L T M M A R K M Q D T D S E E E I R E A F K V F D K D G N G L I	Paramecium
51	M I N E V D A D G N G T I D F P E F L T M M A R K M K D T E E E I R E A F K V F D K D G N G Y I	Drosophila
52	M I N E V D A D G N G N I D F P E F L T M M A R K M K D T E E E I R E A F K V F D K D G N G Y I	Dictyostelium
	60				70				80				90				100

		✝✝		*	✝	x*Δ	✝✝*	Δ			Δ		§	✝✝	✝✝
101	S A A E L R H V M T N L G E K L T D E E V D E M I R E A D I D G D G Q V N Y E E F V Q M M T A K	Vertebrate
101	S A A E L R H V M T N L G E K L T D E E V D E M I R E A D V D G D G Q I N Y E E F V K V M M A K	Spinach
104	S A A E L R H V M T N L G E K L S E E E V D E M I R E A D I D G D G Q V N Y E E F V R M M T S G	Chlamydomonas
101	S A A E L R H V M T N L G E K L T D D E E V D E M I R E A D I D D G D G Q I N Y E E F V R M M T S K	Paramecium
101	S A A E L R H V M T N L G E K L T D E E V D E M I R E A D L D G D G Q V N Y E E F V T M M T S K	Drosophila
103	S A A E L R H V M T S L G E K L T N E E V D E M I R E A D L D G D G Q V N Y D E F V K M M I V R	Dictyostelium
	110				120				130				140			

3. Calmodulin-Regulated Protein Kinases

free Ca^{2+}; and y is bound Ca^{2+} (mol/mol). The overall Ca^{2+}-binding affinity of calmodulin is significantly (up to 200-fold) enhanced when it is complexed with a CaM-binding structure (Haiech et al., 1991; Mirzoeva et al., 1997). It appears that this enhancement is due to an increase in all four macroscopic association constants (Mirzoeva et al., 1997). Moreover, the positive cooperativity of Ca^{2+} binding is also increased. Therefore, the Ca^{2+} sensitivity of a particular enzyme activation represents a nontrivial relationship among the multiple equilibria. For most protein kinases, it is not known for certain how the calcium occupancy state of CaM is related to enzyme activation, although it is generally assumed that the Ca_3:CaM:E or Ca_4:CaM:E state is required for significant enzyme activation.

These relationships among Ca^{2+} and enzyme binding/activation provide an example of how CaM-regulated protein kinases can play a dynamic role in the fine tuning of the Ca^{2+} signal transduction system due to the enhancement of the Ca^{2+} sensitivity of CaM. For example, there is a potential for discrimination in the response of a cell to transient calcium fluxes through the differential calcium sensitivity and subcellular localization of the CaM:enzyme supramolecular complex. The calcium-binding activity of CaM can be enhanced when it is part of a supramolecular complex (Olwin et al., 1984; Haiech et al., 1991; Mirzoeva et al., 1997), and the subcellular localization of CaM can be dictated by the subcellular localization of the protein kinase (Shirinsky et al., 1993; Srinivasan et al., 1994; Heilmeyer, 1991). This theme of selective subcellular targeting, through the properties of the supramolecular association segments of

Fig. 4. Location of peptide contact residues in the amino acid sequences of phylogenetically diverse CaMs. An alignment of CaMs from vertebrate, spinach, the unicellular alga *Chlamydomonas, Paramecium, Drosophila,* and *Dictyostelium* is shown. Shaded areas indicate amino acids that are not phylogenetically conserved among these phylogenetically diverse CaMs. Amino acids in CaM that are proposed as interacting with CaM-binding peptides (i.e., within 4 Å distances based on the three-dimensional structures) are indicated. The plus symbol above an amino acid indicates that it is within this potential interaction distance only in the structure of CaM complexed with the sm/nmMLCK CaM-binding peptide; x, only with CaMPKII; *, with both sm/nmMLCK and CaMPKII; Δ, only with skMLCK; †, with both skMLCK and sm/nmMLCK; §, with both skMLCK and CaMPKII peptides; and ‡, with all three, sm/nmMLCK, skMLCK, and CaMPKII. CaM sequences were obtained from the GenBank or Swiss-Prot databases, and structures used are those from the PDB and discussed in the text.

protein kinases, and differential calcium sensitivity, through interaction with the CaM regulatory segment of protein kinases, is used by multiple CaM-regulated enzymes.

II. MYOSIN LIGHT-CHAIN KINASE

A. Myosin Light-Chain Kinase Function and Cellular Ca^{2+} Signal Transduction

Myosin light-chain kinases are Ca^{2+}/CaM-dependent protein kinases that catalyze the transfer of phosphate to a specific serine residue in the amino-terminal portion of myosin II regulatory light chains (Karaki et al., 1997). Vertebrate myosin II is a hexamer consisting of one pair of heavy chains and pairs of regulatory (RLC) and essential (ELC) light chains. The amino terminus of the heavy chain folds to form a globular head containing the ATPase catalytic site, light chain binding sites, and a site involved in the interaction with actin. Different vertebrate tissue express different myosin isoforms. Similarly, the characterization of vertebrate MLCKs from different tissues and species has led to the identification of at least two types of enzymes that are genetically, physiologically, and biochemically distinct: skMLCK and sm/nmMLCK.

In vertebrates, the skMLCK and sm/nmMLCK are encoded by different genetic loci (Collinge et al., 1992). Each appears to be in the single gene family. Expression of the skMLCK is restricted to striated muscle tissues. The sm/nmMLCK locus has a complex genomic organization, including another gene that encodes a distinct myosin II binding protein, called KRP (Collinge et al., 1992; Shirinsky et al., 1993), that does not bind calmodulin. The protein products of the sm/nmMLCK genetic locus are present in most tissues, although the levels vary among tissues and with development. The theme of a complex genetic organization that includes other proteins involved in the regulation of cellular structure and function, first described for MLCK and the myosin II motor system (Collinge et al., 1992), has now been found with CaMPKII (Bayer et al., 1996) and, most recently, with the focal adhesion kinase (FAK), a tyrosine kinase (Richardson and Parsons, 1996) that is not CaM regulated.

This new form of cellular and organismal regulation may be found in the future with a variety of signal transduction systems.

The net effect of an MLCK response to calcium signals in smooth muscle tissues is the phosphorylation of the regulatory light chain of myosin at a specific serine near the amino terminus, resulting in enhanced interactions with actin and initiation of muscle contraction (for a comprehensive review of calcium and smooth muscle functions, see Karaki et al., 1997). Dephosphorylation of the myosin RLC by a phosphatase is a key event in the completion of the Ca^{2+}-triggered contractile cycle. However, muscle tension declines slowly compared to the return of RLC phosphorylation to basal levels. This finding reflects the involvement of additional regulatory elements that can maintain muscle tension independent of RLC phosphorylation. This theme of a phosphatase activity being only one aspect of the attenuation or prolongation of a physiological event initiated by a CaM-regulated protein kinase is one that is repeated with other CaM-regulated protein kinases (e.g., CaMPKII and the prolongation of a response due to autophosphorylation).

In nonmuscle tissues, the MLCK-mediated phosphorylation of myosin II has been implicated in a diverse array of activities, including cytokinesis, tension development, receptor capping, and cell motility (Karess et al., 1991; Goeckler and Wysolmerski, 1990; Kerrick and Bourguignon, 1984; Fishkind et al., 1991; Wilson et al., 1991). From subcellular immunofluorescence studies, MLCK has been found on actomyosin stress fibers (Feramisco et al., 1981; de Lanerolle et al., 1981) positioning the enzyme for a dynamic role in cellular motility. Ca^{2+}-dependent RLC phosphorylation has been linked to retractile movement in endothelial cells (Wysolmerski and Lagunoff, 1990, 1991) and fibroblasts (Kolodney and Elson, 1993) and is correlated with the generation of tension in these cells (Kolodney and Wysolmerski, 1992; Goeckler and Wysolmerski, 1990). Several characteristics of receptor capping in lymphocyte membranes suggest similarities with mechanisms underlying the control of contraction in smooth muscle fibers. Receptor capping involves the lateral redistribution of cell surface receptors into clusters that contain actin (Bourguignon and Bourguignon, 1984) and myosin (Pasternak et al., 1989). Receptor capping observed in lymphocytes and other cell types is also correlated with Ca^{2+}-mediated increases in myosin RLC phosphorylation and contraction (Majercik and Bourguignon, 1985; Kerrick and Bourguignon, 1984), and MLCK, myosin, and calmodulin were colocalized by immunofluorescent microscopy to receptor caps (Majercik and Bourguignon, 1988). Myosin light-chain kinase has also been implicated in receptor-induced processes in platelets, the blood cells responsible for fibrin-mediated coagulation. Platelets activated by thrombin, ADP, or collagen (Lokeshwar and Bourguignon, 1992) undergo Ca^{2+}-stimulated phosphorylation of myosin RLC that precedes the cellular shape change leading to aggregation (Lokeshwar and Bourguignon, 1992; Hashimoto *et al.*, 1994; Higashihara *et al.*, 1991). During the shape change process, a rearrangement of the platelet cytoskeleton occurs that centralizes the fibrinogen and other cell surface receptors (Kowacsovics and Hartwig, 1996). Consistent with these observations, platelet aggregation is antagonized by actin-disrupting agents and inhibitors of MLCK activity (Yamada *et al.*, 1988; Nakanishi *et al.*, 1990; Kowacsovics and Hartwig, 1996; Lokeshwar and Bourguignon, 1992). Myosin light-chain kinase has also been implicated in the control of cellular motility at neuronal growth cones through the use of pharmacological reagents and colocalization studies of MLCK, CaM, actin, and myosin in growth cone areas (Jian *et al.*, 1996; Berry and Brown, 1995; Bai and Weiss, 1991; Bridgman and Dailey, 1989; Jian *et al.*, 1994).

Pharmacological approaches have also provided insight into potential roles of MLCK in nonmuscle cell function as well as the mechanism of action of MLCK in smooth muscle tissue (Ishikawa *et al.*, 1988; Hagiwara *et al.*, 1987; Ishikawa and Hidaka, 1990; Nakanishi *et al.*, 1990), although there is a lack of specific inhibitors for individual CaM-regulated protein kinases. ML-9, a naphthalene sulfonamide derivative, was initially reported to be a specific inhibitor of MLCK (Ishikawa *et al.*, 1988), but subsequent cell culture studies revealed cellular effects of ML-9 that are usually attributed to other protein kinases (Inoue *et al.*, 1993; Makishima *et al.*, 1991) or inhibition of ion channels (Nakanishi *et al.*, 1989). ML-9, however, does inhibit smooth muscle contraction, a physiological event in which MLCK has been clearly implicated (Ishikawa *et al.*, 1988; Ishikawa and Hidaka, 1990). Staurosporines, a family of small molecule inhibitors derived from Streptomyces, compete *in vitro* with ATP and inhibit a wide variety of protein kinases (Casnellie, 1991), including MLCK (Asano *et al.*, 1995). Staurosporine inhibits smooth muscle contraction (Henrion and Laher, 1993), but it is not clear if MLCK is the only target kinase (Asano *et al.*, 1995). KT5926, a staurosporine derivative, inhibits contraction of rabbit aorta induced by Ca^{2+}-mobilizing agents (Satoh *et al.*, 1995), myosin RLC phosphorylation in platelets (Nakanishi *et al.*, 1990), and nerve growth factor-stimulated neurite elongation (Teng and Greene, 1994). Synthetic peptide inhibitors of MLCK have also been used for altering activity through microinjection into cultured cells (Itoh et al., 1989; Kargacin et al., 1990) or introduction into skinned smooth muscle tissue (Moreland et al., 1992). Most of these peptides, however, also bind to calmodulin so that interpretation of the results must carry this qualification. Clearly, the design of peptide and small molecule reagents that have better selectivity for MLCK would be valuable additions to the tools needed to dissect the exact roles of MLCK and other CaM-dependent protein kinases in intracellular calcium signal transduction.

B. Myosin Light-Chain Kinase as a Calmodulin-Regulated Enzyme

Myosin light-chain kinase has a segmentally organized amino acid sequence, as indicated in Figs. 1 and 5. The experimental detail that relates the various amino acid segments of CaM-regulated protein kinases to function is most extensive with MLCKs, especially the sm/nmMLCK from vertebrate tissues. Therefore, a brief description of the type of data that form the foundations of many assumptions is worthwhile. Limited proteolysis experiments, which demonstrated (Walsh et al., 1982; Foster et al., 1986) the loss of CaM regulation of MLCK with peptide bond cleavage prior to the loss of catalytic activity, suggested that MLCK contains distinct catalytic and CaM regulatory segments. Later, the purification and amino acid sequence characterization of the CaM-binding site from sm/nmMLCK (Lukas et al., 1986) demonstrated that the CaM recognition activity could be found in relatively small (21 amino acids) fragments of a rather large enzyme (M_r 135K). Chemical synthesis of corresponding peptide analogs localized Ca^{2+}-dependent CaM-binding activity to short (20–25 amino acids) peptides (Lukas et al., 1986). Spectroscopic and other studies (Lukas et al., 1986; Roth et al., 1991; Chen et al., 1993; O'Neil et al., 1989) of synthetic peptide analogs revealed that potential secondary and higher order structural features, not primary sequence similarity alone, correlated with the ability to serve as a selective and high-affinity CaM recognition sequence. Concurrent with the studies of synthetic peptides, the amino acid sequences of peptides isolated from the sm/nmMLCK (Lukas et al., 1986) were used to design synthetic deoxyoligonucleotides for the isolation of overlapping cDNA clones that could encode a potential CaM-regulated protein kinase activity (Lukas et al., 1988; Shoemaker et al., 1990). These clones were used to engineer a series of protein kinases, by recombinant

DNA technology in *Escherichia coli,* that were purified to chemical homogeneity and characterized in terms of kinetic properties and regulation by CaM (Lukas *et al.,* 1988; Shoemaker *et al.,* 1990). Other investigators (Foyt *et al.,* 1985) used a fragment analysis approach to show that there is no overlap between the catalytic site and CaM recognition regions of sm/nmMLCK. Similar to sm/nmMLCK, the indication that skMLCK is divided into functional segments came from the partial proteolytic digestion of the enzyme (Srivastava and Hartshorne, 1983; Edelman *et al.,* 1985). Subsequently, a CaM-binding peptide of skMLCK, designated M13, was obtained and characterized (Blumenthal *et al.,* 1985). This peptide corresponds to the extreme C-terminal 27 amino acids of skMLCK and, similar to the CaM-binding peptide from sm/nmMLCK (Lukas *et al.,* 1986), exhibits increased α helicity on binding to CaM (Klevit *et al.,* 1985).

The sm/nmMLCK has at least two isoforms, one of M_r of 135K and the other of M_r 210K (Fig. 5). The M_r 210K MLCK contains a longer amino terminus compared to the M_r 135K isoforms (Watterson *et al.,* 1995). The catalytic and CaM regulatory segments are about 300 amino acid residues in length and phylogenetically conserved. Except for *Drosophila* MLCK, the five known sm/nmMLCK sequences are 80–85% identical throughout the catalytic and CaM regulatory segments (Fig. 6). The amino-terminal supramolecular association domain of sm/nmMLCK is variable in amino acid sequence and length and has repeated sequence motifs found in other proteins (Shoemaker *et al.,* 1990; Olson *et al.,* 1990; Benian *et al.,* 1989). The carboxyl-terminal segment of sm/nmMLCK is identical in amino acid sequence to a M_r 23K protein called KRP (for kinase-related protein) (Shattuck *et al.,* 1988; Lukas *et al.,* 1988; Collinge *et al.,* 1992) or telokin (Ito *et al.,* 1989). The KRP has no kinase or CaM-binding activity (Collinge *et al.,* 1992) but binds to unphosphorylated myosin (Shirinsky *et al.,* 1993). Relatedly, the KRP domain of MLCK is important for myosin binding by MLCK (Shirinsky *et al.,* 1993).

The catalytic segment of sm/nmMLCK is defined by its amino acid sequence similarity to the common catalytic core among serine/threonine kinases (Hunter and Plowman, 1997) and the presence of this core sequence in fully active proteins that contain minimal flanking sequences (Shoemaker *et al.,* 1990). In the case of sm/nmMLCK, the smallest active and CaM-dependent kinases (Bagchi *et al.,* 1992a; Leachman *et al.,* 1992) begin approximately 75 amino acids amino-terminal to the first invariant residue of the catalytic domain (Fig. 6). For skMLCK, a similar results

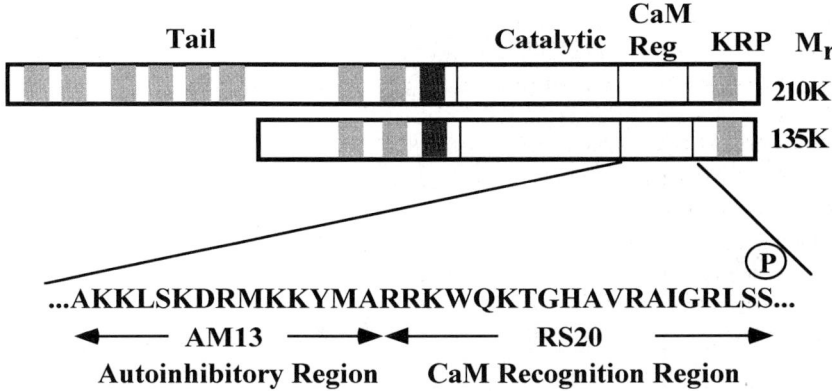

Fig. 5. Segmental organization of MLCK forms found in vertebrate smooth muscle and nonmuscle tissues. Two forms of the vertebrate sm/nmMLCK, which differ at the amino terminus, are made from the same genetic locus. The segments of the M_r 210K and M_r 135K isoforms of MLCK include catalytic, CaM regulatory (CaM Reg), and supramolecular association segments, the latter being implicated in the subcellular targeting of sm/nmMLCK. The carboxyl supramolecular association segment of sm/nmMLCK is called the KRP (kinase related protein) or telokin segment. It has myosin II binding activity and an amino acid sequence that is identical to that of the low molecular weight KRP protein that is made from a separate gene embedded within the MLCK gene. The amino supramolecular segment is called the tail segment based on historical convention. The relative locations of type I (fibronectin III) and type II (IgG superfamily) motifs are indicated by filled boxes. A light gray color indicates type I motifs whereas dark gray boxes represent type II motifs. These motifs are often associated with protein–protein interaction activity. The amino acid sequence of the autoinhibitory (AM13) and CaM recognition (RS20) regions are shown. The phosphorylation site within the RS20 domain is indicated.

was obtained (Takio *et al.*, 1985; Leachman *et al.*, 1992). Thus, the minimal length of an active CaM-dependent protein kinase is generally larger than the core segment, similar to other members of the serine/threonine protein kinase family.

Protein kinase catalytic segments have three regions of conserved residues at the amino half of the catalytic segment. One contains the GxGxxxxV consensus sequence implicated in nucleotide binding. Another region contains the invariant lysine (K). Covalent labeling of this lysine in MLCK (Komatsu and Ikebe, 1993) with an ATP-like affinity reagent FSBA, 5'-*p*-fluorosulfonylbenzoyladenosine, inactivates the enzyme. The third catalytic region contains an invariant glutamic acid.

84

```
chkmlck    IEERLGSKFGQVFRLVEK-KTGKVWAGKFFKKAYSAKKENIRDEITHIMN  569
humlck     IEERLGSKFGQVFRLVEK-KTRKVWAGKFFKKAYSAKKENIRQEITHIMN  592
bovmlck    IEERLGSKFGQVFRLVEK-KTGKIWAGKFFKKAYSAKKENIRQEISHIMN  775
rabmlck    VEDRLGTGKFGQVFHLIEK-STKKVLAGKFIKKAVIHPAERQEISHIMN  746
chkskmlck  SKEILGGKFGEVFHTCTEK-QTGLKL-AAKKVIHKDKEMVLIEVDVMN  520
rabskmlck  SKEALGGGKFGAVCTCTEKSTGLKL-AAAAKVIKKQTPKKDREMEIEVMN  564
dromlckl   IIEELGKGRFGIVYKVQERGQPEQLLAAKVIKSQDRQKVLEISIMR    345
                                                              346

chkmlck    CLHHPKKLVQCVDAFEEEKKANIHHDEDLTERECIK  619
humlck     CLHHPKLLVQCVDAAFEEEKKANIHHDEDLTERECIK  642
bovmlck    CLHHPKKLVQCVDDAFEEEGKTDEIHHDEDLTERECIK 825
rabmlck    DLRHPKLVQCVDDAFEGKTDEIHHDEDLTEREVIK   796
chkskmlck  QLNHRNLLQLYDAIETPHREIVMLFMEYIVEGGELFERIHHDDDYHLTEVDDCMV 614
rabskmlck  ALQHPKLLQLAASFESPREIVMVMEYITGGELFERVVADDFTLTEMDCIL 396
dromlckl                                                      395

chkmlck    YMRQISEGVEYIHKQGIVHLDLKPENIMCVNKTGTRIKLIHDFGLARRLES    669
humlck     YMRQISEGVEYIHKQGIVHLDLKPENIMCVNKTGTRIKLIHDFGLARRLEN    692
bovmlck    YMKQISEGVVEYIHKQGVIVHLDLKPENIMCVNKKTGTRIKLIHDFGLARRLEN 875
rabmlck    YMLQIHVDGVVSFIHHKMRVLHLDLKPENIMCVAATGHMLVKIHDFGLARRYN  846
chkskmlck  FVRQICEEGIREMHHMRVLHLDLKPENIIMCLCVNTTGHHLVKIHDFGLARRYN 664
rabskmlck  FVRQICDGILFMHHGQSVVHLDLKPENIMCHTRTSHQIKIHDFGLAQRLDT    446
dromlckl   FLRQVCDGVAYMHGQSVVHLDLKPENIMCHTRTSHQIKIHDFGLAQRLDT     445
```

```
chkmlck    A G S L K V L F G T P E F V A P E V I N Y E P I G Y E T D M W S I G V I C Y I L V S G L S P F M G D   719
humlck     A G S L K V L F G T P E F V A P E V I N Y E P I G Y A T D M W S I G V I C Y I L V S G L S P F M G D   742
bovmlck    A G S L K V L F G T P E F V A P E V I N Y E P I G Y A T D M W S I G V I H C Y I L V S G L S P F M G D   925
rabmlck    A G S L K V L F G T P E F V A P E V I N Y E P I S Y A T D M W S I G V I C Y I L V S G L S P F M G D   896
gfmlck     A G S L K V V F G T P E F V A P E V V N Y E Q V S Y S T D M W S M G V I T Y M L L S G L S P F L G D   670
chkskmlck  E E K L K V N F G T P E F L S P E V V N Y D Q I S D K T D M W S L G V I T Y M L L S G L S P F L G D   714
rabskmlck  N E K L K V N F G T P E F L S P E V V N Y D Q I S D K T D M W S L G V I T Y M L L S G L S P F L G D   496
dromlck1   K A P V R V L F G T P E F I P P E I I S Y E P I G F Q S D M W S V G V I C Y V L L S G L S P F M G D   495

chkmlck    N D N E T L A N V T S A T W D F D D E A F D E I S N L L K K D M K S R L N C T Q C L   769
humlck     N D N E T L A N V T S A T W D F D D D E A F D E I S N L L K K D M K N R L D C T Q C L   792
bovmlck    N D N E T L A N V T S A T W D F D D D E A F D E I S D D A K D M K N R L D C T Q C L   975
rabmlck    N D N E T L A N V T S A T W D F D D D A K D F I S D D A K D M K N R L D C T Q C L   946
gfmlck     N D N E T L S N N V T S A T W D F E D E A F D E I S D D A K D F I S D D A K D M K A R L S C D Q C F   720
chkskmlck  N D T E T L N N V L A A N W Y F D E E T F E E S V S D E A K D F V S N L L I K E K S A R M S A G Q C L   764
rabskmlck  D D T E T L N N V L S G N W Y F D D E T F E A V S D E A K D F V S N L L I V K E Q G A R M S A A Q C L   546
dromlck1   T D V E T F S N I T R A D Y D Y D D D E A F D C V S Q E A K D F I S Q L L V H R K E D R L T A Q Q C L   545

chkmlck    Q H P W L - - Q K D T K N M E A K K L S K D R - - M K K Y M A R R K W Q K T G H A V R A I G R L S S M   816
humlck     Q H P W L - - M K D T K N M E A K K L S K D R - - M K K Y M A R R K W Q K T G N A V R A I G R L S S M   839
bovmlck    Q H P W L - - M K D T K N M E A K K L S K D R - - M K K Y M A R R K W Q Q K T G N A V R A I G R L S S M   1022
rabmlck    Q H P W L - - M K D T K N M E V K K L S K D R - - M K K Y M A R R K W Q K T G N A V R A I G R L S S M   993
gfmlck     Q H P W L - - K K Q D T K N M E V K K L S K D R - - M L K K Y V L R R K W K K N F I G V C A A N R F K K I   767
chkskmlck  Q H P W L T N L A E K A K R C N R R L K S Q V M L K K Y V M R R R W K K N F I A V S A A N R F K K I   814
rabskmlck  A H P W L N N L A E K A K R C N R R L K S Q H - - L L K K Y L M K R R W K K R F I R R K W Q F I G R M A N L   596
dromlck1   E S K W L S Q R P D D S L S N N K I C T D K - - L L K K F L I R R K W Q F I G R M A N L   593
```

Fig. 6. Alignment of the catalytic and CaM regulatory segments of MLCKs. Amino acid sequences of chicken sm/nmMLCK (chkmlck), human sm/nmMLCK (humlck), bovine sm/nmMLCK (bovmlck), rabbit sm/nmMLCK (rabmlck), goldfish sm/nmMLCK (gfmlck), chicken skMLCK (chkskmlck), rabbit skMLCK (rabskmlck), and *Drosophila* MLCK (dromlck1) are shown. For each vertebrate species, the sm/nmMLCK and skMLCK are made from distinct genetic loci. The fruit fly *Drosophila* uses one genetic locus to make MLCKs for both striated muscle and nonmuscle tissues. Amino acid residues conserved in the catalytic segment of many protein kinases are boxed. The CaM regulatory segments are shaded. Sequences were obtained from the GenBank database.

These three regions of the catalytic segment are followed by regions of comparatively low sequence conservation. The carboxyl half of the catalytic segment has conserved regions that start with the DxKxEN and DFG consensus sequences for protein kinases. Mutation of the glutamate (E) located within the DxKxEN motif to Lys, in both sm/nmMLCK and skMLCK, increased the K_m for the myosin light chain substrate (Herring *et al.*, 1992). The DxKxEN and DFG consensus sequences are followed by the xPE sequence, where x is alanine (A) in many protein kinases. Near the carboxyl end of the catalytic segment is a region that contains residues conserved among serine/threonine kinases. This region contains an invariant DxxxxG and an invariant arginine (R) at the end of the catalytic segment. Myosin light-chain kinases and other CaM-dependent kinases have additional conserved regions located amino-terminal to the DxKxxN motif. One of these, GGELF, has a conserved E residue. Mutation of this amino acid to K weakens peptide substrate interaction in both sm/nmMLCK and skMLCK (Herring *et al.*, 1992).

There are two major models for the mechanism of autoinhibition and activation by CaM. One model is the pseudosubstrate hypothesis (Kemp *et al.*, 1987). It assumes competition between substrate and autoinhibitory region (=pseudosubstrate) with the catalytic site. The binding of CaM to the pseudosubstrate sequence releases it from the active site and allows substrate access. The other mechanism proposes that substrate, autoinhibitory, and CaM recognition segments are distinct sites that are closely coupled (Lukas *et al.*, 1986, 1988; Shoemaker *et al.*, 1990). Calmodulin binding to its recognition segment perturbs the autoinhibitory segment and brings about a relief of autoinhibition (=activation). In contrast to the pseudosubstrate model, the relief-of-autoinhibition mechanism can be achieved in a direct (active site) and/or indirect (allosteric site) mechanism that does not necessarily equate regions of the CaM regulatory segment with autoinhibition (Lukas *et al.*, 1988; Shoemaker *et al.*, 1990).

The pseudosubstrate model of MLCK regulation was based on the amino acid sequence similarity of certain parts of the CaM regulatory segment with the phosphorylation site of the myosin light chain (Kemp *et al.*, 1987). Synthetic peptide analogs of this region of the CaM regulatory segment would inhibit CaM-independent MLCK produced by proteolysis (Kemp *et al.*, 1987; Ikebe *et al.*, 1987). Consistent with the pseudosubstrate model, synthetic peptide analogs gave inhibition kinetics competitive with the light chain substrate. However, one test of pseudosubstrates is that proper placement of a phosphate accepting residue (Ser) should allow the sequence to become a substrate. This strategy was successfully used for PKA and PKC (Glass *et al.*, 1989; Mitchell *et al.*, 1995; House and Kemp, 1987), two protein kinases that use a pseudosubstrate mechanism. This approach has not yet provided good substrates based on the proposed pseudosubstrate region of MLCK. Similarly, mutagenesis of the CaM regulatory segment to generate an intramolecular autophosphorylation site required several modifications to the sequence to enable autophosphorylation (Bagchi *et al.*, 1992b). Mutations of what were thought to be the most important amino acids for the pseudosubstrate, the amino end (RRKWQK) of the RS20 region, did not result in constitutive activity (Shoemaker *et al.*, 1990; Fitzsimons *et al.*, 1992). Examples include mutants 7, 8, and 12 of Fig. 7, where clusters or single basic amino acids (R, K) were changed to acidic or neutral amino acids. With protein kinases that are standards of comparison for a pseudosubstrate mechanism, mutagenesis of the pseudosubstrate domain (in the case of PKC) led to constitutive activity (Pears *et al.*, 1990), and mutations of charged and neutral residues in the pseudosubstrate domain of the regulatory subunit of PKA reduced the potency of inhibition (Buechler and Taylor, 1991; Poteet-Smith *et al.*, 1997; Buechler *et al.*, 1993). Therefore, the proposed pseudosubstrate segment of MLCK does not have some of the properties associated with other protein kinase pseudosubstrate domains.

Lukas *et al.*, (1986, 1988) and Shoemaker *et al.* (1990) proposed that the CaM regulatory segment of MLCK has functionally distinct and coupled centers of autoinhibitory and CaM recognition activities. Mutagenesis within the autoinhibitory/AM13 region (Shoemaker *et al.*, 1990) showed that MLCK can be rendered CaM independent with retention of CaM recognition activity. Coupling of CaM binding at the CaM recognition region to the autoinhibitory segment thus provides for a relief-of-autoinhibition mechanism for MLCK activation (Lukas *et al.*, 1986, 1988; Shoemaker *et al.*, 1990). Deletion of both autoinhibitory/AM13 and CaM-binding/RS20 regions of MLCK resulted in a constitutively active enzyme (Shoemaker *et al.*, 1990), suggesting that most of the elements required for CaM binding and autoinhibition were located within the combined segments. Inversion of the RS20 sequence (mutant 9, Fig. 7) resulted in a kinase that remained Ca^{2+}/CaM dependent, suggesting that the exact sequence of the RS20 region had little influence on kinase autoinhibition (Shoemaker *et al.*, 1990). Therefore, the two center (AM13 + RS20) model proposed by Shoemaker *et al.* (1990) is the most parsimonious for the regulation of MLCK activation and is

Group A. Mutations that result in relief of autoinhibition

	AM13																		RS20														
	A	K	K	L	S	K	D	R	M	K	K	Y	M	A	R	R	K	W	Q	K	T	G	H	A	V	R	A	I	G	R	L	S	S
1.		E	E			E				E	E																							
2.															D							G												
3.															D							G		D										
4.																						G		G										
5.																						G		G	D									
6.																						G		R										

Fig. 7. Summary of amino acid changes in the mutagenesis analysis of the CaM regulatory segment of sm/nmMLCK. The centers of autoinhibitory (AM13) and CaM recognition (RS20) activity are shown. Mutant MLCKs are placed into three groups according to the functional effects of the mutations in the CaM regulatory segment of the enzyme. Only differences in the amino acid sequence between the mutant and the wild-type sm/nmMLCK are shown.

Group B. Mutations that alter Ca^{2+}/CaM recognition or sensitivity

	AM13																RS20																
	A	K	K	L	S	K	D	R	M	K	K	Y	M	A	R	R	K	W	Q	K	T	G	H	A	V	R	A	I	G	R	L	S	S
7.															E	E																	
8.															E	E	D																
9.																L					I	A	R	V	A		G	T	K	Q	W	K	R
10.																					L	K	G	A	I	L	T	T	M	L	A	T	N
11.													A	A	A																		
12.																							E										
13.																							A										
14.																										A							
15.																													A				
16.																												A					
17.																				L													
18.																																	A

Fig. 7. Continued

Group C. Mutations that result in the loss of CaM activation

	AM13																						RS20											
A	K	K	L	S	K	D	R	M	K	K	Y	M	A	R	R	K	W	Q	K	T	G	H	A	V	R	A	I	G	R	L	S	S
19.					E			D																								
20.															A																	
21.															G																	
22.																						A										
23.																														I		
24.																		E			P	M	L	S	E	Q	L	V	D	S		

Fig. 7. Continued

generally applicable with specific modifications to other CaM-dependent protein kinases.

Through site-directed mutagenesis of the CaM recognition of sm/nmMLCK, Shoemaker et al. (1990) demonstrated that CaM recognition sequences among various CaM-dependent protein kinases can be interchangeable. For example, the CaM recognition region of sm/nmMLCK could be replaced with that from skMLCK or CaMPKII (mutant 10, Fig. 7) to produce an enzyme with sm/nmMLCK catalytic activity but the CaM recognition properties skMLCK or CaMPKII (Shoemaker et al., 1990). Similarly, in the complimentary skMLCK mutagenesis experiment, it was shown that its CaM-binding region can be replaced with that from sm/nmMLCK (Leachman et al., 1992). The structural similarity between the CaM recognition regions of skMLCK and sm/nmMLCK also applies to their respective CaM:peptide complexes. Three-dimensional structures of the complexes Ca^{2+}:CaM:RS20 (Meador et al., 1992) and Ca^{2+}:CaM:M13 (Ikura et al., 1992b; Crivici and Ikura, 1995) show that peptides in the complex adopt an α-helical conformation. M13 interacts with many of the same residues of CaM as does RS20, but some CaM-interacting residues are unique for each peptide (see also Fig. 4).

Site-directed mutagenesis of the sm/nmMLCK CaM recognition region (RS20) have demonstrated that selected basic and hydrophobic amino acids contribute to CaM binding and kinase activation. Several different mutations within the RS20 region resulted in alteration of CaM binding/recognition (group B, Fig. 7) and sometimes in an enzyme that could not be activated by CaM (group C, Fig. 7). Charged residue mutagenesis within the M13 region of skMLCK similar to those shown for mutant 8 (Fig. 7) also altered CaM binding/recognition of the enzyme (Herring, 1991) consistent with results obtained by mutations with the CaM recognition/RS20 region of sm/nmMLCK.

Structural features of CaM relevant to activation of sm/nmMLCK have been studied using different mutagenesis strategies. The first strategy was based on the large number of conserved amino acids in CaMs (Fig. 4). For example, results of the charge reversal mutagenesis of conserved acidic clusters at (E82,E83,E84) or (D118,E119,E120) indicated the importance of these residues for the sm/nmMLCK activator activity of CaM (Craig et al., 1987; Weber et al., 1989). Most of the effects of the charge cluster mutations are mimicked by the single mutants E84K or E120K (Shoemaker et al., 1990). Mutant CaMs containing E84K or E120K also fail to activate a chimeric enzyme containing the CaM-binding domain of the skMLCK in the context of the sm/nmMLCK sequence (Shoemaker et al., 1990). Mutation of some other conserved acidic clusters at (E6,E7,Q8) or E45,A46,E47) to lysines did not affect sm/nmMLCK activation (Farrar et al., 1993). Thus, the location of charge perturbation determines whether MLCK activation will be altered. Combinations of mutations generally cause poorer sm/nmMLCK activation as seen in a mutant that has both the 82–84 and the 118–120 clusters changed to lysines (Weber et al., 1989). Poorer sm/nmMLCK activation in double mutants can be found even though the second site mutation by itself may not affect the enzyme. For example, mutants E82K and E83K do not affect MLCK activation, but combining them with E84K reduces MLCK activation compared to E84K alone. A number of CaM mutations have little or no effect on MLCK activation. These results should not be overinterpreted, as other changes at the same site could result in different effects on sm/nmMLCK activation. For example, mutation of a conserved Ca^{2+} ligating residue (E31, E67, E104, or E140) had mixed results depending on the site and nature of the mutation. Mutant CaMs containing E67Q or E67K and E140Q or E140K had significant effects on the sm/nmMLCK activator activity of CaM (Gao et al., 1993). E67A or E140A mutant CaMs had significantly reduced Ca^{2+}-binding stoichiometry in the absence of MLCK, but their Ca^{2+}-binding properties were restored toward normal in the presence of the CaM-binding peptide RS20 (Haiech et al., 1991). Consistent with this observation, both mutants fully activate MLCK (Haiech et al., 1991). Similarly, mutation of E31A, or the mutation of E104A, had little effect on maximal sm/nmMLCK activation (Kilhoffer et al., 1992), whereas mutation of either residue to glutamine perturbs Ca^{2+}/CaM sensitivity, but not maximal activation compared to wild-type CaM (Gao et al., 1993).

A second approach to CaM mutagenesis was to focus on specific amino acids chosen based on studies of chimeric proteins that combine native CaM sequences with those from related proteins that do not activate MLCK, such as yeast CaM or troponin C (TnC). Yeast CaM poorly activates both sm/nmMLCK and skMLCK. However, one of the Ca^{2+}-binding sites of yeast CaM is defective (Starovasnik et al., 1993; Matsuura et al., 1993). Using a series of chimeric constructs, it was found that repair of the Ca^{2+}-binding defect is not sufficient to restore its activator activity toward MLCKs (Lukas et al., 1994; Nakashima et al., 1996; Matsuura et al., 1993). Some of the repaired constructs that poorly activate sm/nmMLCK also function as antagonists of wild-type CaM (Lukas et al., 1994). However, there is one chimeric yeast CaM construct that has completely restored MLCK activator activity (Lukas et al., 1994). This mutant contains a limited amount of yeast CaM amino acid sequence that overlaps with the region that varies among the naturally occurring CaMs that activate MLCK (residues 56–86). Another chimeric approach used TnC, the Ca^{2+} sensor protein responsible for the Ca^{2+}-induced contraction of striated muscle. Characterization of TnC–CaM chimeras established that certain amino acid residues in the first half of CaM were important determinants of sm/nmMLCK activation (George et al., 1990), which led to subsequent investigations of point mutations in CaM based on the limited differences in the chimeric proteins compared to wild-type CaM. For example, mutations at T34, S38, and T110 exhibited reduced sm/nmMLCK activation. A residue that varies among CaMs, S17, can be mutated to glutamic acid without loss of sm/nmMLCK activator activity, but mutation to histidine decreases maximal activation by 30%. Changes of single acidic amino acids such as E14 can also alter sm/nmMLCK activation. As found earlier with charge clusters, combinations of these perturbing mutations cause poorer sm/nmMLCK activation and result in mutant CaMs that can antagonize the activation of wild-type CaM (VanBerkum and Means, 1991). Finally, among the many TnC–CaM chimeric constructs reported (VanBerkum and Means, 1991; Su et al., 1994; George et al., 1993), only one can be classified as a "gain of function" mutant. It has only a small part of the TnC sequence (12 amino acids), including the tripeptide (KGK) insert found in TnC.

In the strategy discussed earlier, mutations of conserved acidic residues such as E14, E84, and E120 were found to alter sm/nmMLCK activation. In the subsequently determined structure of the CaM:peptide complex (Meador et al., 1992), these residues were found to be within 4 Å of basic amino acids of the peptide. E84 makes contacts with the side chains of two of the arginine residues near the carboxyl end of the peptide, E120 interacts with a single lysine, and E14 interacts with an arginine and lysine at the amino-terminal end of the peptide (Figs. 8 and 9). In more recent studies based on the structure, mutation of some of the conserved hydrophobic amino acids in the CaM:peptide interface, such as M124Q, caused reductions in maximal sm/nmMLCK activity (Chin and Means, 1996). Mutation of phylogenetically variable methionine residues in the peptide:CaM interface (e.g., Met-144) did not affect the maximum activity of the CaM:sm/nmMLCK complex, although they did increase the concentration of CaM needed for activation (Chin and Means, 1996). Similarly, the mutation of multiple hydrophobic residues of CaM within the interface to alanines also resulted in CaMs with decreased MLCK activator activity and weaker sm/nmMLCK interaction

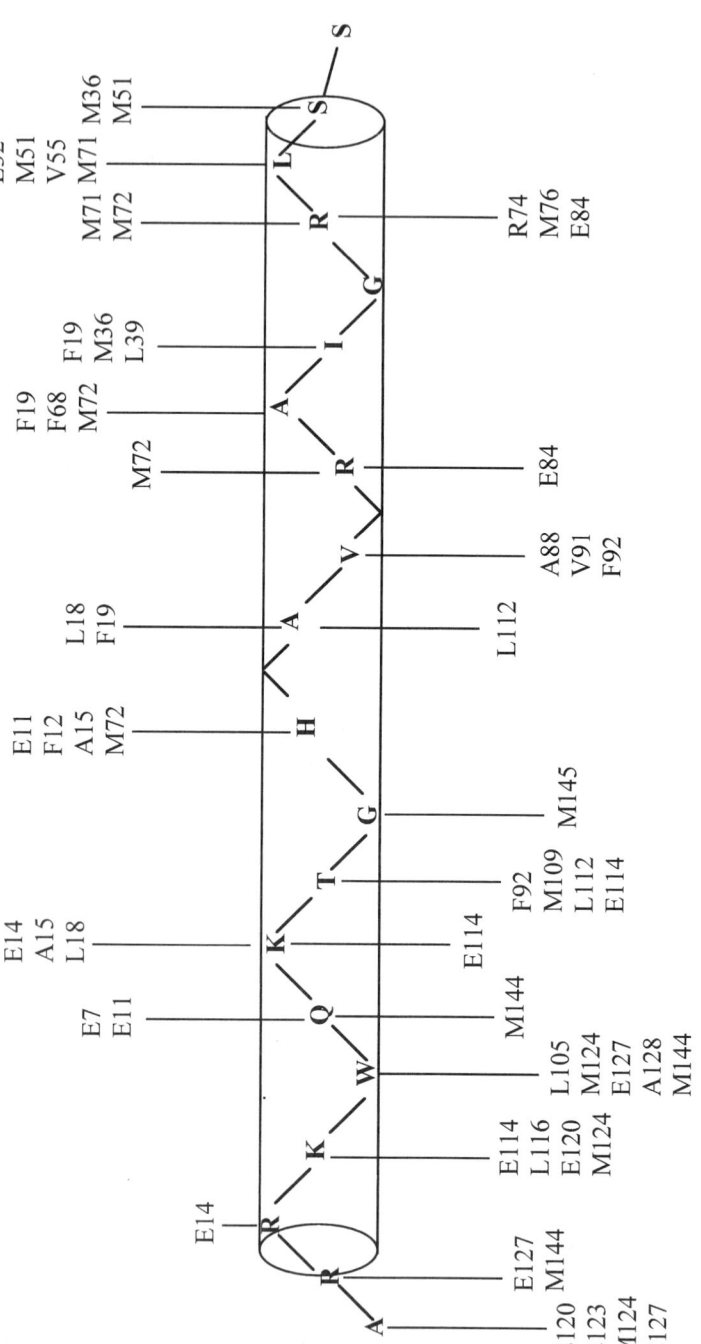

Fig. 8. Interactions between CaM and the CaM recognition peptide of sm/nmMLCK. The amino acid sequence of the RS20 region of sm/nmMLCK is shown within the cylinder. The amino acids shown above and below the cylinder (e.g., E84) are amino acids in CaM that are within 4 Å distances based on the structures of CaM:peptide complexes and, therefore, proposed as interacting residues.

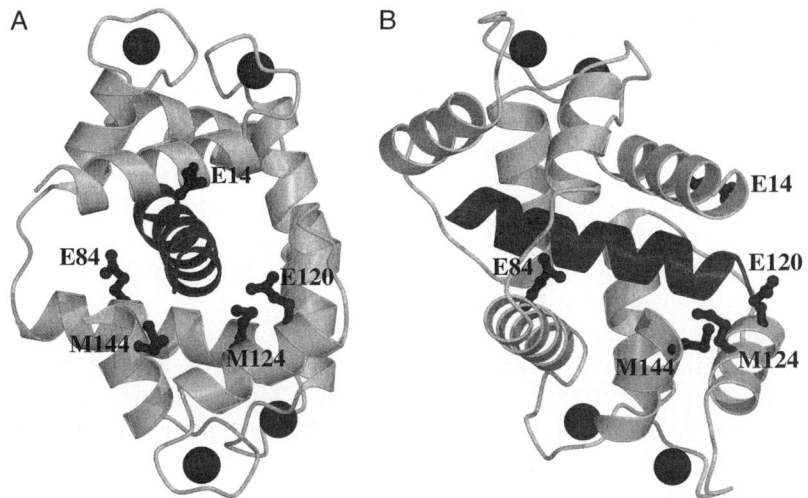

Fig. 9. Three-dimensional structure of the complex between CaM and the CaM recognition peptide of sm/nmMLCK. These ribbon drawings are based on coordinates in the PDB entry 1CDL. Ca^{2+} ions are solid spheres. The CaM recognition peptide is shown as a darkened ribbon. Side chains of selected amino acids from CaM are indicated (e.g., E84). These amino acids are examples of ones within 4 Å as indicated in Fig. 8 (A) View down the long axis of the peptide helix, shown in the center. (B) This view represents a 90° rotation around the axis perpendicular to the peptide.

(Chin *et al.*, 1997). How changes in hydrophobic residues within the interface modify the structure of the CaM:peptide complex is not yet known. However, the crystal structure of one of the charge reversal CaMs complexed with the MLCK peptide (E84K–CaM:RS20) shows that the overall structure of the complex is retained in the mutant complex, and localized perturbation around amino acid 84 and the helix containing the mutation (Weigand *et al.*, 1997). The perturbed Ca^{2+}-binding affinity of the E84K–CaM:RS20 complex (Mirzoeva *et al.*, 1997) can be rationalized from the observed structural changes. However, the crystal structures of the wild-type CaM:sm/nmMLCK and E84K–CaM:sm/nmMLCK complexes are needed in order to explain fully the effect of the E84K mutation on the activator activity of CaM toward sm/nmMLCK.

The CaM recognition segment of skMLCK has not been as extensively studied as sm/nmMLCK. Consistent with the common charged residue interactions observed in the CaM:RS20 and CaM:M13 complexes (Fig. 4), charge reversal mutation of basic amino acids within the M13 segment of skMLCK (KKR → EDE) altered CaM binding, but had no effect on the autoinhibition of the kinase (Herring, 1991), which is essentially the same result found with changing the corresponding region of the RS20 segment of sm/nmMLCK (Shoemaker et al., 1990) (see Fig. 7). A study of the contribution of selected amino acid residues within an M13 peptide analog was done by making synthetic peptides containing alanine replacements (Montigiani et al., 1996). These peptides retained CaM binding activity even when the hydrophobic tryptophan and phenylalanine residues were replaced by alanine. Similarly, replacement of single hydrophilic amino acids such as R16A actually improved CaM binding (Montigiani et al., 1996). Thus, CaM easily tolerates single amino acid replacements in this CaM-binding peptide. A more important conclusion from these studies, however, is that the CaM-binding sequence of skMLCK and probably other CaM-dependent enzymes is not the highest affinity possible, but is sufficient for selective CaM (vs troponin C, and other CaM-like proteins) recognition.

Differences between the M13 and RS20 interactions with CaM observed in the complexes (Fig. 4) appear to be correlated with the differential sensitivity of skMLCK and sm/nm MLCK to mutations in CaM. For example, skMLCK is less sensitive to the presence of the yeast CaM segments in the amino terminus of yeast CaM–vertebrate CaM chimeric molecules (Lukas et al., 1994). Although not directly involved in the CaM:peptide interface, other mutations of E31,E67,E104, and E140, which are Ca^{2+}-ligating residues of CaM, also affect the activation of skMLCK differently from sm/nmMLCK (Gao et al., 1993). In these studies, E31Q CaM exhibited submaximal skMLCK activation, while the maximal activation of sm/nmMLCK was unchanged (Gao et al., 1993). These properties may be due to the specific effects of the mutation on the Ca^{2+}-binding properties of the CaM:skMLCK or CaM:M13 complexes.

Both CaM:RS20 and CaM:M13 complexes also model fairly well the Ca^{2+}-binding properties of the CaM:enzyme complex and exhibit enhanced Ca^{2+}-binding properties compared to CaM alone (Haiech et al., 1991; Yagi et al., 1990; Peersen et al., 1997; Mirzoeva et al., 1997). In studies of the kinetics of dissociation of Ca^{2+} from the CaM:MLCK complex, RS20 and M13 peptides provided nearly the same results as the whole enzymes. The dissociation of Ca^{2+} from the CaM:peptide complexes is biphasic and slower than CaM alone (Persechini et al., 1996;

Peersen *et al.*, 1997; Brown *et al.*, 1997; Johnson *et al.*, 1996). The presence of target peptides not only enhances the Ca^{2+}-binding affinity of CaM, but also restores properties of the mutant CaMs containing Ca^{2+}-binding defects (Maune *et al.*, 1992; Haiech *et al.*, 1991) back toward normal (Findlay *et al.*, 1994; Haiech *et al.*, 1991).

Outside of the conserved catalytic and CaM regulatory segments, sm/nmMLCK has multiple motifs that have been implicated in targeting to contractile filaments such as actin and myosin (Shirinsky *et al.*, 1993; Kanoh *et al.*, 1993). Vertebrate sm/nmMLCK has a motif I (Fig. 5) that contains a pattern related to fibronectin type III motifs (Shoemaker *et al.*, 1990). This motif is also found in cell adhesion proteins such as the N-CAMS. Motif II is an IgG superfamily motif that is also found in cell adhesion proteins and other extracellular receptors. Although the presence of repeated IgG superfamily motifs in titin are associated with contractile filament interactions (Labeit *et al.*, 1992), it is not known whether this function persists in the sm/nmMLCK isoforms. However, the locations of motifs I and II within the amino-terminal supramolecular association region are conserved among M_r 135K and M_r 210K sm/nmMLCK isoforms (Fig. 5). Within the amino terminal segment, actin-binding activity was localized to a 114 amino acid region (Kanoh *et al.*, 1993). The proposed actin-binding region is conserved in other sm/nm-MLCKs. Deletion of this segment, as well as type I and type II motifs in the amino tail segment of sm/nmMLCK, abolished actomyosin interactions (Kin *et al.*, 1997). Thus, the native filament interaction of MLCK is associated with actin- and myosin-binding motifs. The carboxyl end of MLCK has a type II motif that is also associated with myosin II binding (Shirinsky *et al.*, 1993).

C. Molecular Genetics of Myosin Light-Chain Kinase and Related Proteins

In the phylogenetic species for which data are available, the CaM-regulated MLCK is encoded by a single gene. More extensive data are those for the chicken sm/nmMLCK gene (Collinge *et al.*, 1992; Watterson *et al.*, 1995) and the single gene from the fruit fly *Drosophila* (Kojima *et al.*, 1996; Tohtong *et al.*, 1997), which appears to have a single type of MLCK in both muscle and nonmuscle tissues. Less is known about the skMLCK in vertebrates. In the case of the vertebrate sm/nmMLCK gene, the locus is a complex one that produces multiple products, some of which are not protein kinases (Watterson et al., 1995; Collinge et al., 1992; Lukas et al., 1988). Myosin light-chain kinases from earlier phylogenetic species, such as *Dictyostelium* (Tan and Spudich, 1991), are not directly regulated by CaM, and MLCK-like proteins, such as twitchin, are not physiological MLCKs and appear to be regulated by other calcium-binding proteins, such as S100 (Johnson and Quiocho, 1996).

The chicken sm/nmMLCK gene locus encodes at least three size classes (Watterson et al., 1995) of mRNA and proteins (Fig. 10). Two of the size classes of MLCK proteins, M_r 135K and M_r 210K, are isoforms of

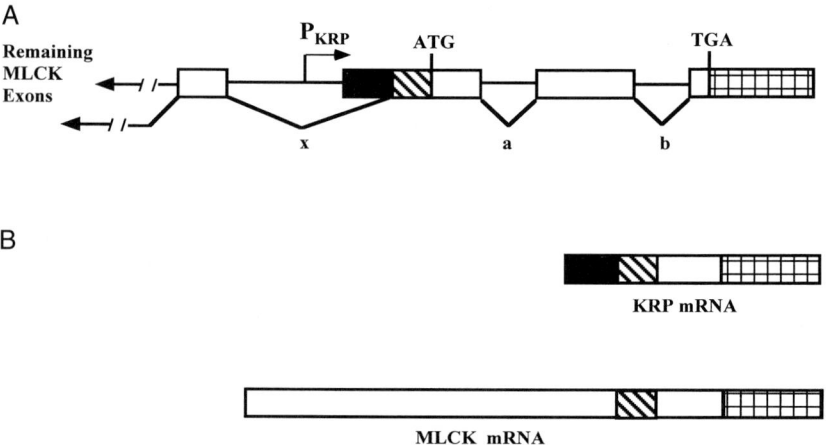

Fig. 10. Diagrammatic summary of the vertebrate KRP gene organization and its relationship to the MLCK gene. The organization shown is that for the chicken KRP/MLCK genetic locus. (A) The chicken KRP transcription unit spans approximately 5.3 kb. The KRP promoter (P_{KRP}) and 109 nucleotides of the 5′ leader sequence (solid bar) are located within a MLCK intron, which results in unique nucleotide sequences at the 5′ end of the KRP mRNA. The remaining portion of the KRP 5′ leader sequence has a nucleotide sequence identical to coding sequences for MLCK (diagonal hatched box), but they are not in frame as far as ORFs until the ATG. The indicated ATG is the KRP translational start site and an internal methionine codon for MLCK. Both MLCK and KRP splice out introns a and b, and have the same stop codon and 3′-nontranslated sequence (cross-hatched bar). The mature MLCK transcript removes MLCK intron x, which contains the KRP promoter and start of transcription. (B) mRNAs for MLCK and KRP that result from the two genes that use a common set of splicing reactions and exons are shown. Note that the same KRP coding sequence is found in MLCK.

sm/nmMLCK encoded by mRNAs of 5.5 and 9 kb, respectively. The proposed reading frames for the overlapping clones indicate that the open reading frame (ORF) for M_r 210K is similar to its apparent molecular chain weight on SDS–PAGE gels (Watterson *et al.*, 1995), but that for the M_r 135K isoform is proposed to be only 108K based on its ORF (Olson *et al.*, 1990). A third size class of mRNA (2.7 kb) and protein, (M_r 20K) are also made from the same locus by a separate gene, the KRP gene, that is imbedded within the MLCK gene (Collinge *et al.*, 1992; Watterson *et al.*, 1995; Shattuck *et al.*, 1988). The KRP/telokin is a myosin-binding protein that lacks protein kinase and CaM-binding activity (Collinge *et al.*, 1992; Shirinsky *et al.*, 1993). The KRP/telokin amino acid sequence is identical to the KRP/telokin domain of MLCK due to the use of a common set of exons and the initiation ATG of KRP/telokin being an internal methionine codon in MLCK (Fig. 10). The mRNAs for the two MLCKs and the KRP/telokin use a common stop codon and 3′-untranslated sequences. The only portion of the KRP/telokin mRNA that can be selectively measured is the 5′ end. This is because the TATA box for the KRP gene and the beginning of KRP exon 1 are derived from genomic sequences that are in an intron for MLCK. A similar relationship has been found more recently with the rabbit KRP/telokin gene (Herring and Smith, 1996). The MLCK-210 differs from the MLCK-108/135 by the presence of an extended amino tail segment (Watterson *et al.*, 1995). The function of this extended tail segment is not known. However, it is potentially a means for differences in subcellular localization, based on the presence of characteristic protein interaction motifs and the postulated roles of the MLCK tail region in interactions with actomyosin. Site-specific antibodies to portions of the extended tail sequence (Watterson *et al.*, 1995) should allow identification of MLCK-210 in future investigations.

A human MLCK gene has been found on chromosome 3 (Van Eldik *et al.*, 1991) and localized to 3q21 (Potier *et al.*, 1995). It produces 2.7 and 6 kb mRNAs (Potier *et al.*, 1995). A KRP/telokin gene is also present in humans, with a conservation of amino acid sequence and genomic organization (D. M. Watterson *et al.*, 1997). However, there does not appear to be a KRP/telokin segment in the single *Drosophila* MLCK gene (Kojima *et al.*, 1996; Tohtong *et al.*, 1997). The *Drosophila* MLCK gene produces multiple mRNA species with ORFs similar to vertebrate sm/nmMLCK-210 (Kojima *et al.*, 1996; Tohtong *et al.*, 1997). The multiple mRNAs are apparently produced by alternative splicing to create unique 5′ ends and give four different mRNA size classes (Kojima *et al.*, 1996;

Tohtong et al., 1997). The higher molecular weight *Drosophila* mRNA can also be spliced toward the 3' end, resulting in an mRNA that lacks the exon containing most of the CaM regulatory sequences. A construct containing the isoform of MLCK missing the putative CaM regulatory sequences did not have significant kinase activity in the presence or absence of CaM (Kojima et al., 1996). Therefore, a mechanism other than CaM regulation may be used to activate this isoform. This is especially interesting because the MLCK that has been characterized from an earlier phylogenetic species, *Dictyostelium*, does not appear to be a CaM-regulated protein kinase and lacks a CaM-binding motif in its primary structure. *Dictyostelium* MLCK also lacks a carboxyl segment that resembles KRP/telokin. Still earlier in phylogeny is the yeast, *Saccharomyces cerevisiae*, whose complete genome has been elucidated. This organism does not encode an MLCK (Hunter and Plowman, 1997). Clearly, much remains to be done with the molecular genetics of MLCK, but there is a trend in the available data. Myosin light-chain kinase has been found in late-evolving eukaryotes, CaM-regulated MLCK has been found as a feature of later evolving multicellular organisms, and MLCKs with a KRP segment and the presence of a KRP gene have been found only in phylogenetically recent organisms.

The novel features of the vertebrate MLCK/KRP gene locus and its regulation of the myosin II motor system have some resemblance to a prokaryotic operon, in which multiple gene products focused on a common pathway are encoded in a clustered manner in the genome. The gene-within-a-gene relationship that was first described for the chicken MLCK/KRP locus (Collinge et al., 1992) appears to be found in other protein kinase signal transduction systems. For example, the CaMPKII gene locus also produces a KAP product that is not a protein kinase or CaM-binding protein (Bayer et al., 1996), and the FAK gene locus produces a distinct product that does not appear to be a tyrosine kinase (Schaller et al., 1993).

In summary, genetic loci for vertebrate sm/nmMLCKs are complex and produce multiple protein products involved in the regulation of distinct aspects of a common cellular structure, the myosin II motor. The vertebrate skMLCK is produced by a different genetic locus and has a different physiological function. Of the various CaM-regulated protein kinases, MLCK is novel in its use of different genes (skMLCK vs sm/nmMLCK) and alternative use of the same gene (sm/nmMLCK-210 vs sm/nmMLCK-108/135) to produce different types of protein kinases (skMLCK, sm/nmMLCK-108/135, and sm/nmMLCK-210) and a nonkinase protein (KRP/telokin) from the same exons.

III. CALMODULIN-DEPENDENT PROTEIN KINASE II

A. Calmodulin-Dependent Protein Kinase II Function and Cellular Ca^{2+} Signal Transduction

Calmodulin-dependent protein kinase II (CaMPKII) represents a family of closely related protein kinases with broad substrate specificity. It is also referred to as multifunctional protein kinase, CaM-dependent multiprotein kinase, CaM kinase II, and type II CaM kinase (Schulman and Hanson, 1993). Calmodulin-dependent protein kinase II can phosphorylate at least 40 proteins *in vitro* (Hanson and Schulman, 1992a). These proteins are involved in cellular metabolism, neurotransmitter release and synthesis, cytoskeletal structure, Ca^{2+} homeostasis, and gene expression. Some of the *in vitro* CaMPKII substrates have been shown to be *in vivo* substrates. These include synapsin I (Schulman and Lou, 1989), MLCK (Stull *et al.*, 1993), and specific ion channels (McGlade-McCulloh *et al.*, 1993; Yakel *et al.*, 1995; Tan *et al.*, 1994).

Calmodulin-dependent protein kinase II is abundant in the brain, where it comprises at least 1% of the total protein, and is enriched in the postsynaptic density (Erondu and Kennedy, 1985). This localization and abundance were early clues that CaMPKII plays a significant role in Ca^{2+} signal transduction in the brain. Calmodulin-dependent protein kinase II is unique among the CaM-regulated kinases because following activation by Ca^{2+}/CaM, it converts to a Ca^{2+}/CaM-independent state as a result of autophosphorylation of its autoinhibitory region (Miller and Kennedy, 1986; Saitoh and Schwartz, 1985). The activity of the Ca^{2+}/CaM-independent CaMPKII is modulated by further autophosphorylation (Lou and Schulman, 1989; Hanson and Schulman, 1992b; Patton *et al.*, 1990; Colbran and Soderling, 1990) and by phosphatases (Strack *et al.*, 1997). These data allowed Lisman and Goldring (1988) to propose that a protein kinase such as CaMPKII has the capability to store signal information that may be part of the encoding process for long-term memory.

The abundance and localization in the postsynaptic density and the ability to make a transient Ca^{2+} stimulus persist through calcium-independent kinase activity led to the hypothesis that CaMPKII is involved in long-term potentiation (LTP) that is associated with memory and learning processes (Lisman, 1994). Long-term potentiation is an electrophysiological characterization of synaptic responses brought about through the activation of voltage-modulated and ligand-gated Ca^{2+} channels, such as (N-methyl-D-aspartate) (NMDA) glutamate receptors (Brown et al., 1988; Davies et al., 1989). The distinguishing feature of LTP in the CA1 region of the hippocampus is that it is a rapid and persistent enhancement of synaptic transmission resulting from pulsed stimulation (Brown et al., 1988; Davies et al., 1989). Because this enhancement can persist *in vivo* for weeks, LTP is a logical candidate as a synaptic mechanism for memory and learning (Brown et al., 1988). It is proposed that LTP results from Ca^{2+} fluxes through Ca^{2+} channels that activate protein kinases such as CaMPKII that can further modulate channel activities through phosphorylation (Yakel et al., 1995; McGlade-McCulloh et al., 1993; Omkumar et al., 1996; Fukunaga et al., 1995). Long-term potentiation also involves the activation of other serine/threonine kinases (Suzuki, 1994; Hvalby et al., 1994; Klann et al., 1991) and tyrosine kinases (Grant et al., 1992; Rosenblum et al., 1996; Rostas et al., 1996). However, a working hypothesis is that CaMPKII is a key protein kinase for the initiation of events that induce LTP (Lisman, 1994). Protein kinase are also involved in maintaining LTP (Lisman, 1994), but the exact role of CaMPKII in the sustaining phase is still under active investigation (Mayford et al., 1997; Otmakhov et al., 1997).

Physiological and pharmacological studies are consistent with this model. In hippocampal CA1 slices, Ca^{2+}-independent CaMPKII activity was increased after induction of LTP and this increased kinase activity persisted for at least 1 hr (Fukunaga et al., 1993; Miyamoto and Fukunaga, 1996). Microinjection of selective (active site directed) peptide inhibitors of CaMPKII blocked the induction of LTP (Malinow et al., 1989; Otmakhov et al., 1997); however, the inhibitor peptides did not affect the maintenance of LTP (Otmakhov et al., 1997). Targeted expression of a minigene containing an inhibitor peptide based on the CaMPKII autoinhibitory domain in *Drosophila* caused alterations in learning and memory in mutant flies (Griffith et al., 1993). Consistent with *in vitro* studies, flies expressing a higher level of inhibitor peptide were more severely affected (Griffith et al., 1993), and characterization of synaptic transmission in neuronal tissues from mutant flies suggested changes in synaptic plasticity (Wang et al., 1994). Therefore, selective inhibition of CaMPKII *in vivo* and *in vitro* results in the same types of effects on neuronal function.

In other studies, KN-62 (1-[N,O-bis(5-isoquinolinosulfonyl)-N-methyl-L-tyrosyl]-4-phenylpiperazine), a relatively selective CaM kinase antagonist (Tokumitsu et al., 1990), suppresses the induction of LTP in hippocampal slices (Ito *et al.* 1991). More extensive discussions of KN-62 pharmacology in smooth muscle, the heart, and the nervous system are available in Hidaka and Okazaki (1996) and will not be reiterated here. Briefly, inhibition data suggest that CaM-regulated protein kinases, including CaMPKII, have a common role in stimulus-secretion cycles in multiple tissues and secretory systems that include insulin release (Niki *et al.*, 1993), gastric acid secretion (Mamiya *et al.*, 1993), and neurotransmitter production (Ishii *et al.*, 1991; Sitges *et al.*, 1995).

The use of targeted mutagenesis of CaMPKII in mice (Silva *et al.*, 1992a; Griffith *et al.*, 1993; Stevens *et al.*, 1994) has also demonstrated CaMPKII involvement in learning, behavior, and neural plasticity. Studies of transgenic mice in which CaMPKIIα was disrupted (Silva *et al.*, 1992a,b; Butler *et al.*, 1995; Waxham *et al.*, 1996; Chen *et al.*, 1994; Stevens *et al.*, 1994; Chapman *et al.*, 1995) illustrated that disruption of this CaMPKII gene is not lethal and the mice do not exhibit major developmental abnormalities. The level of expression of the unaffected CaMPKIIβ, encoded by a different gene, was normal and the level of total CaMPKII activity in brain homogenates decreased relative to control mice. Examination of the hippocampal area of the brains of homozygotes suggested no anatomical abnormalities. However, CaMPKII-deficient mice have altered plasticity of hippocampal synapses, LTP, and short-term potentiation (Stevens *et al.*, 1994; Glazewski *et al.*, 1996). Calmodulin-dependent protein kinase II-deficient mice were also shown to be more sensitive to limbic epileptic seizures than wild-type mice (Butler *et al.*, 1995). This type of behavior is consistent with the decrease in CaMPKII expression found in kindled rat brains (Bronstein *et al.*, 1992) and other models of epileptic seizure (Murray *et al.*, 1995). Also notably, CaMPKII knockout mice exhibit abnormal fear and aggressive behavior (Chen *et al.*, 1994), which appeared to be correlated with a reduction in serotonin synthesis (Chen *et al.*, 1994). This is interesting because tryptophan hydroxylase is a rate-limiting enzyme for serotonin biosynthesis and is activated by CaMPKII phosphorylation (Isobe *et al.*, 1991). Alteration of neurotransmitter biosynthesis could thus be correlated with the behavioral abnormalities seen in the CaMPKII knockout mice.

Homozygous knockout mice also exhibited an ischemic infarct volume nearly twice that of wild-type mice (Waxham *et al.*, 1996). Heterozygous mice developed less ischemic damage than homozygote mice, but the infarct volumes were also elevated compared to wild-type mice (Waxham *et al.*, 1996). In normal animals, CaMPKII partially translocates from cytoplasmic to particulate fractions in neurons after ischemic insult and recovers to the cytoplasm slowly (Hu and Wieloch, 1995; Kolb et al., 1995). Therefore, the enhancement of neuronal damage induced by ischemic insult in mice with reduced levels of CaMPKII is intriguing.

There are also studies of the effects of overexpression of CaMPKII. Tashima et al. (1996) found that overexpression of CaMPKII in cultured PC12 cells inhibited neurite outgrowth that was induced by cyclic nucleotides. Similarly, transient expression of a constitutively active CaMPKII in tadpoles through a viral vector containing CaMPKII altered presynaptic axon arbor growth (Zou and Cline, 1996). Also, transgenic mice containing a mutated CaMPKII (T286D-CaMPKII) that was partially Ca^{2+}/CaM independent showed altered synaptic transmission activity (Mayford et al., 1995; Pettit et al., 1994). The frequency-response coupling of LTP was relatively normal when high-frequency stimulation was used, but at lower frequency stimulation, synapses from CaMPKII mutant animals exhibited little potentiation and tended toward long-term depression. The presence of the transgene apparently shifts the response curve as confirmed by performing synaptic stimulations at multiple frequencies (Mayford et al., 1995). Although it is thought that hippocampal areas are important for explicit memory, changing the levels of CaMPKII activity in other regions of the forebrain may affect memory differently. Additionally, these regions generally express lower levels of CaMPKII compared to the hippocampus and thus may be more sensitive to CaMPKII overexpression. To study the expression in more localized regions of the brain, the CaMPKII mutant (T286D) transgene was introduced under the control of a drug inducible suppressor. A particular transgenic mouse line, BL21, exhibited normal brain CaMPKII activity when the mutant gene was suppressed, but elevated Ca^{2+}/CaM independent activity in brains from rats not given or withdrawn from suppressor drug. Unlike previous transgenic mice, BL21 mice expressed little of the CaMPKII transgene in the hippocampus, but strongly in the striatum and amygdala, a region thought to be important for implicit or perceptual memory (Mayford et al., 1996). BL21 mice had defects in behavioral conditioning that requires implicit memory. The defects were reversed when expression of the transgene was suppressed (Mayford et al., 1996).

In summary, both overexpression and disruption of CaMPKII activity through pharmacological and genetic approaches provide a consistent view of CaMPKII being crucial to normal neuronal function. In the case of the evaluation of higher level activities, such as learning deficits, both underexpression and overexpression of CaMPKII appear to alter normal function. Excluding more technical explanations, the results raise the possibility that a fine balance of CaMPKII and related kinase activities is genetically and physiologically required to maintain a "normal" phenotype. In this regard, CaMPKII may share the role of homeostasis regulator with other CaM-regulated protein kinases such as MLCK and phosphorylase kinase.

B. Calmodulin-Dependent Protein Kinase II as a Calmodulin-Regulated Enzyme

The vertebrate CaMPKII family contains 50–60 K subunit isoforms termed, α, β, γ, and δ that are derived from four closely related genes. The α and β isoforms are present in multiple tissues but are most abundant in brain (Lin et al., 1987; Hanley et al., 1987), whereas the γ and δ isoforms have a broad tissue distribution (Tobimatsu et al., 1988; Tobimatsu and Fujisawa, 1989). The CaMPKII holoenzymes are multimers of 6–12 subunits that total 300–700 kDa as measured by gel filtration or analytical sedimentation methods (Schulman, 1993). The size differences of CaMPKII holoenzymes from different tissues would be consistent with multimers composed of a single isoform. Electron microscopy studies reveal that the CaMPKII holoenzyme has a central core surrounded by globular domains that resemble flower petals (Kanaseki et al., 1991).

In the absence of Ca^{2+}/CaM, CaMPKII is inactive in its dephosphorylated state. Half-maximal activation of the dephosphorylated enzyme occurs at 0.5–1 μM free Ca^{2+} (in the presence of a mole excess of CaM over enzyme) or at 25–100 nM CaM (in the presence of a mole excess of Ca^{2+}) (Schulman, 1988). These *in vitro* properties make CaMPKII less sensitive to CaM than, for example, MLCK. This phenomenon might be important for the timing of sequential phosphorylation events *in vivo*. For example, in response to Ca^{2+} stimulation of smooth muscle, myosin light chain phosphorylation occurs earlier during the transient than MLCK phosphorylation by CaMPKII (Tansey et al., 1994).

The amino acid sequences for all CaMPKII isoforms are closely related and reveal the segmental organization characteristic of CaM-regulated protein kinases. Most differences are outside of the catalytic and CaM regulatory core. Calmodulin-dependent protein kinase II, unlike MLCK, does not have a long amino terminal extension (Fig. 1) so that the amino acid sequence begins with the catalytic segment and, based on the conserved protein kinase sequence motifs, includes approximately the first 265 amino acids that are similar to the catalytic segment of other CaM-regulated kinases.

The stoichiometry of CaM in preparations of holoenzyme and that of exogenous CaM binding to the holoenzyme have not been accurately determined. Therefore, it is not clear whether each catalytic subunit of CaMPKII needs to have a CaM bound to achieve maximal activity. Contributing to this uncertainty is the practice of following activation of CaMPKII by autophosphorylation, which makes the kinase constitutively active (Schulman, 1988). The specific activity of the constitutively active enzyme is comparable to that in the presence of saturating Ca^{2+}/CaM, provided that the ATP concentration is greater than 0.5 mM (Lou et al., 1986; Lou and Schulman, 1989). Dissociation of Ca^{2+}/CaM from the enzyme eventually results in loss of activity and CaM sensitivity (Lou and Schulman, 1989).

Activation of CaMPKII by partial proteolysis (LeVine and Sahyoun, 1987; Kwiatkowski and King, 1989; Yamagata et al., 1991) suggested that the catalytic segment has intrinsic catalytic activity that is inhibited in the basal state. Similar to MLCK, the CaM regulatory segment is on the carboxyl-terminal side of the catalytic segment. Following the CaM regulatory segment is a variable (length and sequence) linker that in some CaMPKII isoforms has a nuclear targeting sequence (Bayer et al., 1996; Srinivasan et al., 1994; Schworer et al., 1993). The remaining carboxyl-terminal segment is an association and oligomerization segment (Lin et al., 1987), and is required for assembly of the subunits into a holoenzyme (LeVine and Sahyoun, 1987; Yamauchi et al., 1989). An expressed fragment of the kinase lacking this segment (e.g., amino acids 1–358 of the α-subunit) is fully Ca^{2+}/CaM dependent but monomeric rather than oligomeric.

The CaM-binding region of CaMPKII was predicted based on homology to other CaM-binding regions such as RS20 in sm/nmMLCK (Lukas et al., 1986; Shoemaker et al., 1990). Synthetic peptides corresponding to the putative CaM-binding region of CaMPKII bind to CaM (Payne et al., 1988; Hanley et al., 1987; Chabbert et al., 1991) and cause conformational changes in CaM similar to those caused by RS20 binding to CaM (Chabbert et al., 1991). Engineered MLCKs with the RS20 region replaced by the 20 residue sequence from CaMPKII respond to CaM activation like CaMPKII, but retain an MLCK substrate specificity (Shoemaker et al., 1990). In addition to supporting the model of this sequence being a CaM recognition sequence in CaMPKII, results with

3. Calmodulin-Regulated Protein Kinases

chimeric kinases (Shoemaker *et al.*, 1990) provide experimental data consistent with the proposed segmental organization of CaM-regulated protein kinases.

The three-dimensional structure of CaM in the complex with a 26 residue synthetic peptide (Figs. 11 and 12) that includes the minimal CaM-binding segment of CaMPKII has been determined (Meador *et al.*, 1993). As with the MLCK RS20 peptide, the CaM molecule wraps around the CaMPKII peptide so that it interacts with carboxyl and amino-terminal domains of CaM (Figs. 11 and 12). Compared to CaM:RS20 complex, a number of sequence differences between the target peptides results in geometrical and contact changes in the binding (Meador *et al.*, 1993). The proposed binding contacts between CaM and the CaMPKII peptide are presented in Fig. 11. Although there are many common interactions, there are some distinct differences (see also Fig. 4). For example, there are specific amino acids in CaM that are found within the distance for MLCK but not CaMPKII, and vice versa. In addition, there are detailed differences among common amino acids. For example,

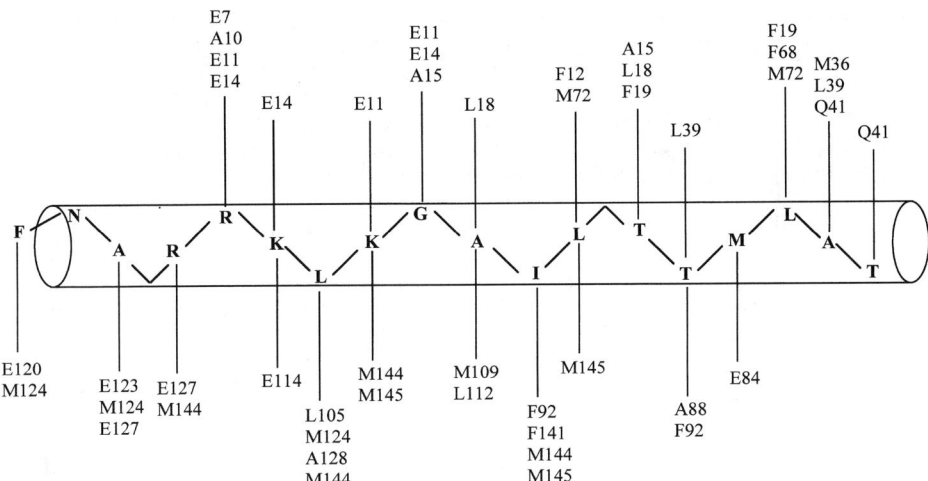

Fig. 11. Interactions between CaM and the CaM recognition peptide of CaMPKII. The amino acid sequence of the CaM recognition region of CaMPKII is shown within the cylinder. Amino acids shown above and below the cylinder (e.g., E120) are amino acids in CaM that are within 4 Å distances based on the structures of CaM:peptide complexes and, therefore, proposed as interacting residues.

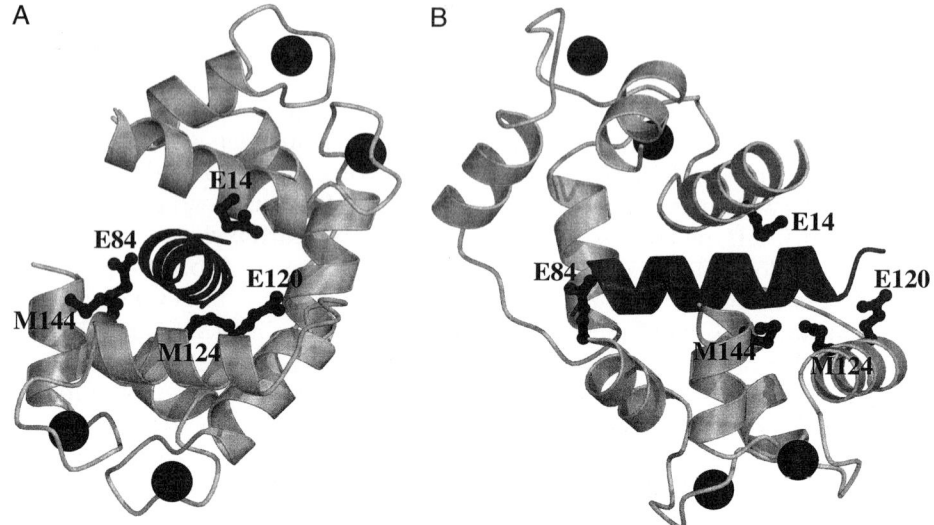

Fig. 12. Three-dimensional structure of the complex between CaM and the CaM recognition peptide of CaMPKII. These ribbon drawings are based on coordinates in the PDB entry 1CDM. Ca^{2+} ions are solid spheres. The CaM recognition peptide is shown as a darkened ribbon. Side chains of selected amino acids from CaM are indicated (e.g., E120). These amino acids are examples of ones within 4 Å as indicated in Fig. 11. (A) View down the long axis of the peptide helix, shown in the center. (B) This view represents a 90° rotation around the axis perpendicular to the peptide. As discussed in the text and noted in Fig. 4, some amino acids in CaM (e.g., E84) are within the 4 Å distance for interaction in both the structure with the CaMPKII CaM-binding peptide and the structure with the MLCK CaM-binding peptide. However, the interactions appear to be different (e.g., more charge–charge features in the CaM:MLCK peptide interactions compared to the CaM:CaMPKII peptide). Thus, unique amino acids in CaM are found within the appropriate distance for interaction as well as common ones, but the common amino acids may use different features of their side chains (e.g., COO^- or $-CH_2-$ of the glutamic acid side chain). The net effect is a unique combination of hydrophobic and charge interactions for each CaM:peptide complex.

E84 in CaM does not make contact with positively charged residues in the CaMPKII peptide as it does in the CaM:RS20 complex. Instead, it makes contact with a hydrophobic methionine residue. Thus, there is less charge–charge interaction in the E84 interaction with CaMPKII compared to MLCK. Structural data with CaM:peptide complexes provide a plausible explanation for differences in the sensitivity of MLCK

3. Calmodulin-Regulated Protein Kinases 109

and CaMPKII to charge-reversal mutants of CaM containing the E84K mutation (Weber *et al.*, 1989; Shoemaker *et al.*, 1990). Similarly, the difference in response of MLCK and CaMPKII to the mutation of M51 and V55 (Chin *et al.*, 1997) can be rationalized from the lack of interactions with the CaMPKII peptide in the three-dimensional structure (Meador *et al.*, 1993). At the same time, the effect of E120K mutation in CaM on the maximal level of activation of both MLCK and CaMPKII (Weber *et al.*, 1989) can also be rationalized from the charge–charge interactions of E120 in both CaM:peptide structures. Thus, as found with the RS20 segment of MLCK, the homologous region of CaMPKII provides a reasonable model for potential interactions of CaM with the kinase. Screening mutagenesis of methionine residues (Chin and Means, 1996) showed that substitution to glutamine can have severe effects on CaMPKII (and MLCK) activation, especially when the mutated contact methionine is phylogenetically conserved (as M124Q, for example) (see Fig. 4). At the same time, when a phylogenetically variant contact methionine was changed (M144Q), mutant CaM was able to activate CaMPKII to 100% maximal activity, with only a slight change in the concentration of CaM required for 50% CaMPKII activation.

Of the CaM-dependent kinases, CaMPKII is one of the more extensively characterized with respect to mutations in the autoinhibitory domain. Calmodulin-dependent protein kinase II presents a more complicated system compared to MLCK because its autoinhibitory domain (Figs. 13 and 14) contains an autophosphorylation site, T286, in the α-isoform. The autophosphorylation of T286 is dependent on Ca^{2+}/CaM and occurs rapidly within the oligomeric complex (Mukherji and Soderling, 1994). On phosphorylation of T286, CaMPKII kinase activity does not require Ca^{2+}/CaM (Schworer *et al.*, 1986). The enzyme undergoes further autophosphorylation in the absence of Ca^{2+}/CaM that results in a loss of CaM activation (Lou and Schulman, 1989; Hanson and Schulman, 1992b; Patton *et al.*, 1990; Colbran and Soderling, 1990). An interesting feature of the autophosphorylation of T286 in CaMPKII is that autophosphorylation results in a large (>1000-fold) increase in apparent CaM affinity (Meyer *et al.*, 1992). The change in affinity for CaM is due to a slowing of the CaM off-rate from the T286 phosphorylated enzyme (Meyer *et al.*, 1992). This phenomenon has been termed "calmodulin trapping" (Meyer *et al.*, 1992) and displays an example of functional coupling between the autoinhibitory and CaM recognition regions of CaMPKII.

As with MLCK, the idea of a pseudosubstrate-like autoinhibitory mechanism was initially based on the ability of selected synthetic peptides from the regulatory segment to inhibit catalytic activity (Colbran *et al.*, 1988, 1989; Payne *et al.*, 1988; Kelly *et al.*, 1988) and bind CaM (Payne *et al.*, 1988; Kelly *et al.*, 1988). For example, the inhibition of CaMPKII by some peptide analogs of the regulatory segment was competitive with ATP and noncompetitive with peptide substrates (Colbran *et al.*, 1988, 1989), whereas other peptides gave inhibition kinetics consistent with competitive substrate inhibition (Smith *et al.*, 1992). Another study showed that labeling of CaMPKII by covalent ATP analogs was enhanced 10-fold by Ca^{2+}/CaM (King *et al.*, 1988), consistent with blockage of the ATP-binding site by the autoinhibitory domain. Together these data suggested that the details of how the autoinhibitory region of CaMPKII interacts with its catalytic segment is distinct from that found in the autoinhibition of MLCK. Results of the synthetic peptide inhibitor studies were consistent with the idea of dual inhibition (ATP and peptide/protein substrate), but it was not clear to what extent the autoinhibitory and CaM-binding regions overlapped. Therefore, refinement of the localization of the CaM-binding and autoinhibitory regions of CaMPKII was done by site-directed mutagenesis studies of the regulatory region. For example, T286, which undergoes autophosphorylation, was mutated to aspartic acid (T286D), which resulted in partial Ca^{2+}/CaM-independent activity (Waldmann *et al.*, 1990; Hagiwara *et al.*, 1991; Brickey *et al.*, 1994). Constitutive activity (67% Ca^{2+}/CaM-independent activity) was also generated with H282D, R283G, and Q284E mutations (Waldmann *et al.*, 1990). Greater disruption of the autoinhibitory region by larger deletions/substitutions caused additional increases in constitutive activity.

Mutations in the CaM recognition region of CaMPKII have been directed at a putative pseudosubstrate sequence in an attempt to generate a CaM-independent kinase. The strategy employed in this regard was

---

Fig. 13. Sequence alignment of the catalytic and CaM regulatory segments of CaMPKIIs. Amino acid sequences of human CaMPKII-α (hucampkiialpha), *Drosophila* CaMPKII-α (drocampkii-alpha), yeast CaMPKII-α (ycampkii2), rat CaMPKII-α (campkiialpha), rat CaMPKII-β (campkiibeta), rat CaMPKII-δ (campkiidelta), and rat CaMPKII-γ (campkiigamma) are shown. The amino acid residues conserved in the catalytic core of most protein kinases are boxed. The CaM regulatory segments are shaded. The threonine residue, which undergoes autophosphorylation, is indicated by an asterisk. Standard single-letter amino acid codes are used throughout. Sequences were obtained from the GenBank database.

```
ratcampkiialpha   Y Q L F E E L G K G A F S V V R R C V K V L A G Q E Y A A K I I N T K K L S A R    52
hucampkii-alpha   - - - - - - - - - - - - - - - - - - - - - - - - - - - - - - - - - - -             1
ysccmpkii2        Y I F G R T L G A G S F G V V R Q A R K L S T N E D V A I H K K A L Q G N         86
drocampkii-alpha  Y D I K E E L G A G S F S I V R R C V Q K S T G H E F F A A K I I N T K K L T A R  53
campkiibeta       Y Q L F E E L G K G A F S V V R R C V K K T S T Q E Y A A K I I N T K K L S A R   53
campkiigamma      Y Q L F E E L G K G A F S V V R R C V K K T P T G Q E Y A A K I I N T K K L S A R 53
campkiidelta      Y Q L F E E L G K G A F S V V R R C M K I P T G Q E Y A A K I I N T K K L S A R  53

ratcampkiialpha   D H Q - - K L E R E - I Y E A R I C R L L K H P N I V R L H D S I S E E G H H Y L I F D    90
hucampkii-alpha   - - - - - - - - - - - - - - - - - - - M K K H P N I V R L H D S I H D S I E E G F H Y L V F D 24
ysccmpkii2        N V Q L Q - - M L Y E E L S I L Q K L S H P N I V S F K D W F E S K D K F Y I V L V F T Q 126
drocampkii-alpha  D F Q - - K L E R E - I Y E A R I C R L L K H P N I V R L H D S I H Q E E N Y H Y L V F D  91
campkiibeta       D H Q - - K L E R E - I Y E A R I C R L L K H P N I V R L H D S I S E E G F H Y L V F D   91
campkiigamma      D H Q - - K L E R E - I Y E A R I C R L L K H P N I V R L H D S I S E E G F H Y L V F D  91
campkiidelta      D H Q - - K L E R E - I Y E A R I C R L L K H P N I V R L H D S I S E E G F H Y L V F D  91

ratcampkiialpha   L V T G G E L F E D I V A R E Y Y S E A D A S H C I Q Q I L E A V L H C H Q M G    130
hucampkii-alpha   L V T G G E L F E D I V A R E Y Y S E A D A S H C H I Q Q H I L E A S V N H I H Q H D  64
ysccmpkii2        L A T G G E L F D R I L S R G K F T E V D D A V K G A V E I H C H L G A V E Y M H S K N 166
drocampkii-alpha  L V T G G E L F E D I V A R E F Y S E A D A S H C I Q Q I L E A V N H C H Q N G       131
campkiibeta       L V T G G E L F E D I V A R E Y Y S E A D A S H C I Q Q I L E A V L H C H Q Q G      131
campkiigamma      L V T G G E L F E D I V A R E Y Y S E A D A S H C I Q Q I L E A S V N H I H Q H D   131
campkiidelta      L V T G G E L F E D I V A R E Y Y S E A D A S H C I Q Q I L E S V N H C H L N G     131

ratcampkiialpha   V V H R D L K P E N L L L A S K L K G A A V K L A D F G L A I E V E G E Q Q A W    170
hucampkii-alpha   I V H R D L K P E N L L L A S K K C K G A A V K L A D F G L A I E V Q G E Q Q A W    104
ysccmpkii2        V V H R D L K P E N V L L Y V D K K S E N S P L V H A D F G L A I E V Q L K G E D L I 206
drocampkii-alpha  V V H R D L K P E N L L L A S K A K G A A V K L A D F G L A I E V Q G D H Q A W       171
campkiibeta       V V H R D L K P E N L L L A S K C K G A A V K L A D F G L A I E V D Q Q A W          171
campkiigamma      I V H R D L K P E N L L L A S K K C K G A A V K L A D F G L A I E V Q G E Q Q A W    171
campkiidelta      I V H R D L K P E N L L L A S K S K G A A V K L A D F G L A I E V Q G D Q Q A W      171

                                                                                          (continued)
```

```
ratcampkiialpha    F G F A G T P G Y L S P E V L R K D P Y G K P V D L W A C G V I L Y I L L L V G Y  210
hucampkii-alpha    F G F A G T P G Y L S P E V L R K D P Y G K P V D H I W A C G V I L Y T L L V G Y  144
ysccmpkii2         Y K A A G S L G Y V A P E V L T Q D G H G K P C D H I W S I G V I T Y T L L C G Y  246
drocampkii-alpha   F G F A G T P G Y L S P E V L K K E P Y G K K S V D H I W A C G V I L Y I L L V G Y  211
campkiibeta        F G F A G T P G Y L S P E V L R K E A Y G K P V D H I W A C G V I L Y I L L V G Y  211
campkiigamma       F G F A G T P G Y L S P E V L R K D P Y G K P V D H I W A C G V I L Y I L L V G Y  211
campkiidelta       F G F A G T P G Y L S P E V L R K D P Y G K P V D M W A C G V I L Y I L L V G Y  211


ratcampkiialpha    P P F W D E D Q H R L Y Q Q I K A G A Y - - D F P S P E W D T V T P E A K D L I   248
hucampkii-alpha    P P F W D E D Q H R L Y Q Q I K A G A Y - - D F P S P E W D T V T P E A K N L I   182
ysccmpkii2         S P F I A E S V E G F M E E C T A S R Y P V T F H M P Y W D N I S I D V K R F I   286
drocampkii-alpha   P P F W D E D Q H R L Y Q Q I K A G A Y - - D Y P S P E W D T V T P E A K N L I   249
campkiibeta        P P F W D E D Q H K L Y Q Q I K A G A Y - - D F P S P E W D T V T P E A K N L I   249
campkiigamma       P P F W D E D Q H K L Y Q Q I K A G A Y - - D F P S P E W D T V T P E A K N L I   249
campkiidelta       P P F W D E D Q H R L Y Q Q I K A G A Y - - D F P S P E W D T V T P E A K D L I   249

                                                                 *
ratcampkiialpha    N K M L T I N P S K R I T A A E A L K H P W I S H R S T V A S C M H R Q E T V D  288
hucampkii-alpha    N Q M L T I N P A K R I T A D Q A L K H P W V C Q R S T V A S M M H R Q E T V E  222
ysccmpkii2         L K A L R L N P A D R P T A T E L L D D P W I T S K R V E T S K R V E T S N I L P D V K - I  325
drocampkii-alpha   N Q M L T V N P N K R I T A H E A L K H P W I C Q R E R T V A S V V H R Q E T V D  289
campkiibeta        N Q M L T I N P A K R I T A D Q A L K H P W V C Q R S T V A S M M H R Q E T V E  289
campkiigamma       N Q M L T I N P A K R I T A D Q A L K H P W V C Q R S T V A S M M H R Q E T V E  289
campkiidelta       N K M L T I N P A K R I T A S E A L K H P W I C Q R S T V A S M M H R Q E T V D  289


ratcampkiialpha    C L K K F N A R R K L K G A I L T T M L A T R N F S   314
hucampkii-alpha    C L R K F N A R R K L K G A I L T T M L V S R N F S   248
ysccmpkii2         - K G F S L R K K L R D A I E I V K L N N R I K R    348
drocampkii-alpha   C L K K F N A R R K L K G A I L T T M L A T R N F S   315
campkiibeta        C L K K F N A R R K L K G A I L T T M L A T R N F S   315
campkiigamma       C L R K F N A R R K L K G A I L T T M L V S R N F S   315
campkiidelta       C L K K F N A R R K L K G A I L T T M L A T R N F S   315
```

Fig. 13. Continued

3. Calmodulin-Regulated Protein Kinases

Group A. Mutations that alter CaM-independent activity

```
         | Autoinhibitory              | Calmodulin Recognition         |
279                *                                                  313
     S C M H R Q E T V D C L K K F N A R R K L K G A I L T T M L A T R N F
1.                 D
2.       D G E
3.                   L
4.                 A
5.       Q
6.             E
7.                 D                                       E
```

Group B. Mutations that alter Ca^{2+}/CaM recognition or sensitivity

```
         | Autoinhibitory              | Calmodulin Recognition         |
279                *                                                  313
     S C M H R Q E T V D C L K K F N A R R K L K G A I L T T M L A T R N F
8.                               S
9.                 S                             S
10.                                              R
11.                                                          R
12.                                                                A
```

Fig. 14. Summary of amino acid changes in mutagenesis analysis of the CaM regulatory segment of CaMPKII. The centers of autoinhibitory and CaM recognition activity are shown. Mutant CaMPKIIs are placed into two groups according to the functional effects of the mutations in the CaM regulatory segment. Only differences in the amino acid sequence between the mutant and the wild-type CaMPKII are shown. The threonine residue, which undergoes autophosphorylation, is indicated by an asterisk.

the substitution of certain amino acids in the CaM-binding region to allow for a possible intramolecular phosphorylation (Mukherji and Soderling, 1995). The mutants N294S, K300S, A302R, and A309R exhibited decreased binding and activation by CaM, but none exhibited enhanced Ca^{2+}-independent kinase activity. Similarly, mutation of K300E combined with T286D resulted in decreased CaM binding, but no increase in constitutive activity beyond that found with the single T286D mutation (Waldmann et al., 1990). Kinase constructs with mutation of R311A, or truncations before this residue, lose CaM-binding activity and the kinases have very low activity (Brickey et al., 1994; Cruzalegui et al., 1992). Therefore, mutation of the CaM recognition region has little effect on autoinhibition of the kinase. Although the mutations studied in CaMPKII are not as extensive as those done with MLCK, apparently the same general theme persists. A short segment (residues 282–294) that precedes a CaM recognition region (residues 296–313) appears to contain most of the elements needed for autoinhibition of kinase activity.

How the mutations in the CaM recognition or autoinhibitory segments affect the general Ca^{2+}/CaM sensitivity of CaMPKII has not been studied extensively. A study of CaMPKII isoforms from *Drosophila* suggested functional differences in Ca^{2+}/CaM sensitivity among the isozymes that differ in the linker region between the CaM regulatory segment and the association domain (GuptaRoy *et al.*, 1996). Additional autophosphorylation of CaMPKII in the absence of Ca^{2+}/CaM can result in an enzyme that has a reduced ability to be regulated by CaM (Lou and Schulman, 1989; Hanson and Schulman, 1992b; Patton *et al.*, 1990). Together the results suggest that the Ca^{2+}/CaM sensitivity of CaMPKII can be modulated in both a positive and a negative direction through autophosphorylation and the structure of the enzyme in regions outside of the CaM regulatory segment.

C. Molecular Genetics of Calmodulin-Dependent Protein Kinase II and Related Proteins

In vertebrates, CaMPKII appears to be encoded by four different genes, each of which encodes a distinct isoform. In *Drosophila*, at least four isoforms of CaMPKII come from a single gene by alternative splicing rather than from separate genes as in rat (Ohsako *et al.*, 1993). Characterization of the *Drosophila* CaMPKII gene indicated that high-level expression of α and β homologs is found in the head, whereas lower expression levels occur in the body of the adult fly (Cho *et al.*, 1991; Ohsako *et al.*, 1993).

The vertebrate α isoform has been cloned from the rat and the intron/exon boundaries characterized (Nishioka *et al.*, 1996). The rat CaMPKII-α gene has 18 exons spanning more than 50 kb. Sequences that comprise the CaM regulatory segment are split between two exons, a phenomenon also seen with the other CaM-dependent kinase genes (Collinge *et al.*, 1992; Karls *et al.*, 1992). Promoter sequences in the rat CaMPKII-α gene have been identified (Olson *et al.*, 1995), and analysis of the promoter region by expression of reporter gene constructs revealed more activity in neural cell lines compared to a control SV40 (simian virus 40) promoter. The CaMPKII-α gene also expresses a nonkinase product (KAP) in skeletal muscle (Bayer *et al.*, 1996). Characterization of the KAP gene from mouse showed that it contains exons encoding the association domain of CaMPKII with a unique 5' end derived from DNA sequences that are a CaMPKII intron. In addition, the KAP cDNA has a 33-bp exon inserted in frame within the first exon of the association domain of CaMPKII (Bayer *et al.*, 1996). This exon insert has also been detected in a CaMPKII-α mRNA from brain (Benson *et al.*, 1991). The significance of this exon is not clear, although it may complete a nuclear targeting sequence (Bayer *et al.*, 1996; Srinivasan *et al.*, 1994; Schworer *et al.*, 1993). The function of the protein product of the KAP gene has not been defined.

A gene for the β isoform of CaMPKII has been identified in the mouse (Karls *et al.*, 1992), and 16 exons that span most of the coding region have been sequenced. The β isoform has alternatively spliced versions of its carboxyl-terminal association domain, but unlike the α isoform, a nonkinase gene product has not been described for this gene.

The γ isoform of CaMPKII is expressed in a variety of neuronal and nonneuronal tissues (Tobimatsu and Fujisawa, 1989; Nghiem *et al.*, 1993). Alternative splicing in the linker between the CaM regulatory and association segments appears to generate three to six variants of the kinase (Nghiem *et al.*, 1993; Kwiatkowski and McGill, 1995). The coding sequences spliced into the variants of CaMPKII-γ are generally unique to this isoform, suggesting that these regions may provide specific targeting or associative elements to these proteins.

There are at least four splice variants of the δ isoform of CaMPKII (δ-A, δ-B, δ-C, and δ-D), which differ in the variable linker region that connects the CaM regulatory segment with the association segment (Schworer *et al.*, 1993). Like the γ isoform, CaMPKII-δ has a wide tissue distribution (Tobimatsu and Fujisawa, 1989). The δ-B form has a consensus nuclear localization signal that targets this variant to the nucleus (Srinivasan *et al.*, 1994). Surprisingly, the context of this sequence is very similar to the proposed nuclear targeting exon of the KAP protein derived from the CaMPKII gene. The sequence in the δ-B form is (K)KRKSSSSVQMM, whereas the sequence in the KAP protein is (K)KRKSSSSVQLM. Srinivasan *et al.* (1994) found that the nuclear targeting exon of δ-B, when placed into a CaMPKII-α construct, will target expression of CaMPKII-α to the nucleus, indicating that the exon is a functional nuclear localization sequence between CaMPKII isoforms.

IV. PHOSPHORYLASE KINASE

A. Phosphorylase Kinase Function and Ca^{2+} Signal Transduction

Phosphorylase kinase (PhosK) is a Ca^{2+}/CaM-dependent protein kinase in which CaM is a tightly bound integral subunit (the δ subunit) of a large multisubunit complex. Other glycogen-metabolizing enzymes such as phosphorylase and glycogen synthase also bind to glycogen, creating the megastructure "glycogen particle" (Heilmeyer, 1991). Phosphorylase kinase is a key regulatory enzyme in glycogen metabolism. Ca^{2+}-dependent phosphorylation of glycogen phosphorylase b by PhosK converts it into the active a form. Calcium-mediated phosphorylation of glycogen phosphorylase b couples glycogenolysis with contraction. Phosphorylase kinase also provides another example of cross-talk among signal transduction pathways at the level of CaM-regulated protein kinases through the ability of PhosK subunits to serve as physiological substrates for PKA.

Phosphorylase kinase is a large multimeric enzyme (Heilmeyer, 1991; Chrisman et al., 1982) consisting of four different subunits, α, β, γ, and δ, that form a hexadecamer $(\alpha, \beta, \gamma, \delta)_4$. The catalytic function of PhosK resides in its γ subunit, whereas the other subunits appear to function in regulatory and subcellular targeting roles. The α and β subunits serve as negative regulatory subunits because they inhibit the isolated γ subunit (Paudel and Carlson, 1987). When α and β are phosphorylated by PKA, the activity and Ca^{2+} sensitivity of phosphorylase kinase increase (Klee and Vanaman, 1982; Paudel and Carlson, 1987). The δ subunit is an integral CaM that requires chelation of calcium and a chaotropic agent to dissociate it from the holoenzyme complex (Cohen et al., 1978; Grand et al., 1981). Thus, PhosK is regulated through two second messenger pathways: Ca^{2+}, through CaM regulation of the γ subunit, and cyclic AMP, through PKA phosphorylation of the α and β subunits (Heilmeyer, 1991). The presence of PhosK in the glycogen particle also provides a subcellular targeting aspect to the physiological regulation.

A deficiency in PhosK activity is associated with several forms of glycogen storage disease (GSD) in humans and animals (Hendrickx and Willems, 1996; van den Berg and Berger, 1990; Schneider et al., 1993; Maichele et al., 1996; Wehner et al., 1994; van den Berg et al., 1995; Burwinkel et al., 1997), diseases that display a variety of phenotypes and inheritance patterns. Generally, PhosK activity or levels are either low or not detected in the relevant tissues of GSD patients with specific

3. Calmodulin-Regulated Protein Kinases

PhosK deficiencies (Wilkinson *et al.*, 1994). Logical candidates for PhosK mutations, based on the known biochemistry of the enzyme, are α subunit genes, γ subunit genes, or the single β subunit gene. The genetic defect in the I-strain of mice, which have a X chromosome-linked muscle PhosK deficiency, has been localized to the α_M subunit expressed in skeletal muscle and is caused by a frameshift mutation in the mouse PHKA1 gene (Schneider *et al.*, 1993). A nonsense mutation in the human PHKA1 gene (α_M subunit) has been identified in a sporadic case of muscle PhosK deficiency (Wehner *et al.*, 1994). A more frequently occurring form of PhosK deficiency in vertebrates affects primarily the liver enzyme and is associated with the X chromosome-linked gene of the α_L subunit, or PHKA2, gene (van den Berg *et al.*, 1995; Hendrickx and Willems, 1996). Mutations in the catalytic γ_L subunit gene (also referred to as the testis, or γ_T, form) are associated with autosomal liver glycogenosis, or liver PhosK deficiency (Maichele *et al.*, 1996). Although not found as a component of a glycogen storage disease, the *Drosophila* DphK γ gene provides insight into other potential functions of the γ subunit. For example, loss-of-function mutations in the *Drosophila* DphK γ gene cause muscle degeneration or abnormal growth of leg musculature (Bahri and Chia, 1994). The DphK γ is interesting because it has ~60% amino acid sequence identity in the catalytic segment to rat and human γ subunits, but contains a 180 amino acid linker sequence between the catalytic core and the putative CaM regulatory segment. The mammalian γ subunit sequences have a linker region of only 8–11 amino acids. It is not known, however, if the linker region in DPhK γ provides an additional association or regulatory domain such that the *Drosophila* kinase is regulated differently from mammalian kinases. Mutations in the single β subunit gene in mammals have been linked to autosomal GSD, which affects both liver and muscle tissues (Burwinkel *et al.*, 1997). In general, phenotypes of the PhosK deficiencies of the liver and muscle enzymes are all consistent with the known biochemical properties of the purified enzymes and their established physiological roles. Although PhosK deficiencies provide one of the better examples of how an inherited mutation in a CaM-regulated protein kinase can result in a disease state, none of the inherited disorders have been mapped to CaM, the δ subunit of the enzyme complex.

B. Phosphorylase Kinase as a Calmodulin-Regulated Enzyme

The primary structure of the skeletal muscle PhosK α subunit (Zander *et al.*, 1988), β subunit (Kilimann *et al.*, 1988; Wüllrich-Schmoll and

Kilimann, 1996), γ subunit (Reimann et al., 1984; Chamberlain et al., 1987; Calalb et al., 1992; Wehner and Kilimann, 1995), and δ subunit (Grand et al., 1981) has been determined. Based on similarity comparisons with MLCK and CaMPKII, the γ subunit can also be displayed as a segmentally organized CaM-regulated protein kinase. Experimental analysis of the association between γ subunit structure and function, however, has been complicated by the tight association of the subunits in the PhosK holoenzyme multimer and by the associated difficulties in obtaining a fully functional reconstituted enzyme after dissociation into individual subunits. Regardless, there has been extensive knowledge gained about the γ subunit as a CaM-regulated protein kinase activity, which shows that it uses themes reminiscent of MLCK and CaMPKII, but is unique among CaM-regulated protein kinases in terms of its regulation and holoenzyme structure.

The γ subunit appears to contain the primary domain for binding the intrinsic CaM because γ δ complexes have been isolated (Chan and Graves, 1982). The δ subunit remains bound to γ in the γ δ complex in the presence of Ca^{2+} chelators (Chan and Graves, 1982). The γ subunit can be obtained by dissociation from the tissue-isolated holoenzyme with a chaotropic agent and then reactivated by a renaturation process (Farrar and Carlson, 1991; Heilmeyer, 1991). The presence of CaM has been beneficial in regenerating enzyme activity during the refolding process (Farrar et al., 1993). More recently, a functional γ subunit has been obtained from bacterial or eukaryotic expression of cloned cDNAs. The bacteria expressed γ subunit was found in inclusion bodies and had to be extracted with 5 M guanidine hydrochloride and then renatured by dilution into a physiological buffer (Huang et al., 1993), similar to the process used with the tissue-isolated enzyme. The γ subunit isolated from recombinant baculovirus expression in Sf9 cells does not require the renaturation step and appears to have kinetic properties comparable to the holoenzyme and specific activity 10-fold greater than the renatured γ subunit (Lanciotti and Bender, 1994). The renatured bacterial and baculovirus expressed proteins, however, exhibit only 1.5- to 2.0-fold stimulation by calmodulin that is not Ca^{2+} dependent (Lanciotti and Bender, 1994; Huang et al., 1993).

The three-dimensional structure of a truncated form of the PhosK γ subunit is shown in Fig. 15 (Owen et al., 1995). Because the putative CaM regulatory segment is missing from the protein used for the crystallography studies, it cannot be directly compared to the CaMPKI (Goldberg et al., 1996) structure that has the CaM regulatory region present

3. Calmodulin-Regulated Protein Kinases

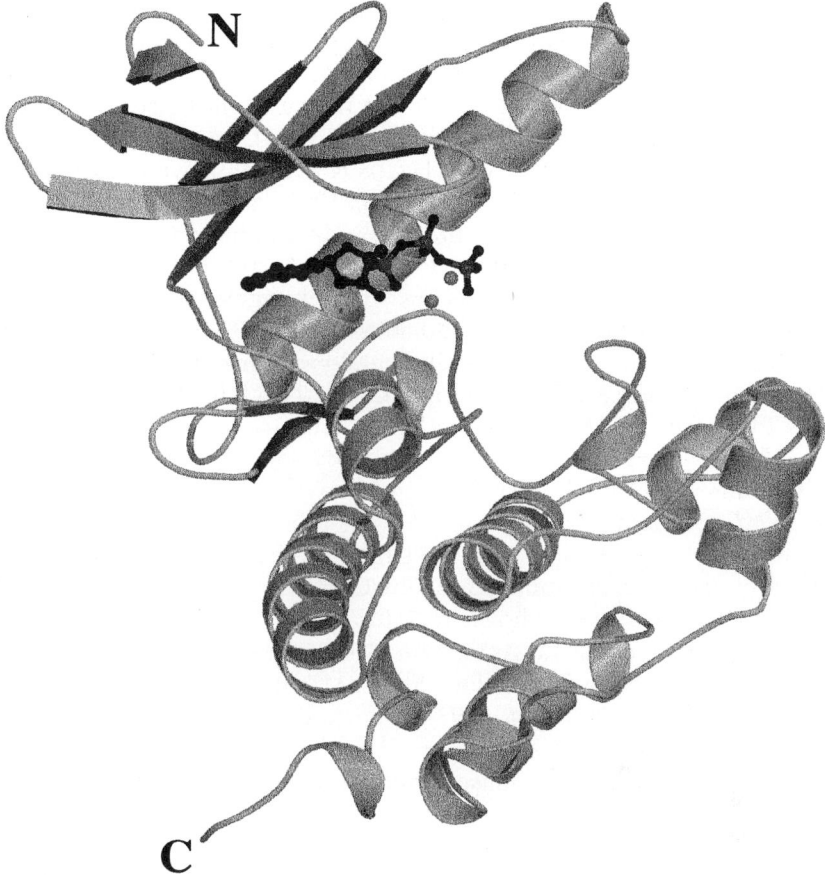

Fig. 15. Ribbon diagram of the PhosK γ structure. This diagram is based on the coordinates kindly provided by the authors (Owen *et al.,* 1995). The peptide backbone trace of the catalytic segment is shown, with the amino end in the upper part of the drawing and the carboxyl end of the catalytic segment at the bottom. The ATP molecule is indicated by the dark gray ball and stick drawing.

and is in the autoinhibited state (see Fig. 18). However, the general scaffold of the catalytic segment of the PhosK γ subunit (Owen *et al.,* 1995) is similar to that for CaMPKI. In addition, site-directed mutations of the catalytic segment of the PhosK γ subunit give results that follow a trend seen with similar mutations in sm/nmMLCK. For example, E110K-PhosKγ and E153R-PhosKγ have increased K_m for ATP and phosphorylase b, indicative of weaker interactions with the active site (Huang et al., 1994). The same conclusions can be drawn from the corresponding mutations in rabbit MLCK (Herring et al., 1992).

Data from limited proteolysis (Harris et al., 1990), synthetic peptide analogs (Dasgupta et al., 1989; James et al., 1991), and deletion mutagenesis (Huang et al., 1993) suggest that the γ subunit has a CaM regulatory segment that appears to extend from residue 302 through 366 at the carboxyl-terminal of the enzyme (Fig. 16). This segment is approximately twice as large as the regulatory domains of CaMPKII and MLCK, but illustrates that the γ subunit is segmentally organized like other CaM-dependent protein kinases (see Fig. 1). Based on synthetic peptide analogs that exhibit CaM-binding activity, there are at least two putative CaM-binding regions within the regulatory segment (Dasgupta et al., 1989; James et al., 1991). These regions have been localized to amino acids 302–326 and 342–366.

Properties of CaM important for γ subunit activation have been investigated by site-directed mutagenesis of CaM. Because the regions of γ subunit that have been shown to interact with CaM are enriched in positive charges, it was suggested that electrostatic interactions are functionally important for interaction with CaM. Charge perturbation mutagenesis of CaM (Farrar et al., 1993) revealed that electrostatic properties of CaM important for γ subunit activation are spread throughout both lobes of CaM, indicating extensive interactions of CaM and the γ subunit of PhosK. As in the case of MLCK and CaMPKII, the combination of electrostatic and hydrophobic interactions was shown to be important for CaM activation. Charge perturbation mutations at E84 or E120 of CaM decrease the maximal Ca^{2+}-dependent activation of the γ subunit, similar to the results with MLCK (Farrar et al., 1993; Shoemaker et al.,

Fig. 16. Sequence alignment of the catalytic and CaM regulatory segments of PhosKs. Amino acid sequences of human PhosK $γ_M$ (humphk), rabbit PhosK $γ_M$ (rabphosk), rat PhosK $γ_M$ (ratphk), *Drosophila* PhosK (drophk), and human PhosK $γ_{T/L}$ (huphkt) are shown. The amino acid residues conserved in the catalytic segment of many protein kinases are boxed. The proposed CaM regulatory segments are shaded. Residues of human PhosK $γ_{T/L}$ found to be mutated in cases of glycogen storage disease are noted with a plus sign. Sequences were obtained from the GenBank database.

```
humphk    EPKEILGRGVSSVVRRCIHKPTSQEYAVKVIDVTGGGSFS    60
rabphosk  EPKKEILGRGVSSVVRRCIHKPTCQEYAVKKIIDVTGGGFS    59
ratphk    EPKKEILGRGVSSVVRRCIHKPTCQEKFAVKKIIDVTGGGFS    59
drophk    EPKKEILGRGISSTVVRRCIEKETGKEFAVKHIIDLGATTESG   63
huphkt    DPKDVIGRGVSSVVRRCVHRATGHEFAVKIMEVTAE-RLSFM   63

humphk    PEEVRELRELRLKEATLKEVDDILRKVSGHPNIHIQLKDTYETNTFF   100
rabphosk  AEEEVQELRELRLKEVDDILRKVSGHPNIHIQQLKDTYETNTFF    99
ratphk    SEEQEELRLKEVDDILRKVSGHPNIHIQQLKDTYETNTFF    99
drophk    ETNPYHMLEATRQEISILRQVMGHPYIHIDLQDVFESDAFV   103
huphkt    PEQLEEVRATRRETHILRLIDSYESSFM    103

humphk    FLVFDLMKRGELFDYLTEKVTLSEKETRKIMRALLEVICT    140
rabphosk  FLVFDLMKKGELFDYLTTEKVTLTEKETRKKIMRALLEVICCA    139
ratphk    FLVFELCPKGELFDYLTEKVTLTEKKETRKIMRQIFEGVEY    139
drophk    FLVFDLMRKGELFDYLTEKVALSEKETRSLLEAVSF    143
huphkt    FLVFDLMRKGELFDYLTEKVALSEKETRSLLEAVSF    143

humphk    LHKLNIVHRDLKPENILLDDNMNIKLTDFGFSCQLEPGER    180
rabphosk  LHKLNIVHRDLKPENILLDDMNIKLTDFGFSCQLDPGEK    179
ratphk    LHKLNIVHRDLKPENILLDENHNIKITDFGFAKAVARGRE    179
drophk    LHAKSIVHRDLKPENIHLLDNMQIIRLSDFGESCHLEPGEK   183
huphkt    LHANNIVHRDLKPENILLDDNMHIKVKITDFGFSCHLEPGEK    183

humphk    LREVCGTPSYLAPEIHECSMNEDHPGYGKEVDMWSTGVIM    220
rabphosk  LREVCGTPSYLAPEIHECSMNEDHPGYGKEVDMWSTGVIM    219
ratphk    LREVCGTPSYLAPEIQCSMDEGHPGYGKEVDHPGYISPGVIM    219
drophk    ITNLCGTPGYLAPETLKCNMFEGSPGYSQEVDIWACGVIL    223
huphkt    LRELCGTPGYLAPEILKCSMDETHPGYGKEVDLWACGVIL   223
```

(continued)

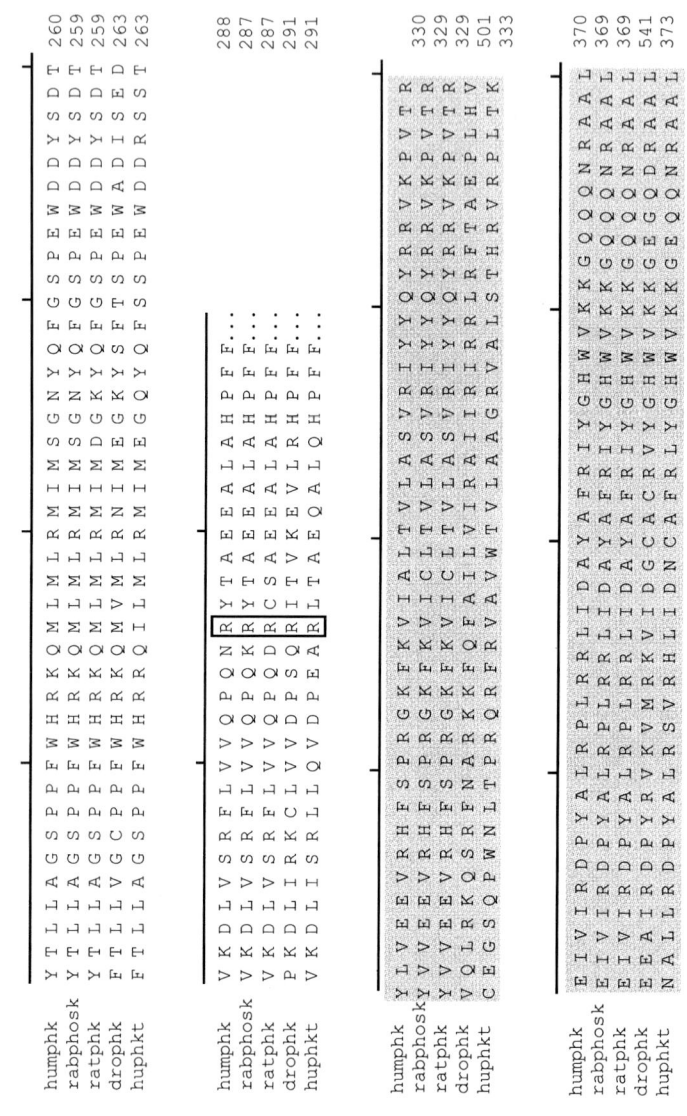

Fig. 16. Continued

1990). In contrast to MLCK, however, the γ subunit of PhosK was shown to be sensitive to mutations of E82, E83, and the cluster of residues 6–8 to lysines (Farrar *et al.*, 1993). This is suggestive of more extensive interactions between CaM and the γ subunit and is consistent with an extended CaM recognition region on the γ subunit.

C. Molecular Genetics of Phosphorylase Kinase

Biochemical and genetic evidence has indicated the existence of several isozymes of PhosK that vary in one or more subunits (Chrisman *et al.*, 1982; Sakai *et al.*, 1979; Daegelen-Proux *et al.*, 1976). There are two isoforms of the PhosK α subunit encoded by two distinct genes, with evidence for differential splicing of the RNA products to form isotypes. The β subunit is found on a single gene with tissue selective RNA splicing used to produce isotypes. There are two isoforms of the γ subunit encoded by two different genes.

The two isoforms of the α subunit of vertebrate PhosK (Davidson *et al.*, 1992), termed muscle ($α_M$) and liver ($α_L$), are encoded by distinct genes. Subtypes of isoforms $α_M$, and $α_L$ subunits are generated by differential mRNA splicing (Harmann *et al.*, 1991; Wullrich *et al.*, 1993). The α subunit isoforms and subtypes show differential tissue distributions. For example, the α subunit typical for fast skeletal muscle (M_r 138K) is replaced in slow skeletal muscle and heart by a smaller subunit. The latter is produced by an internal deletion of 59 amino acids in $α_M$, termed α' (Harmann *et al.*, 1991). The human $α_M$ and $α_L$ genes, also called PHKA1 and PHKA2, have been mapped to the X chromosome (Davidson *et al.*, 1992; Francke *et al.*, 1989).

The gene for the human β subunit, also termed PHKB, has been assigned to chromosome 16 (Francke *et al.*, 1989). The PhosK β subunit gene is more than 140 kb in length and is organized into 33 exons, with 30 exons expressed constitutively (Wüllrich-Schmoll and Kilimann, 1996). Exons 26 and 27 are mutually exclusive exons that encode either muscle-type or nonmuscle-type internal sequences (Wüllrich-Schmoll and Kilimann, 1996). The amino-terminal sequence of the β subunit has a major regulatory phosphorylation site that is modified in some tissues such as brain, uterus, and testis. Exon 2 encodes a brain type of amino-terminal insertion that alters the reading frame encoding the muscle-type amino terminus (Harmann et al., 1991; Wüllrich-Schmoll and Kilimann, 1996).

Isoforms of the γ subunits of PhosK are encoded by distinct genes (Calalb et al., 1992). These isoforms are termed muscle (γ_M), testis (γ_T), or liver (γ_L). The latter two are apparently encoded by the same gene and are sometimes referred to as the testis/liver of $\gamma_{T/L}$. The $\gamma_{T/L}$ and γ_M isoforms have different tissue distribution, as reflected in their names. For example, the mRNA of the γ_M is highly expressed in skeletal muscle and at lower levels in heart and brain, but is undetectable in liver. The $\gamma_{T/L}$ isoform is expressed in the testis and at lower levels in other tissues (Calalb et al., 1992). The human $\gamma_{T/L}$ gene, termed PHKG2, has been assigned to chromosome 16 (Whitmore et al., 1994), whereas γ_M gene PHKG1 was mapped to chromosome 7 (Chamberlain et al., 1987; Jones et al., 1990). The most extensively characterized γ_M subunit gene is that of the rat (Cawley et al., 1993), which spans over 16 kb and contains 10 exons. Unlike CaMPKII and MLCK, most of the CaM regulatory region of the rat γ subunit is on a single exon (Cawley et al., 1993). Characterization of the mouse γ subunit gene suggests the same relationship, although the definitions of the boundaries of the regulatory domain are slightly different (Bender et al., 1993).

Multiple CaM genes in rodents encode the same amino acid sequence. No evidence to date has linked any one of these multiple CaM genes specifically to the δ subunit (CaM) of PhosK. In the I-strain of mice it was shown that the total CaM concentration is decreased by 40% in skeletal muscle compared to normal mice (Shenolikar et al., 1979). This decrease in CaM protein expression is correlated with a comparable decrease in the abundance of the four CaM mRNAs (Bender et al., 1988). As a control, the level of skMLCK was measured and shown to be unaffected in skeletal muscle of the I-strain of mice (Shenolikar et al., 1979).

V. CALMODULIN-DEPENDENT PROTEIN KINASE I

A. Calmodulin-Dependent Protein Kinase I Function and Cellular Ca^{2+} Signal Transduction

Calmodulin-dependent protein kinase I (CaMPKI) is a monomeric protein with an estimated molecular weight of 37K–43K (Picciotto et

3. Calmodulin-Regulated Protein Kinases

al., 1995; DeRemer et al., 1992a,b; Nairn and Greengard, 1987) that is found in multiple tissues and phylogenetic species (Nairn and Greengard, 1987; Picciotto et al., 1993; Haribabu et al., 1995; Picciotto et al., 1995). Three isoforms of the enzyme (CaMPKI-α, CaMPKI-β, and CaMPKI-γ) have been characterized (Picciotto et al., 1993; Yokokura et al., 1997). Calmodulin-dependent protein kiase I α is the more widely distributed isoform. The high level expression of CaMPKI-β and CaMPKI-γ appears to be limited to the brain. Full CaMPKI activity is dependent on both the phosphorylation at a specific threonine residue in a catalytic segment activation loop (by a distinct CaM-dependent kinase kinase) and the activation by Ca^{2+}/CaM at a CaM regulatory domain (Mochizuki et al., 1993a; Picciotto et al., 1993; Lee and Edelman, 1994, 1995). In this regard, CaMPKI is unusual among protein kinases in its dependence on the presence of both regulators (CaM and the CaM-dependent kinase kinase). Calmodulin-dependent protein kinase I might be part of a Ca^{2+}/CaM kinase-activated cascade distinct from the well-known MAP kinases (for reviews see Nishida and Gotoh, 1993; Davis, 1993; Ahn, 1993).

Calmodulin-dependent protein kinase I specifically phosphorylates synapsins I and II (DeRemer et al., 1992a,b; Nairn and Greengard, 1987), the cystic fibrosis transmembrane conductance regulator (Picciotto et al., 1992), the cAMP response element-binding (CREB) protein (Sheng et al., 1991; Sun et al., 1996), and activating transcription factor-1 (ATF-1). Although some of the potential in vivo substrates of CaMPKI overlap those of CaMPKII, CaMPKI phosphorylates synapsin I at a site different from CaMPKII (DeRemer et al., 1992b). Calmodulin-dependent protein kinase I phosphorylation sites also overlap with known PKA phosphorylation sites in the substrate proteins (Lee et al., 1994). However, studies of synthetic peptide substrates indicate that the substrate sequence motif preferred by CaMPKI is Hyd-X-R-X-X-S/T-X-X-X-Hyd, where Hyd is a hydrophobic residue (M.L.I,F,V) (Dale et al., 1995; Lee et al., 1994). This sequence motif is present in all three of the protein substrate phosphorylation sites for which sequence information exists (Nairn and Greengard, 1987; Sun et al., 1996; Sheng et al., 1991) and is distinct from the sequence motifs preferred by CaMPKII and PKA. Although not yet explored in the detail of MLCK and CaMPKII, future studies may reveal that subcellular geography of CaMPKI and its substrates may contribute to in vivo selectivity.

The CaMPKI regulatory pathway might be of greater biological significance than previously appreciated given the broad tissue distribution of the enzyme. Further, initial reports indicate that CaMPKI and its activating kinase coexist in the same cell type and become activated by various calcium-mobilizing stimuli (Aletta et al., 1996). Clearly, a Ca^{2+}/CaM kinase cascade that includes CaMPKI is possible (Aletta et al., 1996; Sugita et al., 1994; Hawley et al., 1995). In addition, a CaM-independent pathway (5' AMP-dependent protein kinase) of CaMPKI activation is also possible (Hawley et al., 1995). Therefore, regulation of cellular function via CaMPKI pathways might involve cross-talk with CaM-independent pathways as well as with other Ca^{2+}/CaM-regulated pathways.

B. Calmodulin-Dependent Protein Kinase I as a Calmodulin-Regulated Enzyme

Based on similarities to the catalytic and CaM regulatory segments of MLCK and CaMPKII, a segmental organization for CaMPKI that includes catalytic, CaM regulatory, and carboxyl-terminal segments can be proposed (Figs. 1 and 17). Although a function for the carboxyl-terminal segment has not been elucidated, it is reasonable to speculate that it might be involved in targeting to subcellular structures, a paradigm found with MLCK and CaMPKII (Shirinsky et al., 1993; Srinivasan et al., 1994), or formation of molecular complexes that include the activating kinase, a paradigm proposed for MAP kinases (Choi et al., 1994). In addition to the three isoforms of CaMPKI that have been characterized based on cloned cDNAs, a kinase known as CaM-dependent protein kinase V (Mochizuki et al., 1993b) may also be a CaMPKI isoform based on amino acid sequence similarity of its ORF to those of the CaMPKI clones (Picciotto et al., 1993; Haribabu et al., 1995; Cho et al., 1994; Yokokura et al., 1997). Electrophoretically distinct species of CaMPKI

Fig. 17. Sequence alignment of the catalytic and CaM regulatory segments of CaMPKIs and CaMPKIV. Amino acid sequences of human CaMPKI-α (hucampki), rat CaMPKI-α (campkialpha), rat CaMPKI-β (campkibeta), rat CaMPKI-γ (campkigamma), and rat CaMPKIV (campkiv) are shown. The amino acid residues conserved in the catalytic segment of many protein kinases are boxed. The proposed CaM regulatory segments are shaded. The threonine residue, which undergoes phosphorylation by CaM-dependent protein kinase kinase, is indicated by an asterisk. Sequences were obtained from the GenBank database.

```
hucampki     F R D V L G T G A F S E V I L A E D K R T Q K L V A I H K C I A K E A L E G K E G   61
campkialpha  F R D V L G T G A F S E V I L A E D K R T Q K L V A I H K C I A K K A L E G K E G   61
campkibeta   F M E V L G T G A F S E V F L V K Q E R V T G S A H L V A L K C I - K K S P A F R D S   63
campkigamma  I R E K L G S G A F S E V M L A Q E R G S A H L V A L K C I P K K A L R G K E A   56
campkiv      V E S E L G R G A T S I V Y R C K Q K G T Q K P Y A L K - V L K K T V D K K -   81

hucampki     S M E N E I A V L H K I K H P N I V A L D D D I Y E S G G H L Y L I M Q L V S G G   101
campkialpha  S M E N E I A V L H K I K H P N I V A L D D D I Y E S G G H L Y L I M Q L V S G G   101
campkibeta   S L E N E I A V L K R I K H E N I V T L E D V H E S P S H L Y L V M Q L V S G G   103
campkigamma  L V E N E I A V L R R I S H P N I V A L E D V H E S P T H L Y L A M E L V T G G   96
campkiv      I V R T E I G V L L R L S H P N I I K L K E I F E T P T E I S L V L E L V T G G   120

hucampki     E L F D R I V E K G F Y T E R D A S R L I F Q V L D A V K Y L H D L G I V H R D   141
campkialpha  E L F D R I V E K G G Y T E R D R S R L I F Q V L D A V K Y L H D L G I V H R D   141
campkibeta   E L F D R I L E R G V Y T E K D A S L V I Q Q V L S A V K Y L H E N G I V H R D   143
campkigamma  E L F D R I M E R G S Y T E K D A S H L V G Q V L G A V S Y L H S L G I V H R D   136
campkiv      E L F D R I V E K G Y Y S E R D A V K Q I L E A V S Y L H E N G I V H R D   160

                                                                    *
hucampki     L K P E N L L Y Y S L D E D S K I M I S D F G L S K M E D P G S V L S T A C G T   181
campkialpha  L K P E N L L Y Y S L D E D S K I M I S D F G L S K M E D P G S V L S T A C G T   181
campkibeta   L K P E N L L Y L T P E E N S K I M I T D F G L S K M E Q N G - V M S T A C G T   182
campkigamma  L K P E N L L Y A T P F E E D S K I M V S D F G L S K K I - Q A G N M L G T A C G T   175
campkiv      L K P E N L L Y A T P A P D A P L K I A D F G L S K I V E H Q V L M K T V C G T   200

                                                                       (continued)
```

```
hucampki    P G Y V A P E V L A Q K P Y S K A V D C W S I G V I A Y I L L C G Y P P F Y D E  221
campkialpha P G Y V A P E V L A Q K P Y S K A V D C W S H I G V I A Y I L L C G Y P P F Y D E  221
campkibeta  P G Y V A P E L L A Q K P Y S K A V D C W S I H G V I T Y H L L C G Y P P F Y E E  222
campkigamma P G Y V A P E L L E Q K P Y G K A V D V W A L G V I S Y I L L C G Y P P F Y D E  215
campkiv     P G Y C A P E I L R G C A Y G P E V D M W S V G I H T Y I L L C G F E P F Y D E  240

hucampki    N D A K - L F E Q I L K A E Y E F D S P Y W D D I S D S A K D F I R H L M E K D  260
campkialpha N D A K - L F E Q I L K A E Y E F D S P Y W D D I S D S A K D F I R H L M E K D  260
campkibeta  T E S K - L F E K I K E G Y Y E F E S P F W D D I S E S A K D F I C H L L E K D  261
campkigamma S D P E - L F S Q I L R A S Y E F D S P F W D D I S E S A K D F I R H L L E R D  254
campkiv     R G D Q F M F R R I L N C E Y Y F I S P W W D E V S L N A K D L V K K L I V L D  280

hucampki    P E K R F T C E Q A L Q H P W I A G D T A L D K N I H Q S V S E Q I K K N F - A  299
campkialpha P E K R F T C E Q A L Q H P W I A G D T A L D K N I H Q S V S S L Q Q I K K N F - A  299
campkibeta  P N E K R Y T C E K A L R H P W I D G N T A L H R D I H Y P S V S L Q Q I H K N F - A  300
campkigamma P Q K R F T C Q Q A L Q H P H L W I S G D A A L - D R D I L G S V S E Q I Q K N F - A  293
campkiv     P K K R L T T F Q A L Q H P W V T G K A A - - N F V H M D T A Q K K L Q E F N A  318

hucampki    K S K W K Q A F N A T A V V R H M  316
campkialpha K S K W K Q A F N R T A V V R H M  316
campkibeta  K S K W R Q A F N . . . S F L R H I  309
campkigamma R T H W K R A F N A T S F L R H I  310
campkiv     R R K L K A A V K A V V A S S R L  336
```

Fig. 17. Continued

(M_r 35K–42K) have been found in tissue extracts (Picciotto et al., 1995). Because the calculated masses of the cDNA ORFs are consistent with the distinct electrophoretic properties of the tissue extracts, it is assumed that the cDNAs of CaMPKI-α encode the M_r 41K–42K proteins and that the cDNAs for CaMPKI-β and CaMPKI-γ encode the smaller proteins (M_r 35K–37K).

Calmodulin-dependent protein kinase I is activated by Ca^{2+}/CaM, but the level of activity is influenced by phosphorylation from an exogenous kinase (a CaMPKK) that is also Ca^{2+}/CaM dependent (Lee and Edelman, 1994; Edelman et al., 1996; Inoue et al., 1995). Studies of bacterially expressed CaMPKI-α indicate that CaMPKI has a comparatively low specific activity in the absence of phosphorylation, even in the presence of Ca^{2+}/CaM (Haribabu et al., 1995; Yokokura et al., 1995). Calmodulin-dependent protein kinase I α remains Ca^{2+}/CaM dependent after phosphorylation by the CaM kinase kinase. Phosphorylation by CaMPKK occurs in a proposed CaMPKI activation loop on T177 (Picciotto et al., 1993). Mutation of this threonine (T177A) prevents phosphorylation and activation of CaMPKI by CaMPKK (Haribabu et al., 1995). Mutation of T177 to aspartic acid (T177D) results in an enzyme that has about a 2.5-fold higher specific activity than the unphosphorylated wild-type enzyme (Haribabu et al., 1995). The mutant T177D–CaMPKI still undergoes activation by Ca^{2+}/CaM. Available data, therefore, indicate that the regulation of the catalytic segment of CaMPKI by the activation loop phosphorylation mechanism and the CaM activation mechanism are distinct but nonexclusive and are functionally coupled through their effects on the active site of the enzyme.

The proposed CaM regulatory segment of CaMPKI-α (Fig. 17) is similar to other CaM-dependent protein kinases such as CaMPKII and MLCK. Deletion mutagenesis studies have been done on bacterially expressed human and rat CaMPKI-α proteins (Tegge et al., 1995; Haribabu et al., 1995). Results of these initial studies and those with synthetic peptide analogs of the proposed CaM recognition region (Yokokura et al., 1995) are consistent with autoinhibitory activity of CaMPKI-α being in the region around residues 286–299 and a CaM recognition region around residues 305–316. Future investigations similar to those done with MLCK should further refine the limits of this region, identify the critical residues for CaM recognition and autoinhibitory activity, and allow a link to the structure of the apoenzyme (Goldberg et al., 1996).

Insight into the structural basis of autoinhibition by CaM-dependent protein kinases has been stimulated by the report (Goldberg et al., 1996) of a X-ray crystallographic structure for CaMPKI-α. The structure (Fig. 18) indicates that part of the proposed CaM regulatory segment interacts with the catalytic segment of the enzyme in a way that would alter the active site, but portions of the CaM regulatory segment do not occupy the active site as predicted by the pseudosubstrate model for CaM-regulated protein kinases. Residues 305–316 of the CaM regulatory segment interact with the outer surface of the upper lobe of the ATP-binding region, whereas residues 290–300 of the regulatory segment interact with a groove in the lower lobe of the enzyme (Goldberg et al., 1996) (Fig. 18). The intramolecular interaction of residues 290–300 is consistent with deletion mutagenesis studies in which constitutive activity results from truncation at residue 294 and CaM-dependent activity requires the sequence through residue 316 (Haribabu et al., 1995). The mode of inhibition of CaMPKI suggested by the structure, where the amino-terminal portion of the CaM regulatory segment associates with the catalytic core, is consistent with the relief-of-autoinhibition model for CaM regulation of protein kinases (Shoemaker et al., 1990). Additional substitutions of amino acid sequences by site-directed mutagenesis by the CaM regulatory segment of CaMPKI-α will be necessary to determine what specific amino acids are necessary for CaM recognition and autoinhibition. However, arguments based on similarity to the well-characterized sm/nmMLCK (Shoemaker et al., 1990) suggest that the putative CaM recognition part of CaMPKI-α (residues 299–316) has the proper exposure of amino acid residues that would facilitate CaM interaction.

Based on the available three-dimensional structures of other protein kinases that are regulated by phosphorylation of a catalytic domain activation loop (DeBondt et al., 1993; Zhang et al., 1994), the proposed activation loop of CaMPKI might function in a similar manner to inhibit its catalytic domain. However, analysis of diffraction data used for the CaMPKI structure indicated that the proposed region of the activation

Fig. 18. Ribbon diagram of the CaMPKI structure. This diagram is based on the coordinates kindly provided by the authors (Goldberg et al., 1996). The peptide backbone of the catalytic segment is shown with the amino end in the upper part of the drawing and the carboxyl end of the catalytic segment at the bottom. The glycine-rich ATP loop is indicated by a dark gray color to the ribbon drawing. The proposed CaM regulatory segment is highlighted with a stick structure.

3. Calmodulin-Regulated Protein Kinases

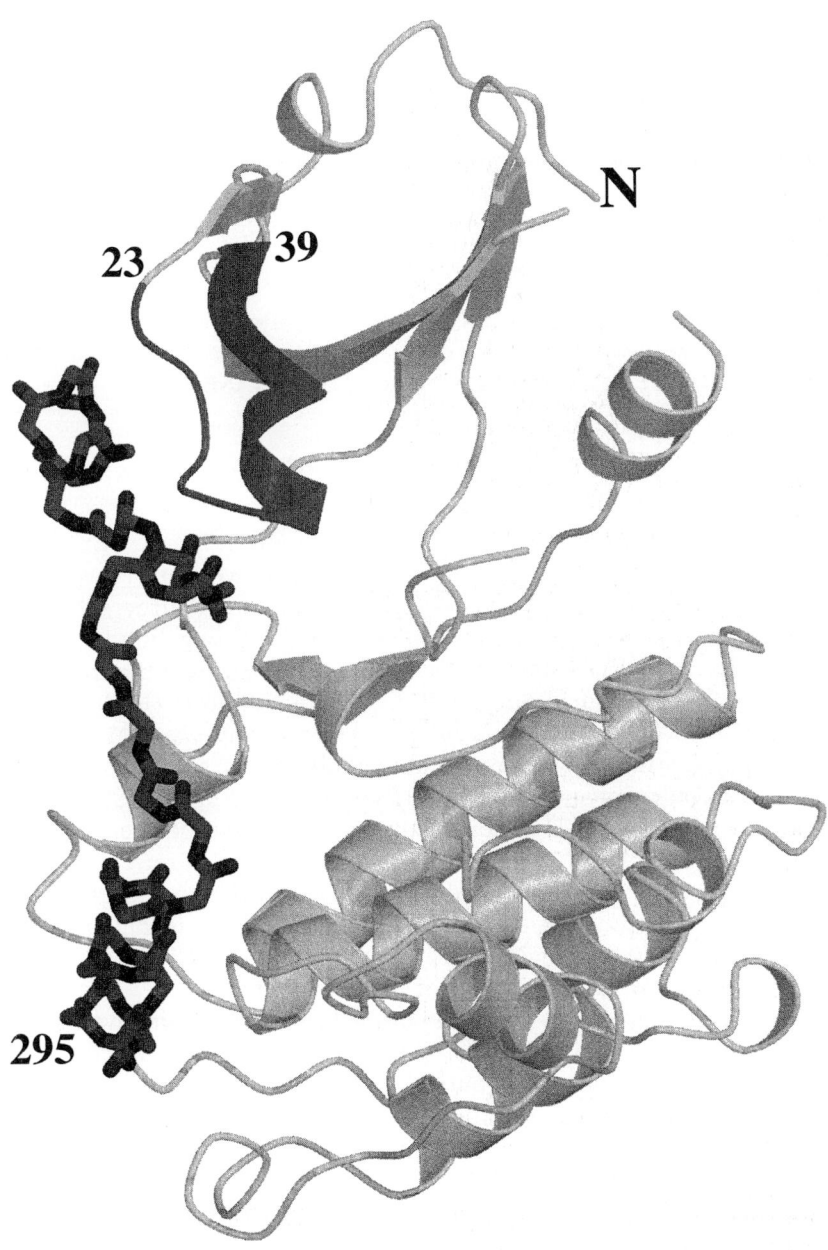

loop was too disordered to allow a structure to be defined with confidence (Goldberg *et al.*, 1996). Regardless, it is reasonable to assume that the activation loop and the CaM regulatory segment could occupy distinct space in the structure such that the activation loop would partially block access to the active site when it is not phosphorylated and the autoinhibitory region of the CaM regulatory segment would affect the active site through interactions similar to that suggested by the structure of CaMPKI (Goldberg *et al.*, 1996).

VI. CALMODULIN-DEPENDENT PROTEIN KINASE III

A. Calmodulin-Dependent Protein Kinase III Function and Cellular Ca^{2+} Signal Transduction

Calmodulin-dependent protein kinase III (CaMPKIII) specifically phosphorylates elongation factor-2 (EF-2) (Nairn and Palfrey, 1987; Mitsui *et al.*, 1993), thus it is also called EF-2 kinase. Elongation factor-2 is a component of eukaryotic ribosomes that is necessary for the translocation step in the elongation phase of protein synthesis (Redpath and Proud, 1994). Phosphorylation of EF-2 (Redpath *et al.*, 1993) reduces its affinity for the ribosome and thus abolishes its ability to perform translocation (Nairn and Palfrey, 1987; Redpath and Proud, 1994). Elongation factor-2 phosphorylation occurs in most eukaryotic cells and has been observed *in vivo* in response to various factors that raise intracellular Ca^{2+} (Brady *et al.*, 1990; Redpath and Proud, 1994; Bagaglio *et al.*, 1993). Although the physiological significance of Ca^{2+}-dependent phosphorylation of EF-2 remains to be elucidated, CaMPKIII seems to be important for the overall control of protein synthesis.

Calmodulin-dependent protein kinase III can be found in diverse phylogenetic species (Hait *et al.*, 1996; Nairn *et al.*, 1985; Ryazanov *et al.*, 1997). The enzyme has been purified from rabbit reticulocytes and rat pancreas and is a single-chain enzyme of about 100K (Nilsson *et al.*, 1991; Mitsui *et al.*, 1993). Calmodulin-dependent protein kinase III is absolutely dependent on Ca^{2+}/CaM for activity and displays nanomolar affinity for CaM (Mitsui *et al.*, 1993), that is comparable to that of MLCK.

The sequence of the consensus phosphorylation site within EF-2 extends from residue 49 to 60 with the amino acid sequence RAGETRFT-

3. Calmodulin-Regulated Protein Kinases

DTRK. The kinase phosphorylates two threonine residues, T56 and T58; phosphorylation of T56 preceding, and apparently being required for, phosphorylation of T58 (Redpath *et al.*, 1993). However, a synthetic peptide based on residues 49–60 of EF-2 is poorly phosphorylated by CaMPKII, suggesting that other features of EF-2 are necessary for efficient phosphorylation (Redpath *et al.*, 1993). Interestingly, both T56 and T58 are conserved in EF-2 from many different species, including yeast *S. cerevisiae*. Moreover, it was shown that yeast EF-2 is an excellent substrate for mammalian CaMPKIII and vice versa (Mitsui *et al.*, 1993), suggesting functional conservation of the kinase and kinase substrate.

B. Calmodulin-Dependent Protein Kinase III as a Calmodulin-Regulated Enzyme

Based on similarities to other CaM-regulated kinases and on the deduced amino acid sequence of an ORF from rat (Redpath *et al.*, 1996), human, mouse, and *Caenorhabditis elegans* (Ryazanov *et al.*, 1997) cDNA clones, CaMPKIII appears to have a segmental organization consisting of catalytic, CaM regulatory, and carboxyl-terminal association segments (Fig. 1). The proposed catalytic segment of CaMPKIII has the least similarity to, and does not contain many of the subdomains present in, the other CaM-regulated protein kinases (Fig. 19). Calmodulin-dependent protein kinase III does have a GxGxxG motif, which is a part of the ATP-binding domain of the other protein kinases. Interestingly, CaMPKIII is homologous to the myosin heavy chain kinase A from *Dictyostelium* (Ryazanov *et al.*, 1997). It is possible that these two kinases might be part of a novel subclass of protein kinases. Clearly, additional investigations that relate structure to function are needed.

Calmodulin-dependent protein kinase III shares some of the regulatory properties elucidated for other CaM-dependent protein kinases. On activation by Ca^{2+}/CaM, CaMPKIII undergoes intramolecular autophosphorylation at multiple unidentified sites that converts it to a partially Ca^{2+}/CaM-independent species (Redpath and Proud, 1993; Mitsui *et al.*, 1993), which is a property shared with CaMPKII. Similar to MLCK and CaMPKI, and CaMPKIII can also serve as a substrate for exogenous kinases such as PKA. Phosphorylation by PKA converts CaMPKIII into a Ca^{2+}/CaM-independent kinase (Redpath and Proud, 1993). Thus, as with other CaM-dependent kinases, CaMPKIII apparently has a dual

```
hucampkiii     QGVGDLYTDPQIHTETGTDF GDNLGV RGMALFFYSHACN  315
mcampkiii      QGVGDLYTDPQIHTEKGTDF GDNLGV RGMALFFYSHACN  314
rcampkiii      QGVGDLYTDPQIHTEKGTDF GDNLGV RGMALFFYSHACN  314
celegcampkiii  QGVGDLYTDPQIHTVVGTDY GDNLGT RGMALFFHSHRCN  298

hucampkiii     RICESMGLAPFDLSPRERDAVNQNTKLLQSAK-------T  348
mcampkiii      RICQSMGLTPFDLSPREQDAVNQSTKLLQSAK-------T  347
rcampkiii      RICQSMGLAPFDLSPREQDAVNQSTKLLQSAK-------T  347
celegcampkiii  DICETMDLSNFELSPPEIEATEVAMEVAAKQKKSCIVPPT  338

hucampkiii     ILRGTEEKCGSPRVRTLSG-SRPPLLRPLSENSGDENMSD  387
mcampkiii      ILRGTEEKCGSPRIRTLSS-SRPPLLRLSENSGDENMSD  386
rcampkiii      ILRGTEEKCGSPRIRTLSG-SRPPLLLRLSENSGDENMSD  386
celegcampkiii  VFEARRNRISSECVHVEHGISMDQLRKRKTLNQSSTDLSA  378

hucampkiii     VTFDSLPSSPSSATPHSQKLDHLHWPVFSDLDNMASRDHD  427
mcampkiii      VTFDSLPSSPSSATPHSQKLDHLHWPVFGDLDNMGPRDHD  426
rcampkiii      VTFDSLPSSPSSATPHSQKLDHLHWPVFGDLDNMGPRDHD  426
celegcampkiii  KSHNEDCVCPECIPV---VEQLCEPCSEDEED-EEEDYP  413

hucampkiii     HLDNHRESENSGDSGYPSEKRGELDDPEPREHGHSYSN--  466
mcampkiii      RMDNHRDSENSGDSGYPSEKRSDLDDPEPREHGHSNGN--  465
rcampkiii      RMDNHRDSENSGDSGYPSEKRSDLDDPEPREHGHSNGN--  465
celegcampkiii  RSEKSGNSQKSRRSRMSISTRSSGDESASRPRKCGFVDLN  453

hucampkiii     ----RKYESDEDSLGS---SGRVCVEK----WNLLNSSR  493
mcampkiii      ----RRHESDEDSLGS---SGRVCVET----WNLLNPSR  492
rcampkiii      ----RRPESDEDSLGS---SGRVCVET----WNLLNPSR  492
celegcampkiii  SLRQRHDSFRSSVGTYSMNSSRQTRDTEKDEFWKVL--R  490
```

```
hucampkiii     L H L P R A S A V A L E V Q R L N A - L D L E K K I G K - - - - - - S I L G K V  526
mcampkiii      L H L P R P S A V A L E V Q R L N A - L D L G R K I G K - - - - - - S V L G K V  525
rcampkiii      L H L P R P S A V A L E V Q R L N A - L D L G R K I G K - - - - - - S V L G K V  525
celegcampkiii  K Q S V P A N I L S L Q L Q Q M A A N L E N D E D V P Q V T G H Q F S V L G Q I  530

hucampkiii     H L A M V R Y H E G G R F C E K G E E - - - - - - - - - - - - - - W D Q E  549
mcampkiii      H L A M V R Y H E G G R F C E K D E E - - - - - - - - - - - - - - W D R E  548
rcampkiii      H L A M V R Y H E G G R F C E K D E E - - - - - - - - - - - - - - W D Q E  548
celegcampkiii  H I D L S R Y H E L G R F V E V D S E H K E M L E G S E N D A R V P I K Y D K Q  570

hucampkiii     S A V F H L E H A A N L G E L E A I V G L G L M Y S Q L P H H I L A D V S L K E  589
mcampkiii      S A I F H L E H A A D L G E L E A I V G L G L M Y S Q L P H H I L A D V S L K E  588
rcampkiii      S A I F H L E H A A D L G E L E A I V G L G L M Y S Q L P H H I L A D V S L E E  588
celegcampkiii  S A I F H L D I A R K C G I L E A V L T S A H I V L G L P H E L L K E V T V D D  610

hucampkiii     - - - - T E E N K T K G F - - - - D Y L L K A A E A G D R Q S M I L V A  617
mcampkiii      - - - - T E E N K T K G F - - - - D Y L L K A A E A G D R H S M I L V A  616
rcampkiii      - - - - T E E N K T K G F - - - - D Y L L K A A E A G D R Q S M I L V A  616
celegcampkiii  L F P N G F G E Q E N G I R D L E E F G S D L M E I A A E M G D K G A M L Y M A  650

hucampkiii     R A F D S G Q N L S P D R C Q D W L E A L H W Y N T A L E - - - - - M T D C  650
mcampkiii      R A F D T G L N L S P D R C Q D W S E A L H W Y N T A L E - - - - - T T D C  649
rcampkiii      R A F D T G L N L S P D R C Q D W S E A L H W Y N T A L E - - - - - T T D C  649
celegcampkiii  H A Y E T G Q H L G P N R R T D Y K K S I D W Y Q R V V G F Q E E E L D S D C  690
```

Fig. 19. Sequence alignment of the proposed catalytic and CaM regulatory segments of CaMPKIIIs. Amino acid sequences of human CaMPKIII (hucampkiii), mouse CaMPKIII (mcampkiii), rat CaMPKIII (rcampkiii), and *C. elegans* CaMPKIII (celegcampkiii) are shown. The GXGXXG nucleotide-binding motif is boxed. The proposed CaM regulatory segments are shaded. Sequences were obtained from the GenBank database.

regulation paradigm that involves autophosphorylation and exogenous phosphorylation.

VII. CALMODULIN-DEPENDENT PROTEIN KINASE IV

A. Calmodulin-Dependent Protein Kinase IV Functions and Calcium Signal Transduction

Calmodulin-dependent protein kinase IV (CaMPKIV) is one of the more recently discovered CaM-regulated protein kinases, so less is known about its structure, function, and molecular mechanisms of Ca^{2+} signal transduction. The tissue and subcellular distribution pattern of CaMPKIV and its substrate selectivity are distinct from that of CaMPKI and CaMPKII, suggesting distinct, and perhaps unique, functions.

Calmodulin-dependent protein kinase IV is abundant in brain, like CaMPKI and CaMPKII, but is also abundant in the thymus. Immunohistochemical staining of brain sections also indicated localizations of CaMPKIV distinct from CaMPKI and CaMPKII. Calmodulin-dependent protein kinase IV immunoreactivity was concentrated in the cerebellum and forebrain and was less abundant in other regions (Sakagami and Kondo, 1996). Calmodulin-dependent protein kinase IV is generally localized in nuclei (Jensen et al., 1991), especially the nuclei of cerebellar granule cells associated with dispersed chromatin (Jensen et al., 1991; Sakagami and Kondo, 1996) and is not found in other particulate fractions or associated with postsynaptic densities or growth cones (Ohmstede et al., 1989). In contrast, CaMPKII is found in both cytosolic and particulate fractions and is particularly abundant in postsynaptic densities.

Because of its nuclear localization and ability to phosphorylate transcription factors, CaMPKIV is a likely mediator of Ca^{2+}/CaM signal transduction that regulates gene expression. Calmodulin-dependent protein kinase IV phosphorylates several transcription factors such as ATF-1, CREB, and CREMt (Cruzalegui and Means, 1993; Sun et al., 1994; Matthews et al., 1994; Sun et al., 1995a, 1996). Transfection of different types of cells with a vector encoding a constitutively active CaMPKIV enhanced expression of a reporter gene through a CRE promoter (Sun *et al.*, 1994, 1996; Enslen *et al.*, 1994; Matthews *et al.*, 1994). Consistent with a model of distinct signal transduction functions, phosphorylation of the CREB protein at one site by CaMPKII leads to a decrease in gene expression, whereas phosphorylation by CaMPKIV at a different site leads to upregulated gene expression (Matthews *et al.*, 1994; Sun *et al.*, 1994).

Specific inhibitors of CaMPKIV are not available, although it is inhibited by KN-62 with about the same potency as this compound inhibits CaMPKII (Enslen *et al.*, 1994). Therefore, clues to the specific *in vivo* processes controlled by CaMPKIV cannot be addressed by pharmacological intervention at this time. Molecular biological and biochemical approaches have been used to address the role of CaMPKIV in specific Ca^{2+}-mediated processes. A combination of these approaches has established a potential role for CaMPKIV in linking Ca^{2+} signaling to the regulation of transcription. In addition, biochemical evidence suggests that type I adenylyl cyclase activity might be regulated by CaMPKIV (Wayman *et al.*, 1996). Regardless of the physiological relevance of the adenyl cyclase regulation, results in the field raise the clear possibility that an additional point of potential cross-talk between the cAMP and Ca^{2+}/CaM signaling pathways may be at the level of CaMPKIV-mediated activities, especially as they link to transcription regulation.

In addition to cross-talk among pathways regulated by different messengers, the cross-talk among different branches of the Ca^{2+} signaling pathways might be through the CaM-dependent protein kinase kinases and their potential signal amplification pathways, or cascades. In this regard, CaMPKIV can be activated by CaM-regulated kinase kinase, and activation of CaMPKIV by the kinase kinase can further augment CaMPKIV-mediated gene expression (Tokumitsu *et al.*, 1995; Park and Soderling, 1995). However, it is not known at this time if this and other CaM-regulated kinase kinases are a true cascade (i.e., is there amplification of the signal at each step of the biological pathway) or if the kinase kinases might be additional regulatory subunits that provide for signal transduction cross-talk (like the subunits of PhosK) but are novel in that they use a phosphotransferase activity among subunits. As discussed in the following sections, the CaMPKIV pathway is a CaM-regulated protein kinase pathway that uses regulatory features found in CaMPKII, CaMPKI, and MLCK and has potential novel aspects of regulation not seen with other classes of CaM-regulated protein kinases.

B. Calmodulin-Dependent Protein Kinase IV as a Calmodulin-Regulated Enzyme

Based on similarities of its catalytic and CaM regulatory segments to MLCK and CaMPKII, CaMPKIV has a segmental organized primary structure (see Fig. 17). However, CaMPKIV is more closely related to CaMPKI than it is to MLCK or CaMPKII when the proposed catalytic and CaM regulatory segments are compared. This is interesting in light of the fact that both CaMPKI and CaMPKIV also use an activation loop mechanism in their catalytic segments. There are two subtypes of CaMPKIV, known as the α (M_r 63–65K) and β (M_r 66–67K) isoforms, that exhibit slightly different electrophoretic mobilities (Miyano et al., 1992; Ohmstede et al., 1989). Two cDNA clones whose ORFs encode proteins of 53K (Means et al., 1991; Ohmstede et al., 1989) and 56K (Sakagami and Kondo, 1993) have been reported. It is possible that these cDNAs and the subtypes of CaMPKIV represent the result of alternative splicing (Sun et al., 1995b). Both subtypes maintain the proposed segmental organization.

As discussed in the previous section, the protein substrate selectivity of CaMPKIV correlates well with subcellular distributions to indicate unique functions for the CaMPKIV regulatory segment. However, CaMPKIV is able to phosphorylate *in vitro* some of the same synthetic peptide substrates as CaMPKII and CaMPKI. This indicates some catalytic similarities that relate to the sequence similarities noted among the catalytic segments for these enzymes. Like CaMPKI, CaMPKIV is a substrate for an exogenous protein kinase that further modulates its phosphotransferase activity (Tokumitsu et al., 1995; Okuno and Fujisawa, 1993; Park and Soderling, 1995; Chatila et al., 1996). The enzyme that phosphorylates CaMPKIV is also a CaM-dependent kinase kinase (Tokumitsu et al., 1994).

Consistent with the proposed CaM regulatory segment, CaMPKIV requires Ca^{2+}/CaM for full activity. Mutagenesis of the autoinhibitory region of the CaM regulatory segment results in constitutive activity, suggesting that CaMPKIV follows a pattern similar to that of CaMPKI. Deletion of residues after L313 in CaMPKIV (Fig. 17) results in an active and CaM-independent kinase (Cruzalegui and Means, 1993). L313 is in the same relative position as I294 of CaMPKI. Site-directed mutagenesis of the CaM regulatory segment of CaMPKIV by the substitution of acidic residues for H305, H306, T308, F316, or N317 results in CaM-independent constitutive activity (Tokumitsu et al., 1994). Experimental results and similarities to MLCK, CaMPKII, and CaMPKI suggest that the autoinhibitory region of CaMPKIV minimally includes residues 305–317. Similar to CaMPKII, CaMPKIV develops some Ca^{2+}/CaM-independent activity that is associated with autophosphorylation (Okuno and Fujisawa, 1993; Cruzalegui and Means, 1993; Selbert et al., 1995; McDonald et al., 1993). In contrast to CaMPKII, CaMPKIV autophosphorylation appears to occur on unidenitified serine residues (Selbert et al., 1995; McDonald et al., 1993), although CaMPKIV has threonine residues located in a position similar to the autophosphorylation site of CaMPKII (Fig. 17). Little has been done to address the CaM recognition features of CaMPKIV.

The function of the carboxyl end of CaMPKIV is not known. Based on the paradigms demonstrated for MLCK and CaMPKII, in which the corresponding carboxyl segment is involved in macromolecular associations, it is reasonable to speculate that a similar function might be found for the carboxyl segment of CaMPKIV. Although CaMPKIV has been found in the nucleus (Jensen et al., 1991), it does not contain the nuclear localization signal found in the δ-B isoform of CaMPKII or the CaMPKII-α gene. However, CaMPKIV contains glutamic acid tracts that are often associated with chromatin-binding proteins (Ohmstede et al., 1989).

C. Molecular Genetics of Calmodulin-Dependent Protein Kinase IV and Related Proteins

The rat gene for CaMPKIV has 12 exons (Sun et al., 1995b) that span 42 kb. By an alternative splicing mechanism, two related isozymes are produced. The smaller α isoform differs from the larger β isoform by the addition at the 5' end of a small coding exon (Sakagami and Kondo, 1993; Sun et al., 1995b). Only during spermatogenesis of postmeiotic cells, does the gene produce a different gene product, called calspermin, through an alternative promotion mechanism (Sun et al., 1995a). Calspermin is missing the catalytic segment but contains the CaM-binding and carboxyl segments (Sun et al., 1995a). The function of calspermin is not known. The CaMPKIV gene is unique among CaM-regulated protein kinase genes in producing protein kinase isoforms by alternative splicing, or nonkinase CaM binding proteins by alternative promotion.

VIII. OTHER CALMODULIN-DEPENDENT PROTEIN KINASES

A. Calmodulin-Dependent Protein Kinase Kinases

Calmodulin-dependent protein kinase kinases (CaMPKK) have been shown to activate other CaM-dependent protein kinases such as CaMPKI and CaMPKIV. Two forms, α (M_r 69K) and β (M_r 73K) (Edelman et al., 1996; Tokumitsu et al., 1995; Kitani et al., 1997), have been identified. Both isoforms phosphorylate CaMPKI and CaMPKIV, but prefer CaMPKI as the substrate (Edelman et al., 1996). The phosphorylation sites of CaMPKI and CaMPKIV are at T177 and T196, respectively, but the limited amino acid sequence identity between these sites precludes definition of a specific recognition motif. It has also been reported that CaMPKK can activate a 5' AMP-dependent protein kinase (Hawley et al., 1995).

Like other CaM-dependent protein kinases described earlier, CaMPKKs appear to be segmentally organized. The kinases are most similar in the kinase catalytic and CaM regulatory segments, but diverge at amino and carboxyl termini (Fig. 20). Within the catalytic segment, CaMPKKs contain an unusual insert of 22 residues (Tokumitsu et al., 1995) that follow the lysine residue essential for catalytic activity of most protein kinases (Fig. 20). The function of this insert is unknown. Only a few protein kinases, such as a yeast homolog of Raf (Rhodes et al., 1990), have an insert in this region. The insert shows a similarity to the proline-rich SH3-binding domain motif (PXXPPPXXP), a motif thought to be important in protein–protein associations (Engel et al., 1993; Ren et al., 1993). SH3 motifs are found (Mayer and Baltimore, 1993) in a variety of proteins, including nonmuscle myosin, fodrin, spectrin, proteins associated with focal adhesions, and tyrosine kinases.

The activity of CaMPKKα and CaMPKKβ requires the presence of Ca^{2+}/CaM, and both enzymes appear to be capable of Ca^{2+}/CaM-dependent autophosphorylation. The CaM dependence of the CaMPKKs was established from their ability to bind to immobilized CaM columns in a Ca^{2+}-dependent fashion and to enhance the phosphotransferase activity of CaMPKI or CaMPKIV preparations whose CaM regulatory capability had been removed (Edelman et al., 1996). The CaM recognition region of CaMPKK has not been identified. One region with some sequence motif similarity has a cluster of basic amino acids at the carboxyl end (Fig. 20) rather than at the amino end of the region, as in other CaM kinases (Fig. 2). However, Shoemaker *et al.* (1990) demonstrated that inversion of a CaM recognition sequence so that the basic cluster was at the carboxyl end of the region resulted in an active CaM-dependent kinase. Thus, if this is the CaM recognition region, then the orientation, although exceptional, would not necessarily mean that CaMPKKs are regulated differently from other CaM-dependent kinases.

B. Death-Associated Protein Kinase

The death-associated protein (DAP) kinase is one of the most recently identified Ca^{2+}/CaM-dependent protein kinases. The DAP kinase is a product of a gene shown to be involved in programmed cell death (Deiss *et al.*, 1995). The DAP kinase is localized to the cytoskeleton (Cohen *et al.*, 1997). Analysis of the deduced amino acid sequence of a human HeLa cell cDNA (Deiss *et al.*, 1995) reveals that the DAP kinase has a potential segmental organization like other CaM regulated kinases. Alignment of the catalytic and CaM regulatory segments of the DAP kinase suggests that it is distantly related to MLCK and other CaM-dependent protein kinases. *In vitro* phosphorylation assays confirmed that the DAP kinase phosphorylates myosin regulatory light chains in a Ca^{2+}/CaM-dependent manner (Cohen *et al.*, 1997). However, *in vivo* substrates and molecular mechanisms have yet to be defined. The DAP kinase also undergoes autophosphorylation (Deiss *et al.*, 1995; Cohen *et al.*, 1997). CaM binds to recombinant expressed DAP kinase (Cohen *et al.*, 1997), and deletion of the putative CaM-binding segment renders the kinase constitutively active, a property shared with other CaM-dependent protein kinases, such as MLCK, CaMPKII, CaMPKIV, and CaMPKI. A stretch of eight ankyrin motif repeats, which are motifs associated with protein–protein interactions, were identified (Cohen *et al.*, 1997) downstream of the CaM-binding region and are followed by a region responsible for the cytoskeleton binding. Thus, the segmental organization of the DAP kinase follows the segmental organization paradigm described for all CaM-dependent kinases (Fig. 1), with a potential supramolecular association segment located on the carboxyl side of the CaM-binding segment. A death domain module was identified at the very carboxyl terminus of the supramolecular association segment of the DAP kinase. The presence of this module is correlated with proteins that are involved in apoptosis (Feinstein *et al.*, 1995).

```
campkk alpha  QYKLQSEIGKGAYGVVRLAYNEREDRHYAMKVLSKKLLK  166
campkk beta   QYTLKDEIGKGSYGVVKLAYNENDNTYYAMKVLSKKLIR  202

campkk alpha  QYGFPRRPPPRGSQAPQGGPAKQLLPLERVYQEIAILKKL  206
campkk beta   QAGFPRRPPPRGTRPAPGGCIQPRGPIEQVYQEIAILKKL  242

campkk alpha  DHVNVVKLIEVLDDPAEDNLYLVFDLLRKGPVMEVPCDKP  246
campkk beta   DHPNVVKLVEVLDDPNEDHLYMVFELVNQGPVMEVPTLKP  282

campkk alpha  FPEEQARLYLRDIILGLEYLHCQKIVHRDIKPSNLLLGDD  286
campkk beta   LSEDQARFYFQDLIKGIEYLHYQKIIHRDIKPSNLLVGED  322

campkk alpha  GHVKIADFGVSNQFEGNDAQLSSTAGTPAFMAPEAISDTG  326
campkk beta   GHIKIADFGVSNEFKGSDALLSNTVGTPAFMAPESLSETR  362
```

```
campkk alpha  Q S F S G K A L D V W A T G V T L Y C F V Y G K C P F I D E Y I L A L H R K I K  366
campkk beta   K I F S G K A L D V W A M G V T L Y C F V F G Q C P F M D E R I M C L H S K I K  402

campkk alpha  N E A V V F P E E P E V S E E L K D L I L K M L D K N P E T R I G V S D I K L H  406
campkk beta   S Q A L E F P D Q P D I A E D L K D L I T R M L D K N P E S R I V V P E I K L H  442

campkk alpha  P W V T K H G E E P L P S E E E H C S V V E V T E E E V K N S V K L I P S W T T  446
campkk beta   P W V T R H G A E P L P S E D E N C T L V E V T E E E V E N S V K H I P S L A T  482

campkk alpha  V I L V K S M L R K R S F G N P F E  464
campkk beta   V I L V K T M I R K R S F G N P F E  500
```

Fig. 20. Sequence alignment of the proposed catalytic and CaM regulatory segments of CaMPKKs. Amino acid sequences of rat CaMPKK α (campkk alpha) and rat CaMPKK β (campkk beta) are shown. The amino acid residues conserved in the catalytic segment of many protein kinases are boxed. The proposed CaM regulatory segments are shaded. The proline-rich insert in the catalytic segment is noted with ●●●. The sequence of CaMPKKα was obtained from the GenBank database, and the sequence of CaMPKK β is from Kitani *et al.* (1997).

C. Caki

Caki is a CaM-regulated protein kinase found in *Drosophila* central nervous system and is associated with a walking defect in flies (Martin and Ollo, 1996). Its sequence reveals the same overall segmental organization pattern described for CaM-regulated protein kinases. Its putative CaM-binding region shares 85%, whereas the catalytic segment shares 41% homology with CaMPKIIα, a CaM-dependent protein kinase that has been implicated in behavioral and developmental plasticity in *Drosophila* (Griffith *et al.*, 1993, 1994).

D. Possible Roles of Calmodulin-Regulated Protein Kinases in G-Protein Signaling and MAP Kinase Pathways

Calmodulin-regulated protein kinases have been implicated in the MAP kinase cascade. In one report (Enslen *et al.*, 1996), the MAP kinase cascade in PC12 cells was activated by transfected CaMPKIV and was further augmented by the presence of CaMPKK. This suggests that Ca^{2+}-dependent gene transcription is one possible mechanism of MAP kinase activation. Alternatively, the activation of the MAP kinase pathway may be direct, involving an unidentified CaM kinase that regulates one or more of the proteins (e.g., Pyk2, src) participating in the MAP kinase cascade (Della Rocca *et al.*, 1997; Van Biesen *et al.*, 1995).

Della Rocca *et al.* (1997) suggest that G-protein-coupled receptors utilize a Ca^{2+}/CaM-dependent phosphorylation event to couple the receptor signal to MAP kinase activation. The involvement of CaM in this process was suspected from studies that showed CaM-binding compounds could block ligand activation of the MAP kinase pathway through G-protein-linked receptors (Della Rocca *et al.*, 1997; Eguchi *et al.*, 1996).

Another potential link between CaM-regulated pathways and G-protein-regulated pathways is through the G-protein receptor kinases (GRKs). G-protein receptor kinases phosphorylate G-protein-coupled receptors (Pronin *et al.*, 1977) and bring about receptor desensitization (Freedman *et al.*, 1995). Calmodulin can mediate autophosphorylation of GRK5 and can inhibit GRK interaction with a G-protein-coupled receptor (Pronin *et al.*, 1997). Thus, CaM or CaM-regulated protein kinases might serve a negative regulatory role for some classes of GRKs, thereby modulating the signaling between receptors and downstream pathways. Another variation of this theme is the report that CaM interacts directly with G-protein subunits (Liu *et al.*, 1997).

Clearly, these initial reports require more extensive analysis in order to determine physiological relevance and to gain insight into molecular mechanism.

IX. SUMMARY

Calmodulin-regulated protein kinases are a diverse family of CaM-regulated enzymes. Among all of the CaM-regulated protein kinases characterized in detail, variations of a common theme are used in the recognition of CaM by the enzyme and in the relief-of-autoinhibition mechanism of activation. This functional theme is reflected in an emerging structural theme of segmental organization of the protein kinase primary structures, which should be represented, presumably, in future reports of three-dimensional structures that are under current investigation. However, the mechanisms for additional regulation are diverse. These additional molecular mechanisms of regulation provide the capability for cross-talk among different branches of the Ca^{2+}/CaM signal transduction pathways, as well as other second and third messenger signaling systems, and for fine tuning and interdigitation of a particular Ca^{2+}/CaM-regulated pathway into cascades and feedback pathways. A particular CaM:protein kinase holoenzyme complex can have its response to Ca^{2+} signals fine tuned by the alteration of the relative affinity of CaM for Ca^{2+} when it is a component of a supramolecular complex and by the subcellular localization of the CaM:enzyme signal transducer complex can also provide a degree of protection to mutations and alterations to the individual components of the supramolecular complex by the functional suppression of untoward effects, such as loss of a critical Ca^{2+} ligation residue. The current state of knowledge about the molecular mechanisms of Ca^{2+} signal transduction mediated by CaM-regulated protein kinases reflects general themes used among protein kinases and other CaM-regulated enzymes, as well as the unique sets of attributes used by each CaM:Enzyme complex in integrating Ca^{2+} signaling with other mechanisms of maintaining cellular and organismal homeostasis.

ACKNOWLEDGMENTS

The authors acknowledge the financial support of National Institutes of Health Grant GM 30861 and colleagues in the field of calmodulin research for helpful comments. We thank Steve Weigand for assistance with the preparation of the molecular graphics pictures.

REFERENCES

Adelstein, R. S., Conti, M. A., Hathaway, D. R., and Klee, C. B. (1978). Phosphorylation of smooth muscle myosin light chain kinase by the catalytic subunit of adenosine 3':5'-monophosphate-dependent protein kinase. *J. Biol. Chem.* **253**, 8347–8350.

Ahn, N. G. (1993). The MAP kinase cascade. Discovery of a new signal transduction pathway. *Mol. Cell. Biochem.* **127–128**, 201–209.

Aletta, J. M., Selberg, M. A., Nairn, A. C., and Edelman, A. M. (1996). Activation of a calcium-calmodulin-dependent protein kinase I cascade in PC12 cells. *J. Biol. Chem.* **271**, 20930–20934.

Asano, M., Matsunaga, K., Miura, M., Ito, K. M., Seto, M., Sakurada, K., Nagumo, H., Sasaki, Y., and Ito, K. (1995). Selectivity of action of staurosporin on Ca^{2+} movements and contractions in vascular smooth muscles. *Eur. J. Pharmacol.* **294**, 693–701.

Babu, Y. S., Bugg, C. E., and Cook, W. J. (1988). Structure of calmodulin refined at 2.2 Å resolution. *J. Mol. Biol.* **204**, 191–204.

Bagaglio, D. M., Cheng, E. H. C., Gorelick, F. S., Mitsui, K., Nairn, A. C., and Hait, W. N. (1993). Phosphorylation of elongation factor 2 in normal and malignant rat glial cells. *Cancer Res.* **53**, 2260–2264.

Bagchi, I. C., Huang, Q., and Means, A. R. (1992a). Identification of amino acids essential for calmodulin binding and activation of smooth muscle myosin light chain kinase. *J. Biol. Chem.* **267**, 3024–3029.

Bagchi, I. C., Kemp, B. E., and Means, A. R. (1992b). Intrasteric regulation of myosin light chain kinase: The pseudosubstrate prototype binds to the active site. *Mol. Endocrinol.* **6**, 621–626.

Bahri, S. M., and Chia, W. (1994). DPhK-gamma, a putative *Drosophila* kinase and homology to vertebrate phosphorylase kinase gamma subunits: molecular characterization of the gene and phenotypic analysis of loss of function mutants. *Mol. Gen. Genet.* **245**, 588–597.

Bai, G., and Weiss, B. (1991). The increase of calmodulin in PC12 cells induced by NGF is caused by differential expression of multiple mRNAs for calmodulin. *J. Cell. Physiol.* **149**, 414–421.

Bayer, K. U., Löhler, J., and Harbers, K. (1996). An alternative, nonkinase product of the brain-specifically expressed Ca^{2+}/calmodulin-dependent kinase II α isoform gene in skeletal muscle. *Mol. Cell. Biol.* **16**, 29–36.

Bender, P. K., Dedman, J. R., and Emerson, C. P. (1988). The abundance of calmodulin in mRNAs is regulated in phosphorylase kinase-deficient skeletal muscle. *J. Biol. Chem.* **263**, 9733–9737.

Bender, P. K., Wang, Z., and Carlson, G. M. (1993). Two exons encode the calmodulin-binding domain in the mouse phosphorylase kinase catalytic subunit gene. *Gene Anal.: Tech. Appl.* **10,** 99–101.

Benian, G. M., Kiff, J. E., Neckelmann, N., Moerman, D. G., and Waterston, R. H. (1989). Sequence of an unusually large protein implicated in regulation of myosin activity in *C. elegans. Nature (London)* **342,** 45–50.

Benson, D. L., Isackson, P. J., Gall, C. M., and Jones, E. G. (1991). Differential effects of monocular deprivation on glutamic acid decarboxylase and type II calcium-calmodulin-dependent protein kinase gene expression in the adult monkey visual cortex. *J. Neurosci.* **11,** 31–47.

Berry, F., and Brown, I. R. (1995). Developmental expression of calmodulin mRNA and protein in regions of the postnatal rat brain. *J. Neurosci. Res.* **42,** 613–622.

Blumenthal, D. K., Takio, K., Edelman, A. M., Charbonneau, H., Titani, K., Walsh, K. A., and Krebs, E. G. (1985). Identification of the calmodulin-binding domain of skeletal muscle myosin light chain kinase. *Proc. Natl. Acad. Sci. U.S.A.* **82,** 3187–3191.

Bourguignon, L. Y., and Bourguignon, G. J. (1984). Capping and the cytoskeleton. *Int. Rev. Cytol.* **87,** 195–224.

Brady, M. J., Nairn, A. C., Wagner, J. A., and Palfrey, H. C. (1990). Nerve growth factor-induced down-regulation of calmodulin-dependent protein kinase III in PC12 cells involves cyclic AMP-dependent protein kinase. *J. Neurochem.* **54,** 1034–1039.

Brickey, D. A., Bann, J. G., Fong, Y.-L., Perrino, L., Brennan, R. G., and Soderling, T. R. (1994). Mutational analysis of the autoinhibitory domain of calmodulin kinase II. *J. Biol. Chem.* **269,** 29047–29054.

Bridgman, R. C., and Dailey, M. E. (1989). The organization of myosin and actin in rapid frozen growth cones. *J. Cell Biol.* **108,** 95–109.

Bronstein, J. M., Micevych, P., Popper, P., Huez, G., Farber, D. B., and Wasterlain, C. G. (1992). Long-lasting decreases of type II calmodulin kinase expression in kindled rat brains. *Brain Res.* **584,** 257–260.

Brown, S. E., Martin, S. R., and Bayley, P. M. (1997). Kinetic control of the dissociation pathway of calmodulin–peptide complexes. *J. Biol. Chem.* **272,** 3389–3397.

Brown, T. H., Chapman, P. F., Kairiss, E. W., and Keenan, C. L. (1988). Long-term synaptic potentiation. *Science* **242,** 724–728.

Buechler, Y. J., and Taylor, S. S. (1991). Mutations in the autoinhibitor site of the regulatory subunit of cAMP-dependent protein kinase I. Replacement of Ala-97 and Ser-99 interferes with reassociation with the catalytic subunit. *J. Biol. Chem.* **266,** 3491–3497.

Buechler, Y. J., Herberg, F. W., and Taylor, S. S. (1993). Regulation-defective mutants of type I cAMP-dependent protein kinase. Consequences of replacing arginine 94 and arginine 95. *J. Biol. Chem.* **268,** 16495–16503.

Burwinkel, B., Maichele, A. J., Aagenaes, O., Bakker, H. D., Lerner, A., Shin, Y. S., Strachan, J. A. and Kilimann, M. W. (1997). Autosomal glycogenosis of liver and muscle due to phosphorylase kinase deficiency is caused by mutations in the phosphorylase kinase β subunit (PHKB). *Hum. Mol. Genet.* **6,** 1109–1115.

Butler, L. S., Silva, A. J., Abeliovich, A., Watanabe, Y., Tonegawa, S., and McNamara, J. O. (1995). Limbic epilepsy in transgenic mice carrying a Ca^{2+}/calmodulin-dependent kinase II α-subunit mutation. *Proc. Natl. Acad. Sci. U.S.A.* **92,** 6852–6855.

Calalb, M. B., Fox, D. T., and Hanks, S. K. (1992). Molecular cloning and enzymatic analysis of the rat homolog of "PhK-gammaT", an isoform of phosphorylase kinase catalytic subunit. *J. Biol. Chem.* **267,** 1455–1463.

Casnellie, J. E. (1991). Protein kinase inhibitors: Probes for the functions of protein phosphorylation. *Adv. Pharmacol.* **22,** 167–203.

Cawley, K. C., Akita, C. G., Angelos, K. L., and Walsh, D. A. (1993). Characterization of the gene for rat phosphorylase kinase catalytic subunit. *J. Biol. Chem.* **268,** 1194–1200.

Chabbert, M., Lukas, T. J., Watterson, D. M., Axelsen, P. H., and Prendergast, F. G. (1991). Fluorescence analysis of calmodulin mutants containing tryptophan: Conformational changes induced by calmodulin-binding peptides from myosin light chain kinase and protein kinase II. *Biochemistry* **30,** 7615–7630.

Chamberlain, J. S., VanTuinen, P., Reeves, A. A., Philip, B. A., and Caskey, C. T. (1987). Isolation of cDNA clones for catalytic "gamma" subunit of mouse muscle phosphorylase kinase: Expression of mRNA in normal and mutant Phk mice. *Proc. Natl. Acad. Sci. U.S.A.* **84,** 2886–2890.

Chan, K.-F. J., and Graves, D. J. (1982). Isolation and physiochemical properties of active complexes of rabbit muscle phosphorylase kinase. *J. Biol. Chem.* **257,** 5939–5947.

Chapman, P. F., Frenguelli, B. G., Smith, A., Chen, C.-M., and Silva, A. J. (1995). The α-Ca^{2+}/calmodulin kinase II: A bidirectional modulator of presynaptic plasticity. *Neuron* **14,** 591–597.

Chatila, T., Anderson, K. A., Ho, N., and Means, A. R. (1996). A unique phosphorylation-dependent mechanism for the activation of Ca^{2+}/calmodulin-dependent protein kinase type IV/GR. *J. Biol. Chem.* **271,** 21542–21548.

Chen, C., Feng, Y., Short, J. H., and Wand, A. J. (1993). The main chain dynamics of a peptide bound to calmodulin. *Arch. Biochem. Biophys.* **306,** 510–514.

Chen, C., Rainnie, D. G., Greene, R. W., and Tonegawa, S. (1994). Abnormal fear response and aggressive behavior in mutant rice deficient for α-calcium-calmodulin kinase II. *Science* **266,** 291–294.

Chin, D., and Means, A. R. (1996). Methionine to glutamine substitutions in the C-terminal domain of calmodulin impair the activation of three protein kinases. *J. Biol. Chem.* **271,** 30465–30471.

Chin, D., Sloan, D. J., Quiocho, F. A., and Means, A. R. (1997). Functional consequences of truncating amino acid side chains located at a calmodulin–peptide interface. *J. Biol. Chem.* **272,** 5510–5513.

Cho, F. S., Phillips, K. S., Bogucki, B., and Weaver, T. E. (1994). Characterization of a rat cDNA clone encoding calcium/calmodulin-dependent protein kinase I. *Biochim. Biophys. Acta* **1224,** 156–160.

Cho, K.-O., Wall, J. B., Pugh, P. C., Ito, M., Mueller, S. A., and Kennedy, M. B. (1991). The α subunit of type II Ca^{2+}/calmodulin-dependent protein kinase is highly conserved in *Drosophila*. *Neuron* **7,** 439–450.

Choi, K.-Y., Satterberg, B., Lyons, D. M., and Elion, E. A. (1994). Ste5 tethers multiple protein kinases in the MAP kinase cascade required for mating in *S. cerevisiae*. *Cell* (*Cambridge, Mass.*) **78,** 499–512.

Chrisman, T. D., Jordan, J. E., and Exton, J. H. (1982). Purification of rat liver phosphorylase kinase. *J. Biol. Chem.* **257,** 10798–10804.

Cohen, O., Feinstein, E., and Kimchi, A. (1997). DAP-kinase is a Ca^{2+}/calmodulin-dependent, cytoskeletal-associated protein kinase, with cell death-inducing functions that depend on its catalytic activity. *EMBO J.* **16,** 998–1008.

Cohen, P., and Klee, C. B. (1988). "Calmodulin." Elsevier, Amsterdam.

Cohen, P., Burchell, A., Foulkes, J. G., Cohen, P. T. W., Vanaman, T. C., and Nairn, A. C. (1978). Identification of the Ca^{2+}-dependent modulator protein as the fourth subunit of rabbit skeletal muscle phosphorylase kinase. *FEBS Lett.* **92,** 287–292.

Colbran, R. J., and Soderling, T. R. (1990). Calcium/calmodulin-independent autophosphorylation sites of calcium/calmodulin-dependent protein kinase II. *J. Biol. Chem.* **265,** 11213–11219.

Colbran, R. J., Fong, Y.-L., Schworer, C. M., and Soderling, T. R. (1988). Regulatory interactions of the calmodulin-binding, inhibitory, and autophosphorylation domains of Ca^{2+}/calmodulin-dependent protein kinase II. *J. Biol. Chem.* **263,** 18145–18151.

Colbran, R. J., Smith, M. K., Schworer, C. M., Fong, Y.-L., and Soderling, T. R. (1989). Regulatory domain of calcium/calmodulin-dependent protein kinase II. Mechanism of inhibition and regulation by phosphorylation. *J. Biol. Chem.* **264,** 4800–4804.

Collinge, M., Matrisian, P. E., Zimmer, W. E., Shattuck, R. L., Lukas, T. J., Van Eldik, L. J., and Watterson, D. M. (1992). Structure and expression of a calcium-binding protein gene contained within a calmodulin-regulated protein kinase gene. *Mol. Cell. Biol.* **12,** 2359–2371.

Craig, T. A., Watterson, D. M., Prendergast, F. G., Haiech, J., and Roberts, D. M. (1987). Site-specific mutagenesis of the α-helices of calmodulin. Effects of altering a charge cluster in the helix that links the two halves of calmodulin. *J. Biol. Chem.* **262,** 3278–3284.

Crivici, A., and Ikura, M. (1995). Molecular and structural basis of target recognition by calmodulin. *Annu. Rev. Biophys. Biomol. Struct.* **24,** 85–116.

Cruzalegui, F. H., and Means, A. R. (1993). Biochemical characterization of the multifunctional Ca^{2+}/calmodulin-dependent protein kinase type IV expressed in insect cells. *J. Biol. Chem.* **268,** 26171–26178.

Cruzalegui, F. H., Kapiloff, M. S., Morfin, J., Kemp, B. E., Rosenfeld, M. G., and Means, A. R. (1992). Regulation of intrasteric inhibition of the multifunctional calcium/calmodulin-dependent protein kinase. *Proc. Natl. Acads. Sci. U.S.A.* **89,** 12127–12131.

Dabrowska, R., Sherry, J. M. F., Aromatorio, D. K., and Hartshorne, D. J. (1978). Modulator protein as a component of the myosin light chain kinase from chicken gizzard. *Biochemistry* **17,** 253–258.

Daegelen-Proux, D., Pierres, M., Alexandre, Y., and Dreyfus, J.-C. (1976). Molecular heterogeneity of rabbit heart phosphorylase kinase. *Biochim. Biophys. Acta* **452,** 398–405.

Dale, S., Wilson, W. A., Edelman, A. M., and Hardie, D. G. (1995). Similar substrate recognition motifs for mammalian AMP-activated protein kinase, higher plant HMG-CoA reductase kinase-A, yeast SNF1, and mammalian calmodulin-dependent protein kinase I. *FEBS Lett.* **361,** 191–195.

Dasgupta, M., Honeycutt, T., and Blumenthal, D. K. (1989). The "gamma"-subunit of skeletal muscle phosphorylase kinase contains two noncontiguous domains that act in concert to bind calmodulin. *J. Biol. Chem.* **264,** 17156–17163.

Davidson, J. J., Ozcelik, T., Hamacher, C., Willems, P. J., Francke, U., and Kilimann, M. W. (1992). cDNA cloning for a liver isoform of the phosphorylase kinase α subunit and mapping of the gene to Xp22.2-p22.1, the region of human X-linked liver glycogenolysis. *Proc. Natl. Acad. Sci. U.S.A.* **89,** 2096–2100.

Davies, S. N., Lester, R. A. J., Reymann, K. G., and Collingridge, G. L. (1989). Temporally distinct pre- and post-synaptic mechanisms maintain long-term potentiation. *Nature (London)* **338,** 500–503.

Davis, R. J. (1993). The mitogen-activated protein kinase signal transduction pathway. *J. Biol. Chem.* **268,** 14553–14556.

De Bondt, H. L., Rosenblatt, J., Jancarik, J., Jones, H. D., Morgan, D. O., and Kim, S.-H. (1993). Crystal structure of cyclin-dependent kinase 2. *Nature (London)* **363**, 595–602.

de Lanerolle, P., Adelstein, R. S., Feramisco, J. R., and Burridge, K. (1981). Characterization of antibodies to smooth muscle myosin kinase and their use in localizing myosin kinase in nonmuscle cells. *Proc. Natl. Acad. Sci. U.S.A.* **78**, 4738–4742.

Deiss, L. P., Feinstein, E., Berissi, H., Cohen, O., and Kimchi, A. (1995). Identification of a novel serine/threonine kinase and a novel 15-kD protein as a potential mediators of the gamma-interferon-induced cell death. *Genes Dev.* **9**, 15–30.

Della Rocca, G. J., Van Biesen, T., Daaka, Y., Luttrell, D. K., Luttrell, L. M., and Lefkowitz, R. J. (1987). Ras-dependent mitogen-activated protein kinase activation by G protein-coupled receptors- Convergence of G_i- and G_q-mediated pathways on calcium/calmodulin, Pyk2, and Src kinase. *J. Biol. Chem.* **272**, 19125–19132.

DeRemer, M. F., Saeli, R. J., Brautigan, D. L., and Edelman, A. M. (1992a). Ca^{2+}-calmodulin-dependent protein kinases Ia and Ib from rat brain. II. Enzymatic characteristics and regulation of activities by phosphorylation and dephosphorylation. *J. Biol. Chem.* **267**, 13466–13471.

DeRemer, M. F., Saeli, R. J., and Edelman, A. M. (1992b). Ca^{2+}-calmodulin-dependent protein kinases Ia and Ib from rat brain. I. Identification, purification, and structural comparisons. *J. Biol. Chem.* **267**, 13460–13465.

Edelman, A. M., Takio, K., Blumenthal, D. K., Hansen, R. S., Walsh, K. A., Titani, K., and Krebs, E. G. (1985). Characterization of the calmodulin-binding and catalytic domains in skeletal muscle myosin light chain kinase. *J. Biol. Chem.* **260**, 11275–11285.

Edelman, A. M., Mitchelhill, K. I., Selbert, M. A., Anderson, K. A., Hook, S. S., Stapleton, D., Goldstein, E. G., Means, A. R., and Kemp, B. E. (1996). Multiple Ca^{2+}-calmodulin-dependent protein kinase kinases from rat brain—Purification, regulation by Ca^{2+}-calmodulin, and partial amino acid sequence. *J. Biol. Chem.* **271**, 10806–10810.

Eguchi, S., Matsumoto, T., Motley, E. D., Utsunomiya, H., and Inagami, T. (1996). Identification of an essential signaling cascade for mitogen-activated protein kinase activation by angiotensin II in cultured rat vascular smooth muscle cells—Possible requirement of G_q-mediated $p21^{ras}$ activation coupled to a Ca^{2+}/calmodulin-sensitive tyrosine kinase. *J. Biol. Chem.* **271**, 14169–14175.

Engel, K., Plath, K., and Gaestel, M. (1993). The MAP kinase-activated protein kinase 2 contains a proline-rich SH3-binding domain. *FEBS Lett.* **336**, 143–147.

Enslen, H., Sun, P., Brickey, D., Soderling, S. H., Klamo, E., and Soderling, T. R. (1994). Characterization of Ca^{2+}/calmodulin-dependent protein kinase IV. *J. Biol. Chem.* **269**, 15520–15527.

Enslen, H., Tokumitsu, H., Stork, P. J. S., Davis, R. J., and Soderling, T. R. (1996). Regulation of mitogen-activated protein kinases by a calcium/calmodulin-dependent protein kinase cascade. *Proc. Natl. Acad. Sci. U.S.A.* **93**, 10803–10808.

Erondu, N. E., and Kennedy, M. B. (1985). Regional distribution of type II Ca^{2+}/calmodulin-dependent protein kinase in rat brain. *J. Neurosci.* **5**, 3270–3277.

Farrar, Y. J. K., and Carlson, G. M. (1991). Kinetic characterization of the calmodulin-activated catalytic subunit of phosphorylase kinase. *Biochemistry* **30**, 10274–10279.

Farrar, Y. J., Lukas, T. J., Craig, T. A., Watterson, D. M., and Carlson, G. M. (1993). Features of calmodulin that are important in the activation of the catalytic subunit of phosphorylase kinase. *J. Biol. Chem.* **268**, 4120–4125.

Feinstein, E., Kimchi, A., Wallach, D., Boldin, M., and Varfolomeev, E. (1995). The death domain: A module shared by proteins with diverse cellular functions. *Trends Biochem. Sci.* **20**, 342–344.

Feramisco, J. R., Burridge, K., de Lanerolle, P., and Adelstein, R. S. (1981). Localization of myosin light-chain kinase in nonmuscle cells. *Cold Spring Harbor Conf. Cell Proliferation* **8,** 855–868.

Findlay, W., Martin, S., Beckingham, K., and Bayley, P. (1994). Recovery of native structure by calcium binding site mutants of calmodulin upon binding of sk-MLCK target peptides. *Biochemistry* **34,** 2087–2094.

Fishkind, D., Cao, L. G., and Wang, Y. L. (1991). Microinjection of the catalytic fragment of myosin light chain kinase into dividing cells: Effects on mitosis and cytokinesis. *J. Cell Biol.* **114,** 967–975.

Fitzsimons, D. P., Herring, B. P., Stull, J. T., and Gallagher, P. J. (1992). Identification of basic residues involved in activation and calmodulin binding of rabbit smooth muscle myosin light chain kinase. *J. Biol. Chem.* **267,** 23903–23909.

Foster, C., Van Fleet, M., and Marshak, A. (1986). Tryptic digestion of myosin light chain kinase produces an inactive fragment that is activated on continued digestion. *Arch. Biochem. Biophys.* **251,** 616–623.

Foyt, H. L., Guerriero, V., Jr., and Means, A. R. (1985). Functional domains of chicken gizzard myosin light chain kinase. *J. Biol. Chem.* **260,** 7765–7774.

Francke, U., Darras, B. T., Zander, N. F., and Kilimann, M. W. (1989). Assignment of human genes for phosphorylase kinase subunits α (PHKA) to Xq12-q13 and β (PKKB) to 16q12-q13. *Am. J. Hum. Genet.* **45,** 276–282.

Freedman, N. J., Liggett, S. B., Drachman, D. E., Pei, G., Caron, M. G., and Lefkowitz, R. J. (1995). Phosphorylation and desensitization of the human β_1-adrenergic receptor. *J. Biol. Chem.* **270**(30), 17953–17961.

Fukunaga, K., Stoppini, L., Miyamoto, E., and Muller, D. (1993). Long-term potentiation is associated with an increased activity of Ca^{2+}/calmodulin-dependent protein kinase II. *J. Biol. Chem.* **268,** 7863–7867.

Fukunaga, K., Muller, D., and Miyamoto, E. (1995). Increased phosphorylation of Ca^{2+}/calmodulin-dependent protein kinase II and its endogenous substrates in the induction of long term potentiation. *J. Biol. Chem.* **270,** 6119–6124.

Gao, Z. H., Krebs, J., VanBerkum, M. F. A., Tang, W.-J., Maune, J. F., Means, A. R., Stull, J. T., and Beckingham, K. (1993). Activation of four enzymes by two series of calmodulin mutants with point mutations in individual Ca^{2+} binding sites. *J. Biol. Chem.* **268,** 20096–20104.

George, S. E., VanBerkum, M. F. A., Ono, T., Cook, R., Hanley, R. M., Putkey, J. A., and Means, A. R. (1990). Chimeric calmodulin-cardiac troponin C proteins differentially activate calmodulin target enzymes. *J. Biol. Chem.* **265,** 9228–9235.

George, S. E., Su, Z., Fan, D., and Means, A. R. (1993). Calmodulin-cardiac troponin C chimeras. Effects of domain exchange on calcium binding and enzyme activation. *J. Biol. Chem.* **268,** 25213–25220.

Glass, D. B., Cheng, H.-C., Mende-Mueller, L., Reed, J., and Walsh, D. A. (1989). Primary structural determinants essential for potent inhibition of cAMP-dependent protein kinase by inhibitory peptides corresponding to the active portion of the heat-stable inhibitor protein. *J. Biol. Chem.* **264,** 8802–8810.

Glazewski, S., Chen, C.-M., Silva, A., and Fox, K. (1996). Requirement for α-CaMKII in experience-dependent plasticity of the barrel cortex. *Science* **272,** 421–423.

Goeckler, Z. M., and Wysolmerski, R. B. (1990). Myosin light chain kinase-regulated endothelial cell contraction: The relationship between isomeric tension, actin polymerization, and myosin phosphorylation. *Proc. Natl. Acad. Sci. U.S.A.* **87,** 16–20.

Goldberg, J., Nairn, A. C., and Kuriyan, J. (1996). Structural basis for the autoinhibition of calcium/calmodulin-dependent protein kinase I. *Cell (Cambridge, Mass.)* **84,** 875–887.

Grand, R. J., Shenolikar, S., and Cohen, P. (1981). The amino acid sequence of the δ subunit (calmodulin) of rabbit skeletal muscle phosphorylase kinase. *Eur. J. Biochem.* **113,** 359–367.

Grant, S. G. N., O'Dell, T. J., Karl, K. A., Stein, P. L., Soriano, P., and Kandel, E. R. (1992). Impaired long-term potentiation, spatial learning, and hippocampal development in *fyn* mutant mice. *Science* **258,** 1903–1910.

Griffith, L. C., Verselis, L. M., Aitken, K. M., Kyriacou, C. P., Danho, W., and Greenspan, R. J. (1993). Inhibition of calcium/calmodulin-dependent protein kinase in *Drosophila* disrupts behavioral plasticity. *Neuron* **10,** 501–509.

Griffith, L. C., Wang, J., Zhong, Y., Wu, C.-F., and Greenspan, R. J. (1994). Calcium/calmodulin-dependent protein kinase II and potassium channel subunit *Eag* similarly affect plasticity in *Drosophila. Proc. Natl. Acad. Sci. U.S.A.* **91,** 10044–10048.

GuptaRoy, B., Beckingham, K., and Griffith, L. C. (1996). Functional diversity of alternatively spliced isoforms of *Drosophila* Ca^{2+}/calmodulin-dependent protein kinase II—A role for the variable domain in activation. *J. Biol. Chem.* **271,** 19846–19851.

Hagiwara, M., Inagaki, M., Watanabe, M., Ito, M., Onoda, K., Tanaka, T., and Hidaka, H. (1987). Selective modulation of calcium-dependent myosin phosphorylation by novel protein kinase inhibitors, isoquinolinesulfonamide derivatives. *Mol. Pharmacol.* **32,** 7–12.

Hagiwara, T., Ohsako, S., and Yamauchi, T. (1991). Studies on the regulatory domain of Ca^{2+}/calmodulin-dependent protein kinase II by expression of mutated cDNAs in *Escherichia coli. J. Biol. Chem.* **266,** 16401–16408.

Haiech, J., Kilhoffer, M.-C., Lukas, T. J., Craig, T. A., Roberts, D. M., and Watterson, D. M. (1991). Restoration of the calcium binding activity of mutant calmodulins toward normal by the presence of a calmodulin binding structure. *J. Biol. Chem.* **266,** 3427–3431.

Hait, W. N., Ward, M. D., Trakht, I. N., and Ryazanov, A. G. (1996). Elongation factor-2 kinase: Immunological evidence for the existence of tissue-specific isoforms. *FEBS Lett.* **397,** 55–60.

Hanley, R. M., Means, A. R., Ono, T., Kemp, B. E., Burgin, K. E., Waxham, N., and Kelly, P. T. (1987). Functional analysis of a complementary DNA for the 50-kilodalton subunit of calmodulin kinase II. *Science* **237,** 293–297.

Hanson, P. I., and Schulman, H. (1992a). Neuronal Ca^{2+}/calmodulin-dependent protein kinases. *Annu. Rev. Biochem.* **61,** 559–601.

Hanson, P. I., and Schulman, H. (1992b). Inhibitory autophosphorylation of multifunctional Ca^{2+}/calmodulin-dependent protein kinase analyzed by site-directed mutagenesis. *J. Biol. Chem.* **267,** 17216–17224.

Hanson, P. I., Kapiloff, M. S., Lou, L. L., Rosenfeld, M. G., and Schulman, H. (1989). Expression of a multifunctional Ca^{2+}/calmodulin dependent protein kinase and mutational analysis of its autoregulation. *Neuron* **3,** 59–70.

Haribabu, B., Hook, S. S., Selberg, M. A., Goldstein, E. D., Tomhave, E. D., Edelman, A. M., Snyderman, R., and Means, A. R. (1995). Human calcium-calmodulin dependent protein kinase I: cDNA cloning, domain structure and activation by phosphorylation at threonine-177 by calcium-calmodulin dependent protein kinase I kinase. *EMBO J.* **14,** 3679–3686.

3. Calmodulin-Regulated Protein Kinases

Harmann, B., Zander, N. F., and Kilimann, M. W. (1991). Isoform diversity of phosphorylase kinase α and β subunits generated by alternative RNA splicing. *J. Biol. Chem.* **266,** 15631–15637.

Harris, W. R., Malencik, D. A., Johnson, C. M., Carr, S. A., Roberts, G. D., Byles, C. A., Anderson, S. R., Heilmeyer, L. M. G., Jr., Fischer, E. H., and Crabb, J. W. (1990). Purification and characterization of catalytic fragments of phosphorylase kinase gamma subunit missing a calmodulin-binding domain. *J. Biol. Chem.* **265,** 11740–11745.

Hashimoto, Y., and Soderling, T. R. (1990). Phosphorylation of smooth muscle myosin light chain kinase by Ca^{2+}/calmodulin-dependent protein kinase II: Comparative study of the phosphorylation sites. *Arch. Biochem. Biophys.* **278,** 41–45.

Hashimoto, Y., Sasaki, H., Togo, M., Tsukamoto, K., Horie, Y., Fukata, H., Watanabe, T., and Kurokawa, K. (1994). Roles of myosin light-chain kinase in platelet shape change and aggregation. *Biochim. Biophys. Acta* **1223,** 163–169.

Hawley, S. A., Selbert, M. A., Goldstein, E. G., Edelman, A. M., Carling, D., and Hardie, D. G. (1995). 5'-AMP activates the AMP-activated protein kinase cascade, and Ca^{2+}/calmodulin activates the calmodulin-dependent protein kinase I cascade, via three independent mechanisms. *J. Biol. Chem.* **270**(45), 27186–27191.

Heilmeyer, L. M. G. (1991). Molecular basis of signal intergration in phosphorylase kinase. *Biochim. Biophys. Acta* **1094,** 168–174.

Hendrickx, J., and Willems, P. J. (1996). Genetic deficiencies of the glycogen phosphorylase system. *Hum. Genet.* **97,** 551–556.

Henrion, D., and Laher, I. (1993). Effects of staurosporine and calphostin C, two structurally unrelated inhibitors of protein kinase C, on vascular tone. *Can. J. Physiol. Pharmacol.* **71,** 521–524.

Herring, B. P. (1991). Basic residues are important for Ca^{2+}/calmodulin binding and activation but not autoinhibition of rabbit skeletal muscle myosin light chain kinase. *J. Biol. Chem.* **266,** 11838–11841.

Herring, B. P., and Smith, A. F. (1996). Telokin expression is mediated by a smooth muscle cell-specific promoter. *Am. J. Physiol. Cell Physiol.* **270,** C1656–C1665.

Herring, B. P., Gallagher, P. J., and Stull, J. T. (1992). Substrate specificity of myosin light chain kinases. *J. Biol. Chem.* **267,** 25945–25950.

Hidaka, H., and Okazaki, K. (1996). KN-62: A specific Ca^{2+}/calmodulin-dependent protein kinase inhibitor as a putative function-searching probe for intracellular signal transduction. *Cardiovascular Drug Rev.* **14**(1), 84–95.

Higashihara, M., Takahata, K., and Kurokawa, K. (1991). Effect of phosphorylation of myosin light chain by myosin light chain kinase and protein kinase C on conformational change and ATPase activities of human platelet myosin. *Blood* **78,** 3224–3231.

House, C., and Kemp, B. E. (1987). Protein kinase C contains a pseudosubstrate prototope in its regulatory domain. *Science* **238,** 1726–1728.

Hu, B.-R., and Wieloch, T. (1995). Persistent translocation of Ca^{2+}/calmodulin-dependent protein kinase II to synaptic junctions in the vulnerable hippocampal CA1 region following transient ischemia. *J. Neurochem.* **64,** 277–284.

Huang, C.-Y. F., Yuan, C.-J., Livanova, N. B., and Graves, D. J. (1993). Expression, Purification, characterization, and deletion mutations of phosphorylase kinase gamma subunit: Identification of an inhibitory domain in the gamma subunit. *Mol. Cell. Biochem.* **127–128,** 7–18.

Huang, C.-Y. F., Yuan, C.-J., Luo, S., and Graves, D. J. (1994). Mutational analyses of the metal ion and substrate binding sites of phosphorylase kinase gamma subunit. *Biochemistry* **33,** 5877–5883.

Hunter, T., and Plowman, G. D. (1997). The protein kinases of budding yeast: Six score and more. *Trends Biochem. Sci.* **22**, 18–22.

Hvalby, O., Hemmings, H. C., Jr., Paulsen, O., Czernik, A. J., Nairn, A. C., Godfraind, J.-M., Jensen, V., Raastad, M., Storm, J. F., Andersen, P., and Greengard, P. (1994). Specificity of protein kinase inhibitor peptides and induction of long-term potentiation. *Proc. Natl. Acad. Sci. U.S.A.* **91**, 4761–4765.

Ikebe, M., Stepinska, M., Kemp, B. E., Means, A. R., and Hartshorne, D. J. (1987). Proteolysis of smooth muscle myosin light chain kinase. Formation of inactive and calmodulin-independent fragments. *J. Biol. Chem.* **262**, 13828–13834.

Ikura, M., Clore, G. M., Gronenborn, A. M., Zhu, G., Klee, C. B., and Bax, A. (1992a). Solution structure of a calmodulin-target peptide complex by multidimensional NMR. *Science* **256**, 632–638.

Ikura, M., Barbato, G., Klee, C. B., and Bax, A. (1992b). Solution structure of calmodulin and its complex with a myosin light chain kinase fragment. *Cell Calcium* **13**, 391–400.

Inoue, G., Kuzuya, H., Hayashi, T., Okamoto, M., Yoshimasa, Y., Kosaki, A., Kono, S., Maeda, I., Kubota, M., and Imura, H. (1993). Effects of ML-9 on insulin stimulation of glucose transport in 3T3-L1 adipocytes. *J. Biol. Chem.* **268**, 5272–5278.

Inoue, S., Mizutani, A., Sugita, R., Sugita, K., and Hidaka, H. (1995). Purification and characterization of a novel protein activator of Ca^{2+}/calmodulin-dependent protein kinase I. *Biochem. Biophys. Res. Commun.* **215**, 861–867.

Ishii, A., Kiuchi, K., Kobayashi, R., Sumi, M., Hidaka, H., and Nagatsu, T. (1991). A selective Ca^{2+}/calmodulin-dependent protein kinase II inhibitor, KN-62, inhibits the enhanced phosphorylation and the activation of tyrosine hydroxylase by 56 mM K^+ in rat pheochromocytoma PC12h cells. *Biochem. Biophys. Res. Commun.* **176**, 1051–1056.

Ishikawa, T., and Hidaka, H. (1990). Molecular pharmacology of calcium, calmodulin-dependent myosin phosphorylation in vascular smooth muscle. *Am. J. Hypertens.* **3**, 231–234.

Ishikawa, T., Chijiwa, T., Hagiwara, M., Mamiya, S., Saitoh, M., and Hidaka, H. (1988). ML-9 inhibits the vascular contraction via the inhibition of myosin light chain phosphorylation. *Mol. Pharmacol.* **33**, 598–603.

Isobe, T., Ichimura, T., Sunaya, T., Okuyama, T., Takahashi, N., Kuano, R., and Takahashi, Y. (1991). Distinct forms of the protein kinase-dependent activator of tyrosine and tryptophan hydroxylases. *J. Mol. Biol.* **217**, 125–132.

Ito, I., Hidaka, H., and Sugiyama, H. (1991). Effects of KN-62, a specific inhibitor of calcium/calmodulin-dependent protein kinase II, on long-term potentiation in the rat hippocampus. *Neurosci. Lett.* **121**, 119–121.

Ito, M., Dabrowska, R., Guerriero, V., Jr., and Hartshorne, D. J. (1989). Identification in turkey gizzard of an acidic protein related to the C-terminal portion of smooth muscle myosin light chain kinase. *J. Biol. Chem.* **264**, 13971–13974.

Itoh, T., Ikebe, M., Kargacin, G. J., Hartshorne, D. J., Kemp, B. E., and Fay, F. S. (1989). Effects of modulators of myosin light-chain kinase activity in single smooth muscle cells. *Nature (London)* **338**, 164–167.

James, P., Cohen, P., and Carafoli, E. (1991). Identification and primary structure of calmodulin binding domains in the phosphorylase kinase holoenzyme. *J. Biol. Chem.* **266**, 7087–7091.

Jensen, K. F., Ohmstede, C.-A., Fisher, R. S., and Sahyoun, N. (1991). Nuclear and axonal localization of Ca^{2+}/calmodulin-dependent protein kinase type Gr in rat cerebellar cortex. *Proc. Natl. Acad. Sci. U.S.A.* **88**, 2850–2853.

3. Calmodulin-Regulated Protein Kinases

Jian, X., Hidaka, H., and Schmidt, J. T. (1994). Kinase requirement for retinal growth cone motility. *J. Neurobiol.* **25,** 1310–1328.

Jian, X. Y., Szaro, B. G., and Schmidt, J. T. (1996). Myosin light chain kinase. Expression in neurons and upregulation during axon regeneration. *J. Neurobiol.* **31,** 379–391.

Johnson, J. D., Snyder, C., Walsh, M., and Flynn, M. (1996). Effects of myosin light chain kinase and peptides on Ca^{2+} exchange with the N- and C-terminal Ca^{2+} binding sites of calmodulin. *J. Biol. Chem.* **271,** 761–767.

Johnson, K. A., and Quiocho, F. A. (1996). Protein kinases—Twitching worms catch S100. *Nature (London)* **380,** 585–587.

Jones, T. A., Da Cruz e Silva, E. F., Spurr, N. K., Sheer, D., and Cohen, P. T. W. (1990). Localisation of the gene encoding the catalytic gamma subunit of phosphorylase kinase to human chromosome bands 7p12-q21. *Biochim. Biophys. Acta* **1048,** 24–29.

Kanaseki, T., Ikeuchi, Y., Sugiura, H., and Yamauchi, T. (1991). Structural features of Ca^{2+}/calmodulin-dependent protein kinase II revealed by electron microscopy. *J. Cell Biol.* **115,** 1049–1060.

Kanoh, S., Ito, M., Niwa, E., Kawano, Y., and Hartshorne, D. J. (1993). Actin-binding peptide from smooth muscle myosin light chain kinase. *Biochemistry* **32,** 8902–8907.

Karaki, H., Ozaki, H., Hori, M., Mitsui-Saito, M., Amano, K., Harada, K., Miyamoto, S., Nakazawa, H., Won, K., and Sato, K. (1997). Calcium movements, distribution and functions in smooth muscle. *Pharmacol. Rev.* **49,** 157–229.

Karess, R. E., Chang, X., Edwards, K. A., Kulkarni, S., Aguilera, I., and Kiehart, D. P. (1991). The regulatory light chain of nonmuscle myosin is encoded by *Spaghetti-squash*, a gene required for cytokinesis in *Drosophila*. *Cell (Cambridge, Mass.)* **65,** 1177–1189.

Kargacin, G. J., Ikebe, M., and Fay, F. S. (1990). Peptide modulators of myosin light chain kinase affect smooth muscle cell contraction. *Am. J. Physiol.* **259,** 315–324.

Karls, U., Muller, U., Gilbert, D. J., Copeland, N. G., Jenkins, N. A., and Harbers, K. (1992). Structure, expression and chromosome location of the gene for the β subunit of brain-specific Ca^{2+}/calmodulin-dependent protein kinase II identified by transgene integration in an embryonic lethal mouse mutant. *Mol. Cell. Biol.* **12,** 3644–3652.

Kelly, P. T., Weinberger, R. P., and Waxham, M. N. (1988). Active site-directed inhibition of Ca^{2+}/calmodulin-dependent protein kinase type II by a bifunctional calmodulin-binding peptide. *Proc. Natl. Acad. Sci. U.S.A.* **85,** 4991–4995.

Kemp, B. E., Pearson, R. B., Guerriero, V., Bagchi, I. C., and Means, A. R. (1987). The calmodulin binding domain of chicken smooth muscle myosin light chain kinase contains a pseudosubstrate sequence. *J. Biol. Chem.* **262,** 2542–2548.

Kerrick, W. G. L., and Bourguignon, L. Y. W. (1984). Regulation of receptor capping in mouse lymphoma T cells by Ca^{2+}-activated myosin light chain kinase. *Proc. Natl. Acad. Sci. U.S.A.* **81,** 165–169.

Kilhoffer, M.-C., Lukas, T. J., Watterson, D. M., and Haiech, J. (1992). The heterodimer calmodulin: Myosin light-chain kinase as a prototype vertebrate calcium signal transduction complex. *Biochim. Biophys. Acta Protein Struct. Mol. Enzymol.* **1160,** 8–15.

Kilimann, M. W., Zander, N. F., Kuhn, C. C., Crabb, J. W., Meyer, H. E., and Heilmeyer, L. M. G., Jr. (1988). The α and β subunits of phosphorylase kinase are homologous: cDNA cloning and primary structure of the β subunit. *Proc. Natl. Acad. Sci. U.S.A.* **85,** 9381–9385.

Kin, P.-J., Luby-Phelps, K., and Stull, J. T. (1997). Binding of myosin light chain kinase to cellular actin-myosin filaments. *J. Biol. Chem.* **272,** 7412–7420.

King, M. M., Shell, D. J., and Kwiatkowski, A. P. (1988). Affinity labeling of the ATP-binding site of type II calmodulin-dependent protein kinase by 5'-*p*-fluorosulfonylbenzoyl adenosine. *Arch. Biochem. Biophys.* **267,** 467–473.

Kitani, T., Okuno, S., and Fujisawa, H. (1997). Molecular cloning of Ca^{2+}/calmodulin-dependent protein kinase kinase β. *J. Biochem. (Tokyo)* **122,** 243–250.

Klann, E., Chen, S.-J., and Sweatt, J. D. (1991). Persistent protein kinase activation in the maintenance phase of long-term potentiation. *J. Biol. Chem.* **266,** 24253–24256.

Klee, C. B., and Vanaman, T. C. (1982). Calmodulin. *Adv. Protein Chem.* **35,** 213–321.

Klevit, R. E., Blumenthal, D. K., Wemmer, D. E., and Krebs, E. G. (1985). Interaction of calmodulin and a calmodulin-binding peptide from myosin light chain kinase: Major spectral changes in both occur as the result of complex formation. *Biochemistry* **24,** 8152–8157.

Kojima, S., Mishima, M., Mabuchi, I., and Hotta, Y. (1996). A single *Drosophila melanogaster* myosin light chain kinase gene products multiple isoforms whose activities are differently regulated. *Genes Cells* **1,** 855–871.

Kolb, S. J., Hudmon, A., and Waxham, M. N. (1995). Ca^{2+}/calmodulin kinase II translocates in a hippocampal slice model of ischemia. *J. Neurochem.* **64,** 2147–2152.

Kolodney, M. S., and Elson, E. L. (1993). Correlation of myosin light chain phosphorylation with isometric contraction of fibroblasts. *J. Biol. Chem.* **268,** 23850–23855.

Kolodney, M. S., and Wysolmerski, R. B. (1992). Isometric contraction by fibroblasts and endothelial cells in tissue culture: A quantitative study. *J. Cell Biol.* **117,** 73–82.

Komatsu, H., and Ikebe, M. (1993). Affinity labelling of smooth-muscle myosin light-chain kinase with 5'-[*p*-(fluorosulphonyl)benzoyl]adenosine. *Biochem. J.* **296,** 53–58.

Kowacsovics, T. J., and Hartwig, J. H. (1996). Thombin-induced GpIb-IX centralization on the platelet surface requires actin assembly and myosin II activation. *Blood* **87,** 618–629.

Krebs, E. G. (1993). Protein phosphorylation and cellular regulation I. *Biosci. Rep.* **13,** 127–142.

Kuboniwa, H., Tjandra, N., Grzesiek, S., Ren, H., Klee, C. B., and Bax, A. (1995). Solution structure of calcium-free calmodulin. *Nat. Struct. Biol.* **2,** 768–776.

Kwiatkowski, A. P., and King, M. M. (1989). Autophosphorylation of the type II calmodulin-dependent protein kinase is essential for formation of a proteolytic fragment with catalytic activity. Implications for long-term synaptic potentiation. *Biochemistry* **28,** 5380–5385.

Kwiatkowski, A. P., and McGill, J. M. (1995). Human biliary epithelial cell line Mz-ChA-1 expresses new isoforms of calmodulin-dependent protein kinase II. *Gastroenterology* **109,** 1316–1323.

Labeit, S., Gautel, M., Lakey, A., and Trinick, J. (1992). Towards a molecular understanding of titin. *EMBO J.* **11,** 1711–1716.

Lai, Y., Nairn, A. C., and Greengard, P. (1986). Autophosphorylation reversibly regulates the Ca^{2+}/calmodulin- dependence of Ca^{2+}/calmodulin-dependent protein kinase II. *Proc. Natl. Acad. Sci. U.S.A.* **83,** 4253–4257.

Lanciotti, R. A., and Bender, P. K. (1994). Baculovirus-directed expression of the gamma-subunit of phosphorylase kinase: Purification and calmodulin dependence. *Biochem. J.* **299,** 183–189.

Leachman, S. A., Gallagher, P. J., Herring, B. P., McPhaul, M. J., and Stull, J. T. (1992). Biochemical properties of chimeric skeletal and smooth muscle myosin light chain kinases. *J. Biol. Chem.* **267,** 4930–4938.

Lee, J. C., and Edelman, A. M. (1994). A protein activator of Ca^{2+}/calmodulin-dependent protein kinase Ia. *J. Biol. Chem.* **269,** 2158–2164.

Lee, J. C., and Edelman, A. M. (1995). Activation of Ca^{2+}-calmodulin-dependent protein kinase Ia is due to direct phosphorylation by its activator. *Biochem. Biophys. Res. Commun.* **210,** 631–637.

Lee, J. C., Kwon, Y., Lawrence, D. S., and Edelman, A. M. (1994). A requirement of hydrophobic and basic amino acid residues for substrate recognition by Ca^{2+}/calmodulin-dependent protein kinase Ia. *Proc. Natl. Acad. Sci. U.S.A.* **91,** 6413–6417.

Levine, H., and Sahyoun, N. E. (1987). Characterization of a soluble Mr = 30000 catalytic fragment of the neuronal calmodulin-dependent protein kinase II. *Eur. J. Biochem.* **168,** 481–486.

Lin, C. R., Kapiloff, M. S., Durgerian, S., Tatemoto, K., Russo, A. F., Hanson, P., Schulman, H., and Rosenfeld, M. G. (1987). Molecular cloning of a brain-specific calcium/calmodulin-dependent protein kinase. *Proc. Natl. Acad. Sci. U.S.A.* **84,** 5962–5966.

Lisman, J. (1994). The CaM kinase II hypothesis for the storage of synaptic memory. *Trends Neurosci.* **17,** 406–412.

Lisman, J. E., and Goldring, M. A. (1988). Feasibility of long-term storage of graded information by the Ca^{2+}/calmodulin-dependent protein kinase molecules of the postsynaptic density. *Proc. Natl. Acad. Sci. U.S.A.* **85,** 5320–5324.

Liu, M. Y., Yu, B., Nabanishi, O., Wieland, T., and Simon, M. (1997). The Ca^{2+}-dependent binding of calmodulin to an N-terminal motif of the heterotrimeric G protein β subunit. *J. Biol. Chem.* **272,** 18801–18807.

Lokeshwar, V. B., and Bourguignon, L. Y. W. (1992). The involvement of Ca^{2+} and myosin light chain kinase in collagen-induced platelet activation. *Cell Biol. Int. Rep.* **16,** 883–897.

Lou, L. L., and Schulman, H. (1989). Distinct autophosphorylation sites sequentially produce autonomy and inhibition of the multifunctional Ca^{2+}/calmodulin-dependent protein kinase. *J. Neurosci.* **9,** 2020–2032.

Lou, L. L., Lloyd, S. J., and Schulman, H. (1986). Activation of the multifunctional Ca^{2+}/calmodulin-dependent protein kinase by autophosphorylation: ATP modulates production of an autonomous enzyme. *Proc. Natl. Acad. Sci. U.S.A.* **83,** 9497–9501.

Lukas, T. J., Burgess, W. H., Prendergast, F. G., Lau, W., and Watterson, D. M. (1986). Calmodulin binding domains: Characterization of a phosphorylation and calmodulin binding site from myosin light chain kinase. *Biochemistry* **25,** 1458–1464.

Lukas, T. J., Haiech, J., Lau, W., Craig, T. A., Zimmer, W. E., Schattuck, R. L., Shoemaker, M. O., and Watterson, D. M. (1988). Calmodulin and calmodulin-regulated protein kinases as transducers of intracellular calcium signals. *Cold Spring Harbor Symp. Quant. Biol.* **53,** 185–193.

Lukas, T. J., Collinge, M., Haiech, J., and Watterson, D. M. (1994). Gain of function mutations for yeast calmodulin and calcium dependent regulation of protein kinase activity. *Biochim. Biophys. Acta* **1223,** 341–347.

McDonald, O. B., Merrill, B. M., Bland, M. M., Taylor, L. C. E., and Sahyoun, N. (1993). Site and consequences of the autophosphorylation of Ca^{2+}/calmodulin- dependent protein kinase type "Gr". *J. Biol. Chem.* **268,** 10054–10059.

McGlade-McCulloh, E., Yamamoto, H., Tan, S.-E., Brickey, D. A., and Soderling, T. R. (1993). Phosphorylation and regulation of glutamate receptors by calcium/calmodulin-dependent protein kinase. II. *Nature (London)* **362,** 640–642.

Maichele, A. J., Burwinkel, B., Maire, I., Sovik, O., and Kilimann, M. W. (1996). Mutations in the testis/liver isoform of the phosphorylase kinase gamma subunit (PHKG2) causes autosomal liver glycogenosis in the GSD rat and in humans. *Nat. Genet.* **14,** 337–340.

Majercik, M. H., and Bourguignon, L. Y. W. (1985). Insulin receptor capping and its correlation with calmodulin-dependent myosin light chain kinase. *J. Cell. Physiol.* **124,** 403–410.

Majercik, M. H., and Bourguignon, L. Y. W. (1988). Insulin-induced myosin light-chain phosphorylation during receptor capping in IM-9 human B-lymphoblasts. *Biochem. J.* **252,** 815–823.

Makishima, M., Honma, Y., Hozumi, M., Sampi, K., Hattori, M., and Motoyoshi, K. (1991). Induction of differentiation of human leukemia cells by inhibitors of myosin light chain kinase. *FEBS Lett.* **287,** 175–177.

Malinow, R., Schulman, H., and Tsien, R. W. (1989). Inhibition of postsynaptic PKC or CaMKII blocks induction but not expression of LTP. *Science* **245,** 862–866.

Mamiya, N., Goldenring, J. R., Tsunoda, Y., Modlin, I. M., Yasui, K., Usuda, N., Ishikawa, T., Natsume, A., and Hidaka, H. (1993). Inhibition of acid secretion in gastric parietal cells by the Ca^{2+}/calmodulin-dependent protein kinase II inhibitor KN-93. *Biochem. Biophys. Res. Commun.* **195,** 608–615.

Martin, J. R., and Ollo, R. (1996). A new *Drosophila* Ca^{2+} calmodulin-dependent protein kinase (Caki) is localized in the central nervous system and implicated in walking speed. *EMBO J.* **15,** 1865–1876.

Matsuura, I., Kimura, E., Tai, K., and Yazawa, M. (1993). Mutagenesis of the fourth calcium-binding domain of yeast calmodulin. *J. Biol. Chem.* **268,** 13267–13273.

Matthews, R. P., Guthrie, C. R., Wailes. L. M., Zhao, X., Means, A. R., and McKnight, G. S. (1994). Calcium/calmodulin-dependent protein kinase types II and IV differentially regulate CREB-dependent gene expression. *Mol. Cell. Biol.* **14,** 6107–6116.

Maune, J. F., Beckingham, K., Martin, S. R., and Bayley, P. M. (1992). Circular dichroism studies on calcium binding to two series of Ca^{2+} binding site mutants of *Drosophila melanogaster* calmodulin. *Biochemistry* **31,** 7779–7786.

Mayer, B. J., and Baltimore, D. (1993). Signalling through SH2 and SH3 domains *Trends Cell Biol.* **3,** 8–13.

Mayford, M., Wang, J., Kandel, E. R., and O'Dell, T. J. (1995). CaMKII regulates the frequency-response function of hippocampal synapses for the production of both LTD and LTP. *Cell (Cambridge, Mass.)* **81,** 891–904.

Mayford, M., Bach, M. E., Huang, Y.-Y., Wang, L., Hawkins, R. D., and Kandel, E. R. (1996). Control of memory formation through regulated expression of a CaMKII transgene. *Science* **274,** 1678–1683.

Mayford, M., Mansuy, I. M., Muller, R. V., and Kandel, E. R. (1997). Memory and behavior: A second generation of genetically modified mice. *Curr. Biol.* **7,** R580–R589.

Meador, W. E., Means, A. R., and Quiocho, F. A. (1992). Target enzyme recognition by calmodulin: 2.4 Å Structure of a calmodulin–peptide complex. *Science* **257,** 1251–1255.

Meador, W. E., Means, A. R., and Quiocho, F. A. (1993). Modulation of calmodulin plasticity in molecular recognition on the basis of X-ray structures. *Science* **262,** 1718–1721.

Means, A. R., Cruzalegui, F., LeMagueresse, B., Needleman, D. S., Slaughter, G. R., and Ono, T. (1991). A novel Ca^{2+}/calmodulin-dependent protein kinase and a male germ cell-specific calmodulin-binding protein are derived from the same gene. *Mol. Cell. Biol.* **11,** 3960–3971.

Meyer, T., Hanson, P. I., Stryer, L., and Schulman, H. (1992). Calmodulin trapping by calcium-calmodulin-dependent protein kinase. *Science* **256,** 1199–1202.

Miller, S. G., and Kennedy, M. B. (1986). Regulation of brain type II Ca^{2+}/calmodulin-dependent protein kinase by autophosphorylation: A Ca^{2+}-triggered molecular switch. *Cell (Cambridge, Mass.)* **44,** 861–870.

Mirzoeva, S., Lukas, T. J., Weigand, S., Anderson, W. F., and Watterson, D. M. (1997). Alteration of the coupling between calcium binding and peptide recognition properties in E84K-calmodulin, submitted.

Mitchell, R. D., Glass, D. B., Wong, C.-W., Angelos, K. L., and Walsh, D. A. (1995). Heat-stable inhibitor protein derived peptide substrate analogs: Phosphorylation by cAMP-dependent and cGMP-dependent protein kinases. *Biochemistry* **34,** 528–534.

Mitsui, K., Brady, M., Palfrey, H. C., and Nairn, A. C. (1993). Purification and characterization of calmodulin-dependent protein kinase III from rabbit reticulocytes and rat pancreas. *J. Biol. Chem.* **268,** 13422–13433.

Miyamoto, E., and Fukunaga, K. (1996). A role of Ca^{2+}/calmodulin-dependent protein kinase II in the induction of long-term potentiation in hippocampal CA1 area. *Neurosci. Res.* **24,** 117–122.

Miyano, O., Kameshita, I., and Fujisawa, H. (1992). Purification and characterization of a brain-specific multifunctional calmodulin-dependent protein kinase from rat cerebellum. *J. Biol. Chem.* **267,** 1198–1203.

Mochizuki, H., Sugita, R., Ito, T., and Hidaka, H. (1993a). Phosphorylation of Ca^{2+}/calmodulin-dependent protein kinase V and regulation of its activity. *Biochem. Biophys. Res. Commun.* **197,** 1595–1600.

Mochizuki, H., Ito, T., and Hidaka, H. (1993b). Purification and characterization of Ca^{2+}/calmodulin-dependent protein kinase V from rat cerebrum. *J. Biol. Chem.* **268,** 9143–9147.

Montigiani, S., Neri, G., Neri, P., and Neri, D. (1996). Alanine substitutions in calmodulin-binding peptides results in unexpected affinity enhancement. *J. Mol. Biol.* **258,** 6–13.

Moreland, S., Ikebe, M., Hunt, J. T., and Moreland, R. S. (1992). Peptide analogs of the pseudosubstrate domain of smooth muscle myosin light chain kinase inhibit actomyosin ATPase activity at concentrations that do not inhibit superprecipitation. *Biochem. Biophys. Res. Commun.* **187,** 1279–1284.

Mukherji, S., and Soderling, T. R. (1994). Regulation of Ca^{2+}/calmodulin-dependent protein kinase II by inter- and intrasubunit-catalyzed autophosphorylations. *J. Biol. Chem.* **269,** 13744–13747.

Mukherji, S., and Soderling, T. R. (1995). Mutational analysis of Ca^{2+}-independent autophosphorylation of calcium/calmodulin-dependent protein kinase II. *J. Biol. Chem.* **270,** 14062–14067.

Murray, K. D., Gall, C. M., Benson, D. L., Jones, E. G., and Isackson, P. J. (1995). Decreased expression of the alpha subunit of Ca^{2+}/calmodulin-dependent protein kinase type II mRNA in the adult rat CNS following recurrent limbic seizures. *Mol. Brain Res.* **32,** 221–232.

Nairn, A. C., and Greengard, P. (1987). Purification and characterization of Ca^{2+}/calmodulin-dependent protein kinase I from bovine brain. *J. Biol. Chem.* **262,** 7273–7281.

Nairn, A. C., and Palfrey, H. C. (1987). Identification of the major Mr 100,000 substrate for calmodulin-dependent protein kinase III in mammalian cells as elongation factor-2. *J. Biol. Chem.* **262,** 17299–17303.

Nairn, A. C., Bhagat, B., and Palfrey, H. C. (1985). Identification of calmodulin-dependent protein kinase III and its major Mr 100,000 substrate in mammalian tissue. *Proc. Natl. Acad. Sci. U.S.A.* **82,** 7939–7943.

Nakanishi, A., Yoshizumi, M., Hamano, S., Morita, K., and Oka, M. (1989). Myosin light-chain kinase inhibitor, 1-(5-chlornaphthalene-1-sulfonyl)-1H-hexahydro-1,4-diazepine (ML-9), inhibits catecholamine secretion from adrenal chromaffin cells by inhibiting Ca^{2+} uptake into the cells. *Biochem. Pharmacol.* **38,** 2615–2619.

Nakanishi, S., Yamada, K., Iwahashi, K., Kuroda, K., and Kase, H. (1990). KT5926, a potent and selective inhibitor of myosin light chain kinase. *Mol. Pharmacol.* **37,** 482–488.

Nakashima, K., Maekawa, H., and Yazawa, M. (1996). Chimeras of yeast and chicken calmodulin demonstrate differences in activation mechanisms of target enzymes. *Biochemistry* **35,** 5602–5610.

Nghiem, P., Saati, S. M., Martens, C. L., Gardner, P., and Schulman, H. (1993). Cloning and anaylsis of two new isoforms of multifunctional Ca^{2+}/calmodulin-dependent protein kinase. Expression in multiple human tissues. *J. Biol. Chem.* **268,** 5471–5479.

Niki, I., Okazaki, K., Saitoh, M., Niki, A., Niki, H., Tamagawa, T., Iguchi, A., and Hidaka, H. (1993). Presence and possible involvement of Ca/calmodulin-dependent protein kinases in insulin release from the rat pancreatic β cell. *Biochem. Biophys. Res. Commun.* **191,** 255–261.

Nilsson, A., Carlberg, U., and Nygård, O. (1991). Kinetic characterisation of the enzymatic activity of the eEF-2-specific Ca^{2+}- and calmodulin-dependent proteinkinase III purified from rabbit reticulocytes. *Eur. J. Biochem.* **195,** 377–383.

Nishida, E., and Gotoh, Y. (1993). The MAP kinase cascade is essential for diverse signal transduction pathways. *Trends Biochem. Sci.* **18,** 128–131.

Nishioka, N., Shiojiri, M., Kadota, S., Morinaga, H., Kuwahara, J., Arakawa, T., Yamamoto, S., and Yamauchi, T. (1996). Gene of rat Ca^{2+}/calmodulin-dependent protein kinase II α isoform—its cloning and whole structure. *FEBS Lett.* **396,** 333–336.

O'Neil, K. T., Wolfe, H. R., Jr., Erickson-Viitanen, S., and deGrado, W. F. (1987). Fluorescence properties of calmodulin-binding peptides reflect alpha-helical periodicity. *Science* **236,** 1454–1456.

O'Neil, K. T., Erickson-Viitanen, S., and deGrado, W. F. (1989). Photolabeling of calmodulin with basic, amphiphilic "alpha"-helical peptides containing p-benzoylphenylalanine. *J. Biol. Chem.* **264,** 14571–14578.

Ohmstede, C. A., Jensen, K. F., and Sahyoun, N. E. (1989). Ca^{2+}/calmodulin-dependent protein kinase enriched in cerebellar granule cells. *J. Biol. Chem.* **264,** 5866–5875.

Ohsako, S., Nishida, Y., Ryo, H., and Yamauchi, T. (1993). Molecular characterization and expression of the *Drosophila* Ca^{2+}/calmodulin-dependent protein kinase II gene. *J. Biol. Chem.* **268,** 2052–2062.

Okuno, S., and Fujisawa, H. (1993). Requirement of brain extract for the activity of brain calmodulin-dependent protein kinase IV expressed in *Escherichia coli*. *J. Biochem. (Tokyo)* **114,** 167–170.

Olson, N. J., Pearson, R. B., Needleman, D. S., Hurwitz, M. Y., Kemp, B. E., and Means, A. R. (1990). Regulatory and structural motifs of chicken gizzard myosin light chain kinase. *Proc. Natl. Acad. Sci. U.S.A.* **87,** 2284–2288.

Olson, N. J., Masse, T., Suzuki, T., Chen, J., Alam, D., and Kelly, P. T. (1995). Functional identification of the promoter for the gene encoding the α-subunit of calcium/calmodulin-dependent protein kinase II. *Proc. Natl. Acad. Sci. U.S.A.* **92,** 1659–1663.

3. Calmodulin-Regulated Protein Kinases 161

Olwin, B. B., Edelman, A. M., Krebs, E. G., and Storm, D. R. (1984). Quantitation of energy coupling between Ca^{2+}, calmodulin, skeletal muscle myosin light chain kinase, and kinase substrates. *J. Biol. Chem.* **259**, 10949–10955.

Omkumar, R. V., Kiely, M. J., Rosenstein, A. J., Min, K.-T., and Kennedy, M. B. (1996). Identification of a phosphorylation site for calcium/calmodulin-dependent protein kinase II in the NR2B subunit of the N-methyl-D-aspartate receptor. *J. Biol. Chem.* **271**, 31670–31678.

Otmakhov, N., Griffith, L. C., and Lisman, J. E. (1997). Postsynaptic inhibitors of calcium/calmodulin-dependent protein kinase type II block induction but not maintanance of pairing-induced long-term potentiation. *J. Neurosci.* **17**, 5357–5365.

Owen, D. J., Noble, M. E., Garman, E. F., Papageorgiou, A. C., and Johnson, L. N. (1995). Two structures of the catalytic domain of phosphorylase kinase: An active protein kinase complexed with substrate analogue and product. *Structure* **3**, 467–482.

Park, I. K., and Soderling, T. R. (1995). Activation of Ca^{2+}/calmodulin-dependent kinase (CaM-kinase) IV by CaM-kinase kinase in Jurkat T lymphocytes. *J. Biol. Chem.* **270**, 30464–30469.

Pasternak, C., Spudich, J. A., and Elson, E. L. (1989). Capping of surface receptors and concomitant cortical tension are generated by conventional myosin. *Nature (London)* **341**, 549–551.

Patton, B. L., Miller, S. G., and Kennedy, M. B. (1990). Activation of type II calcium/calmodulin-dependent protein kinase by Ca^{2+}/calmodulin is inhibited by autophosphorylation of threonine within the calmodulin-binding domain. *J. Biol. Chem.* **265**, 11204–11212.

Paudel, H. K., and Carlson, G. M. (1987). Inhibition of the catalytic subunit of phosphorylase kinase by its α/β subunits. *J. Biol. Chem.* **262**, 11912–11915.

Payne, M. E., Fong, Y.-L., Ono, T., Colbran, R. J., Kemp, B. E., Soderling, T. R., and Means, A. R. (1988). Calcium/calmodulin-dependent protein kinase II. *J. Biol. Chem.* **263**, 7190–7195.

Pears, C. J., Kour, G., House, C., Kemp, B. E., and Parker, P. J. (1990). Mutagenesis of the pseudosubstrate site of protein kinase C leads to activation. *Eur. J. Biochem.* **194**, 89–94.

Peersen, O. B., Madsen, T. S., and Falke, J. J. (1997). Intermolecular tuning of calmodulin by target peptides and proteins: Differential effect on Ca^{2+} binding and implications for kinase activation. *Protein Sci.* **6**, 794–807.

Persechini, A., White, H. D., and Gansz, K. J. (1996). Different mechanisms for Ca^{2+} dissociation from complexes of calmodulin with nitric oxide synthase or myosin light chain kinase. *J. Biol. Chem.* **271**, 62–67.

Pettit, D. L., Perlman, S., and Malinow, R. (1994). Potentiated transmission and prevention of further LTP by increased CaMKII activity in postsynaptic hippocampal slice neurons. *Science* **266**, 1881–1885.

Picciotto, M. R., Cohn, J. A., Bertuzzi, G., Greengard, P., and Nairn, A. C. (1992). Phosphorylation of the cystic fibrosis transmembrane conductance regulator. *J. Biol. Chem.* **267**, 12742–12752.

Picciotto, M. R., Czernik, A. J., and Nairn, A. C. (1993). Calcium/calmodulin-dependent protein kinase I. cDNA cloning and identification of autophosphorylation site. *J. Biol. Chem.* **268**, 26512–26521.

Picciotto, M. R., Zoli, M., Bertuzzi, G., and Nairn, A. C. (1995). Immunochemical localization of calcium/calmodulin-dependent protein kinase I. *Synapse* **20**, 75–84.

Pinna, L. A., and Ruzzene, M. (1996). How do protein kinases recognize their substrates? *Biochim. Biophys. Acta Mol. Cell Res.* **1314,** 191–225.

Poteet-Smith, C. E., Shabb, J. B., Francis, S. H., and Corbin, J. D. (1997). Identification of critical determinants for autoinhibition in the pseudosubstrate region of type I-alpha cAMP-dependent protein kinase. *J. Biol. Chem.* **272,** 379–388.

Potier, M. C., Chelot, E., Pekarsky, Y., Gardiner, K., Rossier, J., and Turnell, W. G. (1995). The human myosin light chain kinase (MLCK) from hippocampus: Cloning, sequencing, expression, and localization to 3qcen-q21. *Genomics* **29,** 562–570.

Pronin, A. N., Satpaev, D. K., Slepak, V. Z., and Benovic, J. L. (1997). Regulation of G protein-coupled receptor kinases by calmodulin and localization of the calmodulin binding domain. *J. Biol. Chem.* **272,** 18273–18280.

Redpath, N. T., and Proud, C. G. (1993). Cyclic AMP-dependent protein kinase phosphorylates rabbit reticulocyte elongation factor-2 kinase and induces calcium-independent activity. *Biochem. J.* **293,** 31–34.

Redpath, N. T., and Proud, C. G. (1994). Molecular mechanisms in the control of translation by hormones and growth factors. *Biochim. Biophys. Acta* **1220,** 147–162.

Redpath, N. T., Price, N. T., Severinov, K. V., and Proud, C. G. (1993). Regulation of elongation factor-2 by multisite phosphorylation. *Eur. J. Biochem.* **213,** 689–699.

Redpath, N. T., Price, N. T., and Proud, C. G. (1996). Cloning and expression of cDNA encoding protein synthesis elongation factor-2 kinase. *J. Biol. Chem.* **271,** 17547–17554.

Reimann, E. M., Titani, K., Ericsson, L. H., Wade, R. D., Fischer, E. H., and Walsh, K. A. (1984). Homology of the "gamma" subunit of phosphorylase b kinase with cAMP-dependent protein kinase. *Biochemistry* **23,** 4185–4192.

Ren, R., Mayer, B. J., Cicchetti, P., and Baltimore, D. (1993). Identification of a ten-amino acid proline-rich SH3 binding site. *Science* **259,** 1157–1161.

Rhodes, N., Connell, L., and Errede, B. (1990). STE11 is a protein kinase required for cell-type-specific transcription and signal transduction in yeast. *Genes Dev.* **4,** 1862–1874.

Richardson, A., and Parsons, J. T. (1996). A mechanism for regulation of the adhesion-associated protein tyrosine kinase pp125[FAK]. *Nature (London)* **380,** 538–540.

Rosenblum, K., Dudai, Y., and Richter-Levin, G. (1996). Long-term potentiation increases tyrosine phosphorylation of the N-methyl-D-aspartate receptor subunit 2B in rat dentate gyrus *in vivo. Proc. Natl. Acad. Sci. U.S.A.* **93,** 10457–10460.

Rostas, J. A. P., Brent, V. A., Voss, K., Errington, M. L., Bliss, T. V. P., and Gurd, J. W. (1996). Enhanced tyrosine phosphorylation of the 2B subunit of the N-methyl-D-aspartate receptor in long-term potentiation. *Proc. Natl. Acad. Sci. U.S.A.* **93,** 10452–10456.

Roth, S. M., Schneider, D. M., Strobel, L. A., VanBerkum, M. F. A., Means, A. R., and Wand, A. J. (1991). Structure of the smooth muscle myosin light-chain kinase calmodulin-binding domain peptide bound to calmodulin. *Biochemistry* **30,** 10078–10084.

Ryazanov, A. G., Ward, M. D., Mendola, C. E., Pavur, K. S., Dorovkov, M. V., Wiedman, M., Erdjument-Bromage, H., Tempst, P., Gestone Parmer, T., Prostko, C. R., Germino, F. J., and Hait, W. N. (1997). Identification of a new class of protein kinases represented by eukaryotic elongation factor-2 kinase. *Proc. Natl. Acad. Sci. U.S.A.* **94,** 4884–4889.

Saitoh, T., and Schwartz, J. H. (1985). Phosphorylation-dependent subcellular translocation of a Ca^{2+}/calmodulin-dependent protein kinase produces an autonomous enzyme in Aplysia neurons. *J. Cell Biol.* **100,** 835–842.

3. Calmodulin-Regulated Protein Kinases

Sakagami, H., and Kondo, H. (1993). Cloning and sequencing of a gene encoding the β polypeptide of Ca^{2+}/calmodulin-dependent protein kinase IV and its expression confined to the mature cerebellar granule cells. *Mol. Brain Res.* **19**, 215–218.

Sakagami, H., and Kondo, H. (1996). Immunohistochemical localization of Ca^{2+}/calmodulin-dependent protein kinase type IV in the mature and developing rat retina. *Brain Res.* **719**, 154–160.

Sakai, K., Matsumura, S., Okimura, Y., Yamamura, H., and Nishizuka, Y. (1979). Liver glycogen phosphorylase kinase. Partial purification and characterization. *J. Biol. Chem.* **254**, 6631–6637.

Satoh, M., Kobuku, N., Matsuo, K., and Takayanagi, I. (1995). Alpha 1A-adrenoreceptor subtype effectively increases Ca^{2+}-sensitivity for contraction in rabbit thoracic aorta. *Gen. Pharmacol.* **26**, 357–362.

Schaller, M. D., Borgman, C. A., and Parsons, J. T. (1993). Autonomous expression of a noncatalytic domain of the focal adhesion-associated protein tyrosine kinase pp125^{FAK}. *Mol. Cell Biol.* **13**, 785–791.

Schneider, A., Davidson, J. J., Wüllrich, A., and Kilimann, M. W. (1993). Phosphorylase kinase deficiency in I-strain mice is associated with a frameshift mutation in the α subunit muscle isoform. *Nat. Genet.* **5**, 381–385.

Schulman, H. (1988). The multifunctional Ca^{2+}/calmodulin-dependent protein kinase. *Adv. Second Messenger Phosphoprotein Res.* **22**, 39–112.

Schulman, H. (1993). The multifunctional Ca^{2+}/calmodulin-dependent protein kinases. *Curr. Opin. Cell Biol.* **5**, 247–253.

Schulman, H., and Hanson, P. I. (1993). Multifunctional Ca^{2+}/calmodulin-dependent protein kianse. *Neurochem. Res.* **18**, 65–77.

Schulman, H., and Lou, L. L. (1989). Multifunctional Ca^{2+}/calmodulin-dependent protein kinase: Domain structure and regulation. *Trends Biochem. Sci.* **14**, 62–66.

Schworer, C. M., Colbran, R. J., and Soderling, T. R. (1986). Reversible generation of a Ca^{2+}-independent form of Ca^{2+} (calmodulin)-dependent protein kinase II by an autophosphorylation mechanism. *J. Biol. Chem.* **261**, 8581–8584.

Schworer, C. M., Rothblum, L. I., Thekkumkara, T. J., and Singer, H. A. (1993). Identification of novel isoforms of the δ subunit of Ca^{2+}/calmodulin-dependent protein kinase II. Differential expression in rat brain and aorta. *J. Biol. Chem.* **268**, 14443–14449.

Selbert, M. A., Anderson, K. A., Huang, Q-H., Goldstein, E. G., Means, A. R., and Edelman, A. M. (1995). Phosphorylation and activation of Ca^{2+}-calmodulin-dependent protein kinase IV by Ca^{2+}-calmodulin-dependent kinase Ia kinase. *J. Biol. Chem.* **270**, 17616–17621.

Shattuck, R. L., Zimmer, W. E., Lukas, T. J., and Watterson, D. M. (1988). Characterization of a calcium binding protein which is probably related to smooth muscle myosin light chain kinase by alternative promotion. *J. Cell Biol.* **107**, 747a.

Sheng, M., Thompson, M. A., and Greenberg, M. E. (1991). CREB: A Ca^{2+}-regulated transcription factor phosphorylated by calmodulin-dependent kinases. *Science* **252**, 1427–1430.

Shenolikar, S., Cohen, P. T. W., Cohen, P., Nairn, A. C., and Perry, S. V. (1979). The role of calmodulin in the structure and regulation of phosphorylase kinase from rabbit skeletal muscle. *Eur. J. Biochem.* **100**, 329–337.

Shirinsky, V. P., Vorotnikov, A. V., Birukov, K. G., Nanaev, A. K., Collinge, M., Lukas, T. J., Sellers, J. R., and Watterson, D. M. (1993). A kinase related protein stabilizes unphosphorylated smooth muscle myosin minifilaments in the presence of ATP. *J. Biol. Chem.* **268,** 16578–16583.

Shoemaker, M. O., Lau, W., Shattuck, R. L., Kwiatkowski, A. P., Matrisian, P. E., Guerra-Santos, L., Wilson, E., Lukas, T. J., Van Eldik, L. J., and Watterson, D. M. (1990). Use of DNA sequence and mutant analyses and antisense oligodeoxynucleotides to examine the molecular basis of non-muscle myosin light chain kinase autoinhibition, calmodulin recognition, and activity. *J. Cell Biol.* **111,** 1107–1125.

Silva, A. J., Paylor, R., Wehner, J. M., and Tonegawa, S. (1992). Impaired spatial learning in α-calcium-calmodulin kinase II mutant mice. *Science* **257,** 206–211.

Silva, A. J., Stevens, C. F., Tonegawa, S., and Wang, Y. (1992b). Deficient hippocampal long-term potentiation in α-calcium-calmodulin kinase II mutant mice. *Science* **257,** 201–206.

Sitges, M., Dunkley, P. R., and Chiu, L. M. (1995). A role for calcium/calmodulin kinase(s) in the regulation of GABA exocytosis. *Neurochem. Res.* **20,** 245–252.

Smith, M. K., Colbran, R. J., Brickey, D. A., and Soderling, T. R. (1992). Functional determinants in the autoinhibitory domain of calcium/calmodulin-dependent protein kinase II. Role of His^{282} and multiple basic residues. *J. Biol. Chem.* **267,** 1761–1768.

Srinivasan, M., Edman, C. F., and Schulman, H. (1994). Alternative splicing introduces a nuclear localization signal that targets multifunctional CaM kinase to the nucleus. *J. Cell. Biol.* **126,** 839–852.

Srivastava, S., and Hartshorne, D. J. (1983). Conversion of a Ca^{2+}-dependent myosin light chain kinase from skeletal muscle to a Ca^{2+}-independent form. *Biochem. Biophys. Res. Commun.* **110,** 701–708.

Starovasnik, M. A., Davis, T. N., and Klevit, R. E. (1993). Similarities and differences between yeast and vertebrate calmodulin: An examination of the calcium-binding and structural properties of calmodulin from the yeast. *Saccharomyces cerevisiae. Biochemistry* **32,** 3261–3270.

Stevens, C. F., Tonegawa, S., and Wang, Y. (1994). The role of calcium-calmodulin kinase II in three forms of synaptic plasticity. *Curr. Biol.* **4,** 687–693.

Strack, S., Barban, M. A., Wadzinski, B. E., and Colbran, R. J. (1997). Differential inactivation of postsynaptic density-associated and soluble Ca^{2+}/calmodulin-dependent protein kinase II by protein phosphatases 1 and 2A. *J. Neurochem.* **68,** 2119–2128.

Stull, J. T., Tansey, M. G., Tang, D.-C., Word, R. A., and Kamm, K. E. (1993). Phosphorylation of myosin light chain kinase: A cellular mechanism for Ca^{2+} desensitization. *Mol. Cell. Biochem.* **127–128,** 229–237.

Su, Z., Fan, D., and George, S. E. (1994). Role of domain 3 of calmodulin in activation of calmodulin-stimulated phosphodiesterase and smooth muscle myosin light chain kinase. *J. Biol. Chem.* **269,** 16761–16765.

Sugita, R., Mochizuki, H., Ito, T., Yokokura, H., Kobayashi, R., and Hidaka, H. (1994). Ca^+/calmodulin-dependent protein kinase kinase cascade. *Biochem. Biophys. Res. Commun.* **203,** 694–701.

Sun, P., Enslen, H., Myung, P., and Maurer, R. (1994). Differential activation of CREB by Ca^{2+}/calmodulin-dependent protein kinases type II and type IV involves phosphorylation of a site that negatively regulates activity. *Genes Dev.* **8,** 2527–2539.

Sun, Z., Sassone-Corsi, P., and Means, A. R. (1995a). Calspermin gene transcription is regulated by two cyclic AMP response elements contained in an alternative promoter in the calmodulin kinase IV gene. *Mol. Cell. Biol.* **15,** 561–571.

3. Calmodulin-Regulated Protein Kinases

Sun, Z. M., Means, R. L., LeMagueresse, B., and Means, A. R. (1995b). Organization and analysis of the complete rat calmodulin-dependent protein kinase IV gene. *J. Biol. Chem.* **270,** 29507–29514.

Sun, P., Lou, L., and Maurer, R. A. (1996). Regulation of activating transcription factor-1 and the cAMP response element-binding protein by Ca^{2+}/calmodulin-dependent protein kinases type I, II and IV. *J. Biol. Chem.* **271,** 3066–3073.

Suzuki, T. (1994). Protein kinases involved in the expression of long-term potentiation. *Int. J. Biochem.* **26,** 735–744.

Takio, K., Blumenthal, D. K., Edelman, A. M., Walsh, K. A., Krebs, E. G., and Titani, K. (1995). Amino acid sequence of an active fragment of rabbit skeletal muscle myosin light chain kinase. *Biochemistry* **24,** 6028–6037.

Tan, J. L., and Spudich, J. A. (1991). Characterization and bacterial expression of the *Dictyostelium* myosin light chain kinase cDNA. *J. Biol. Chem.* **266,** 16044–16049.

Tan, S., Wenthold, R. J., and Soderling, T. R. (1994). Phosphorylation of AMPA-type glutamate receptors by calcium/calmodulin-dependent protein kinase II and protein kinase C in cultured hippocampal neurons. *J. Neurosci.* **14,** 1123–1129.

Tansey, M. G., Luby-Phelps, K., Kamm, K. E., and Stull, J. T. (1994). Ca^{2+}-dependent phosphorylation of myosin light chain kinase decreases the Ca^{2+} sensitivity of light chain phosphorylation within smooth muscle cells. *J. Biol. Chem.* **269,** 9912–9920.

Tashima, K., Yamamoto, H., Setoyama, C., Ono, T., and Miyamato, E. (1996). Overexpression of Ca^{2+}/calmodulin-dependent protein kinase II inhibits neurite outgrowth of PC12 cells. *J. Neurochem.* **66,** 57–64.

Tegge, W., Frank, R., Hofmann, F., and Dostmann, W. R. G. (1995). Determination of cyclic nucleotide-dependent protein kinase substrate specificity by the use of peptide libraries on cellulose paper. *Biochemistry* **34,** 10569–10577.

Teng, K. K., and Greene, L. A. (1994). KT5926 selectivity inhibits nerve growth factor-dependent neurite elongation. *J. Neurosci.* **14,** 2624–2635.

Thiel, G., Czernik, A. J., Gorelick, F., Nairn, A. C., and Greengard, P. (1988). Ca^{2+}/calmodulin-dependent protein kinase II: Identification of threonine-286 as the autophosphorylation site in the alpha subunit associated with the generation of Ca^{2+}-independent activity. *Proc. Natl. Acad. Sci. U.S.A.* **85,** 6337–6341.

Tobimatsu, T., and Fujisawa, H. (1989). Tissue-specific expression of four types of rat calmodulin-dependent protein kinase II mRNAs. *J. Biol. Chem.* **264,** 17907–17912.

Tobimatsu, T., Kameshita, I., and Fujisawa, H. (1988). Molecular cloning of the cDNA encoding the third polypeptide (gamma) of brain calmodulin-dependent protein kinase II. *J. Biol. Chem.* **263,** 16082–16086.

Tohtong, R., Rodriguez, D., Maughan, D., and Simcox, A. (1997). Analysis of cDNAs encoding *Drosophila melanogaster* myosin light chain kinase. *J. Muscle Res. Cell Motil.* **18,** 43–56.

Tokumitsu, H., Chijiwa, T., Hagiwara, M., Mizutani, A., Terasawa, M., and Hidaka, H. (1990). KN-62, 1-[N,O-bis(5-isoquinolinesulfonyl)-N-methyl-L-tyrosyl]-4-phenylpiperazine, a specific inhibitor of Ca^{2+}/calmodulin-dependent protein kinase II. *J. Biol. Chem.* **265,** 4315–4320.

Tokumitsu, H., Brickey, D. A., Glod, J., Hidaka, H., Sikela, J., and Soderling, T. R. (1994). Activation mechanisms for Ca^{2+}/calmodulin-dependent protein kinase IV. Identification of a brain CaM-kinase IV kinase. *J. Biol. Chem.* **269,** 28640–28647.

Tokumitsu, H., Enslen, H., and Soderling, T. R. (1995). Characterization of a Ca^{2+}/calmodulin-dependent protein kinase cascade. Molecular cloning and expression of calcium/calmodulin-dependent protein kinase kinase. *J. Biol. Chem.* **270,** 19320–19324.

VanBerkum, M. F. A., and Means, A. R. (1991). Three amino acid substitutions in domain I of calmodulin prevent the activation of chicken smooth muscle myosin light chain kinase. *J. Biol. Chem.* **266,** 21488–21495.

Van Biesen, T., Hawes, B. E., Luttrell, D. K., Krueger, K. M., Touhara, K., Porfiri, E., Sakaue, M., Luttrell, L. M., and Lefkowitz, R. J. (1995). Receptor-tyrosine-kinase- and G$\beta\gamma$-mediated MAP kinase activation by a common signaling pathway. *Nature (London)* **376,** 781–784.

van den Berg, I. E. T., and Berger, R. (1990). Phosphorylase *b* kinase deficiency in man: A review. *J. Inherited Metab. Dis.* **13,** 442–451.

van den Berg, I. E. T., van Beurden, E. A. C., Malingre, H. E. M., van Amstel, H. K. P., Poll-The, B. T., Smeitink, J. A. M., Lamers, E. H., and Berger, R. (1995). X-linked liver phosphorylase kinase deficiency is associated with mutations in the human liver phosphorylase kinase α subunit. *Am. J. Hum. Genet.* **56,** 381–387.

Van Eldik, L. J., Kwiatkowski, A. P., and Watterson, D. M. (1991). Molecular genetics of a human protein kinase: Chromosomal localization and analysis of the catalytic and calmodulin-regulatory regions. *FASEB J.* **5,** A448 (Abstract).

Waldmann, R., Hanson, P. I., and Schulman, H. (1990). Multifunctional Ca^{2+}/calmodulin-dependent protein kinase made Ca^{2+} independent for functional studies. *Biochemistry* **29,** 1679–1684.

Walsh, M. P., Dabrowska, R., Hinkins, S., and Hartshorne, D. J. (1982). Calcium-independent myosin light chain kinase of smooth muscle. Preparation by limited chymotryptic digestion of the calcium ion dependent enzyme, purification, and characterization. *Biochemistry* **21,** 1919–1925.

Wang, J., Renger, J. J., Griffith, L. C., Greenspan, R. J., and Wu, C.-F. (1994). Concomitant alterations of physiological and developmental plasticity in *Drosophila* CaM Kinase II-inhibited synapses. *Neuron* **13,** 1373–1384.

Watterson, D. M., and Vincenzi, F. F., eds (1980). Calmodulin and cell functions. *Ann. N.Y. Acad. Sci.* **356,** 1–446.

Watterson, D. M., Collinge, M., Lukas, T. J., Van Eldik, L., Birukov, K. G., Stepanova, O. V., and Shirinsky, V. S. (1995). Multiple gene products are produced from a novel protein kinase transcription region. *FEBS Lett.* **373,** 217–220.

Waxham, M. N., Aronowski, J., Westgate, S. A., and Kelly, P. T. (1990). Mutagenesis of Thr-286 in monomeric Ca^{2+}/calmodulin-dependent protein kinase II eliminates Ca^{2+}/calmodulin-independent activity. *Proc. Natl. Acad. Sci. U.S.A.* **87,** 1273–1277.

Waxham, M. N., Grotta, J. C., Silva, A. J., Strong, R., and Aronowski, J. (1996). Ischemia-induced neuronal damage: A role for calcium/calmodulin-dependent protein kinase II. *J. Cerebral Blood Flow Metab.* **16,** 1–6.

Wayman, G. A., Wei, J., Wong, S., and Storm, D. R. (1996). Regulation of type I adenylyl cyclase by calmodulin kinase IV *in vivo*. *Mol. Cell. Biol.* **16,** 6075–6082.

Weber, P. C., Lukas, T. J., Craig, T. A., Wilson, E., King, M. M., Kwiatkowski, A. P., and Watterson, D. M. (1989). Computational and site-specific mutagenesis analyses of the asymmetric charge distribution on calmodulin. *Proteins: Struct. Funct. Genet.* **6,** 70–85.

Wehner, M., and Kilimann, M. W. (1995). Human cDNA encoding the muscle isoform of the phosphorylase kinase gamma subunit (PHKG1). *Hum. Genet.* **96,** 616–618.

Wehner, M., Clemens, P. R., Engel, A. G., and Kilimann, M. W. (1994). Human muscle glycogenois due to phosphorylase kinase deficiency associated with a nonsense mutation in the muscle isoform of the α subunit. *Hum. Mol. Genet.* **3,** 1983–1987.

3. Calmodulin-Regulated Protein Kinases

Weigand, S., Shuvalova, L., Lukas, T. J., Mirzoeva, S., Watterson, D. M., and Anderson, W. F. (1997). The 1.9 Å resolution structure of the E84K-calmodulin peptide complex, submitted.

Whitmore, S. A., Apostolous, S., Lane, S., Nancarrow, J. K., Phillips, H. A., Richards, R. I., Sutherland, G. R., and Callen, D. F. (1994). Isolation and characterization of transcribed sequences from a chromosome 16 hn-cDNA library and the physical mapping of genes and transcribed sequences using a high-resolution somatic cell panel of human chromosome 16. *Genomics* **20,** 169–175.

Wilkinson, D. A., Tonin, P., Shanske, S., Lombes, A., Carlson, G. M., and DiMauro, S. (1994). Clinical and biochemical features of 10 adult patients with muscle phosphorylase kinase deficiency. *Neurology* **44,** 461–466.

Wilson, A. K., Gorgas, G., Claypool, W. D., and de Lanerolle, P. (1991). An increase or a decrease in myosin II phosphorylation inhibits macrophage motility. *J. Cell Biol.* **114,** 277–283.

Wüllrich-Schmoll, A., and Kilimann, M. W. (1996). Structure of the human gene encoding the phosphorylase kinase β subunit (PHKB). *Eur. J. Biochem.* **238,** 374–380.

Wullrich, A., Hamacher, C., Schneider, A., and Kilimann, M. W. (1993). The multiphosphorylation domain of the phosphorylase kinase α_M and α_L subunits is a hotspot of differential mRNA processing and of molecular evolution. *J. Biol. Chem.* **268,** 23208–23214.

Wysolmerski, R. B., and Lagunoff, D. (1990). Involvement of myosin light-chain kinase in endothelial cell retraction. *Proc. Natl. Acad. Sci. U.S.A.* **87,** 16–20.

Wysolmerski, R. B., and Lagunoff, D. (1991). Regulation of permeabilized endothelial cell retraction by myosin phosphorylation. *Am. J. Physiol. Cell Physiol.* **261,** C32–C40.

Yagi, K., Yazawa, M., Ikura, M., and Hikichi, K. (1990). Interaction between calmodulin and target proteins. *Adv. Exp. Med. Biol.* **255,** 147–154.

Yakel, J. L., Vissavajjhala, P., Derkach, V. A., Brickey, D. A., and Soderling, T. R. (1995). Identification of a Ca^{2+}/calmodulin-dependent protein kinase II regulatory phosphorylation site in non-*N*-methyl-D-aspartate glutamate receptors. *Proc. Natl. Acad. Sci. U.S.A.* **92,** 1376–1380.

Yamada, K., Iwahashi, K., and Kase, H. (1988). Parallel inhibition of platelet-activating factor-induced protein phosphorylation and serotonin release by K-252a, a new inhibitor of protein kinases, in rabbit platelets. *Biochem. Pharmacol.* **37,** 1161–1166.

Yamagata, Y., Czernik, A. J., and Greengard, P. (1991). Active catalytic fragment of Ca^{2+}/calmodulin-dependent protein kinase II. Purification, characterization, and structural analysis. *J. Biol. Chem.* **266,** 15391–15397.

Yamauchi, T., Ohsako, S., and Deguchi, T. (1989). Expression and characterization of calmodulin-dependent protein kinase II from cloned cDNAs in chinese hamster ovary cells. *J. Biol. Chem.* **264,** 19108–19116.

Yokokura, H., Picciotto, M. R., Nairn, A. C., and Hidaka, H. (1995). The regulatory region of calcium calmodulin-dependent protein kinase I contains closely associated autoinhibitory and calmodulin-binding domains. *J. Biol. Chem.* **270,** 23851–23859.

Yokokura, H., Terada, O., Naito, Y., and Hidaka, H. (1997). Isolation and comparison of rat cDNAs encoding Ca^{2+}/calmodulin-dependent protein kinase I isoforms. *Biochim. Biophys. Acta Protein Struct. Mol. Enzymol.* **1338,** 8–12.

Zander, N. F., Meyer, H. E., Hoffman-Posorske, E., Crabb, J. W., Heilmeyer, L. M. G., and Kilimann, M. W. (1988). cDNA cloning and complete primary structure of skeletal muscle phosphorylase kinase (α-subunit). *Proc. Natl. Acad. Sci. U.S.A.* **85,** 2929–2933.

Zhang, F., Strand, A., Robbins, D., Cobb, M. H., and Goldsmith, E. J. (1994). Atomic structure of the MAP kinase ERK2 at 2.3 angstrom resolution. *Nature (London)* **367,** 704–711.

Zhang, M., Tanaka, T., and Ikura, M. (1995). Calcium-induced conformational transition revealed by solution structure of apo calmodulin. *Nat. Struct. Biol.* **2,** 758–767.

Zou, D.-J., and Cline, H. T. (1996). Expression of constitutively active CaMKII in target tissue modifies presynaptic axon arbor growth. *Neuron* **16,** 529–539.

4

Biochemistry and Pharmacology of Calmodulin-Regulated Phosphatase Calcineurin

BRIAN A. PERRINO* AND THOMAS R. SODERLING[†]

*Department of Physiology and Cell Biology
University of Nevada School of Medicine
Reno, Nevada 89557

[†]Vollum Institute for Advanced Biomedical Research
Oregon Health Sciences University
Portland, Oregon 97201

I. Introduction	170
II. Subunit Composition and Functional Domain Organization	171
A. Activation by Ca^{2+}/Calmodulin	171
B. Activation by Ca^{2+}/Calcineurin Subunit B	175
C. Regulation by Autoinhibitory Domain	177
D. Interaction of Autoinhibitory Domain with Catalytic Core	180
III. Catalytic Properties	183
A. Catalytic Mechanism	183
B. Deactivation by Ca^{2+}/Calmodulin and Oxidation	184
C. Substrate Specificity	185
D. Inhibition by Immunophilin/Immunosuppressant Complexes	186
E. Subunit Isoforms	190
IV. Physiological Roles of Calmodulin-Regulated Serine Threonine Phosphatase	192
A. Subunit Isoforms: Expression and Distribution	192
B. Subcellular Localization and Targeting Mechanisms	193
C. Antagonism of cAMP Signal	199
D. Regulation of Postsynaptic Potential	201
E. Regulation of Neurotransmitter Release	202
F. Regulation of Cytoskeleton	203
G. Regulation of Ion Channels and Transporters	205
H. Regulation of Gene Expression	211

I. Regulation of Ca^{2+}-Activated Cell Death............................. 218
J. Summary and Perspective 220
   References 221

I. INTRODUCTION

Calcineurin (CaN) was initially identified in mammalian brain as a "calmodulin (CaM)-dependent cyclic nucleotide phosphodiesterase inhibitor protein" by virtue of its ability to bind CaM and inhibit phosphodiesterase (Wang and Desai, 1977; Klee and Krinks, 1978). Independently, a Ca^{2+}/CaM-regulated serine threonine phosphatase was discovered in skeletal muscle and named phosphoprotein phosphatase 2B (PP2B) based on the Ingebritsen and Cohen classification scheme (Stewart et al., 1982). SDS–PAGE of purified skeletal muscle PP2B showed that its subunit structure was identical to calcineurin, the major CaM-binding protein in the brain (Klee and Krinks, 1978). Subsequent studies confirmed that CaN is a Ca^{2+}/CaM-regulated Ser/Thr phosphatase with a similar substrate specificity to skeletal muscle PP2B (Stewart et al., 1982). In the Ingebritsen and Cohen classification scheme, type I phosphatases perferentially dephosphorylate the β subunit of phosphorylase kinase and are sensitive to the inhibitor-1 and inhibitor-2 proteins (Ingebritsen and Cohen, 1983). Calcineurin is a type 2B phosphatase because it preferentially dephosphorylates the α subunit of phosphorylase kinase, requires a divalent cation and cofactor for activity, and is insensitive to inhibitor-1 and inhibitor-2 (Ingebritsen and Cohen, 1983). More recently, several isoforms of the type I and type II phosphatases have been described and several new phosphatase genes have been identified that are difficult to assign by this classification scheme. Initially, the term calcineurin was used to refer to the enzyme present in the brain, whereas the nonneuronal enzyme was referred to as PP2B or the CaM-regulated phosphatase. Nonneuronal Ca^{2+}/CaM-regulated protein phosphatases having similar subunit structure and functional domain organization to neuronal calcineurin have been referred to as calcineurin. In this chapter the term CaN will be used when referring to either neuronal calcineurin or nonneuronal type 2B phosphatases to avoid unnecessary complication of the nomenclature.

The biochemical characterization of neuronal CaN proceeded much more rapidly than that of the nonneuronal enzyme and more rapidly than studies of its physiological roles for several reasons. Because it is one of the major CaM-binding proteins in the brain and comprises 1% of the total protein, CaN is easily purified in large quantities from brain (Krinks et al., 1985). In contrast, CaN is present at levels of only 0.1–0.01% of total protein in nonneuronal tissues (Krinks et al., 1985). The proteolytic sensitivity of CaN, particularly in the presence of Ca^{2+}, made accurate assays of its activity in tissue homogenates difficult (Klee et al., 1988). Physiological studies were also hampered due to the fact that specific, cell-permeable inhibitors of the enzyme were unknown until 1991 when it was discovered that CaN is the target of the immunophilin/immunosuppressant complexes cyclophilin/cyclosporin, or FKBP12/FK506 (J. Liu et al., 1991).

II. SUBUNIT COMPOSITION AND FUNCTIONAL DOMAIN ORGANIZATION

A. Activation by Ca^{2+}/Calmodulin

All isoforms of CaN from yeast to human are composed of a 1:1 molar complex of a 56- to 61-kDA catalytic subunit (CnA) and a 19-kDa Ca^{2+}-binding regulatory subunit (CnB). Like CaM, the CnB subunit is an EF-hand Ca^{2+}-binding protein having four Ca^{2+}-binding loops (Klee et al., 1988). Calmodulin and CnB share considerable structural homology within the Ca^{2+}-binding loops, and similar to CaM, CnB contains a long central helix connecting Ca^{2+}-binding loops II and III (Klee et al., 1988). However, CaM and CnB are distinct proteins with no functional cross-reactivity (Klee et al., 1988). Calmodulin does not bind to the CnB-binding helix in CnA, and CnB does not bind to the CaM-binding domain nor does it inhibit the activation of cyclic nucleotide phosphodiesterase by CaM (Klee et al., 1985; Merat et al., 1985). Calmodulin binding to CaN is absolutely dependent on Ca^{2+}, whereas CnB remains tightly associated with CnA even in the presence of Ca^{2+} chelators (Kincaid et al., 1982; Stewart et al., 1982). However, using the proteolyzed form of brain CaN, Stemmer and Klee (1994) showed that 2 mol of Ca^{2+} bound per mol of CaN at a Ca^{2+} concentration of 0.07 μM. This finding suggests that the CnB subunit may still contain some bound Ca^{2+} in the presence of Ca^{2+} chelators.

The crystal structures of Ca^{2+}/CaM complexed with a synthetic peptide corresponding to its binding domain and of CnB complexed with CnA reveal the basis for the lack of functional cross-reactivity between CnB and CaM (Meador et al., 1992; Griffith et al., 1995; Kissinger et al., 1995). Ca^{2+}-binding loops I and II of both free CaM and CnB are separated from loops III and IV by an α-helical central linker segment (Griffith et al., 1995). Calmodulin bounds to its binding domain shows two globular lobes containing each pair of Ca^{2+}-binding loops wrapped around its binding domain opposite each other, with the CaM central helix encircling the α-helical binding domain (Meador et al., 1992). In contrast, the CnB central α helix is kinked by Gly-85 (Griffith et al., 1995). Thus, the two globular domains containing each pair of Ca^{2+}-binding loops are arranged linearly on top of the CnB-binding α helix of CnA (Griffith et al., 1995). In Griffith et al. (1995), which shows the electrostatic potential surface of CnB bound to its α-helical binding domain, the shape of CnB is not unlike that of a baseball catcher's mitt.

The CaN catalytic subunit contains a CaM-binding domain (see Fig. 1). Like most other CaM-regulated enzymes, calcineurin is so activated by limited proteolysis (Klee and Cohen, 1988). The CnB subunit is not dissociated from the CnA subunit and is unaffected during limited

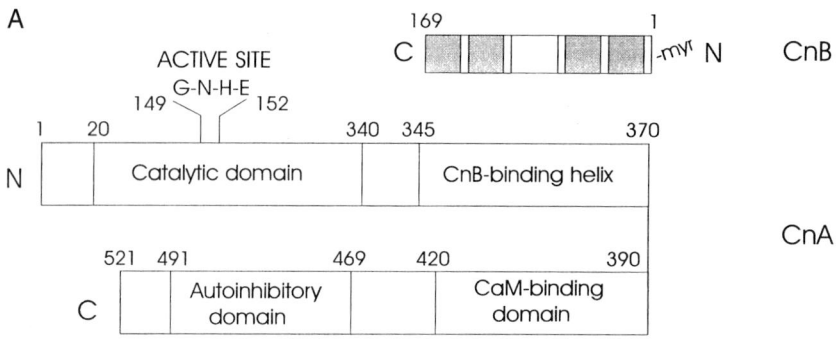

Fig. 1. (A) Schematic diagram of the CaN domain structure. The numbering for CnA and CnB is based on mammalian brain CnAα and CnB amino acid sequences, respectively. The CnB EF-hand Ca^{2+}-binding loops are shaded. (B) Ribbon representation of recombinantly expressed human brain CaN. The CnB subunit (dark gray) binds to an extended α helix in the CnA subunit (light gray). The Fe and Zn ion pair (light gray spheres) is bound in a cleft at the active site in the CnA subunit. The CnB subunit contains four Ca^{2+} ions (light gray spheres). (Figure 1B kindly provided by Charles R. Kissinger and colleagues, Agouron Pharmaceuticals.) (Reprinted, with permission from Villafranca et al., 1996, and Current Biology Ltd.).

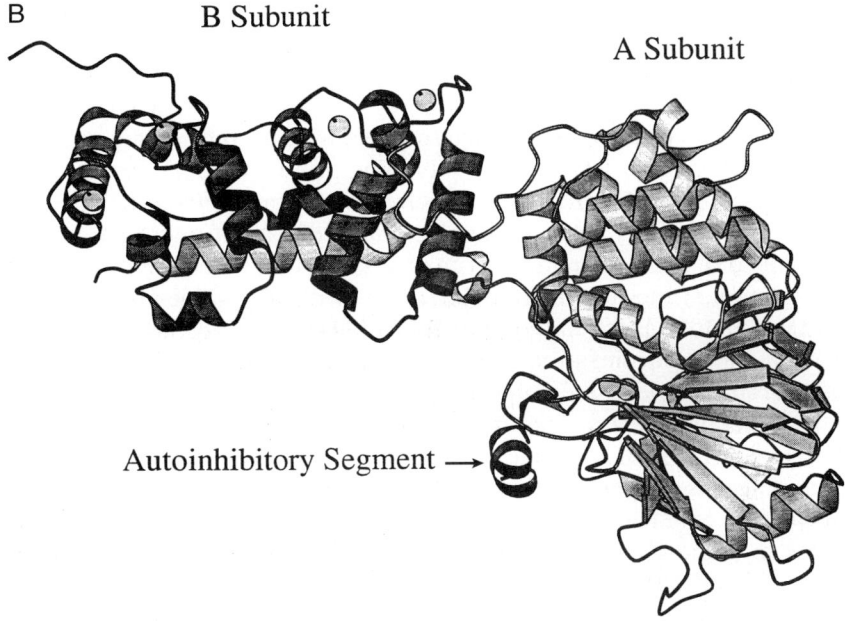

Fig. 1. Continued

proteolysis (Klee and Cohen, 1988). Limited proteolysis of CaN in the absence of CaM converts the 61-kDa CnA subunit to a 43-kDa proteolytic fragment that no longer binds CAM but is constitutively active (Klee and Cohen, 1988). These limited proteolysis studies localized the CaM-binding domain to the carboxyl-terminal third of the CnA subunit. By homology to other known CaM-binding domains, a putative CaM-binding amphipathic α helix was identified between residues 390 and 420 and was confirmed with synthetic CaM-binding peptides containing this sequence (Kincaid et al., 1988). As several other CaM-activated enzymes contain autoinhibitory domains whose proteolytic removal results in constitutive activity, the proteolysis experiments suggested a putative autoinhibitory domain near the COOH terminus of CaN (Hashimoto et al., 1990; Soderling, 1990). As detailed in a later section, subsequent studies identified an autoinhibitory domain in CaN.

The interaction between CaN and CaM is absolutely dependent on Ca^{2+}. In the absence of Ca^{2+}, binding is still not detected at CaN and

CaM concentrations $>10^{-5}$ M, resulting in a lowest estimated dissociation constant of 10^{-4} M (Hubbard and Klee, 1987). In contrast, in the presence of Ca^{2+}, CaM has an extremely high affinity for CaN. Calcineurin and CaM form a tight 1:1 molar complex with a dissociation constant of 10^{-9}–10^{-10} M (Klee et al., 1988). Interestingly, despite its high affinity for CaM, the other EF-hand Ca^{2+}-binding proteins tested—troponin C, parvalbumin, S-100, and the CnB subunit—do not substitute for CaM in stimulating CaN activity (Yang et al., 1982; Wolff and Sved, 1985). Similar to other CaM-regulated enzymes, the Ca^{2+} dependence of CaM stimulation is the low micromolar range, with K_{act} values between 0.3 and 1.0×10^{-6} M (Stemmer and Klee, 1994). Thus, in the presence of 10^{-6} M CaM concentrations, cytosolic Ca^{2+} concentrations in stimulated cells (100–500 μM) are sufficient to activate CaN. Stemmer and Klee (1994) detailed several important characteristics of the interaction between CaM and CaN. Like other CaM-regulated enzymes, the Ca^{2+} stimulation is cooperative with a Hill coefficient of 2.8–3, indicating that occupancy of at least three Ca^{2+}-binding loops are required for activation (Stemmer and Klee, 1994). However, the presence of the Ca^{2+}-binding CnB subunit complicates the analysis of the Ca^{2+}/CaM stimulation of CaN. Similar to the Ca^{2+}/CaM-dependent stimulation of cyclic nucleotide phosphodiesterase, the inverse plots of CaM-stimulated activity vs CaM concentration are linear (Stemmer and Klee, 1994). However, unlike the cyclic nucleotide phosphodiesterase plots, the CaN plots did not intercept on the ordinate (Stemmer and Klee, 1994). A plot of the extrapolated values of Ca^{2+}-dependent phosphatase activity at infinite CaM concentrations shows that the Ca^{2+} dependence of the CaM stimulation of CaN is very similar to the Ca^{2+} dependence of CnB-mediated CaN activation (measured in the absence of CaM) (Stemmer and Klee, 1994). Thus, Ca^{2+} binding to the CnB subunit is required for CaM stimulation of CaN, and the three Ca^{2+}-binding loops required for activation by CaM may reside on both the CnB subunit and CaM. As has been observed for other CaM-regulated enzymes, the Ca^{2+} concentration required for half-maximal activation of CaN by CaM decreases as the CaM concentration increases (Stemmer and Klee, 1994). Also, in the presence of the CaM-binding peptide from CaN and 6 mM Mg^{2+}, the affinities of all four Ca^{2+}-binding loops of CaM for Ca^{2+} are increased (Stemmer and Klee, 1994). Because amino acid sequences of CaN CaM-binding domains from yeast to humans are highly conserved, it is expected that, similar to the brain isoform, all CaNs will have a very high affinity for CaM.

B. Activation by Ca^{2+}/Calcineurin Subunit B

The requirement of an additional Ca^{2+}-binding protein in the regulation of Ca^{2+}/CaM-dependent enzyme activity is unique to CaN. Although the regulation of CaN by Ca^{2+}/CaM has been extensively studied, the role of the B subunit is still being characterized. *In vitro*, in the absence of CaM, Ca^{2+} binding to the B subunit results in a small V_{max} increase (Stewart *et al.*, 1982). Circular dichroism analysis revealed that the B and A subunits both undergo conformational changes on Ca^{2+} binding to the B subunit (Wolff and Sved, 1985). In the presence of Ca^{2+}, CnB and CaM give synergistic activation of the CnA subunit, suggesting different mechanisms of activation for the two proteins. Studies have confirmed that this is indeed the case. The K_m of CaM-activated A subunit alone for the RII peptide substrate is fivefold higher than CaM-activated native brain CaN, and the V_{max} is less than 10% of the V_{max} value attained by the native enzyme (Perrino *et al.*, 1992). Following reconstitution with the purified B subunit, the phosphatase activity is characterized by low K_m (20 μM) and high V_{max} values (0.6–2.0 μmol/min/mg) obtained with the purified brain enzyme (Blumenthal *et al.*, 1986; Perrino *et al.*, 1992). Although proteolyzed CaN or CaN mutants lacking the autoinhibitory domain have high V_{max} values in the absence or presence of Ca^{2+}, the K_m in the absence of Ca^{2+} is fivefold higher than the K_m in the presence of Ca^{2+}. Thus, Ca^{2+}/CaM binding increases the V_{max} and, although Ca^{2+} binding to the B subunit also slightly increases the V_{max}, the major effect is to lower the K_m and increase the affinity of the catalytic domain for substrate (Stemmer and Klee, 1994; Perrino *et al.*, 1995).

Some important questions still to be resolved are the mechanisms by which Ca^{2+} binding to CnB affects the catalytic domain to (i) decrease the K_m and increase the affinity of the catalytic domain for substrate, (ii) cause the characteristic small V_{max} increase in the absence of CaM, and (iii) affect the stimulation of phosphatase activity by Ca^{2+}/CaM. In the current model of CaN activation, Ca^{2+}/CaM binding to the CaM-binding domain induces conformational changes in the autoinhibitory domain that cause the autoinhibitory domain to be displaced from the catalytic domain and result in the V_{max} increase. Similarly, could Ca^{2+} binding to CnB transmit a conformational change in the B subunit-binding domain to the autoinhibitory domain and cause a partial displacement of the autoinhibitory domain from the catalytic domain to account for the slight V_{max} increase?

Like several other proteins involved in signal transduction (c-*src*, protein kinase A, MARCKS), the B subunit is myristoylated on its aminoterminal Gly (Aitken *et al.*, 1982). For many proteins, myristoylation facilitates membrane association or serves a structural role. At present, the available information about the functional significance to CaN of this important posttranslational modification is unclear. No difference has been found in the subunit composition or kinetic parameters of purified CaN containing the myristoylated or nonmyristoylated B subunit (Sikkink *et al.*, 1995; Watanabe *et al.*, 1995). The crystal structure of proteolyzed CaN (removal of the CaM-binding and autoinhibitory domains from the A subunit) shows the myristyl group lying on top of the B subunit, away from the catalytic domain and the interface between the A and the B subunits, arguing against a structural role (Griffith *et al.*, 1995). However, a structural role for the B subunit myristic acid is suggested by the finding that the thermal stability of *in vitro* reconstituted CaN containing myristoylated B subunit is increased relative to CaN containing nonmyristoylated B subunit (Kennedy *et al.*, 1996). The myristic acid lays against and parallel to the N-terminal hydrophobic helix of CnB and is connected to this helix by a 15-residue loop (Griffith *et al.*, 1995). The limited proteolysis removed the first 23 residues of the CnA subunit. The crystal structure of recombinant mammalian brain CaN expressed in *Escherichia coli* shows that an amino-terminal segment of CnA containing residues 14–23 comprises an additional CnB-binding interface by forming part of a binding cleft for the carboxy-terminal lobe of CnB (Kissinger *et al.*, 1995). The CnB subunit of the *E. coli* expressed CaN is not myristoylated. Thus, the crystal structure of the proteolyzed brain CaN contains the myristic acid but lacks the first 23 residues of the CnA subunit, whereas the recombinant CaN lacks the myristic acid. Interestingly, residues 1–13 and 1–4 of CnA and CnB, respectively, are not visualized in the crystal structure of recombinant CaN and are presumed to be disordered. Flexible regions of crystallized proteins are often disordered. It is interesting to speculate whether the aminoterminal myristate of CnB could interact with the CnA amino terminus via these flexible disordered segments.

Previous reports have demonstrated a Ca^{2+}-dependent association of CaN with lymphocyte plasma membranes and with membrane phospholipids *in vitro* (Chantler, 1985; Politino and King, 1987). Immunohistochemical studies of brain suggest an equal distribution of CaN between the cytosol and the plasma membrane, and CaN has been purified from placenta membrane preparations (Pallen *et al.*, 1985; Klee *et al.*, 1988).

4. Calmodulin-Regulated Phosphatase Calcineurin

Some reports suggest that several neuronal ion channels and membrane-associated proteins are regulated by CaN, suggesting that CaN may be targeted to the plasma membrane to dephosphorylate membrane proteins (Aperia *et al.*, 1992; Liu *et al.*, 1994a; Chen *et al.*, 1995; Tong *et al.*, 1995; Marrion, 1996). The CaN B subunit has been shown to contain a site of interaction for phospholipids, suggesting a role for the myristic acid in CaN membrane association (Politino and King, 1990). However, Zhu *et al.* (1995) used genetically altered *Sascharomyces cerevisiae* lacking the CnB amino-terminal Gly to generate a yeast strain containing CaN composed of the nonmyristoylated CnB subunit. In assays of CaN function in yeast, recovery from α-factor-induced growth arrest, survival during cation stress, and viability of a CaN-dependent strain, the biological activity of CaN containing the nonmyristoylated CnB subunit was indistinguishable from that of the wild-type enzyme (Zhu *et al.*, 1995). In addition, both the wild-type CnB subunit and the nonmyristoylated CnB subunit associated with the particulate fraction are detergent extractable, indicating that myristoylation is not required for cytoskeletal or membrane association. However, a smaller fraction of the nonmyristoylated CnB subunit was recovered by detergent extraction, suggesting that the myristoyl group is partially involved in membrane association of yeast CaN.

C. Regulation by Autoinhibitory Domain

Similar to the CaM-regulated kinases, Ca^{2+}/CaM binding to the CaM-binding domain of CaN is thought to induce conformational changes in an autoinhibitory domain that cause the autoinhibitory domain to be displaced from the catalytic domain and result in the V_{max} increase without affecting the K_m for substrate (Klee *et al.*, 1988). The CaM effect on the V_{max} is seen with both native heterodimeric CaN and the isolated CnA subunit, although the Ca^{2+}/CaM-induced V_{max} increase of the isolated CnA subunit is 13-fold lower (Perrino *et al.*, 1992). As is the case for CaM-regulated kinases, CaN is activated by limited proteolysis (Manalan and Klee, 1983; Soderling, 1990). Excellent reviews of these studies have appeared (Klee and Cohen, 1988; Klee *et al.*, 1988), so their findings will only be summarized briefly here. Limited proteolysis in the absence of Ca^{2+}/CaM results in removal of the carboxyl-terminal third of the CnA subunit, including the CaM-binding domain, resulting in a fully active proteolytic fragment. Likewise, limited proteolysis in the presence of Ca^{2+}/CaM results in a fully active proteolytical fragment, but this fragment still contains the CaM-binding domain and binds CaM. These results indicated the presence of a carboxy-terminal autoinhibitory domain in the CnA subunit that was distinct from the CaM-binding domain (Stemmer and Klee, 1991). Searches for protein kinase autoinhibitory domains are aided by the fact that these motifs generally have sequence homology to consensus phosphorylation site sequences for the kinase but lack the phosphorylatable Ser or Thr (e.g., pseudosubstrates) (Soderling, 1990). However, it appears that the substrate specificity determinants for the phosphatases, including CaN, do not reside in a well-defined linear stretch of primary sequence surrounding the dephosphorylatable Ser or Thr (Pinna and Donella-Deana, 1994). Thus, the CaN autoinhibitory domain was identified by testing the abilities of overlapping synthetic peptides corresponding to regions of the CnA carboxy terminus to inhibit CaN (Hashimoto *et al.*, 1990). Using this approach, a 25-residue peptide with an IC_{50} of 10–15 μM was identified, corresponding to a region of the CnA subunit (residues 467–492; numbering is based on the CnAα isoform) located 43 residues carboxy terminal of the CaM-binding domain (Hashimoto *et al.*, 1990). The CaN crystal structure shows this sequence interacting with the catalytic core, providing direct evidence for its autoinhibitory function (Kissinger *et al.*, 1995). The isolation and sequence determination of cDNA clones for the CnA catalytic subunit from various mammalian tissues, yeast, fungi, and *Drosophila* indicates a high degree of sequence conservation within the autoinhibitory domains, although not as high as for other subdomains of the catalytic subunit (Guerini and Klee, 1991). In fact, the CnAβI isoform lacks a sequence similar to that of residues 467–492, suggesting that this isoform has a different mode of regulation (Guerini and Klee, 1989). However, in the tissues examined, the CnAβI mRNA levels are too low to be detected by standard Northern blot analysis (Guerini and Klee, 1991). Interestingly, except for CnAβI, all CaN autoinhibitory domains contain a conserved Ser or Thr residue followed by a casein kinase II recognition site (Guerini and Klee, 1991). The physiological or functional significance to CaN of this casein kinase II phosphorylation site is unknown, although purified CaN contains substoichiometric amounts of phosphate (King and Huang, 1984).

Using a combination of synthetic autoinhibitory peptides with substituted amino acids and site-directed mutagenesis of the autoinhibitory domain, some residues involved in the regulation of autoinhibitory function have been identified. Synthetic autoinhibitory peptides with D477A or R486A/R487A amino acid substitutions lose the ability to inhibit CaN (Perrino et al., 1995). At the IC_{50} concentration of the parent autoinhibitory peptide (10–15 μM) these two peptides each result in only 5% inhibition of activity (Perrino et al., 1995). These results suggest a role for D477 and R486/R487 in the autoinhibitory peptide for inhibitory function. However, the corresponding autoinhibitory domain mutants did not have increased Ca^{2+}-independent phosphatase activity as they were still inactive in the absence of Ca^{2+} (Perrino et al., 1995). From similar studies of the CaM kinase II autoinhibitory domain, it appears that the intermolecular interaction of exogenously added peptides with the catalytic domain is weaker than the intrasubunit interaction between catalytic and autoinhibitory domains (Fong and Soderling, 1990; Brickey et al., 1994). The native structure of the interacting domains would allow multiple interactions between residues, stabilizing the intrasubunit interaction and making single amino acid substitutions less effective at disrupting the interaction. Although the autoinhibitory domain D^{477}A and R^{486}A/R^{487}A mutants are inactive in the absence of Ca^{2+}, their Ca^{2+}-dependent and Ca^{2+}/CaM-dependent V_{max} values are both elevated two- to threefold relative to wild-type CaN (Perrino et al., 1995). In addition, these CaN autoinhibitory domain mutants were more sensitive to activation by limiting Ca^{2+} concentrations (0.05–0.5 μM). Interestingly, CaN containing a $D^{477}N$ mutation from a human B-cell lymphoma cell line has elevated activity at low cytosolic Ca^{2+} concentrations (Fruman et al., 1995).

As mentioned previously, Ca^{2+} binding to the CnB subunit causes a modest V_{max} increase, and CnB from proteolyzed CaN has a higher affinity for Ca^{2+} than native enzyme CnB (Stemmer and Klee, 1994). These findings suggest that mutual regulatory interactions occur between the autoinhibitory domain and the CnB subunit. Based on their crystal structure data of proteolyzed brain CaN and the wealth of biochemical data from studies of CaN regulation, Griffith et al. (1995) proposed a model of CaN regulation by Ca^{2+}, CaM, and its autoinhibitory domain. In this model, the amino acid sequence immediately carboxy terminal of the CnB-binding helix (residues 374–389) bends back to bring the amphipathic α-helical CaM-binding domain up under and against the CnB-binding helix. This arrangement would place the autoinhibitory domain in the active site and result in inhibition of phosphatase activity. The Ca^{2+}/CaM-induced changes in the conformation of the CaM-binding domain would then disrupt its interaction with the CnB-binding helix and simultaneously cause release of the autoinhibitory domain from the active site. The binding of autoinhibitory domain residues 469–486 to the active site in the crystal structure of native, recombinant CaN supports this model and indicates that the autoinhibitory interaction is stabilized by the interaction of autoinhibitory domain residues with active site residues. These findings do not preclude additional stabilizing interactions between the CaM-binding and CnB-binding domains. However, in the crystal structure of native recombinant CaN, residues 374–389, 390–414 (the CaM-binding domain), 415–468, and 487–521 are disordered and not resolved. Thus direct evidence for association between CaM-binding and CnB-binding domains is lacking. However, an interaction between the CaM-binding domain and the CnB-binding helix proposed in this model would represent a novel aspect of the structure/function regulation of CaN. Such a model is not inconsistent with biochemical data on the regulation of CaN activation. If Ca^{2+}/CaM binding to its α-helical binding domain disrupts its interaction with the CnB-binding helix, it is equally plausible that Ca^{2+} binding to CnB could also disrupt the interaction between the two binding helices. However, the Ca^{2+}/CnB-induced disruption would be expected to cause only a partial displacement of the autoinhibitory domain, as indicated by the modest V_{max} increase caused by Ca^{2+} binding to CnB. Also, could disruption of a CaM-binding domain interaction with the CnB-binding domain by removal of the CaM-binding domain cause subtle conformational changes in the CnB-binding domain that affect CnB and contribute to the increased affinity for Ca^{2+} for CnB from proteolyzed CaN? Mechanistically, an association between CaM-binding and the CnB-binding domains might be expected to consist primarily of electrostatic interactions, as the hydrophobic side of the amphipathic CnB-binding helix is embedded in a hydrophobic-binding groove of the CnB subunit (Griffith *et al.*, 1995). This predicts that hydrophobic interactions would be important for Ca^{2+}/CaM binding to the CnA CaM-binding domain and is consistent with data indicating a hydrophobic patch on CaM is involved in CaM binding to its amphipathic α-helical binding domains.

D. Interaction of Autoinhibitory Domain with Catalytic Core

Crystal structure data show autoinhibitory domain residues 469–486 interacting with the catalytic core (Kissinger *et al.*, 1995). The structure of the autoinhibitory segment consists of two short amino-terminal α helices followed by five residues in an extended conformation (Kissinger *et al.*, 1995). Ten residues of the autoinhibitory domain participate with 20 residues of the catalytic domain to form the binding interface (Kissinger *et al.*, 1995). One of the key residues implicated in the autoinhibitory domain peptide study, Asp-477 forms one of seven hydrogen bonds. Interestingly, extensive contacts with the substrate-binding cleft are precluded by the conformations of the autoinhibitory domain amino-terminal α helices. Also, because of the similarity of the autoinhibitory domain sequence R486-R-D-A489 to the consensus protein kinase A (PKA) phosphorylation motif R-R-X-S/T-PO$_4$, it was thought that perhaps the autoinhibitory domain may function as a "pseudosubstrate" inhibitor (Kissinger *et al.*, 1995). However, the R486-R-D-A489 sequence is not observed in the CaN crystal structure, suggesting that this sequence is disordered and not critical for autoinhibitory function. Sequence E481-R-M-P484 is the most highly conserved segment of the autoinhibitory domain and makes the most extensive contacts with the substrate-binding cleft. Water molecules bound to the active site Fe and Zn ions form hydrogen bonds with E481.

Studies of the kinetic mechanism of inhibition by the 25-residue (467–492) autoinhibitory peptide have indicated competitive inhibition with the RII phosphopeptide, although noncompetitive inhibition with this substrate and with the phosphomyosin light chain has also been reported (Hashimoto *et al.*, 1990; Perrino *et al.*, 1990; Parsons *et al.*, 1994). It is unclear why these differences in the mechanism of inhibition have been obtained. Different substrates (RII peptide vs myosin light chain) were used in these studies. Although the K_m of CaN toward the RII peptide is similar to the K_m value for native RII protein, small peptide substrates may interact differently with the catalytic domain than full-length proteins. Also, the sequence surrounding the dephosphorylation site of the myosin light chain is quite different from the sequence of the RII peptide and other CaN substrates (Hashimoto *et al.*, 1990). As noted previously, the affinity of the catalytic domain for the RII peptide substrate is increased by Ca^{2+} binding to the CnB subunit (Stemmer and Klee, 1994; Perrino *et al.*, 1995). These findings suggest that a conformational change in the catalytic domain may result from Ca^{2+}-induced conformational changes in the CnB subunit. In contrast to the peptide substrate, the IC$_{50}$ value of the autoinhibitory peptide is the same with or without Ca^{2+} (Perrino *et al.*, 1995). Thus, the autoinhibitory domain may interact with the catalytic domain in a manner distinct from that of substrates.

However, the CaN crystal structure indicates that residues 469–486 bind to the active site, providing strong evidence for a competitive mechanism of inhibition by the autoinhibitory domain.

Compared to protein kinase autoinhibitory peptides, the CaN autoinhibitory peptide is a relatively weak inhibitor (Hashimoto *et al.*, 1990). Also, unlike CaM kinase II and MLCK, which have partially overlapping CaM-binding and autoinhibitory domains, the CaN autoinhibitory domain is separated from the CaM-binding domain by about 40 residues (Hashimoto *et al.*, 1990). These findings raise the question of the function of this region between CaM-binding and autoinhibitory domains. Could the autoinhibitory segment from residues 467–492 represent a minimum inhibitory sequence? Removal of the entire region carboxy-terminal of the CaM-binding domain results in levels of Ca^{2+}-independent activity similar to the Ca^{2+}/CaM-stimulated activity of the native enzyme. These findings provided strong evidence of the presence of an autoinhibitory domain distinct from the CaM-binding domain. Subsequent to the identification of residues 467–492 as an autoinhibitory domain, stepwise truncation of the CnA carboxy-terminal region was used to further define the autoinhibitory segment. Two reports have shown that removal of the region carboxy-terminal of Lys-441 or Lys-466 results in phosphatase activity that is only partially Ca^{2+} independent and still requires Ca^{2+}/CaM to attain maximum activity (Parsons *et al.*, 1994; Perrino *et al.*, 1995). In contrast, further amino-terminal truncation to Lys-399 or Val-420 resulted in Ca^{2+}-independent activity similar to the Ca^{2+}/CaM-stimulated activity of the native enzyme (Parsons *et al.*, 1994; Perrino *et al.*, 1995). These findings suggest that the region between the CaM-binding domain and the autoinhibitory domain (i.e., residues 467–492) contains additional autoinhibitory elements. However, the presence of only residues 469–486 binding to the active site in the crystal structure suggests that additional residues are not important for inhibitory function (Kissinger *et al.*, 1995). It was concluded that the CaM-binding domain, the region between the CaM-binding domain and the autoinhibitory domain (residues 420–468), and 487–521 were disordered due to their flexibility (Kissinger *et al.*, 1995). However, these regions may have been disordered due to the presence of Ca^{2+} bound to the CnB subunit (Kissinger *et al.*, 1995). It is possible that CaN was crystalized in a partially active state, as Ca^{2+} binding to the CnB subunit results in measurable phosphatase activity (Stewart *et al.*, 1982). It would be interesting to test the effect of crystallizing CaN in EGTA on the interaction of the autoinhibitory domain with the catalytic domain, as well as on the disordered regions observed in the presence of Ca^{2+}.

III. CATALYTIC PROPERTIES

A. Catalytic Mechanism

As a Ser/Thr phosphoprotein phosphatase, CaN is a member of a set of enzymes that hydrolyze phosphate esters. For a detailed description of the catalytic mechanism of CaN, the reader is referred to the original reports of the PP1, CaN, and purple-acid phosphatase crystal structures, as well as the excellent review by Lohse *et al.* (1995), which summarizes the findings from the crystal structures of Ser/Thr phosphoprotein phosphatases as they relate to the catalytic mechanism of CaN (Goldberg *et al.*, 1995; Griffith *et al.*, 1995; Kissinger *et al.*, 1995). Only the major points will be summarized here. The crystal structures of purple-acid phosphatase, PP1, and CaN show that Ser/Thr phosphatases contain similar core elements within the catalytic site (Lohse *et al.*, 1995). The shared core elements of these enzymes reveal essential residues involved in catalysis (Lohse *et al.*, 1995). Currently, data strongly indicate that dephosphorylation by CaN (like all the Ser/Thr phosphatases) is catalyzed in a single step by a metal-activated water molecule (Griffith *et al.*, 1995). The core of the CaN catalytic domain consists of two mixed β sheets, flanked by a mixed $\alpha\beta$ structure and an α helix (Griffith *et al.*, 1995). The two β sheets form a β sandwich, with an open and closed end. The active site is located above the closed end of this β sandwich, in the middle of a curved channel in which substrates presumably bind (Griffith *et al.*, 1995). The active site contains two metal ions, Fe^{3+} and Zn^{2+}, confirming earlier atomic absorption experiments that showed substoichiometric amounts of Fe and Zn in the CnA subunit (Griffith *et al.*, 1995; Kissinger *et al.*, 1995). The active site consists of Fe^{3+} and Zn^{2+} ions and sequence G149-N-H-E/D152, which is present in the active site of all homologous Ser/Thr phosphatases (Lohse *et al.*, 1995). Asn-150 and His-151 are required for phosphate ester hydrolysis, and Glu-152 forms a pair of hydrogen bonds around Asp-118, the bridging ligand to Fe^{3+} and Zn^{2+} ions (Griffith *et al.*, 1995). The current model for the mechanism of CaN phosphatase activity involves a water or hydroxide molecule acting as a bridge ligand between the Fe^{3+} and Zn^{2+}. The water of hydroxide ion would then be positioned within 2.9 Å of the substrate phosphorous and directly in line with the P—O scissile bond of the substrate. The water or hydroxide would then be positioned to carry out a nucleophilic attack on the phosphorous. His-151 is positioned to donate a proton to the leaving Ser side chain. This mechanism of attacking and leaving groups is consistent with an S_n2-type reaction as suggested by mechanistic studies of CaN and purple-acid phosphatase showing phosphate ester hydrolysis without the formation of a phosphoenzyme intermediate (Griffith et al., 1995).

B. Deactivation by Ca^{2+}/Calmodulin and Oxidation

It was observed early on that following activation by Ca^{2+}/CaM in vitro, CaN undergoes a time-dependent inactivation that requires the continued presence of Ca^{2+}/CaM (King and Huang, 1984; Pallen and Wang, 1984). It was reported that following the removal of low molecular weight, heat-stable molecules, CaN in cell lysates is 10–20 times more active than purified CaN (Stemmer et al., 1995). However, CaN in cell lysates is also susceptible to time- and Ca^{2+}/CaM-dependent inactivation (Stemmer et al., 1995). These findings suggested the presence of an endogenous cofactor responsible for the high activity of unpurified CaN in cell extracts, which is lost on purification of CaN, and whose activity is inhibited by low molecular weight, heat-stable molecules (Wang et al., 1996). Such a factor that protects CaN from inactivation and is responsible for the high endogenous activity of unpurified CaN was demonstrated to be superoxide dismutase (Wang et al., 1996). It was also shown that superoxide dismutase is inhibited by small heat-stable molecules, indicating that inhibition of CaN by these molecules is due to inhibition of superoxide dismutase (Wang et al., 1996). As confirmed by the CaN crystal structures, CaN is a metalloenzyme that requires the presence of Fe^{3+} and Zn^{2+} in the active site for phosphatase activity. Based on these findings it was proposed that superoxide dismutase protects CaN by preventing oxidation of the Fe^{3+} at the catalytic core (Wang et al., 1996). Thus, following activation of CaN by Ca^{2+}/CaM binding, the autoinhibitory domain is displaced from the catalytic domain, allowing substrates access to the active site but also exposing the active site Fe^{3+} to oxidation (Wang et al., 1996). The demonstration of the sensitivity of CaN phosphatase activity to oxidation suggests that CaN in vivo may be regulated by both Ca^{2+}/CaM and the intracellular redox potential, which couples Ca^{2+}/CaN-dependent signaling pathways to the oxidation state of the cell (Wang et al., 1996).

C. Substrate Specificity

In contrast to the Ser/Thr protein kinases, the substrate specificity determinants of the Ser/Thr phosphoprotein phosphatases do not appear to rely on a primary " consensus-dephosphorylation sequence" of amino acids surrounding the dephosphorylated Ser/Thr. Thus, what makes a protein a "good" substrate for any Ser/Thr phosphoprotein phosphatase, including CaN, is still largely unknown. The fact that the kinases outnumber the known phosphatases suggests that the phosphatases must be able to dephosphorylate a number of different proteins that have been phosphorylated by several different kinases. As the primary amino acid sequences surrounding the phosphorylated Ser/Thr can be quite different, the substrate specificity determinants of the phosphatases cannot be as stringent as the kinases. Substrates of CaN have mostly been determined empirically, almost by default. Prior to the discovery in 1991 that CaN is inhibited by immunophilin/immunosuppressant complexes, there was no specific cell-permeable inhibitor of CaN. Thus, CaN activity was defined as okadaic acid-insensitive, Ca^{2+}/CaM-dependent phosphatase activity (Cohen *et al.*, 1989; Stemmer and Klee, 1991). Despite the lack of experimental data on the mechanism of substrate selectivity of CaN toward whole proteins, the use of short synthetic peptides containing a phosphorylated Ser or Thr has provided some insight into the substrate specificity determinants of CaN.

In contrast to PP2A and PP2C, peptide substrates shorter than 10 amino acid residues are poor substrates of CaN (Pinna and Donella-Deana, 1994). However, the 19-residue RII, peptide (corresponding to residues 81–99 of the RII subunit) and the 18-residue DARPP32 peptide (corresponding to residues 20–38 of DARPP32) are dephosphorylated as efficiently as the native proteins *in vitro,* suggesting that critical substrate determinants are present within these peptides (Blumenthal *et al.*, 1986; Hemmings *et al.*, 1990). In fact, removal of the first 4 residues from the RII peptide increased the K_m fivefold and reduced the V_{max} sixfold (Blumenthal *et al.*, 1986). It was concluded that the tendency of these residues to form an amphipathic β sheet was a strong positive determinant. On the one hand, several substrates of CaN do not appear to contain such secondary structures amino-terminal of the target Ser/Thr. On the other hand, studies with other peptides have shown that similar to PP2C, the carboxy-terminal acidic residue(s) adjacent to the phosphorylated target is a strong negative determinant for CaN (Pinna and

Donella-Deana, 1994). Like PP2A, CaN perceives basic residues aminoterminal of the phosphorylated Ser/Thr as positive determinants (Pinna and Donella-Deana, 1994). Finally, in contrast to other Ser/Thr phosphatases, the Sp/Tp-P motif is not a negative determinant for CaN (Pinna and Donella-Deana, 1994). It is possible that some substrates of CaN will contain substrate-specificity determinants within the vicinity of the dephosphorylated residue, whereas other substrates will have determinants present elsewhere in the native proteins that are positioned close to the CaN active site during substrate binding. Using the structural information provided by CaN crystal structures, the amino acid sequences of different CaN substrates, and enzymatic assays using site-directed mutants of CaN and its substrates, computer modeling of the CaN catalytic domain with phosphoprotein or peptide substrates may be the most efficient approach in determining the substrate specificity determinants of CaN.

D. Inhibition by Immunophilin/Immunosuppressant Complexes

Cyclosporin A (CsA) and FK506 are fungal compounds that are used clinically to suppress the immune response following organ transplantation and to prevent rejection of the foreign tissue. These structurally unrelated drugs bind to immunophilins and inhibit their peptidylprolylisomerase activities (Emmel et al., 1989; Fisher et al., 1989; Harding et al., 1989; Siekierka et al., 1989; Takahashi et al., 1989). Surprisingly, although cyclophilin A (CyP) and FKBP12, the intracellular binding proteins for CsA and FK506, respectively, are both peptidylprolylisomerases, they have no similarity in sequence or three-dimensional structure (Fisher et al., 1989; Harding et al., 1989; Jin et al., 1991; Price et al., 1991; Kieffer et al., 1992). Several studies have shown that inhibition of interleukin (IL)-2 gene induction does not involve inhibition of peptidylprolylisomerase activity, but it does involve the binding of CsA or FK506 to its respective immunophilin (Schreiber, 1991; Sigal et al., 1991). These findings led to a model of action in which the immunophilin/immunosuppressant complex inhibits some other cellular process by binding to another target protein (Schreiber, 1991). Early studies indicated that CsA blocks T-cell activation primarily by inhibiting IL-2 gene induction (Ringe, 1991; Schreiber, 1991). Antigen recognition and signaling by the T-cell receptor complex induces IL-2 gene transcription (Tocci et al.,

4. Calmodulin-Regulated Phosphatase Calcineurin

1989; Schreiber, 1991). Antigen-induced IL-2 gene transcription results from the PKC- or phorbol ester-dependent induction of the nuclear component of the T-cell transcription factor NFAT, along with a Ca^{2+}-dependent translocation of the preexisting cytosolic component of NFAT (NFATc) into the nucleus (Flanagan et al., 1991). This study also showed that CsA and FK506 block the Ca^{2+}-dependent translocation of NFATc into the nucleus (Flanagan et al. 1991). Liu et al. (1991a) showed that CsA/CyP or FK506/FKBP12, but not the individual components, bind to and inhibit the phosphatase activity of CaN. Several subsequent reports rapidly established that inhibition of CaN activity in T cells by CsA or FK506 resulted in inhibition of IL-2 gene induction (Clipstone and Crabtree, 1992; Liu et al., 1992; O'Keefe et al., 1992). Finally, several groups demonstrated that NFATc translocates into the nucleus only after it is dephosphorylated by CaN (Tocci et al., 1989; Liu et al., 1992; Jain et al., 1992; McCaffrey et al., 1993; Ruff and Leach, 1995; Shaw et al., 1995).

Initial biochemical studies of the mechanism of binding and inhibition of CaN by CsA/CyP or FK506/FKBP12 showed (i) binding requires only Ca^{2+} binding to the CnB subunit, but is enhanced by Ca^{2+}/CaM (Liu et al., 1991a), (ii) binding to proteolyzed CaN does not require Ca^{2+} (Liu et al., 1991a), (iii) although structurally unrelated, CsA/CyP and FK506/FKBP12 bind to the same site on CaN as indicated by the competitive displacement of FK506/FKBP12 by CsA/CyP (Liu et al., 1991a), (iv) the IC_{50} in vitro is 10^{-9} M with phosphopeptide substrates (Liu et al., 1992), (v) the inhibition is noncompetitive with substrate (Kissinger et al., 1995), and (vi) the phosphatase activity of the CaN/immunophilin/immunosuppressant complex toward the same organic substrate p-nitrophenyl phosphate is actually increased about threefold (Liu et al., 1991a). Findings that the binding of the immunophilin/immunosuppressant complex is Ca^{2+} dependent for native CaN and Ca^{2+} independent for proteolyzed CaN indicated that displacement or removal of the CaM-binding and autoinhibitory domains is necessary for the interaction between CaN and the immunophilin/immunosuppressant complex (Liu et al., 1991a, 1992; Kissinger et al., 1995). However, several lines of evidence suggested that the autoinhibitory domain and immunophilin/immunosuppressant complexes do not share a common binding site within the catalytic domain. The autoinhibitory peptide, unlike the immunophilin/immunosuppressant complex, inhibits CaN phosphatase activity toward both the peptide substrate and the p-nitrophenyl phosphate (Hashimoto et al., 1990). This finding suggested that the immunophilin/immunosuppressant complexes bound away from the catalytic site (Liu *et al.*, 1991a). Furthermore, binding studies, cross-linking studies, and site-directed mutagenesis studies indicated a requirement for the CnB subunit in the interaction between immunophilin/immunosuppressant complexes and the CnA subunit (Haddy *et al.*, 1992; Li and Handschumacher, 1993; Clipstone *et al.*, 1994; Husi *et al.*, 1994; Milan *et al.*, 1994; Kawamura and Su, 1995). In fact, mutations within the CnB subunit significantly reduce binding of the immunophilin/immunosuppressant complex to CaN and inhibition of phosphatase activity (Milan *et al.*, 1994). Similar results were obtained using CnA subunits containing mutated amino acid residues near the CnB-binding helix (Kawamura and Su, 1995). Together, these studies led to the hypothesis that the immunophilin/immunosuppressant complexes bound to a surface formed by the interface between the CnA and the CnB subunits, which was confirmed by the crystal structures of the ternary complex of CaN/FK506/FKBP12 (Griffith *et al.*, 1995; Kissinger *et al.*, 1995).

The crystal structures of the CaN/FK506/FKBP12 ternary complex confirmed much of the data from the biochemical and genetic studies of the interaction between immunophilin/immunosuppressant complexes and CaN. The FKBP12/FK506 complex binds to CaN at the amino-terminal end of the CnB-binding helix, with His^{87}-Ile^{90}, Arg^{42}-Phe^{46}, and Asp^{37}-Asp^{41} of FKBP12 making contacts with the CnB-binding helix, the CnB subunit, and the catalytic domain, respectively (Griffith *et al.*, 1995). These points of contact between FKBP12 and CaN that surround FK506 have previously been identified as critical for FK506/FKBP12/CaN binding by site-directed mutagenesis studies (Aldape *et al.*, 1992; Yang *et al.*, 1993; Futer *et al.*, 1995). A hydrophobic cleft formed by the interface of the CnB subunit with the CnB-binding helix of CnA is the principal site of interaction between CaN and FK506 (Griffith *et al.*, 1995). Side chains of residues from both the CnB-binding helix and the CnB subunit form this cleft (Griffith *et al.*, 1995). The C_{15-17}, and C_{21} allyl groups of FK506 extend into a deep pocket within the hydrophobic cleft and are stabilized by several van der Waals contacts with main-chain and side-chain atoms (Griffith *et al.*, 1995). These findings confirmed earlier studies using chemical modifications of FK506 that indicated that changes in the FK506 cyclohexyl moiety do not affect function, whereas alterations of the C_{21} allyl group result in greatly diminished CaN inhibition (Goulet *et al.*, 1994). The crystal structures of the ternary complex of CaN/FKBP12/FK506 indicate that large-scale changes in the conformation of the FKBP12/FK506 binary complex do not occur following binding to CaN (Griffith *et al.*, 1995; Kissinger *et al.*, 1995). They also show that the composite surface of the FKBP12/FK506 binary complex contains the determinants for CaN binding. However, slight changes to FK506 and FKBP12 do occur at the CaN-binding interface. The His^{87}-Ile^{90} loop of FKBP12 shifts, and Ile-90 adopts a different side-chain conformation (Griffith *et al.*, 1995). In addition, FK506 rotates slightly away from FKBP12, displacing the C_{21} allyl group. Together, these changes allow the allyl group to interact more tightly with the CnB-binding helix (Griffith *et al.*, 1995).

As mentioned previously, although structurally unrelated, CsA/CyP and FK506/FKBP12 most likely bind to the same region of CaN, as indicated by the competitive displacement of FK506/FKBP12 by CsA/CyP (Liu *et al.*, 1991a). Also, mutations of residues 118–124 of CnB result in loss of binding and inhibition of CaN by Cyp/CsA (Milan *et al.*, 1994). As shown in the CaN/FKBP12/FK506 crystal structure, the majority of the surface area contact between CnB and FKBP12/FK506 occurs at this region (Griffith *et al.*, 1995; Kissinger *et al.*, 1995). However, evidence suggests that the specific residues of the CnA and CnB subunits involved in the interactions between the two different immunophilin/immunosuppressant complexes are not the same. Kawamura and Su (1995) showed that certain mutations within the CnB-binding helix greatly increase the IC_{50} values for FKBP12/FK506 inhibition while having little or no effect on the IC_{50} values of the CyP/CsA complex. Also, as shown by Zhu *et al.* (1995), yeast CaN containing myristoylated or nonmyristoylated CnB binds equally well to both CyP/CsA and FKBP12/FK506. However, yeast containing the nonmyristoylated CnB subunit is partially resistant to inhibition by CyP/CsA but not FKBP12/FK506 (Zhu *et al.*, 1995). These results show that although CnB myristoylation is not involved in the binding of Cyp/Csa or FKBP12/FK506 to CaN, it may be involved in the inhibition of CaN by the CyP/CsA complex (Zhu *et al.*, 1995).

Although both FK506/FKBP12 and CsA/CyP inhibit CaN noncompetitively with respect to phosphopeptide substrates, the nature of the noncompetitive inhibition is still a matter of debate. In one proposed model of inhibition, the noncompetitive inhibition of the FKBP12/FK506 complex toward phosphopeptide and phosphoprotein substrates arises from a physical occlusion of the active site by the FKBP12 molecule (Griffith *et al.*, 1995). This model is consistent with observations that CaN dephosphorylation of phosphopeptides and phosphoproteins, but not the small organic molecule *p*-nitrophenyl phosphate, is inhibited by the FKBP12/

FK506 complex (Liu et al., 1991a). Because noncompetitive inhibition usually implies the formation of an inactive enzyme–substrate–inhibitor complex, another model proposes that the FKBP12/FK506 complex does not prevent phosphopeptide binding, but does result in changes in the conformation of the active site that cause inhibition of catalysis (Kissinger et al., 1995). Data from two biochemical and site-directed mutagenesis studies of Cyp/Csa or FKBP12/FK506 inhibition of CaN show that the CnB residues involved in binding to Cyp/Csa or FKBP12/FK506 are also involved in activation of the phosphatase activity of the CnA subunit (Milan et al., 1994; Watanabe et al., 1996). Several CnB mutants were identified that were poor activators of the phosphatase activity of the CnA subunit and also poorly interacted with CyP/Csa (Milan et al., 1994). These CnB mutants bound to CnA as tightly as native CnB, thus the poor activation was not due to weak binding of CnB to CnA (Milan et al., 1994). These results support the proposal that the Cyp/Csa and FKBP12/FK506 complexes inhibit CaN by affecting regions involved in the allosteric activation of CaN phosphatase activity. However, the elevation of phosphatase activity toward p-nitrophenyl phosphate in the presence of the immunophilin/immunosuppressant complex suggests that the active site retains catalytic activity.

E. Subunit Isoforms

Similar to PP1 and PP2A, several isoforms of the CnA and CnB subunits have been described (Guerini and Klee, 1991; Mumby and Walter, 1993). The CnA subunits of CaN exhibit the extreme conservation of sequence characteristic of the Ser/Thr phosphoprotein phosphatases. The sequences of the CnA subunits from fungi and yeast still have 54–62% identity to the major human brain CnA subunit isoform (Cyert et al., 1991; Higuchi et al., 1991; Liu et al., 1991b; Mumby and Walter, 1993). In mammalian brain, two different CnA clones are present, termed CnAα and CnAβ (these clones have also been called PP2Bα1 and PP2Bα2) (Guerini and Klee, 1989; Ito et al., 1989; Kuno et al., 1989; Kincaid et al., 1990; Rathna Giri et al., 1992). The CnAα and CnAβ isoforms are 74 and 81% identical in their cDNA and amino acid sequences, respectively, indicating that CnAα and CnAβ are encoded by two different genes (Guerini and Klee, 1991). The two isoforms exhibit 90% sequence identity between Asp-15 and leu-376 (Guerini and Klee,

1991). The major sequence differences reside at the amino and carboxy termini. The amino terminus of CnAβ, but not of CnAα, contains an unusual stretch of 12 prolines within the first 23 amino acid residues (Guerini and Klee, 1989). In addition, the last 20 amino acid residues of each isoform (carboxy-terminal of the autoinhibitory domain) contain multiple amino acid differences (Guerini and Klee, 1991). Alternative splicing generates an additional isoform from CnAα and two additional isoforms from CnAβ (Guerini and Klee, 1989; Kincaid et al., 1990). For both CnAα and CnAβ, an isoform exists that lacks a specific 30-bp sequence in the carboxy-terminal region (Guerini and Klee, 1989; Kuno et al., 1989; Kincaid et al., 1990). This deletion results in the loss of a 10-residue amino acid segment between CaM-binding and autoinhibitory domains. An additional CnAβ isoform (CnAβ2) from human brain containing a 54-bp insertion in the amino-terminal half of the coding sequence and a carboxy-terminal region with no sequence homology to the previously characterized autoinhibitory region has also been described (Guerini and Klee, 1989). In mammalian brain, total CaN is composed of 80% CnAα and 20% CnAβ, with 50–80% of the CnAα being composed of the isoform containing the carboxy-terminal 30-bp deletion (Kincaid et al., 1990). Similarly, CnAα is the predominant isoform in thymus and T cells (Zhang et al., 1996).

A third CnA gene has been described from mouse and human that encodes a testes-specific catalytic subunit (Muramatsu et al., 1992; Muramatsu and Kincaid, 1992). The deduced amino acid sequence of the testes-specific gene is 80% identical to both CnAα and CnAβ, although the testes-specific gene product does not contain an amino-terminal polyproline sequence (Muramatsu et al., 1992; Muramatsu and Kincaid, 1992). The mouse gene is developmentally expressed, with a large increase in expression during the later stages of spermatogenesis (Muramatsu et al., 1992). This correlation suggests that CaN in the testis may be involved in germ-cell maturation, which is consistent with the report of a flagellum-associated form of CaN that may regulate sperm motility (Tash et al., 1988).

As reported in the phylogenetic analysis of CnA subunit genes carried out by Rathna Giri et al. (1992), Northern blots of mammalian brain mRNA showed transcripts for both CnAα and CnAβ, whereas in two nonmammalian vertebrates (chicken and lizard), only CnAα transcripts were detected. In addition, the cloned CnA sequences from fungi, yeast, *Dictyostelium,* and *Drosophila melanogaster* do not contain an amino-terminal polyproline sequence and are more similar to the mammalian

CnAα isoform (Rathna Giri *et al.*, 1992). These findings led to the proposal that CnAβ might have evolved in response to the appearance of a different substrate(s) or a higher substrate concentration in mammals (Rathna Giri *et al.*, 1992). Similarly, the 30-bp insertion of the CnAα isoform has not been described in nonmammalian species.

In contrast to the CnA subunit, CnB appears to be the product of two genes. One CnB isoform is found in all organisms and tissues examined from yeast to human brain (Guerini and Klee, 1991). Mammalian brain CnB amino acid sequences are 99% identical (Guerini and Klee, 1991). The consensus myristoylation sequence composed of the amino-terminal glycine and the serine at position 5 is present in all cloned CnB sequences (Guerini and Klee, 1991; Guerini *et al.*, 1992; Ueki *et al.*, 1992; Perrino *et al.*, 1995; Kuno *et al.*, 1996). The phylogenetic sequence conservation of CnB is greater than that of the CnA subunit, with 82 and 88% amino acid sequence homology, respectively, of the yeast and *Drosophila* CnB subunit with mammalian CnB (Guerini *et al.*, 1992). Unlike all other tissues examined, testis contains the mRNA for the brain CnB isoform and an additional testis-specific isoform having 80% sequence identity with the brain isoform (Ueki *et al.*, 1992). Interestingly, only the mRNA expression of the testis-specific isoform is increased along with the testis-specific CnA during germ-cell maturation (Ueki *et al.*, 1992).

IV. PHYSIOLOGICAL ROLES OF CALMODULIN-REGULATED SERINE/THREONINE PHOSPHATASE

A. Subunit Isoforms: Expression and Distribution

The functional and physiological significance of the multiple CnA subunit isoforms is unknown. Only mammalian brain CaN containing the CnAα isoform has been crystallized, thus a structural comparison to CnAβ and the polyproline amino-terminal has not been carried out. Studies of CaN immunoreactivity in the brain have shown that CaN is present in neuronal cell bodies, dendrites, postsynaptic densities, spines, axons, and terminals (Klee and Cohen, 1988; Klee *et al.*, 1988). The highest concentrations of CaN are found in the hippocampus, caudate nucleus, and putamen, with lower levels observed in cerebral and cerebellar neocortex (Klee and Cohen, 1988; Klee *et al.*, 1988). In the developing rat brain, CaN immunoreactivity is not present until postnatal day 5 and then increases to adult levels between postnatal days 7 and 20 (Polli *et al.*, 1991). The regional expression of the CnAα and CnAβ isoforms in brain shows areas of overlapping expression, but also shows areas of uneven distribution and expression (Goto *et al.*, 1986; Takaishi *et al.*, 1991). Calcineurin immunoreactivity is not present in glial cells (Klee *et al.*, 1988). Earlier studies using biochemical and immunohistochemical techniques have reported the presence of CaN in the cell nucleus in both brain and liver (Bosser *et al.*, 1993; Pujol *et al.*, 1993). An immunohistochemical study by Usuda *et al.* (1996) demonstrated a nuclear localization of CnAα, but not CnAβ, in human neurons. These findings, along with the findings of CaN-regulated gene transcription, as well as the reports of a physical association between CaN and the T-cell transcription factor NFAT, suggest that the scope of the regulatory roles of CaN is broader than previously thought (Jain *et al.*, 1995; Loh *et al.*, 1996a; Shibasaki *et al.*, 1996; Wesselborg *et al.*, 1996).

The existence of multiple isozymes of CaN raises important questions about their functional and physiological significance. The mechanisms regulating their developmental expression, differential tissue expression, and subcellular localization remain to be determined. In addition, it is unknown whether the different isozymes display differing substrate specificities and/or different Ca^{2+}/CaM regulation. Continuing studies of substrate regulation by CaN in fungi, yeast, and *Drosophila* should provide clues to the substrate preferences and signal transduction pathways regulated by CaN in the more complex vertebrate mammalian systems.

B. Subcellular Localization and Targeting Mechanisms

A fundamental question in signal transduction is how the activation of multifunctional kinases and phosphatases by diffusible second messengers leads to temporally and spatially restricted phosphorylation or dephosporylation of specific substrates. The targeting hypothesis proposed by Hubbard and Cohen (1993) proposes that a phosphorylation-dependent signal depends not only on the balance of kinase and phosphatase activity, but also where these enzymes are located within the cell. Different localization mechanisms restrict multifunctional kinases and phosphatases to sites where their activity toward a particular substrate is required to produce a specific physiological response. Examples of well-characterized targeting mechanisms are the association of PP1 in striated muscle with glycogen via the Gm (glycogen-binding subunit) subunit of PP1 and the localization of PKA by AKAP79 at the plasma membrane of postsynaptic densities to regulate the activity of AMPA/kainate receptors (Cohen, 1992; Rosenmund *et al.*, 1994). Biochemical and immunohistochemical studies have shown that brain CaN is evenly distributed between the cytosol and the membrane (Klee *et al.*, 1988). In addition, CaN has been purified from human placental membrane and has been shown to be associated with purified sarcolemmal membrane preparations (Pallen *et al.*, 1985; Walsh *et al.*, 1995). In B lymphocytes CaN has been shown to translocate to the plasma membrane in a Ca^{2+}-dependent manner (Chantler, 1985). The activities of several ion channels and the membrane-associated protein dynamin 1 have been shown to be regulated by CaN (see later). In addition, the membrane-associated protein dystrophin is dephosphorylated *in vitro* by CaN (Walsh *et al.*, 1995). In rat hippocampus and cortex, kindling causes an increase in CaN immunoreactivity in the membrane fraction (Moia *et al.*, 1994). Together, these findings suggest that CaN is targeted to the plasma membrane to dephosphorylate membrane proteins. A number of CaN–protein interactions have been described, providing additional mechanisms to regulate CaN phosphatase toward a particular substrate(s) within a specific subcellular component (see later).

As shown in earlier studies. CaN binds to acidic phospholipids in a Ca^{2+}-sensitive manner *in vitro* (Politino and King, 1987, 1990). Photoaffinity labeling of CaN with [^{125}I]ASA-PtdEtn (a phosphatidylethanolamine derivative) is Ca^{2+} dependent and results in predominant labeling of the B subunit (Politino and King, 1990). These findings indicate that the Ca^{2+}-binding CaN B subunit is responsible for the Ca^{2+}-dependent association of CaN with acidic phospholipids *in vitro* and suggest a role for the amino-terminal myristic acid in CaN membrane association (Politino and King, 1987, 1990). Recoverin, a myristoylated EF-hand Ca^{2+}-binding protein of vertebrate photoreceptors, associates with membranes in a Ca^{2+}-dependent manner (Dizhoor *et al.*, 1993). Ca^{2+} binding induces extrusion of the myristic acid and exposure of a hydrophobic region, allowing recoverin to bind to neutral phospholipids (Tanaka *et al.*, 1995). In contrast, nonmyristoylated recoverin, even in the presence of Ca^{2+}, does not bind to membrane phospholipids, indicating the requirement of the myristic acid for the Ca^{2+}-dependent binding of recoverin to membrane phospholipids (Dizhoor *et al.*, 1993). The amino-terminal myristic acid and basic residue cluster of the CaN B

4. Calmodulin-Regulated Phosphatase Calcineurin

subunit suggest an analogous mechanism by which CaN may interact with acidic membrane phospholipids.

Presently, the physiological function of membrane-associated CaN is unknown. However, as noted previously, the finding that CaN regulates several ion channels and the membrane-associated proteins dynamin 1 and dystrophin implicates membrane-associated CaN in their regulation. Why would membrane-associated CaN be a better regulator of membrane-bound substrates than cytosolic CaN? One obvious reason is that the diffusion of a protein is restricted to the two-dimensional surface of the plasma membrane instead of through the three-dimensional surface of the cytoplasm following membrane binding (McLaughlin and Aderem, 1995). Thus, membrane-associated CaN would have a higher probability of encountering and dephosphorylating membrane-bound substrates. For a sperical cell with a radius r of 10 μm, it has been calculated that translocation of a protein from the cytoplasm to the plasma membrane increases its effective concentration by a factor of 10^3 (Mclaughlin and Aderem, 1995). Furthermore, the Ca^{2+} concentration increases very rapidly in the immediate vicinity of voltage-activated Ca^{2+} channels and glutamate receptors (Tong et al., 1995; Dolphin, 1996). In addition, for CaN, Ca^{2+}/phospholipid binding to the CnB subunit has been shown to increase the V_{max} to the same extent as Ca^{2+}/CaM and elevates Ca^{2+}/CaM-stimulated activity an additional threefold (Politino and King, 1987). Thus, the combination of increased effective concentration and elevated V_{max} values in the presence of Ca^{2+}/CaM could allow for extremely rapid regulation of substrate activity by membrane-associated CaN.

Several CaN–protein interactions have been described that function to target the phosphatase activity of CaN toward a specific substrate. These CaN–protein interactions include association of CaN with sperm flagellum (Tash et al., 1988), regulation of interleukin-2 gene transcription via a physical interaction of CaN with NFAT (Loh et al., 1996a; Shibasaki et al., 1996; Wesselborg et al., 1996), downregulation of the IP_3 and ryanodine receptors by the association of CaN with FKBP (Cameron et al., 1995), and submembranous localization of CaN in postsynaptic densities via interaction with AKAP79 (Coghlan et al., 1995). Similar to PKC-targeting proteins, CaN–protein interactions are characterized by different regulatory characteristics.

The dephosphorylation of NFAT1 by CaN causes translocation of NFAT1 into the nucleus and also increases the DNA binding of NFAT1 (Ruff and Leach, 1995). The CaN/NFAT1 interaction is Ca^{2+} dependent, but independent of the NFAT1 phosphorylation state, and thus resembles the stable interaction of JNK with phosphorylated or dephosphorylated c-*jun* (Derijard et al., 1994; Loh et al., 1996a; Wesselborg et al., 1996). This stable CaN/NFAT1 interaction is different from the PKC substrate/binding proteins interaction, which only occurs with the phosphorylated substrate/binding protein (Liao et al., 1994). Two reports showed that even though immunophilin/immunosuppressant complexes inhibit NFAT1 dephosphorylation by CaN, they do not disrupt the CaN/NFAT1 interaction (Loh et al., 1996a; Wesselborg et al., 1996). These findings (1) imply that NFAT1 is still bound to the substrate-binding site of the catalytic domain in the presence of the immunophilin/immunosuppressant complex and (2) support the proposal that FKBP12/FK506 and CyP/CsA inhibit the allosteric activation of CaN phosphatase activity (Kissinger et al., 1995). However, *in vitro* the preformed immunophilin/immunosuppressant CaN complex blocks the CaN/NFAT1 interaction (Loh et al., 1996a; Wesselborg et al., 1996). The latter finding suggests that FKBP12/FK506 and CyP/CsA inhibit CaN activity by blocking the access of large substrates to the substrate-binding site in the catalytic domain (Griffith et al., 1995). These findings also raise the exciting possibility that new immunosuppressive drugs could be developed that specifically disrupt the CaN/NFAT1 interaction without inhibiting CaN phosphatase activity (Luo et al., 1996). The clinical side effects of FK506 and Cyp/CsA include hypertension, nephrotoxicity, and neurotoxicity (Luo et al., 1996). It is likely that NFAT1 is not the primary target of CaN in these other tissues. Thus, drugs that inhibit the CaN/NFAT1 interaction but not CaN activity may lack the toxic side effects of FK506 and Cyp/CsA (Luo et al., 1996). The nature of the CaN/NFAT1 interaction is under investigation. Mice containing T cells that lack the CnAα gene have defective antigen-induced T-cell signaling. However, they are still sensitive to FKBP12/FK506 and CyP/CsA inhibition of T-cell activation, suggesting that both CnAα and CnAβ are involved in CaN-dependent signaling in T cells (Zhang et al., 1996). Using a 415 amino acid residue fragment of the amino-terminal portion of NFAT1, Luo *et al.* (1996) localized the CaN-binding region to a distinct regulatory region amino-terminal to the DNA-binding domain. In contrast, the NFAT1-binding site on CaN is unidentified. However, because NFAT1 is a CaN substrate, the interaction between CaN and NFAT1 most likely includes the substrate-binding region of the catalytic domain.

The interaction of CaN with IP_3 and ryanodine receptors resembles that of PKC with receptors for activated C-kinase (RACKs) proteins.

RACKs bind PKC but are not substrates for the kinase (Mochly-Rosen et al., 1991). Similarly, FKBP12, which is a component of IP$_3$ and ryanodine receptors but is not a CaN substrate, binds CaN to the receptor complex (Cameron et al., 1995). Using rat cerebellar membranes, Cameron et al. (1995) showed that IP$_3$ receptor activity is modulated by receptor-associated CaN. Similar to CaN binding to the FKBP12/FK506 complex, CaN binding to the FKBP12/IP$_3$ complex is Ca^{2+} dependent (Cameron et al., 1995). However, CaN associated with FKBP12/IP$_3$ still retains its phosphatase activity toward exogenous substrates and, more importantly, it also dephosphorylates the IP$_3$ receptor complex (Cameron et al., 1995). What accounts for the different effects on CaN phosphatase activity in the two complexes? FK506 disrupts the interaction between FKBP12 and the IP3 receptor, suggesting that the IP3 receptor competes with FK506 for the FK506-binding domain on FKBP12 (Cameron et al., 1995). The crystal structures of FKBP12/FK506 and FKBP12/FK506/CaN may provide a model for investigating the IP3 receptor/KFBP12/CaN interaction. To promote the interaction with FKBP12, a region of the IP3 receptor may mimic the structure of FK506 or form a surface similar to the composite surface formed by the CnB subunit, the CnB-binding domain, and FK506. Thus, the binding of CaN to the IP3 receptor/FKBP12 complex may be quite different from its binding to the FKBP12/FK506 complex. In the FKBP12/FK506 complex, a region of FKBP12 having sequence similarity with AKAP79 (see later) interacts with CaN, although the contribution of this region to FKBP12 inhibition of CaN activity is unknown (Kissinger et al., 1995). Perhaps this region does not interact with CaN in the FKBP12/IP3 receptor complex.

Colocalization of a kinase or a phosphatase with a particular substrate not only provides specificity of enzyme activity, but it also allows for rapid regulation of substrate activity, as diffusion of the enzyme to the substrate is no longer necessary or very limited. Thus, it is not hard to imagine that colocalization of both a kinase and a phosphatase with their substrates would provide highly specific and rapid substrate regulation. A new class of multivalent adapter proteins that organize multienzyme signaling complexes has been identified (Faux and Scott, 1996). A well-characterized example is the yeast scaffold protein STE 5, which coordinates the yeast MAP/kinase pathway (Faux and Scott, 1996). Initiation of the pheromone mating response via a G-protein-linked receptor activates the STE 20 kinase (Faux and Scott, 1996). The next three kinases in the pathway, STE 11 (a MEKK homolog), STE 7 (a MEK homolog), and FUS 3 or KSS 11 (MAP-kinase homologs), are all associated with

STE 5, allowing for efficient sequential kinase phosphorylation and activation (Faux and Scott, 1996). In mammalian brain, AKAP79, originally characterized as a targeting locus of PKA via its binding to the RII subunit of PKA, appears to function as a multivalent adapter protein (Rubin, 1994; Faux and Scott, 1996).

As a targeting locus, AKAP79 contains targeting domains that interact with a particular subcellular structure (Faux and Scott, 1996). A portion of neuronal PKA is localized to the postsynaptic density (PSD) by AKAP79 where it regulates the activity of AMPA/kainate receptors (Rosenmund et al., 1994). The yeast–dihybrid system was used to identify proteins that bound and localized AKAP79 in the PSD and surprisingly yielded a positive clone that contained the cDNA of CnAβ (Coghlan et al., 1995). However, the interaction of CaN with AKAP79 is apparently not isoform specific, as the biochemical characterization of the CaN/AKAP79/II interaction was carried out using recombinant CaN containing the CnAα subunit (Coghlan et al., 1995). A subsequent report using similar biochemical methods demonstrated the simultaneous association of PKA, PKC, and CaN with AKAP79 (Klauck et al., 1996). These findings have led to the proposal that AKAP79 serves as a mammalian scaffold protein in coordinating the activity of three major signaling proteins in the PSD in response to three distinct activation signals (Klauck et al., 1996). Calcineurin bound to AKAP79 is inactive, and Ca^{2+}/CaM does not relieve the inhibition by AKAP79. This suggests that release from AKAP79 and activation of CaN are not due to CaM binding to CaN or AKAP79 (Coghlan et al., 1995). Furthermore, AKAP79 inhibits a truncation mutant of CaN that lacks the autoinhibitory domain, indicating that the binding and inhibition of CaN is Ca^{2+} independent (Coghlan et al., 1995). Substrates of AKAP79-regulated CaN are unknown and cannot be identified until the mechanism of release from AKAP79 is determined. A 14 amino acid residue peptide corresponding to AKAP79 residues 88–102 inhibits CaN activity, thus localizing the CaN-binding site of AKAP79 to a region distinct from both PKA- and PKC-binding domains (Coghlan et al., 1995). In contrast, the AKAP79-binding site on CaN is unidentified. However, the amino acid sequence of AKAP79 residues 88–102 has a high degree of sequence similarity to FKBP12 residues 32–47, which interact with the catalytic domain of CaN, suggesting that AKAP79 and FKBP12/FK506 bind to and inhibit CaN by a similar mechanism (Kissinger et al., 1995).

C. Antagonism of cAMP Signal

Prior to the discoveries of cell-permeable inhibitors of PP1 and PP2A (okadaic acid) and of CaN (cyclosporin A and FK506), the physiological roles of these protein phosphatases were inferred from biochemical data. One initial strategy employed the dual approach of determining the regional colocalization of CaN with well-characterized phosphoproteins and then determining the catalytic efficiency (k_{cat}/K_m ratios) of purified CaN toward a particular phosphoprotein substrate as an indicator of possible physiological dephosphorylation by CaN *in vivo*. Using this approach, Inhibitor-1 and DARPP32, the RII subunit of PKA, microtubule-associated protein 2 (MAP2), tau proteins, and myosin light chains were identified as physiological substrates of CaN (King *et al.*, 1984; Williams *et al.*, 1986; Klee *et al.*, 1988; Shenolikar and Nairn, 1991).

DARPP32 and Inhibitor-1 are small, heat-stable inhibitors of PP1 having a high degree of sequence homology (Williams *et al.*, 1986). In brain, Inhibitor-1 is ubiquitously expressed whereas DARPP32 expression is restricted to neurons possessing D1 dopamine receptors (Hemmings *et al.*, 1992). It is well established that DARPP32 (as well as Inhibitor-1) becomes a potent, specific inhibitor of PP1 when phosphorylated on a specific Thr residue (Thr-34 or Thr-35 for DARPP32 or inhibitor-1, respectively) by PKA and that PP1 inhibition is completely reversed by CaN dephosphorylation (Stemmer and Klee, 1991). As the functional regulation of these two proteins depends on the stimulation of two enzymes with opposing actions, it has been hypothesized that DARPP32 and Inhibitor-1 play important roles in cellular processes where cAMP and Ca^{2+} act antagonistically (Walaas and Greengard, 1991). The report by Halpain *et al.* (1990) provides strong support for this hypothesis. Using striatal slices from rat brain, it was demonstrated that the forskolin-induced phosphorylation of DARPP32 on Thr residues was abolished by simultaneous treatment of the slices with *N*-methyl-D-aspartate (NMDA) (Halpain *et al.*, 1990). Because the neurons in these slices contain both D1 dopamine receptors and NMDA receptors, it has been proposed that dopamine and glutamate have antagonistic effects in these neurons (Halpain *et al.*, 1990). In this model the dopamine-induced increase in cAMP activates PKA and results in phosphorylation of DARPP32 and inhibition of PP1 (Halpain *et al.*, 1990). The action of PKA is facilitated by the general decrease in phosphatase activity brought about by PP1 inhibition. Conversely, the glutamate-induced increase in Ca^{2+} (via NMDA channels) activates CaN, resulting in DARPP32 dephosphorylation and activation of PP1 (Halpain et al., 1990). The stimulation of CaN activity and the indirect activation of PP1 by Ca^{2+} result in dephosphorylation of proteins phosphorylated by PKA. This model most likely can be extended to cells where responses to different signals are mediated by cAMP or Ca^{2+} and Inhibitor-1 is present.

Ca^{2+} antagonism of the cAMP signal by CaN activity is reinforced by additional mechanisms. Phosphorylation of RII by the PKA catalytic subunit inhibits reassociation and inactivation of the catalytic subunit in the absence of cAMP (Hubbard and Cohen, 1993). Dephosphorylation of RII by CaN could lead to an increase in the rate of association of the catalytic subunit with RII, resulting in a more rapid inhibition of PKA activity (Hubbard and Cohen, 1993). Previous studies have shown feedback inhibition of cAMP production in AtT20 cells by cAMP-induced enhancement of Ca^{2+} influx through dihydropyridine-sensitive Ca^{2+} channels (Shipston et al., 1994; Cooper et al., 1995). A report by Antoni et al. (1995) showed that this Ca^{2+}-induced inhibition of corticotropin-releasing factor (CRF)-induced cAMP formation in AtT20 cells is mediated by CaN and is associated with the expression of a novel adenylyl cyclase (adenylate cyclase) isotype. FK506 enhanced the CRF-induced cAMP production, whereas pretreatment of the cells with A_{23187} ionophore inhibited CRF-induced cAMP formation in AtT20 cells (Antoni et al., 1995). The target of CaN action is unknown, but the previous findings implicate the novel adenylyl cyclase (Antoni et al., 1995). This type of antagonism of the cAMP signal by Ca^{2+} appears to be cell type specific, as another study reported that CaN activity stimulates cAMP production in bovine adrenal cortical cells (Baukal et al., 1994).

The finding that PP1 activity can be regulated by CaN dephosphorylation of DARPP32 or Inhibitor-1 has relevance to the use of phosphatase inhibitors to identify the phosphatase responsible for the regulation of a specific protein. Observing that a Ca^{2+}-dependent effect on the activity of a protein is blocked by treatment of the cell or tissue with FK506 or CsA is not sufficient evidence to conclude a direct effect on activity by CaN dephosphorylation, particularly when coupled to treatments that increase cAMP levels. Inhibition of CaN by FK506 or CsA could favor the net phosphorylation of DARPP32 or Inhibitor-1, leading to PP1 inhibition. In this case, the protein would be directly regulated by PP1 and indirectly by CaN via its dephosphorylation of DARPP32 or Inhibitor-1. Thus, an effect that is FK506 or CsA sensitive, and also okadaic acid-sensitive, would suggest the previous mechanism of regulation. An okadaic acid-sensitive and FK506- or CsA-insensitive effect would suggest an effect due to PP1 or PP2A with no involvement by CaN. Conversely, an effect that is FK506 or CsA sensitive and okadaic acid insensitive suggests direct regulation by CaN. An example of a cellular process that is both okadaic acid and FK506 sensitive is the long-term depression (LTD) of postsynaptic potentials that is induced by repetitive low frequency, presynaptic stimulation of CA1 pyramidal neurons in the hippocampus (Malenka, 1994).

D. Regulation of Postsynaptic Potential

A characteristic of neuronal connections in many regions of the brain is their plasticity. The number and position of synaptic connections can change as well as the strength of individual synaptic transmission. Certain types of presynaptic activity lead to long-lasting changes in the strength of the postsynaptic response (Bear and Malenka, 1994). In area CA1 of the hippocampus, short, repetitive bursts of high-frequency stimulation give rise to a long-term potentiation of the postsynaptic action potential, whereas longer, repetitive bursts of low-frequency stimulation induce a long-term depression or weakening of postsynaptic potential (Lisman, 1989; Malenka, 1994). Surprisingly, LTP and LTD share some common elements for induction. Glutamate and depolarization-induced Ca^{2+} influx through NMDA receptors is required for both LTP and LTD induction (Dudek and Bear, 1992; Malenka, 1994). However, LTP induction may also require cAMP production by a Ca^{2+}/CaM-dependent adenylyl cyclase and the opening of mGluRs and non-NMDA receptors (Bashir *et al.*, 1993; Blitzer *et al.*, 1995; Malenka, 1994). Findings that LTP requires protein kinase activation led to the hypothesis that LTD may require stimulation of protein phosphatase activity (Lisman, 1989; Malenka, 1994). This hypothesis was confirmed by findings that okadaic acid, thiophosphorylated inhibitor-1, or FK506 could each, by itself, prevent the induction of LTD (Mulkey *et al.*, 1993, 1994). These observations also demonstrated the requirement of CaN and PP1 for induction and maintenance of hippocampal LTD (Mulkey *et al.*, 1994). Based on these findings, a model for LTD has been proposed in which the depolarization-induced Ca^{2+} influx through NMDA receptors activates CaN by Ca^{2+}/CaM. The dephosphorylation of inhibitor-1 by CaN then results in the activation of PP1 and dephosphorylation of the appropriate substrates (Mulkey et al., 1994). The substrates of PP1 relevant to LTD induction and maintenance are unknown. In addition, it is unknown whether Inhibitor-1 is the only CaN substrate that must be dephosphorylated to induce and maintain LTD (Mulkey et al., 1994). Two potential PP1 substrates are autophosphorylated CaMKII, which is dephosphorylated by PP1 *in vitro*, and the AMPA receptor, which has elevated responses in the presence of okadaic acid and CaM kinase II (Fukunaga et al., 1989; McGlade-McCulloh et al., 1993). A potential additional CaN substrate is the NMDA receptor itself. Lieberman and Mody (1994) demonstrated that high (10 μM) but not low concentrations of okadaic acid increased the duration of NMDA receptor openings. This okadaic acid-induced increase in NMDA channel opening did not occur when Ca^{2+} was prevented from entering the channel (Lieberman and Mody, 1994). FK506 also increased the duration of NMDA receptor openings (Lieberman and Mody, 1994). In addition, Tong et al. (1995) showed that synaptic desensitization of NMDA receptors was prevented by CaN inhibitors or intracellular BAPTA but not by inhibitors of PP1 and PP2A or tyrosine phosphatases. Together these results indicate that CaN phosphatase activity decreases Ca^{2+} influx through the NMDA receptor, and they also suggest that the NMDA receptor is directly regulated by CaN (Lieberman and Mody, 1994; Tong et al., 1995).

E. Regulation of Neurotransmitter Release

The GTP-binding protein dynamin 1 plays a key role in the regulation of synaptic vesicle recycling in the presynaptic terminus (De Camilli and Takei, 1996). Dynamin 1 GTPase activity is required for coated pit invagination and detachment (De Camilli and Takei, 1996). In the temperature-sensitive *Drosophila* mutant *shibire*, endocytosis of empty synaptic vesicles is defective (Robinson et al., 1994). The *shibire* mutation has been mapped to the GTPase domain of dynamin 1, indicating that dynamin 1 GTPase activity is important for synaptic vesicle endocytosis following depolarization-induced neurotransmitter release (Robinson et al., 1994). Dynamin 1 GTPase activity is increased by PKC phosphorylation and decreased by CaN dephosphorylation, implicating PKC and CaN in the regulation of synaptic vesicle endocytosis (Robinson et al., 1994). During the depolarization-induced Ca^{2+} influx into the presynaptic terminal through voltage-gated Ca^{2+} channels, dynamin 1 is rapidly dephosphorylated while other PKC substrates (MARCKS, neuromodulin) are phosphorylated (Liu et al., 1994b). Using purified rat brain synaptosomes, Liu et al. (1994b) showed that the depolarization-induced dynamin 1 dephosphorylation and inhibition of its GTPase activity are prevented by CsA, suggesting an involvement of CaN in regulating the phosphorylation state and GTPase activity of dynamin 1 during depolarization-induced Ca^{2+} influx into the presynaptic terminal. The finding that CsA treatment of cultured rat neurons increases the rate, but not the amplitude, of glutamate receptor-mediated excitatory postsynaptic currents suggests glutamate release from the presynaptic terminal is increased (Victor et al., 1995). These findings support the conclusion that CaN activity is involved in the regulation of neurotransmitter release. Nichols et al. (1994) showed that inhibition of CaN activity in purified synaptosomes prevents dynamin 1 dephosphorylation and increases glutamate release, providing strong support for a role of CaN phosphatase activity toward dynamin 1 in the regulation of synaptic vesicle endocytosis and neurotransmitter release.

F. Regulation of Cytoskeleton

Early studies identified MAP2 and tau proteins as CaN substrates *in vitro*, suggesting that CaN played a role in the regulation of microtubule assembly (Goto et al., 1985). *In vitro* studies have demonstrated that the phosphorylation state of MAP2 and tau regulate microtubule assembly (Jameson et al., 1980; Murthy and Flavin, 1983; Yamamoto et al., 1983). Dephosphorylation of MAP2 or tau by CaN promotes microtubule assembly and stabilizes preformed microtubules, whereas phosphorylation has the opposite effect (Murthy and Flavin, 1983; Yamauchi and Fujisawa, 1983). A role for CaN in stabilizing microtubule structures in neurons has been suggested by several studies. In dendritic spines, immunohistochemical studies showed that MAP2 and CaN are colocalized on microtubules (Wood et al., 1980; Matus et al., 1981). Using rat brain slices, Halpain and Greengard (1990) showed that activation of NMDA receptors induces dephosphorylation of MAP2. *N*-Methyl-D-aspartate receptor antagonists blocked MAP2 dephosphorylation, whereas non-NMDA receptor antagonists had no effect (Halpain and Greengard, 1990). In addition, kainate and quisqualate were less effective in promoting MAP2 dephosphorylation. These results suggest that CaN, activated by Ca^{2+} influx through NMDA receptors, regulates microtubule dynamics by dephosphorylating MAP2 (Halpain and Greengard, 1990). However, the effects of protein phosphatase inhibitors on NMDA receptor-dependent MAP2 dephosphorylation have not been tested. The localization of CaN in growth cones of developing neurons is sensitive to cytochalasin or nocodazole, indicating that the integrity of microfilaments and microtubules is required for CaN localization within the growth cone (Ferreira *et al.*, 1993). Inhibition of CaN activity prevented axonal elongation and increased the amount of phosphorylated tau protein (Ferreira *et al.*, 1993). Using cultured DRG chick neurons, Chang *et al.* (1995) showed that the CaN inhibitors CsA and FK506 inhibit neuritogenesis and neurite extension. In addition, localized photochemical inactivation of CaN within growth cones by malachite-green conjugated anti-CaN antibodies caused filopodial retraction (Chang *et al.*, 1995). These studies provide compelling evidence for a role of CaN in the regulation of cytoskeletal dynamics during neuronal development and also during activity-dependent changes in neuronal connections.

In smooth muscle, calponin inhibits the actin-activated MgATPase of smooth muscle myosin and inhibits actin filament movement over immobilized myosin in the *in vitro* motility assay (Fraser and Walsh, 1995). Calponin phosphorylated *in vitro* by PKC or CaM kinase II no longer binds to actin or inhibits the actomyosin MgATPase (Fraser and Walsh, 1995). It has been shown that PKC or CaM kinase II phosphorylated calponin is dephosphorylated by CaN and that CaN dephosphorylation restores actin binding and MgATPase inhibition by calponin *in vitro* (Fraser and Walsh, 1995). These findings suggest that CaN phosphatase activity is involved in the association of calponin with actin and in the regulation of smooth muscle contraction. In addition to an involvement in the regulation of smooth muscle contraction, CaN activity has been reported to be involved in the regulation of neutrophil motility on vitronectin (Hendey *et al.*, 1996).

Several studies have implicated a loss of CaN activity in the formation of paired helical filaments (PHFs), a hallmark of Alzheimer's disease (AD) (Drewes *et al.*, 1993; Harris *et al.*, 1993; Gong *et al.*, 1994; Garver *et al.*, 1995; Kayyali *et al.*, 1995). The major component of PHFs is hyperphosphorylated tau protein (Gong *et al.*, 1994). Tau proteins normally contain 2–3 mol phosphate/mol of tau; however, nine phospho-Ser residues in PHF tau have been identified that are not phosphorylated in normal tau (Gong *et al.*, 1994). A report by Harris *et al.* (1993) showed that only concentrations of okadaic acid that inhibit CaN (5 μM) induced the formation of hyperphosphorylated, PHF-like tau in fresh brain slices. In addition, these PHF-like tau were dephosphorylated by CaN. Gong *et al.* (1994) demonstrated that CaN dephosphorylated the abnormal phosphorylation sites of authentic AD tau to much greater extents than PP1 and PP2A. Together, the results from these studies suggest that CaN phosphatase activity in AD brains might be reduced. Support for this proposal comes from a study using knockout mice lacking CnAα, the major neuronal CnA subunit isoform (Kayyali *et al.*, 1995). Elevated levels of hyperphosphorylated tau were present in brains of the CnA$\alpha^{-/-}$ mice, occurring primarily in hippocampal mossy fibers (Kayyali *et al.*, 1995). Because the CnAα isoform accounts for the majority of CaN catalytic subunit activity in the brain, these findings indicate that reduced CaN levels can lead to the accumulation of hyperphosphorylated tau and may be relevant to the mechanism of AD progression (Kayyali *et al.*, 1995).

G. Regulation of Ion Channels and Transporters

As mentioned in the previous section, the use of FK506 and CsA has provided evidence indicating regulation of the NMDA receptor by CaN. Biochemical studies, as well as studies of channel activity in the absence or presence of these CaN inhibitors, has implicated CaN in the regulation of additional ion channels from yeast to human. In yeast, CaN appears to play a central role in regulating intracellular ion homeostasis (Engele-Garrett *et al.*, 1995). Yeast mutants lacking the CnB subunit gene exhibit impaired salt tolerance as evidenced by vanadate hypersensitivity, slow growth in alkaline pH, and accumulation of high intracellular levels of NaCl or LiCl (Li$^+$ is a toxic analog of Na$^+$) (Nakamura *et al.*, 1993). Nakamura *et al.* (1993) showed that although there was no difference between wild-type and CnB-deficient yeast mutants in the rate of Na$^+$ accumulation, wild-type cells lowered their intracellular Na$^+$ levels more rapidly. In addition, FK506 treatment of wild-type cells decreased the rate of Na$^+$ export. This study suggested a role for CaN in enhancing Na$^+$ export through a membrane channel. However, the involvement of gene transcription or posttranslational modification of existing proteins was not addressed (Nakamura *et al.*, 1993). A more recent study by Mendoza *et al.* (1994) presented evidence indicating that CaN regulation of both gene expression and posttranslational modifications of ion transporters is involved in NaCl tolerance of yeast. In yeast, Na^+ enters the cell through a K^+ transport system (Mendoza et al., 1994). In the presence of high extracellular NaCl, the affinity of the K^+ uptake system for K^+ increases, resulting in a decreased Na^+ influx. The high K^+ affinity state is due to expression of TRK1, a gene that encodes a putative K^+ transporter (Mendoza et al., 1994). Na^+ export is mediated by a P-type ATPase encoded by ENA1, an essential gene for NaCl tolerance, which is induced by Li^+, Na^+, or alkaline pH. Using CnB null mutants, it was shown that the accumulation of high intracellular levels of Li^+ in CnB-deficient mutants was due to reduced Li^+ efflux (Mendoza et al., 1994). Furthermore, Northern blot analysis showed reduced levels of ENA1 expression in the CnB mutants. These findings suggest that CaN phosphatase activity is required for normal levels of ENA1 gene transcription (Mendoza et al., 1994). In addition to affecting Na^+ and Li^+ efflux, the CnB null mutation prevented the transition of the K^+ uptake system from the low-affinity to the high-affinity state. This finding, along with the finding that this transition does not require new protein synthesis, suggests that CaN dephosphorylation regulates, directly or indirectly, the TRK1 K^+ channel (Mendoza et al., 1994).

In nonexcitable mammalian cells, Na^+ and K^+ transport mainly occurs through the ouabain-sensitive Na^+,K^+-ATPase of the plasma membrane, which exports Na^+ and imports K^+ (Kopp and DiBona, 1992). In renal tubule cells, α-adrenergic receptor activation by norepinephrine is associated with elevated cytosolic Ca^{2+} levels and increased Na^+ export (Minneman, 1988; Michel et al., 1990). These findings led to Aperia et al. (1992) showing that α-adrenergic stimulation of proximal convoluted renal tubules increased the activity of the Na^+,K^+-ATPase. Furthermore, incubation of the tubules with the Ca^{2+} ionophore A_{23187} also activated the enzyme. The stimulation of Na^+,K^+-ATPase activity by α-adrenergic receptor activation was prevented by the CaN autoinhibitory peptide and also by FK506, indicating that the increase in Na^+,K^+-ATPase activity mediated by α-adrenergic stimulation involves activation of CaN (Aperia et al., 1992). The CaN regulation of the renal Na^+,K^+-ATPase appears to involve indirect and direct mechanisms. DARPP32 is present in renal tubules, and a fragment of phospho-DARPP32 has been shown to inhibit Na^+,K^+-ATPase activity in permeabilized renal tubules (Aperia et al., 1991). Similar to the opposing actions of dopamine and glutamate in the brain, dopamine and norepinephrine have opposing actions on renal Na^+ homeostasis (Michel et al., 1990; Marcaida et al., 1996). Dopamine

4. Calmodulin-Regulated Phosphatase Calcineurin

inhibits renal Na^+,K^+-ATPase, leading to increased cytosolic Na^+ levels (Aperia et al., 1991). Thus, in renal tubule cells, dopamine-induced increases in cAMP would lead to phosphorylation of DARPP32 by PKA and inhibition of PP1 activity (Aperia et al., 1992). In contrast, the norepinephrine-induced elevation of cytosolic Ca^{2+} levels would cause activation of CaN, dephosphorylation of DARPP32, and activation of PP1 (Aperia et al., 1992). The opposing actions of cAMP and Ca^{2+} on the renal tubule cell Na^+,K^+-ATPase represents another example of Ca^{2+} and cAMP antagonism. Similar findings were reported using primary cultures of rat cerebellar neurons. (Marcaida et al., 1996). Glutamate treatment of neurons resulted in activation of the Na^+, K^+-ATPase, which was prevented by MK-801 or reversed by PMA treatment (Marcaida et al., 1996). These findings indicate that Na^+, K^+-ATPase activation was mediated by Ca^{2+} influx through the NMDA receptor and was reversed by PKC phosphorylation. The activation of Na^+,K^+-ATPase by glutamate was prevented by CsA, suggesting that Na^+,K^+-ATPase activity is regulated by PKC and CaN in cerebellar neurons (Marcaida et al., 1996).

The suppression of the $GABA_A$ receptor-mediated inward Cl^- current by NMDA receptor activation represents another example of modulation of ion channel activity by NMDA receptor activity (Chen and Wong, 1995). Studies have shown that the $GABA_A$ receptor-mediated response is modified by changes in cytosolic Ca^{2+} levels (Mulle et al., 1992). Ca^{2+} influx through the nicotinic receptor channel has been shown to suppress the GABA responses in habenular cells (Mulle et al., 1992). Also, caffeine-induced Ca^{2+} release from intracellular stores strongly inhibits the GABA response (Mouginot et al., 1991). Both studies also demonstrated that Ca^{2+} entry through voltage-gated channels had no effect on GABA responses (Mouginot et al., 1991; Mulle et al., 1992). Early studies indicated that regulation of $GABA_A$ receptor-mediated responses involved phosphorylation (Chen et al., 1990). Based on these findings, it was shown, using acutely isolated hippocampal neurons from adult guinea pig, that the suppression of $GABA_A$ receptor-mediated current by NMDA receptor activation was partially inhibited by okadaic acid ($6 \mu M$) and completely reversed by the CaN autoinhibitory peptide ($50 \mu M$) (Chen and Wong, 1995). The CaN autoinhibitory peptide ($50 \mu M$, in vitro) results in 80–90% inhibition of CaN activity, whereas the IC_{50} concentration for inhibition of CaN by okadaic acid is $5 \mu M$, suggesting that CaN may not have been completely inhibited by $6 \mu M$ okadaic acid in these experiments (Chen and Wong, 1995). Unfortunately, lower concentrations of okadaic acid were not used and thus it is not known whether the regulation of $GABA_A$ receptor-mediated responses involves a direct or indirect CaN mechanism. Intracellular perfusion of constitutively active CaN completely suppressed the $GABA_A$ receptor-mediated response, providing additional evidence for CaN regulation of $GABA_A$ receptor activity (Chen and Wong, 1995). However, similar experiments using phosphorylated inhibitor-1 were not carried out, so a role for PP1 in $GABA_A$ receptor regulation cannot be ruled out.

In higher plants, regulation of guard cell stomatal pores by K^+ channels controls CO_2 uptake and transpirational H_2O loss (Gilroy et al., 1990). K^+ influx through guard cell K^+ channels results in opening of stomatal pores, whereas elevation of cytosolic Ca^{2+} levels prevents stomatal pore opening by inactivating the guard cell K^+ channel (Schroeder and Hagiwara, 1989; Mansfield et al., 1990). Using patch-clamp techniques, it has been shown that CsA or FK506 prevents the Ca^{2+}-induced inactivation of K^+ current in fava bean guard cells (Luan et al., 1993). In addition, the constitutively active proteolytic fragment of CaN inhibited K^+ channel activity in the absence of Ca^{2+} (Luan et al., 1993). Furthermore, an endogenous Ca^{2+}-sensitive and FKBP/FK506- or CyP/CsA-sensitive phosphatase activity was identified in guard cell extracts (Luan et al., 1993). These findings provide evidence that K^+ channel activity in guard cells is regulated by a Ca^{2+}/CaM-regulated, CaN-like protein phosphatase.

Several studies have shown that the activity of the voltage-sensitive Na^+ channel of excitable cells is modulated by phosphorylation (Numann et al., 1991; Li et al., 1992; Chen et al., 1995). Protein kinase A or PKC phosphorylation of the rat brain Na^+ channel α subunit decreases Na^+ current during depolarization, whereas PKC phosphorylation also slows channel inactivation (Numann et al., 1991; Li et al., 1992). Protein kinase A phosphorylates the α subunit on four serine residues located in the intracellular loop between domains I and II (Murphy et al., 1993). Serine-610 is a common phosphorylation site of PKA and PKC, whereas Ser-1506 on the loop between domains III and IV is the PKC site that mediates the effect of PKC on channel inactivation and augments the inhibitory effect of PKA (Murphy et al., 1993). These in vitro studies showed that PP2A and CaN, but not PP1, dephosphorylated purified Na^+ channels (Murphy et al., 1993). Using rat brain extracts, Chen et al. (1995) showed that inhibition of phosphatase activity toward Na^+ channels by 500 nM okadaic acid did not significantly increase the Na^+ channel dephosphorylation observed with 10 nM okadaic acid. This finding implicates PP2A in Na^+ channel dephosphorylation. However, incubation of the extracts with 10 nM okadaic acid and 2 mM EGTA resulted in a further 12% decrease in phosphatase activity, suggesting that CaN also dephosphorylates Na^+ channels (Chen et al., 1995). Finally, treatment of synaptosomes with okadaic acid or CsA increased Na^+ channel phosphorylation at PKA sites, providing further evidence that PP2A and CaN are involved in the regulation of neuronal voltage-sensitive Na^+ channels (Chen et al., 1995).

The M current is a noninactivating K^+ current that regulates the excitability of several neuronal cell types (Brown, 1988). M current activity causes hyperpolarization and decreased cell excitability, whereas inhibition of the M current leads to depolarization and increased cell excitability (Brown, 1988). It has been shown that the M current is enhanced by small increases in cytosolic Ca^{2+} (50 to 150 nM), but is inhibited by higher Ca^{2+} concentrations (> 200 nM) (Marrion, 1996). The findings that basal M current activity and the activity enhanced by the small Ca^{2+} increase require ATP suggested the requirement of kinase activity for these effects and also suggested that dephosphorylation was required for M current inhibition by higher Ca^{2+} concentrations (Yu et al., 1994). In support of this proposal, Marrion (1996) showed that intracellular application of constitutively active CaN into bullfrog sympathetic neurons inhibits the M current. The Ca^{2+}-dependent inhibition of the M current was blocked by the CaN autoinhibitory peptide, or CsA. Okadaic acid or microcystin LR had no effect on the M current, suggesting that the effect of CaN on the M current is not due to activation of PP1 by Inhibitor-1 dephosphorylation. Incubation of excised inside-out patches with constitutively active CaN reduced the high open probability activity of the M current, suggesting amplitude regulation by CaN (Marrion, 1996). These findings are consistent with the results of previous studies of the effects of elevated cytosolic Ca^{2+} levels on the macroscopic M current (Marrion, 1996).

Studies have shown that cloned 5-HT_{1C} receptors transfected into A9 cells are desensitized by PMA treatment, suggesting a role for PKC in 5-HT_{1C} receptor desensitization (Boddeke et al., 1993a). In addition, it was reported that 5-HT treatment of A9 cells expressing 5-HT_{1C} receptors increased the cytosolic Ca^{2+} concentration and activated a Ca^{2+}-dependent K^+ current, which was decreased by phorbol esters (Boddeke et al., 1993b). These findings are consistent with the findings that 5-HT_{1C} receptor activation causes increased phosphoinositide hydrolysis (Boddeke et al., 1993b). Based on these studies, the involvement of CaN in the regulation of 5-HT_{1C} receptors transfected into A9 cells was investigated (Boddeke et al., 1993b). It was shown that 5-HT evoked a Ca^{2+}-dependent K^+ current of 109 pA and that PMA treatment reduced this current by 46% (Boddeke et al., 1993b). Calphostin C completely inhibited PMA-induced 5-HT_{1C} receptor desensitization, providing further evidence for the involvement of PKC. Intracellular injection of CaN into A9 cells pretreated with PMA resulted in a partial reversal of PMA-induced 5-HT_{1C} receptor desensitization, as shown by the 34% increase in the 5-HT-induced K^+ currrent. In addition, the K^+ current of 84 pA induced by IP_3 was not decreased by PMA pretreatment, suggesting that the effects of PKC were reversed by Ca^{2+}. Together, these results suggest that the desensitization of 5-HT_{1C} receptors by PKC is reversed by CaN phosphatase activity (Boddeke et al., 1993b). However, the effects of CsA or FK506 on the endogenous CaN of A9 cells were not investigated. Additional support for the conclusion that 5-HT receptor desensitization is reversed by CaN phosphatase activity is provided by a study showing that intracellular application of the CsA/CyP complex or the CAN autoinhibitory peptide, but not okadaic acid, prevented the loss of 5-HT_3 receptor responsiveness in NG108 cells (Boddeke et al., 1996). In contrast, 5-HT_{1C} receptor-mediated responses in pyramidal neurons of the rat piriform cortex are resistant to desensitization; thus the physiological role of CaN in 5-HT receptor regulation may be cell type specific (Sheldon and Aghajanian, 1991).

It has long been known that Ca^{2+} influx through voltage-activated Ca^{2+} channels of excitable cells is modulated by PKA and cytosolic Ca^{2+} levels (Hadley and Lederer, 1991). The Ca^{2+} regulation occurs through a negative feedback mechanism (Eckert and Chad, 1984). The original report by Chad and Eckert (1986) showed that perfusion of molluscan neurons with CaN and CaM increases the rate of Ca^{2+} channel inactivation, suggesting that the Ca^{2+}-induced inactivation of voltage-activated Ca^{2+} channels is mediated by a Ca^{2+}-stimulated phosphatase. Although there has been debate on the matter, it appears that PKA phosphorylation of the channel is required for its voltage-dependent activation (Hadley and Lederer, 1991). Support for this conclusion is provided from a study of the mechanisms of the voltage- and Ca^{2+}-dependent inactivation of voltage-activated Ca^{2+} channels in guinea pig ventricular myocytes (Hadley and Lederer, 1991). It was shown that voltage-dependent inactivation immobilizes the gating charge. In contrast, the Ca^{2+}-dependent inactivation was severely inhibited by isoproteronol-induced, cAMP-dependent phosphorylation of the channel. These results support the hypotheses that phosphorylation is required for activity and that Ca^{2+}

4. Calmodulin-Regulated Phosphatase Calcineurin

-dependent inactivation of voltage-activated Ca^{2+} channels requires activation of CaN (Hadley and Lederer, 1991).

Similarly, CaN activity appears to negatively modulate the regulation of N-type Ca^{2+} channels in sympathetic neurons of the male rat major pelvic ganglion (Zhu and Yakel, 1998). Dialysis of the cultured sympathetic neurons with the CaN autoinhibitory peptide or the CsA/CyP complex greatly reduced the α_2-noradrenergic- or somatostatin receptor-induced inhibition of N-type Ca^{2+} channels. Okadaic acid (1 μM) had no effect, suggesting a direct role of CaN. The findings that CaN inhibition or dialysis of the sympathetic neurons with constitutively active CaN had no effect on N-type Ca^{2+} current suggests that CaN is not affecting channel function directly (Zhu and Yakel, 1988). Furthermore, muscarinic receptor-induced inhibition of N-type Ca^{2+} channel activity is unaffected by CaN inhibition (Zhu and Yakel, 1998). The α_2-noradrenergic- or somatostatin receptor-induced inhibitions of N-type Ca^{2+} channels are mediated by pertussis toxin-sensitive G-proteins, whereas the effects of muscarin are modulated by a different G-protein (Hille, 1994). These findings indicate that in sympathetic neurons of the major pelvic ganglion, CaN directly regulates the pertussis toxin-sensitive G-protein-mediated inhibition of N-type Ca^{2+} channels (Zhu and Yakel, 1997).

H. Regulation of Gene Expression

Prior to 1991 it was known that the immunosuppressant drugs CsA and FK506 blocked T-cell activation by primarily inhibiting IL-2 gene induction (Schreiber, 1991). CsA and FK506 bind to and inhibit the peptidylprolyl isomerase activity of distinct members of the immunophilin protein family (Bierer et al., 1990; Kallen et al., 1991; Schreiber, 1991). However, several studies had shown that the immunosuppressive effects of CsA and FK506 are not due to their inhibitory effect on the enzymatic activity of the immunophilins (Bierer et al., 1990; Heitman et al., 1991; Kallen et al., 1991; Koltin et al., 1991). It was known that these drugs inhibit a Ca^{2+}-dependent step required for IL-2 gene induction (Kay et al., 1989; Mattila et al., 1990). One of several transactivating factors that bind to the transcriptional regulatory region of the IL-2 gene is NFAT (Muegge and Durum, 1990). Antigen binding by the T-cell receptor activates signal transduction pathways involving tyrosine phosphorylation and phosphoinositide breakdown (Schreiber, 1991). The findings that phorobl esters and Ca^{2+} ionophores increase NFAT activity and induce IL-2 gene expression indicated that NFAT induction requires PKC activity and is regulated by cytosolic Ca^{2+} levels (Rao, 1995). These findings supported the hypothesis that CsA and FK506 inhibited IL-2 gene induction by blocking the induction of NFAT, although the molecular mechanism of inhibition of NFAT induction was unknown (Schreiber, 1991).

In 1991 two key observations were made concerning the mechanism of immunosuppression by CsA and FK506. The findings that the immunosuppressive effects of CsA and FK506 were not due to the inhibition of the enzymatic activity of the immunophilins supported a model in which the drug/protein complexes bound to and affected another cellular target protein (Schreiber, 1991). This hypothesis was directly tested and confirmed by Liu *et al.* (1991a) showing that the CyP/CsA or FK506/FKBP12 complexes bound to and inhibited the phosphatase activity of CaN. A simultaneous report by Flanagan *et al.* (1991) appeared showing that transcriptionally competent NFAT is formed on antigen binding to the T-cell receptor or by PMA and ionomycin treatment, which induce the cytoplasmic subunit of NFAT to translocate into the nucleus and combine with the newly synthesized nuclear subunit of NFAT. This report further showed that CsA or FK506 prevented the PMA- and ionomycin-dependent translocation of the cytoplasmic component of NFAT into the nucleus. These two reports showed that CsA/Cyp or FK506/FKBP12 inhibit CaN and that they block the Ca^{2+}-dependent translocation of the cytosolic subunit of NFAT into the nucleus. The stage was set to test the hypothesis suggested by these data that CsA/Cyp or FK506/FKBP12 block the dephosphorylation of a substrate by CaN that is required for translocation of the cytosolic subunit of NFAT into the nucleus.

Within 4 years this hypothesis was essentially confirmed and the likely CaN substrate identified. O'Keefe *et al.* (1992) showed that overexpression of CaN in the Jurkat T-cell line increased the IC_{50} concentrations of CsA and FK506 required to inhibit the IL-2 promoter and that PMA-treated cells expressing a constitutively active CaN did not require Ca^{2+} for IL-2 promoter activity. Using structural variants of CsA and FK506, an excellent correlation was found between the inhibitory potency of the drug/immunophilin complexes toward CaN and their ability to inhibit NFAT-mediated gene transcription (Fruman *et al.*, 1992a; Liu *et al.*, 1992). In addition, it was shown that the concentrations of CsA and FK506 that inhibit CaN activity in T-cell lysates also prevent IL-2 expression. It was also shown that CsA/CyP or FK506/FKBP12 do not inhibit the activities of PP1 or PP2A (Liu et al., 1992). Together, these results indicated that CaN phosphatase activity is required for NFAT-mediated gene transcription and that the immunosuppressive effects of CsA/Cyp and FK506/FKBP12 are due to their inhibition of CaN. A number of studies showed that the DNA-binding ability of the cytosolic subunit of NFAT (NFATp) is increased by CaN dephosphorylation and that the dephosphorylation and DNA binding of NFAT is inhibited in cell extracts from CsA- or FK506-treated T cells (June, 1991; Rao, 1995). It has also been demonstrated that NFATp dephosphorylation and nuclear translocation are inhibited by CsA and FK506 (June, 1991). These findings provided strong evidence that NFATp is dephosphorylated by CaN and that this dephosphorylation is required for nuclear translocation. Furthermore, as mentioned previously, a physical interaction between NFATp and CaN has been described, providing additional support for the hypothesis that CaN dephosphorylates NFATp.

The findings that CaN phosphatase activity is required for NFATp activation suggest that kinase activity may be the "off" signal for NFAT-mediated gene induction. The termination of NFAT-dependent signaling pathways induced by T-cell receptor activation is enhanced by subsequent treatment with PMA, suggesting a role for PKC in the deactivation of NFAT (Loh et al., 1996b). Interleukin-2 transcriptional inhibition also results from Ca^{2+} influx into T cells in the absence of PMA (Ngheim et al., 1996). To investigate the mechanism of Ca^{2+}-dependent inhibition of IL-2 gene expression, Ngheim et al. (1996) showed that PMA- and ionomycin-induced IL-2 receptor gene activity was inhibited 90% by transfection of constitutively active CaM kinase II, but not by active mutants of CaM kinase IV, CaN, PKC, or the inactive mutant of CaM kinase II. These findings suggest that CaM kinase II phosphorylation may antagonize CaN-dependent T-cell activation. Together, these two reports indicate that phosphorylation may downregulate CaN-dependent T-cell activation pathways, although the target(s) is unknown. The simplest hypothesis is that a kinase phosphorylates NFATp that contains consensus phosphorylation sites for CaM kinase II (Ngheim et al., 1996).

Similarly, elevation of cAMP levels inhibits CaN-dependent T-cell activation pathways (Paliogianni et al., 1993). Immunosuppression by prostaglandin E_2 (PGE_2) involves inhibition of CaN-dependent IL-2 gene expression (Paliogianni et al., 1993). Interleukin-2 promoter activity in Jurkat T cells induced by PMA and ionomycin is specifically reduced by PGE_2, and overexpression of constitutively active CaN decreased the inhibitory potency of PGE_2 toward IL-2 promoter activity (Paliogianni et al., 1993). These findings suggest that PKA activity is involved in counteracting CaN phosphatase activity. However, a cAMP-inducible transcriptional repressor protein (ICER), whose expression is under circadian control, was described in the hypothalamic–pituitary–gonadal axis (Molina et al., 1993). Bodor et al. (1996) demonstrated a role of ICER in PGE_2-mediated inhibition of IL-2 gene induction in T cells. It was shown that elevation of cAMP levels by PGE_2 induced ICER expression and that ICER expression suppressed ionomycin and PMA-induced IL-2 promoter activity. Thus, these findings suggest that ICER induction is involved in the antagonism of CaN-dependent T-cell activation pathways by cAMP (Bodor et al., 1996).

A growing body of literature now exists reporting CsA- or FK506-sensitive regulation of NFAT-mediated gene transcription in immune cells. In addition to IL-2 gene induction, CsA- or FK506-sensitive induction of IL-3, IL-4, IL-6, GM-CSF, tumor necrosis factor α, and c-*rel* has been reported (Rao, 1995; Rao et al., 1997). Furthermore, not only has NFAT DNA-binding activity induced by T-cell receptor activation by antigen binding been shown to be CsA or FK506 sensitive, but similar findings have been found for FcεRI receptor activation by IgE binding (Hutchinson and McCloskey, 1995). In contrast, it has been shown that CsA treatment of the erythropoietin-responsive murine erythroleukemia cell line ELM-I-1 results in inhibition of CaN activity and in prevention of Ca^{2+}-dependent suppression of c-*myb* mRNA synthesis (Schaefer et al., 1996). These results indicate that CaN activation is involved in the Ca^{2+}-dependent inhibition of c-*myb* gene transcription and suggest that CaN activity in erythroid precursor cells is involved in the regulation of growth and differentiation via inhibition of c-*myb* expression.

In nonimmune cells, depending on the system or gene being studied. CaN can activate or inhibit gene transcription. In a whole animal study of the nephrotoxic effects of CsA treatment, it was shown that renal PEPCK gene expression is specifically reduced, whereas hepatic PEPCK mRNA levels were unaffected (Morris et al., 1992). In addition, FK506 treatment has been reported to cause a similar reduction in renal PEPCK mRNA levels (Morris et al., 1991). In contrast to PEPCK mRNA, a 50% increase in procollagen mRNA levels in rat renal cortex was observed during CsA treatment (Nast et al., 1991). The sensitivity of the levels of renal PEPCK and procollagen mRNa to CsA implicates CaN in the regulation of their gene expression; however, renal CaN phosphatase activity was not measured in these experiments.

A study of preproendothelin-1 (PreproET-1) gene expression in human umbilical vein endothelial cells showed, by Northern blot analysis, that KN-62 treatment (the specific inhibitor of CaM kinase II) inhibited thrombin-mediated preproET-1 gene expression (Marsen et al., 1995). In contrast, CsA treatment in the absence of prior thrombin exposure increased preproET-1 mRNA levels to almost the same extent as levels attained with thrombin (Marsen et al., 1995). These results suggest the involvement of CaN phosphatase activity in inhibition of preproEt-1 gene expression. Incubation with CsA and thrombin did not result in further increase in preproET-1 gene expression, suggesting that CaN activity is inhibited during thrombin-mediated signaling pathways leading to preproEt-1 gene induction; however, the effect of thrombin treatment on CaN activity was not determined (Marsen et al., 1995).

A detailed study of the roles of CaM-dependent kinases and CaN in the Ca^{2+}-dependent transcription of immediate early genes (IEG) in PC12 cells demonstrated differential sensitivities of three IEGs to CsA or FK506 (Enslen and Soderling, 1994). As shown by Northern blot analysis and nuclear run-on assays, KN-62 pretreatment strongly inhibited c-*fos*, NGF1-A, and NGF1-B gene expression induced by the Ca^{2+} ionophore A23187. In contrast, CsA or FK506 preincubation strongly enhanced NGF1-A expression, blocked NGF1-B expression, and had no effect on c-*fos* expression. The effects of KN-62 and FK506 were specific for Ca^{2+}-stimulated transcription, as transcription of the same IEGs in response to forskolin or PMA was unaffected. These results indicate that CaN is involved in Ca^{2+}-mediated gene expression in PC12 (rat adrenal pheochromocytoma) cells, but the target of CaN phosphatase activity is unknown.

Several reports have demonstrated in pancreatic islet cells that both cAMP-stimulated and Ca^{2+}-stimulated expression of genes regulated by a CRE (cAMP-responsive element) require CaN activity (Schwaninger et al., 1993, 1995; Eckert et al., 1996). Using pancreatic islet cells transiently transfected with a reporter gene fused to the CRE-containing promoter region of the rat glucagon gene, it was shown that depolarization-induced glucagon gene transcription was strongly inhibited by $10^{-9} M$ concentrations of FK506 or CsA (Schwaninger et al., 1993). It was also shown that CaN phosphatase activity was inhibited by FK506 or CsA treatment and that rapamycin reversed the effects of FK506, but not CsA, on depolarization-induced, CRE-mediated gene transcription. Moreover, the CRE-mediated gene transcription became resistant to inhibition by FK506 or CsA by overexpression of CaN. Similar results were demonstrated for CRE-mediated gene transcription stimulated by increases in cytosolic Ca^{2+} levels caused by treatment with the insulin secretagogues arginine vasopressin, bombesin, or acetylcholine (Eckert et al., 1996). Finally, stimulation of CRE-mediated gene transcription by 8-bromo-cAMP was also demonstrated to be inhibited by FK506 or CsA (Schwaninger et al., 1995). Rapamycin reversed the effects of FK506, but not CsA, on 8-bromo-cAMP induced, CRE-mediated gene transcription, and CRE-mediated gene transcription became resistant to inhibition by FK506 or CsA by overexpression of CaN (Schwaninger et al., 1995). Together, these findings provide strong evidence that CRE-mediated gene transcription requires CaN phosphatase activity in pancreatic islet cells. It was also shown in these experiments that phosphorylation of CREB on Ser-119 (Ser-119 in CREB-327 corresponds to Ser-133 in CREB-341) is not affected by FK506 or CsA, suggesting that CRE-mediated gene transcription depends on CREB phosphorylation at Ser-119 as well as CaN acting elsewhere. This conclusion implies the existence of an inhibitory phosphorylation site, which is dephosphorylated by CaN, that interferes with the transcriptional ability of Ser-119 phosphorylated CREB (Eckert et al., 1996). The kinase involved in this inhibitory phosphorylation, as well as the substrate, is unknown (Eckert et al., 1996).

Finally, it should be noted that not all FK506- or CsA-sensitive gene expression is mediated by CaN-dependent signaling pathways. It has been shown that FK506 or CsA treatment of L929 mouse fibroblasts stably transfected with the MMTV-CAT reporter plasmid enhances glucocorticosteroid receptor-mediated gene expression (Renoir et al., 1995). Cyclosporin A or FK506 by themselves had no effect on CAT gene expression in the absence of dexamethasone, and the antisteroid RU486 prevented the CsA- or FK506-enhanced gene expression obtained in the presence of dexamethasone (Renoir et al., 1995). However, rapamycin, as well as FK506 and CsA analogs that do not inhibit CaN phosphatase activity, also potentiated dexamethasone-induced increases in CAT gene expression (Renoir et al., 1995). These results indicate that the potentiation of dexamethasone-induced gene expression by CsA or FK506 does not occur through a CaN-dependent signaling pathway. Western blots of immunoprecipitated glucocorticoid receptor complexes showed co-precipitation of heat shock protein 90 and the FK506-binding protein FKBP59 and the CsA-binding protein CyP40 (Renoir et al., 1995). These findings suggest that modulation of steroid receptor activity by FK506 or CsA binding to the steroid receptor complex may be involved in the enhancement of dexamethasone-induced gene expression (Renoir et al., 1995).

An important question raised by the findings implicating CaN in the regulation of gene transcription is in which subcellular compartments is CaN activity required for transcriptional regulation? In the case of NFAT, dephosphorylation of NFATp by CaN most likely takes place in both the cytoplasm and nucleus as phosphorylated NFATp is not found in the nucleus (Rao et al., 1998). As reported by Shibasaki et al. (1996), NF-AT4 expressed in BHK cells translocates to the nucleus following A_{23187} treatment. Cyclosporin A blocked NF-AT4 translocation to the nucleus and the dephosphorylation-dependent mobility shift of NF-AT4 in SDS-PAGE. In addition, coexpression of catalytically inactive CaN mutants with NF-AT4 prevented NF-AT4 import into the nucleus. These results indicate that cytoplasmic CaN activity is required for nuclear translocation of NF-AT4 (Shibasaki et al., 1996). Coexpression of CaN and NF-AT4 showed cotranslocation of CaN and NF-AT4 into the nucleus on Ca^{2+} ionophore treatment (Shibasaki et al., 1996). Interestingly, CsA treatment caused both CaN and NF-AT4 to exit the nucleus and reenter the cytoplasm, indicating that CaN phosphatase activity in the nucleus is required for its nuclear localization. Immunoblot analysis and immunohistochemical studies have demonstrated the presence of CaN in the nucleus of rat brain neurons and rat liver cells (Bosser et al., 1993; Pujol et al., 1993; Usuda et al., 1996). Furthermore, CaN phosphatase activity was demonstrated in nuclear extracts of rat liver cells as shown by the Ca^{2+}/CaM-dependent dephosphorylation of three endogenous nuclear phosphoproteins (Bosser et al., 1993). In addition, the developmentally regulated localization of CaN in the nucleus of murine elongating spermatids has been demonstrated by immunofluorescence analysis (Moriya et al., 1995). These results suggest that CaN phosphatase activity in the nucleus is involved in the transcriptional regulation of certain genes. With regard to this possibility, it is intriguing that low micromolar concentrations (0.1 μM) of histone H1 in vitro inhibit the enzymatic activities of CaM kinase II and CaN (Rasmussen and Garen, 1993). In addition, the inhibitory effect of histone H1 was abolished by DNA binding to histone H1. These results suggest an additional mechanism of regulation of CaN phosphatase activity and suggest a dual regulation of nuclear CaN phosphatase activity by Ca^{2+}/CaM and histone H1 (Rasmussen and Garen, 1993). However, evidence for nuclear localization of CaN in undifferentiated or differentiated PC12 cells was not observed by immunofluorescence (Farber et al., 1987).

I. Regulation of Ca^{2+}-Activated Cell Death

It is well known that glutamate is the major excitatory neurotransmitter of the central nervous system. It is also known that glutamate binding to NMDA receptors can induce neuronal cell death (Choi, 1992). The findings that NMDA receptor antagonists prevent glutamate-induced elevations in cytosolic Ca^{2+} levels and protect against neuronal damage caused by neurodegenerative diseases or focal brain ischemia implicate high levels of cytosolic Ca^{2+} in mediating glutamate neurotoxicity (Choi, 1992). However, it has been shown that nitric oxide synthase (NOS) inhibitors block glutamate neurotoxicity due to NMDA receptor activation, indicating that nitric oxide (NO) is involved in NMDA neurotoxicity (Dawson et al., 1992). The finding that phosphorylation of NOS by PKA, PKC, PKG, or CaM kinase II inhibits its catalytic activity suggested that decreased phosphorylation of NOS might potentiate NMDA neurotoxicity (Bredt et al., 1992). The findings that NMDA receptor activation results in MAP2 and DARPP-32 dephosphorylation in vivo suggested that a similar mechanism of CaN activation may be involved in NMDA neurotoxicity (Dawson et al., 1993). To investigate this hypothesis, Dawson et al. (1993) showed that FK506 treatment of primary rat neuronal cultures prior to NMDA treatment significantly inhibited cell death (as measured by trypan blue exclusion). These results suggested that protection against neurotoxicity involved inhibition of CaN and increased NOS phosphorylation. These findings also suggested that NOS is a CaN substrate. Thus, it was also demonstrated that PKC phosphorylated NOS is dephosphorylated by CaN, and FKBP12/FK506 enhanced NOS phosphorylation by PKC in the presence of CaN in vitro (Dawson et al., 1993). In addition, using 293 cells stably transfected with NOS, it was shown that phorbol 12-tetradecanoate 13 acetate (TPA) treatment increased NOS phosphorylation, and FK506 further increased TPA-stimulated NOS phosphorylation (Dawson et al., 1993). Finally, the increased cGMP production in cultured neurons exposed to NMDA was almost completely prevented by FK506, indicating that FK506 inhibits NOS catalytic activity. These data strongly suggest that NOS catalytic activity is increased by CaN dephosphorylation and suggest the involvement of CaN in glutamate neurotoxicity mediated by NOS and the generation of NO.

As noted earlier, prior to 1991 it was known that CsA and FK506 inhibited T-cell growth by inhibiting a Ca^{2+}-dependent step in the signal-

4. Calmodulin-Regulated Phosphatase Calcineurin   219

ing pathway activated by antigen binding to the T-cell receptor. Concurrent studies of T-cell hydridomas and immature T cells showed that anti-CD3 antibodies could also cause cell death in a Ca^{2+}-dependent manner and that CsA inhibited T-cell activation-induced cell death (Smith et al., 1989; Fruman et al., 1992b). It is thought that programmed cell death in the immune system is one mechanism by which tolerance to self-antigens is generated during development (Kabelitz et al., 1993). Following the demonstration that inhibition of CaN phosphatase activity by CsA or FK506 is responsible for the inhibition of T-cell growth, it was demonstrated that CaN activity is required for activation-induced cell death in T-cell hydribomas (Shi et al., 1989; Ucker et al., 1989). Thus, stimulation of T-cell hybridomas by anti-CD3 antibodies results in Ca^{2+} mobilization, NFATp translocation into the nucleus, and cytokine gene transcription (Fruman et al., 1992b; Kabelitz et al., 1993; Krammer et al., 1994). Cell death occurs 6 to 8 hr later and is cycloheximide sensitive, indicating the requirement for new protein synthesis for T-cell hybridoma programmed cell death (Fruman et al., 1992b; Kabelitz et al., 1993). It has been demonstrated that activation-induced cell death in T cells also involves binding of FasL to the Fas receptor (Izquierdo et al., 1996). In addition, it has been shown that Fas-regulated cell death is Ca^{2+} dependent and that CsA blocks activated T-cell receptor-induced FasL gene induction (Singer and Abbas, 1994). These findings suggest that CaN activity is involved in activation-induced cell death in T cells and is required for FasL gene expression. However, the molecular events involved in CaN-dependent FasL gene expression are unknown.

Shibasaki and McKeon (1995) presented evidence that CaN is also involved in the programmed cell death of nonimmune cells. Using BHK fibroblasts cotransfected with CnA and CnB subunits, it was demonstrated that greater than 60% of serum-deprived cells treated with ionomycin entered apoptosis, whereas only 3% of transfected cells were apoptotic in the presence of serum (Shibasaki and McKeon, 1995). These results are consistent with previous findings showing that growth factors inhibit apoptosis (Evan et al., 1992). In contrast to the findings with T cells, it was also shown that new protein synthesis is not required for the CaN-mediated programmed cell death in BHK fibroblasts (Shibasaki and McKeon, 1995). Finally, the cotransfection of the CnA carboxy-terminal deletion mutant with the CnB subunit induced apoptosis in serum-deprived cells in the absence of ionomycin (Shibasaki and McKeon, 1995). These findings provide strong evidence suggesting that CaN phosphatase activity toward a preexisting substrate is involved in Ca^{2+}-

dependent programmed cell death in nonimmune cells. It was also shown that coexpression of BcL-2 with CnA and CnB blocks apoptosis (Shibasaki and McKeon, 1995). BcL-2 expression has been implicated in preventing the generation of reactive oxygen species associated with cell death and in the inhibition of Ca^{2+} efflux from the endoplasmic reticulum (Hockenberry *et al.*, 1993; Lam *et al.*, 1994). Reactive oxygen species can react with NO and produce additional oxidants (Crow and Beckman, 1995). As noted earlier, CaN dephosphorylation of NOS results in increased NO generation, suggesting that BcL-2 may act to decrease the levels of oxidants produced by CaN-stimulated NO generation. However, the demonstration by Wang *et al.* (1996) that CaN is inactivated by an oxidative attack suggests that an antagonism of oxidation by BcL-2 might protect CaN from oxidative attack. However, the effects of antioxidants on CaN-mediated apoptosis have not been reported. Further studies will be required to define the molecular events associated with CaN-mediated apoptosis in nonimmune cells.

J. Summary and Perspective

A more detailed and complex picture has emerged of the functional regulation of CaN by Ca^{2+} and CaM and of its varied physiological roles. The identification of different CaN isoforms and the determination of their phylogenetic, tissue, and subcellular distributions raised questions about such potential functional differences as Ca^{2+}/CaM regulation, substrate specificity, and subcellular targeting mechanisms. The crystal structures of CaN and the CaN/FK506/FKBP12 complex confirmed much of the biochemical data on the structural regulation of CaN by its own Ca^{2+}-binding B subunit, the autoinhibitory domain, and Ca^{2+}/CaM. However, important unresolved questions still remain, such as the mechanism by which the B subunit regulates the K_m of the catalytic domain for substrates, and the determinants of substrate specificity within the catalytic domain and within substrates. A new question arising from the crystal structure is the mechanism by which Ca^{2+}/CaM binding to the CaM-binding domain disrupts the association between autoinhibitory and catalytic domains.

An exciting emerging concept is that similar to the targeting of Ser/Thr protein kinase, protein phosphatases, including CaN, are also regulated by targeting mechanisms. The reversible targeting of CaN to different subcellular sites or proteins provides an additional regulatory mechanism for specifying temporal and spatial aspects of CaN phosphatase activity toward a particular substrate.

The discovery that cyclosporin A/cyclophilin A and FK506/FKBP12 are specific inhibitors of CaN has provided the pharmacological tools needed to investigate its physiological roles. The use of these drugs has revealed the involvement of CaN in the regulation of a wide variety of physiological processes and signaling pathways. It is clear that the scope of the regulatory roles of CaN is broader than previously thought. It has become apparent that the enzymes and proteins of the major metabolic pathways are not regulated by CaN. Experimental data overwhelmingly indicate that the overall physiological role of CaN is to regulate the activities of enzymes and proteins in Ca^{2+}-activated signal transduction pathways. Defining and further characterizing the functions of CaN will involve continuing studies of its structural regulation, identifying substrate specificity determinants, and identifying substrates and determining how dephosphorylation by CaN regulates their activities.

ACKNOWLEDGMENTS

We thank our colleagues for their helpful discussions during the preparation of this chapter and express our appreciation to those who generously provided us with manuscripts prior to publication. We sincerely apologize to those whose work we have not cited because of space limitations.

REFERENCES

Aitken, A., Cohen, P., Santikarn, S., Williams, D. H., Calder, A. G., and Klee, C. B. (1982). Identification of the NH_2-terminal blocking group of calcineurin as myristic acid. *FEBS Lett.* **150,** 314–318.

Aldape, R. A., Futer, O., DeCenzo, M. T., Jarrett, B. P., Murcko, M. A., and Livingston, D. J. (1992). Charged surface residues of FKBP12 participate in formation of the FKBP12-FK506–calcineurin complex. *J. Biol. Chem.* **267,** 16029–16032.

Antoni, F. A., Barnard, R. J. O., Shipston, M. J., Smith, S. M., Simpson, J., and Paterson, J. M. (1995). Calcineurin feedback inhibition of agonist-evoked cAMP formation. *J. Biol. Chem.* **270,** 28055–28061.

Aperia, A., Meister, B., and Hokfelt, T. (1991). "Contemporary Issues in Nephrology" (S. Goldfarb and F. N. Ziyadeh, eds.), pp. 315–338. Churchill Livingstone, New York.

Aperia A., Ibarra, F., Svensson, L. B., Klee, C. B., and Greengard, P. (1992). Calcineurin mediates α-adrenergic stimulation of Na^+,K^+-ATPase activity in renal tubule cells. *Proc. Natl. Acad. Sci. U.S.A.* **89,** 7394–7397.

Bashir, Z. I., Bortolotto, Z. A., Davies, C. H., Beretta, N., Irving, A. J., Seal, A. J., Henley, J. M., Jane, D. E., Watkins, J. C., and Collingridge, G. L. (1993). Induction of LTP in the hippocampus needs synaptic activation of glutamate metabotropic receptors. *Nature (London)* **363,** 347–350.

Baukal, A. J., Hunyady, L., Catt, K. J., and Balla, T. (1994). Evidence for participation of calcineurin in potentiation of agonist-stimulated cyclic AMP formation by the calcium-mobilizing hormone angiotensin II. *J. Biol. Chem.* **269,** 24546–24549.

Bear, M. F., and Malenka, R. C. (1994). Synaptic plasticity: LTP and LTD. *Curr. Opin. Neurobiol.* **4,** 389–399.

Bierer, B. E., Somers, P. K., Wandless, T. J., Burakoff, S. J., and Schreiber, S. L. (1990). Probing immunosuppressant action with a nonnatural immunophilin ligand. *Science* **250,** 556–559.

Blitzer, R. D., Wong, T., Nouranifar, R., Iyengar, R., and Landau, E. M. (1995). Postsynaptic cAMP pathway gates early LTP in hippocampal CA1 region. *Neuron* **15,** 1403–1414.

Blumenthal, D. K., Takio, Hansen, R. S., and Krebs, E. G. (1986). Dephosphorylation of cAMPCa^{2+}dependent protein kinase regulatory subunit (Type II) by calmodulin-dependent protein phosphatase. *J. Biol. Chem.* **261,** 8140–8145.

Boddeke, H. W. G. M., Hoffman, B. J., Palacios, J. M., and Hoyer, D. (1993a). Characterization of functional responses in A9 cells transfected with cloned rat 5-HT$_{1C}$ receptors. *Naunyn-Schmiedeberg's Arch. Pharmacol.* **346,** 119–124.

Boddeke, H. W. G. M., Hoffman, B. J., Palacios, J. M., and Hoyer, D. (1993b). Calcineurin inhibits desensitization of cloned rat 5-HT1C receptors. *Naunyn-Schmiedeberg's Arch. Pharmacol.* **348,** 221–224.

Boddeke, H. W. G. M., Meigel, I., Boeijinga, P., Arbuckle, J., and Docherty, R. J. (1996). Modulation by calcineurin of 5-HT$_3$ receptor function in NG108-15 neuroblastoma x glioma cells. *Br. J. Pharmacol.* **118,** 1836–1840.

Bodor, J., Spetz, A., Strominger, J. L., and Habener, J. F. (1996). cAMP inducibility of transcriptional repressor ICER indeveloping and mature human T lymphocytes. *Proc. Natl. Acad. Sci. U.S.A.* **93,** 3536–3541.

Bosser, R., Aligue, R., Guerini, D., Agell, N., Carafoli, E., and Bachs, O. (1993). Calmodulin can modulate protein phosphorylation in rat liver cells nuclei. *J. Biol. Chem.* **268,** 15477–15483.

Bredt, D. S., Ferris, C. D., and Snyder, S. H. (1992). Nitric oxide synthase regulatory sites. *J. Biol. Chem.* **267,** 10976–10981.

Brickey, D. A., Bann, J. G., Fong, Y. L., Perrino, L., Brennan, R. G., and Soderling, T. R. (1994). Mutational analysis of the autoinhibitory domain of calmodulin kinase II. *J. Biol. Chem.* **269,** 29047–29054.

Brown, D. A. (1988). M-currents. *In* "Ion Channels" (T. Narahashi, ed.), pp. 55–94. Plenum, New York.

Cameron, A. M., Steiner, J. P., Roskams, A. J., Ali, S. M., Ronnet, G. V., and Snyder, S. H. (1995). Calcineurin associated with the inositol 1,4,5-trisphosphate receptor–FKBP12 complex modulates Ca^{2+} flux. *Cell (Cambridge, Mass.)* **83,** 463–472.

Chad, J. E., and Ecker, R. (1986). An enzymatic mechanism for calcium current inactivation in dialysed Helix neurones. *J. Physiol. (London)* **378,** 31–51.

Chang, H. Y., Takei, K., Sydor, A. M., Born, T. L., Rusnak, F., and Jay, D. G. (1995). Asymmetric retraction of growth cone filopodia following focal inactivation of calcineurin. *Nature (London)* **376,** 686–690.

Chantler, P. D. (1985). Calcium-dependent association of a protein complex with the lymphocyte plasma membrane: Probable indentity with calmodulin–calcineurin. *J. Cell Biol.* **101,** 207–216.

Chen, Q. X., and Wong, R. K. S. (1995). Suppression of $GABA_A$ receptor responses by NMDA application in hippocampal neurones acutely isolated from the adult guinea pig. *J. Physiol. (London)* **482,** 353–362.

Chen, Q. X., Stelzer, A., Kay, A. R., and Wong, R. K. S. (1990). $GABA_A$ receptor function is regulated by phosphorylation in acutely dissociated guinea-pig hippocampal neurons. *J. Physiol. (London)* **420,** 207–221.

Chen, T., Law, B., Kondratyuk, T., and Rossie, S. (1995). Identification of soluble protein phosphatases that dephosphorylate voltage-sensitive sodium channels in rat brain. *J. Biol. Chem.* **270,** 7750–7756.

Choi, D. W. (1992). Bench to bedside: The glutamate connection. *Science* **258,** 241–243.

Clipstone, N. A., and Crabtree, G. R. (1992). Identification of calcineurin as a key signalling enzyme in T-lymphocyte activation. *Nature (London)* **357,** 695–697.

Clipstone, N. A., Fiorentino, D. F., and Crabtree, G. R. (1994). Molecular analysis of the interaction of calcineurin with drug–immunophilin complexes. *J. Biol. Chem.* **269,** 26431–26437.

Coghlan, V. M., Perrino, B. A., Howard, M., Langeberg, L. K., Hicks, J. B., Gallatin, W. M., and Scott, J. D. (1995). Association of protein kinase A and protein phosphatase 2B with a common anchoring protein. *Science* **267,** 108–111.

Cohen, P. (1992). Signal integration at the level of protein kinases, protein phosphatases and their substrates. *Trends Biochem. Sci.* **17,** 408–413.

Cohen, P., Klumpp, S., and Schelling, D. L. (1989). An improved procedure for identifying and quantitating protein phosphatases in mammalian tissues. *FEBS Lett.* **250,** 596–600.

Cooper, D. M. F., Mons, N., and Karpen, J. W. (1995). Adenylyl cyclases and the interaction between calcium and cAMP signalling. *Nature (London)* **374,** 421–424.

Crow, J. P., and Beckman, J. S. (1995). Reactions between nitric oxide, superoxide, and peroxynitrite: Footprints of peroxynitrite in vivo. *Adv. Pharmacol.* **34,** 17–43.

Cyert, M. S., Kunisawa, D., Kaim, D., and Thorner, J. (1991). Yeast has homologs (CNA1 and CNA2 gene products) of mammalian calcineurin, a calmodulin-regulated phosphoprotein phosphatase. *Proc. Natl. Acad. Sci. U.S.A.* **88,** 7376–7380.

Dawson, T. M., Dawson, V. L., and Snyder, S. H. (1992). A novel neuronal messenger molecule in brain: The free radical, nitric oxide. *Ann. Neurol.* **32,** 297–311.

Dawson, T. M., Steiner, J. P., Dawson, V. L., Dinerman, J. L., Uhl, G. R. and Snyder, S. H. (1993). Immunosuppressant FK506 enhances phosphorylation of nitric oxide synthase and protects against glutamate neurotoxicity. *Proc. Natl. Acad. Sci. U.S.A.* **90,** 9808–9812.

De Camilli, P., and Takei, K. (1996). Molecular mechanisms in synaptic vesicle endocytosis and recycling. *Neuron* **16,** 481–486.

Derijard, B., Hibi, M., Wu, I.-H., Barret, T., Su, B., Deng, T., Karin, M., and Davis, R. J. (1994). JNK1: A protein kinase stimulated by UV light and Ha-Ras that binds and phosphorylates the c-jun activation domain. *Cell (Cambridge, Mass.)* **76,** 1025–1037.

Dizhoor, A. M., Chen, C. K., Olshevskaya, E., Sinelnikova, V. V., Phillipov, V., and Hurley, J. B. (1993). Role of the acylated amino terminus of recoverin in Ca^{2+}-dependent membrane interaction. *Science* **259,** 829–832.

Dolphin, A. C. (1996). Facilitation of Ca^{2+} current in excitable cells. *Trends Neurosci.* **19**, 35–43.
Drewes, G., Mandelkow, E.-M., Baumann, K., Goris, J., Merlevede, W., and Mandelkow, E. (1993). Dephosphorylation of tau protein and Alzheimer paired helical filaments by calcineurin and phosphatase-2A. *FEBS Lett.* **336**, 425–432.
Dudek, S. M., and Bear, M. F. (1992). Homosynaptic long-term depression in area CA1 of the hippocampus and effects of *N*-methyl-D-aspartate receptor blockade. *Proc. Natl. Acad. Sci. U.S.A.* **89**, 4363–4367.
Eckert, B., Schwaninger, M., and Knepel, W. (1996). Calcium-mobilizing insulin secretagogues stimulate transcription that is directed by the cyclic adenosine 3',5'-monophosphate/calcium response element in a pancreatic β-cell line. *Endocrinology (Baltimore)* **137**, 225–233.
Eckert, R., and Chad, J. E. (1984). Inactivation of Ca channels. *Prog. Biophys. Mol. Biol.* **44**, 215–267.
Emmel, E. A., Verweij, C. L., Durand, D. B., Higgins, K. M., Lacy, E., and Crabtree, G. R. (1989). Cyclosporin A specifically inhibits function of nuclear proteins involved in T cell activation. *Science* **246**, 1617–1620.
Engele-Garrett, P., Moilanen, B., and Cyert, M. S. (1995). Calcineurin, the Ca^{2+}/calmodulin-dependent protein phosphatase, is essential in yeast mutants with cell integrity defects and in mutants that lack a functional vacuolar H^+-ATPase. *Mol. Cell. Biol.* **15**, 4103–4114.
Enslen, H., and Soderling, T. R. (1994). Roles of calmodulin-dependent protein kinases and phosphatases in calcium-dependent transcription of immediate early genes. *J. Biol. Chem.* **269**, 20872–20877.
Evan, G. I., Wyllie, A. H., Gilbert, C. S., Littlewood, T. D., Land, H., Brooks, M., Waters, C. M., Penn, L. Z., and Hancock, D. C. (1992). Induction of apoptosis in fibroblasts by c-Myc protein. *Cell (Cambridge, Mass.)* **68**, 119–128.
Farber, L. H., Wilson, F. J., and Wolff, D. J. (1987). Calmodulin-dependent phosphatases of PC12, GH_3 and C_6 cells: Physical, kinetic, and immunochemical properties. *J. Neurochem.* **49**, 404–414.
Faux, M. C., and Scott, J. D. (1996). More on target with protein phosphorylation: Conferring specificity by location. *Trends Biochem. Sci.* **21**, 312–315.
Ferreira, A., Kincaid, R. L., and Kosik, K. S. (1993). Calcineurin is associated with the cytoskeleton of cultured neurons and has a role in the acquisition of polarity. *Mol. Biol. Cell* **4**, 1225–1238.
Fisher, G., Wittman-Liebold, B., Lang, K., Kiefhaber, T., and Schmid, F. X. (1989). Cyclophilin and peptidyl-prolyl *cis-trans* isomerase are probably identical proteins. *Nature (London)* **337**, 476–478.
Flanagan, M. W., Corthesy, B., Bram, R. J., and Crabtree, G. R. (1991). Nuclear association of a T-cell transcription factor blocked by FK-506 and cyclosporin A. *Nature (London)* **352**, 803–807.
Fong, Y. L., and Soderling, T. R. (1990). Studies on the regulatory domain of Ca^{2+}/calmodulin-dependent protein kinase II. *J. Biol. Chem.* **265**, 11091–11097.
Fraser, E. D., and Walsh, M. P. (1995). Dephosphorylation of calponin by Type 2B protein phosphatase. *Biochemistry* **34**, 9151–9158.
Fruman, D. A., Klee, C. B., Bierer, B. E., and Burakoff, S. J. (1992a). Calcineurin phosphatase activity in T lymphocytes is inhibited by FK 506 and cyclosporin A. *Proc. Natl. Acad. Sci. U.S.A.* **89**, 3686–3690.

Fruman, D. A., Mather, P. E., Burakoff, S. J., and Bierer, B. E. (1992b). Correlation of calcineurin phosphatase activity and programmed cell death in T cell hybridomas. *Eur. J. Immunol.* **22,** 2513–2517.

Fruman, D. A., Pai, S. Y., Burakoff, S. J., and Bierer, B. E. (1995). Characterization of a mutant calcineurin Aα gene expressed by EL4 lymphoma cells. *Mol. Cell. Biol.* **15,** 3857–3863.

Fukunaga, K., Rich, D. P., and Soderling, T. R. (1989). Generation of the Ca^{2+}-independent form of Ca^{2+}/calmodulin-dependent protein kinase II in cerebellar granule cells. *J. Biol. Chem.* **264,** 21830–21836.

Futer, O., DeCenzo, M. T., Aldape, R. A., and Livingston, D. J. (1995). FK506 binding protein mutational analysis. *J. Biol. Chem.* **270,** 18935–18940.

Garver, T. D., Oyler, G. A., Harris, K. A., Polavarapu, R. G., Damuni, Z., Lehman, R. A. W., and Billingsley, M. L. (1995). Tau phosphorylation in brain slices: Pharmacological evidence for convergent effects of protein phosphatases on tau and mitogen-activated protein kinase. *Mol. Pharmacol.* **47,** 745–756.

Gilroy, S., Read, N. D., and Trewavas, A. J. (1990). Elevation of cytoplasmic calcium by caged calcium or carged inositol trisphosphate initiates stomatal closure. *Nature (London)* **346,** 769–771.

Goldberg, J., Huang, H., Kwon, Y., Greengard, P., Nairn, A. C., and Kuriyan, J. (1995). Three-dimensional structure of the catalytic subunit of protein serine/threonine phosphatase-1. *Nature (London)* **376,** 745–753.

Gong, C.-X., Singh, T. S., Grundke-Iqbal, K. (1994). Alzheimer's disease abnormally phosphorylated τ is dephosphorylated by protein phosphatase-2B (calcineurin). *J. Neurochem.* **62,** 803–806.

Goto, S., Yamamoto, H., Fukunaga, K., Iwasa, T., Matsukado, Y., and Miyamoto, E. (1985). Dephosphorylation of microtubule-associated protein 2, τ factor, and tubulin by calcineurin. *J. Neurochem.* **45,** 276–283.

Goto, S., Matsukado, Y., Mihara, Y., Inoue, N., and Miyamoto, E. (1986). The distribution of calcineurin in rat brain by light and electron microscopic immunohistochemistry and enzyme-immunoassay. *Brain Res.* **397,** 161–172.

Goulet, M. T., Rupprecht, K. M., Sinclair, P. J., Wyvratt, M. J., and Parsons, W. H. (1994). The medicinal chemistry of FK-506. *Perspect. Drug Discovery Design* **2,** 145–162.

Griffith, J. P., Kim, J. L., Kim, E. E., Sintchak, M. D., Thomson, J. A., Fitzgibbon, M. J., Fleming M. A., Carson, P. R., Hsiao, K., and Navia, M. A. (1995). X-ray structure of calcineurin inhibited by the immunophilin–immunosuppressant FKBP12-FK506 complex. *Cell (Cambridge, Mass.)* **82,** 507–522.

Guerini, D., and Klee, C. B. (1989). Cloning of human calcineurin A: Evidence for two isozymes and identification of a polyproline structural domain. *Proc. Natl. Acad. Sci. U.S.A.* **86,** 9183–9187.

Guerini, D., and Klee, C. B. (1991). Structural diversity of calcineurin, a Ca^{2+} and calmodulin-stimulated protein phosphatase. *Adv. Protein Phosphatases* **6,** 391–410.

Guerini, D., Montell, C., and Klee, C. B. (1992). Molecular cloning and characterization of the genes encoding the two subunits of *Drosophila melanogaster* calcineurin. *J. Biol. Chem.* **267,** 22542–22549.

Haddy, A., Swanson, S. K. H., Born, T. L., and Rusnak, F. (1992). Inhibition of calcineurin by cyclosporin A-cyclophilin requires calcineurin B. *FEBS Lett.* **314,** 37–40.

Hadley, R. W., and Lederer, W. J. (1991). Ca^{2+} and voltage inactivate Ca^{2+} channels in guinea-pig ventricular myocytes through independent mechanisms. *J. Physiol. (London)* **444,** 257–268.

Halpain, S., and Greengard, P. (1990). Activation of NMDA receptors induces rapid dephosphorylation of the cytoskeletal protein MAP2. *Neuron* **5,** 237–246.

Halpain, S., Girault, J. A., and Greengard, P. (1990). Activation of NMDA receptors induces dephosphorylation of DARPP-32 in rat striatal slices. *Nature (London)* **343,** 369–372.

Harding, M. W., Galat, A., Uehling, D. E., and Schreiber, S. L. (1989). A receptor for the immunosuppressant FK506 is a *cis-trans* peptidyl-prolyl isomerase. *Nature (London)* **341,** 758–760.

Harris, K. A., Oyler, G. A., Doolittle, G. M., Vincent, I., Lehman, R. A. W., Kincaid, R. L., and Billingsley, M. L. (1993). Okadaic acid induces hyperphosphorylated forms of tau protein in human brain slices. *Ann. Neurol.* **33,** 77–87.

Hashimoto, Y., Perrino, B. A., and Soderling, T. R. (1990). Identification of an autoinhibitory domain in calcineurin. *J. Biol. Chem.* **265,** 1924–1927.

Heitman, J., Movva, N. R., Hiestand, P. C., and Hall, M. N. (1991). FK506-binding protein proline rotamase is a target for the immunosuppressive agent FK506 in Saccharomyces cerevisiae. *Proc. Natl. Acad. Sci. U.S.A.* **88,** 1948–1952.

Hemmings, H. C. J., Nairn, A. C., Elliot, J. I., and Greengard, P. (1990). Synthetic peptide analogs of DARPP-32 (Mr 32,000 dopamine- and cAMP-regulated phosphoprotein), an inhibitor of protein phosphatase-1. *J. Biol. Chem.* **265,** 20369–20376.

Hemmings, H. C. J., Girault, J. A., Nairn, A. C., Bertuzzi, G., and Greengard, P. (1992). Distribution of protein phosphatase inhibitor-1 in brain and peripheral tissues of various species: Comparison with DARPP-32. *J. Neurochem.* **59,** 1053–1061.

Hendey, B., Lawson, M., Marcantonio, E. E., and Maxfield., F. R. (1996). Intracellular calcium and calcineurin regulate neutrophil motility on vitronectin through a receptor identified by antibodies to integrins αv and $\beta 3$. *Blood* **87,** 2038–2048.

Higuchi, S., Tamura, J., Rathna Giri, P., Polli, J. W., and Kincaid, R. L. (1991). Calmodulin-dependent protein phosphatase from *Neurospora crassa*. Molecular cloning and expression of recombinant catalytic subunit. *J. Biol. Chem.* **266,** 18104–18112.

Hille, B. (1994). Modulation of ion-channel function by G-protein-coupled receptors. *Trends Neurosci.* **17,** 531–536.

Hockenberry, D. M., Oltvai, Z. N., Yin, X., Milliman, C. L., and Korsmeyer, S. J. (1993). BcL-2 functions in an antioxidant pathway to prevent apoptosis. *Cell (Cambridge, Mass.)* **75,** 241–251.

Hubbard, M. J., and Cohen, P. (1993). On target with a new mechanism for the regulation of protein phosphorylation. *Trends Biochem. Sci.* **18,** 172–177.

Hubbard, M. J., and Klee, C. B. (1987). Calmodulin binding by calcineurin. Ligand-induced renaturation of protein immobilized on nitrocellulose. *J. Biol. Chem.* **262,** 15062–15070.

Husi, H., Luyten, M. A., and Zurini, M. G. M. (1994). Mapping of the immunophilin–immunosuppressant site of interaction on calcineurin. *J. Biol. Chem.* **269,** 14199–14204.

Hutchinson, L. E., and McCloskey, M. E. (1995). Fcε RI-mediated induction of nuclear factor of activated T cells. *J. Biol. Chem.* **270,** 16333–16338.

Ingebritsen, T. S., and Cohen, P. (1983). Protein phosphatases: Properties and role in cellular regulation. *Science* **221,** 331–338.

Ito, A., Hashimoto, T., Hirai, M., Takeda, T., Shuntoh, H., Kuno, T., and Tanaka, C. (1989). The complete primary structure of calcineurin A, a calmodulin-binding protein homologous with protein phosphatases 1 and 2A. *Biochem. Biophys. Res. Commun.* **163,** 1492–1497.

Izquierdo, M., Ruiz-Ruiz, M. C., and Lopez-Rivas, A. (1996). Stimulation of phosphatidylinositol turnover is a key event for Fas-dependent, activation-induced apoptosis in human T lymphocytes. *J. Immunol.* **157,** 21–28.

Jain, J., McCaffrey, P. G., Valge-Arthur, V. E., and Rao, A. (1992). Nuclear factor of activated T cells contains fos and jun. *Nature (London)* **356,** 801–804.

Jain, J., Loh, C., and Rao, A. (1995). Transcriptional regulation of the *IL-2* gene. *Curr. Opin. Immunol.* **7,** 333–342.

Jameson, L., Frey, T., Zeeberg, B., Dalldorf, F., and Caplow, W. (1980). Inhibition of microtubule assembly by phosphorylation of microtubule-associated proteins. *Biochemistry* **19,** 2472–2479.

Jin, Y. J., Albers, M. W., Lane, W. S., Bierer, B. E., Schreiber, S. L., and Burakoff, S. J. (1991). Molecular cloning of a membrane-associated human FK506- and rapamycin-binding protein, FKBP13. *Proc. Natl. Acad. Sci. U.S.A.* **88,** 6677–6681.

June, C. H. (1991). Signal transduction in T cells. *Curr. Opin. Immunol.* **3,** 287–293.

Kabelitz, D., Pohl, T., and Pechhold, K. (1993). Activation-induced cell death (apoptosis) of mature peripheral T lymphocytes. *Immunol. Today* **14,** 338–339.

Kallen, J., Spitzfaden, C., Zurini, M. G. M., Wider, G., Widmer, H., Wuthrich, K., and Walkinshaw, M. D. (1991). Structure of human cyclophilin and its binding site for cyclosporin A determined by X-ray crystallography and NMR spectroscopy. *Nature (London)* **353,** 276–279.

Kawamura, A., and Su, M. S.-S. (1995). Interaction of FKBP12-FK506 with calcineurin A at the B subunit-binding domain. *J. Biol. Chem.* **270,** 15463–15466.

Kay, J. E., Doe, S. E., and Benzie, C. R. (1989). The mechanism of action of the immunosuppressive drug FK506. *Cell Immunol.* **124,** 175–181.

Kayyali, U. S., Zhang, W., Yee, A. G., Seidman, J. G, and Potter, H. (1995). Accumulation of hyperphosphorylated tau in knock-out mice lacking calcineurin. *Soc. Neurosci.* **21,** 742.

Kennedy, M. T., Brockman, H., and Rusnak, F. (1996). Contributions of myristoylation to calcineurin structure/function. *J. Biol. Chem.* **271,** 26517–26521.

Kieffer, L. J., Thalhammer, T., and Handschumacher, R. E. (1992). Isolation and characterization of a 40-kDa cyclophilin-related protein. *J. Biol. Chem.* **267,** 5503–5507.

Kincaid, R. L., Vaughan, M., Osborne, J. C., Jr., and Tkachuk, V. A. (1982). Ca^{2+}-dependent interaction of 5-dimethylaminonaphthalene-1-sulfonyl-calmodulin with cyclic nucleotide phosphodiesterase, calcineurin, and troponin I. *J. Biol. Chem.* **257,** 10638–10643.

Kincaid, R. L., Nightingale, M. S., and Martin, B. M. (1988). Characterization of a cDNA clone encoding the calmodulin binding domain of mouse brain calcineurin. *Proc. Natl. Acad. Sci. U.S.A.* **85,** 8983–8987.

Kincaid, R. L., Rathna Giri, P., Higuchi, S., Tamura, J., Dixon, S. C., Marietta, C. A., Amorese, D. A., and Martin, B. M. (1990). Cloning and characterization of molecular isoforms of the catalytic subunit of calcineurin using nonisotopic methods. *J. Biol. Chem.* **265,** 11312–11319.

King, M. M., and Huang, C. Y. (1984). Calmodulin-dependent activation and deactivation of the protein phosphatase, calcineurin, and the effects of nucleotides, pyrophosphate, and divalent metal ions. *J. Biol. Chem.* **259,** 8847–8856.

King, M. M., Huang, C. Y., Boon Chock, P., Nairn, A. C., Hemmings, H. C. J., Jesse-Chan, K. F., and Greengard, P. (1984). Mammalian brain phosphoproteins as substrates for calcineurin. *J. Biol. Chem.* **259,** 8080–8083.

Kissinger, C. R., Parge, H. E., Knighton, D. R., Lewis, C. T., Pelletier, L. A., Tempczyk, A., Kalish, V. J., Tucker, K. D., Showalter, R. E., Moomaw, E. W., Gastinel, L. N., Habuka, N., Chen, X., Maldonado, F., Barker, J. E., Bacquet, R., and Villafranca, J. E. (1995). Crystal structures of human calcineurin and the human FKBP12-FK506 complex. *Nature (London)* **378,** 641–644.

Klauck, T. M., Faux, M. C., Labudda, K., Langeberg, L. K., Jaken, S., and Scott, J. D. (1996). Coordination of three signaling enzymes by AKAP79, a mammalian scaffold protein. *Science* **271,** 1589–1592.

Klee, C. B., and Cohen, P., eds. (1988). The calmodulin-regulated protein phosphatase. *In* "Calmodulin," pp. 225–248. Elsevier, Amsterdam.

Klee, C. B., and Krinks, M. H. (1978) Purification of cyclic 3′,5′-nucleotide phosphodiesterase inhibitory protein by affinity chromatography on activator protein coupled to Sepharose. *Biochemistry* **17,** 120–126.

Klee, C. B., Krinks, M. H., Manalan, A. S., Draetta, G. F., and Newton, D. L. (1985). "Advances in Protein Phosphatases 1" (W. Merleverde and J. Di Salvo, eds.), pp. 135–146. Leuven Univ. Press, Belgium.

Klee, C. B., Draetta, G. F., and Hubbard, M. J. (1988). Calcineurin. *Adv. Enzymol.* **61,** 149–200.

Koltin, Y., Faucette, L., Bergsma, D. J., Levy, M. A., Cafferkey, R., Koser, P. L., Johnson, R. K., and Livi, G. P. (1991). Rapamycin sensitivity in *Saccharomyces cerevisiae* is mediated by a peptidyl-prolyl cis-trans isomerase related to human FK506-binding protein. *Mol. Cell. Biol.* **11,** 1718–1723.

Kopp, U. C., and DiBona, G. F. (1992). "The Kidney: Physiology and Pathophysiology" (D. W. Seldin and G. Giebisch, eds.), pp. 1157–1204. Raven, New York.

Krammer, P. H., Dhein, J., Walczack, H., Behrmann, I., Mariani, S., Matiba, B., Fath, M., Daniel, P. T., Knipping, E., Westendorp, M. O., Stricker, K., Baumler, C., Hellbardt, S., Germer, M., Peter, M., and Debatin, K.-M. (1994). The role of Apo 1-mediated apoptosis in the immune system. *Immunol. Rev.* **142,** 175–181.

Krinks, M. H., Manalan, A. S., and Klee, C. B. (1985). Calcineurin: A brain-specific isozyme of protein phosphatase-2 B. *Fed. Proc.* **44,** 707a.

Kuno, T., Takeda, T., Hirai, M., Ito, A., Mukai, H., and Tanaka, C. (1989). Evidence for a second isoform of the catalytic subunit of calmodulin-dependent protein phosphatase (calcineurin A). *Biochem. Biophys. Res. Commun.* **165,** 1352–1358.

Kuno, T., Tanaka, H., Mukai, H., Chang, C. D., Hiraga, K., Miyakawa, T., and Tanaka, C. (1996). cDNA cloning of a calcineurin B homolog in Saccharomyces cerevisiae. *Biochem. Biophys. Res. Commun.* **180,** 1159–1163.

Lam, M., Dubyak, G., Chen, L., Nunez, G., Miesfeld, R. L., and Distelhorst, C. W. (1994). Evidence that BcL-2 represses apoptosis by regulating endoplasmic-reticulum-associated Ca^{2+} fluxes. *Proc. Natl. Acad. Sci. U.S.A.* **91,** 6569–6573.

Li, M., West, J. W., Lai, Y., Scheuer, T., and Catterwall, W. A. (1992). Functional modulation of brain sodium channels by cAMP-dependent phosphorylation. *Neuron* **8,** 1151–1159.

Li, M., and Handschumacher, R. E. (1993). Specific interaction of the cyclophilin–cyclosporin complex with the B subunit of calcineurin. *J. Biol. Chem.* **268,** 14040–14044.

Liao, L., Hyatt, S. L., Chapline, C., and Jaken, S. (1994). Protein kinase C domains involved in interactions with other proteins. *Biochemistry* **33,** 1229–1233.

Lieberman, D. N., and Mody, I. (1994). Regulation of NMDA channel function by endogenous Ca^{2+}-dependent phosphatase. *Nature (London)* **369,** 235–239.

4. Calmodulin-Regulated Phosphatase Calcineurin

Lisman, J. (1989). A mechanism for the Hebb and the anti-Hebb processes underlying learning and memory. *Proc. Natl. Acad. Sci. U.S.A.* **86,** 9574–9578.

Liu, J., Farmer, J. D., Jr., Lane, W. S., Friedman, J., Weissman, I., and Schreiber, S. L. (1991a). Calcineurin is a common target of cyclophilin–cyclosporin A and FKBP–FK506 complexes. *Cell (Cambridge, Mass.)* **66,** 807–815.

Liu, J., Albers, M. W., Wandless, T. J. Luan, S., Alberg, D. G., Belshaw, P. J., Cohen, P., MacKintosh, C., Klee, C. B. and Schreiber, S. L. (1992). Inhibition of T-cell signaling by immunophilin–ligand complexes correlates with loss of calcineurin phosphatase activity. *Biochemistry* **31,** 3896–3901.

Liu, J. P., Powell, K. A., Sudhof, T. C., and Robinson, P. J. (1994a). Calcineurin inhibition of dynamin 1 GTPase activity coupled to nerve terminal depolarization. *Science* **265,** 970–973.

Liu, J. P., Powell, K. A., Sudhof, T. C., and Robinson, P. J. (1994b). Dynamin 1 is a Ca^{2+}-sensitive phospholipid-binding protein with very high affinity for protein kinase C. *J. Biol. Chem.* **269,** 21043–21050.

Liu, Y. S., Ishii, M., Tokai, H., Tsutsumi, O., Ohki, R., Akada, K., Tanaka, E., Tsuchiya, S., Fuku, I., and Miyakawa, T. (1991b). The *Saccharomyces cerevisiae* genes (CMP1 and CMP2) encoding calmodulin-binding proteins homologous to the catalytic subunit of mammalian brain protein phosphatase 2B. *Mol. Gen. Genet.* **227,** 52–59.

Loh, C., Shaw, K. T. Y., Carew, J., Viola, J. P. B., Luo, C., Perrino, B. A., and Rao, A. (1996a). Calcineurin binds the transcription factor NFAT1 and reversibly regulates its activity. *J. Biol. Chem.* **271,** 10884–10891.

Loh, C., Carew, J. A., Kim, J., Hogan, P. G., and Rao, A. (1996b). T-cell receptor stimulation elicits an early phase of activation and a later phase of deactivation of the transcription factor NFAT1. *Mol. Cell. Biol.* **16,** 3945–3954.

Lohse, D. L., Denu, J. M., and Dixon, J. E. (1995). Insights derived from the structures of the Ser/Thr phosphatases calcineurin and protein phosphatase 1. *Structure* **15,** 987–990.

Luan, S., Li, W., Rusnak, F., Assman, S. M., and Schreiber, S. L. (1993). Immunosuppressants implicate protein phosphatase regulation of K^+ channels in guard cells. *Proc. Natl. Acad. Sci. U.S.A.* **90,** 2202–2206.

Luo, C., Shaw, K. T. Y., Raghavan, A., Aramburu, J., Garcia-Cozar, F., Perrino, B. A., Hogan, P. G., and Rao, A. (1996). Interaction of calcineurin with a domain of the transcription factor NFAT1 that controls nuclear import. *Proc. Natl. Acad. Sci. U.S.A.* **93,** 8907–8912.

McCaffrey, P. G., Perrino, B. A., Soderling, T. R., and Rao, A. (1993). NF-ATp, a T lymphocyte DNA-binding protein that is a target for calcineurin and immunosuppressive drugs. *J. Biol. Chem.* **268,** 3747–3752.

McGlade-McCulloh, E., Yamamoto, H., Tan, S.-E., Brickey, D. A., and Soderling, T. R. (1993). Phosphorylation and regulation of glutamate receptors by calcium/calmodulin-dependent protein kinase II. *Nature (London)* **362,** 640–642.

McLaughlin, S., and Aderem, A. (1995). The myristoyl-electrostatic switch: A modulator of reversible protein–membrane interactions. *Trends Biochem. Sci.* **20,** 272–276.

Malenka, R. C. (1994). Synaptic plasticity in the hippocampus: LTP and LTD. *Cell (Cambridge, Mass.)* **78,** 535–538.

Manalan, A. S., and Klee, C. B. (1983). Activation of calcineurin by limited proteolysis. *Proc. Natl. Acad. Sci. U.S.A.* **80,** 4291–4295.

Mansfield, T. A., Hetherington, A. M., and Atkinson, C. J. (1990). Some current aspects of stomatal physiology. *Annu. Rev. Plant Physiol. Mol. Biol.* **41,** 55–75.

Marcaida, G., Kosenko, E., Minana, M., Grisolia, S., and Felipo, V. (1996). Glutamate induces a calcineurin-mediated dephosphorylation of Na$^+$,K$^+$-ATPase that results in its activation in cerebellar neurons in culture. *J. Neurochem.* **66**, 99–104.

Marrion, N. V. (1996). Calcineurin regulates M channel modal gating in sympathetic neurons. *Neuron* **16**, 163–173.

Marsen, T. A., Simonson, M. S., and Dunn, M. J. (1995). Thrombin-mediated ET-1 gene regulation involves CaM kinases and calcineurin in human endothelial cells. *J. Cardiovasc. Pharmacol.* **26**(Suppl. 3), S1–S4.

Mattila, P. S., Ullman, K. S., Fiering, S., Emmel, E. A., McCutcheon, M., Crabtree, G. R., and Herzenberg, L. A. (1990). The actions of cyclosporin A and FK506 suggest a novel step in the activation of T lymphocytes. *EMBO J.* **9**, 4425–4433.

Matus, A., Bernhardt, R., and Hugh-Jones, T. (1981). High molecular weight microtubule-associated proteins are preferentially associated with dendritic microtubules in brain. *Proc. Natl. Acad. Sci. U.S.A.* **78**, 3010–3014.

Meador, W. E., Means, A. R., and Quiocho, F. A. (1992). Target enzyme recognition by calmodulin: 2.4 Å structure of a calmodulin–peptide complex. *Science* **257**, 1251–1255.

Mendoza, I., Rubio, F., Rodriguez-Navarro, A. and Pardo, J. M. (1994). The protein phosphatase calcineurin is essential for NaCl tolerance of *Saccharomyces cerevisiae*. *J. Biol. Chem.* **269**, 8792–8796.

Merat, D. L., Hu, Z. Y., Carter, T. C., and Cheung, W. Y. (1985). Bovine brain calmodulin-dependent protein phosphatase regulation of subunit A activity by calmodulin and subunit B. *J. Biol. Chem.* **260**, 11053–11059.

Michel, M. C., Brodde, O.-E., and Insel, P. A. (1990). Peripheral adrenergic receptors in hypertension. *Hypertension* **16**, 107–120.

Milan, D., Griffith, J., Su, M., Price, E. R., and McKeon, F. D. (1994). The latch region of calcineurin B is involved in both immunosuppressant–immunophilin complex docking and phosphatase activation. *Cell (Cambridge, Mass.)* **79**, 437–447.

Minneman, K. P. (1988). al-Adrenergic receptor subtypes, inositol phosphates, and sources of cell Ca^{2+}. *Pharmacol. Rev.* **40**, 87–119.

Mochly-Rosen, D., Khaner, H., and Lopez, J. (1991). Identification of intracellular receptor proteins for activated protein kinase C. *Proc. Natl. Acad. Sci. U.S.A.* **88**, 3997–4000.

Moia, L. J. M. P., Matsui, H., deBarros, G. A. M., Tomizawa, K., Miyamoto, K., Kuwata, Y., Tokuda, M., Itano, T., and Hatase, O. (1994). Immunosuppressants and calcineurin inhibitors, cyclosporin A and FK506, reversibly inhibit epileptogenesis in amygdaloid kindled rat. *Brain Res.* **648**, 337–341.

Molina, C. A., Foulkes, N. S., Lalli, E., and Sassone-Corsi, P. (1993). Inducibility and negative autoregulation of CREM: An alternative promoter directs the expression of ICER, an early response repressor. *Cell (Cambridge, Mass.)* **75**, 875–886.

Moriya, M., Fujinaga, K., Yazawa, M., and Katagiri, C. (1995). Immunohistochemical localization of the calcium/calmodulin-dependent protein phosphatase, calcineurin, in the mouse testis: Its unique accumulation in spermatid nuclei. *Cell Tissue Res.* **201**, 273–281.

Morris, S. M. J., Kepka-Lenhart, D., Curthoys, N. P., McGill, R. L., Marcus, R. J., and Adler, S. (1991). Disruption of renal function and gene expression by FK506 and cyclosporin. *Transplant Proc.* **23**, 3116–3118.

Morris, S. M. J., Kepka-Lenhart, D., McGill, R. L., Curthoys, N. P., and Adler, S. (1992). Specific disruption of renal function and gene transcription by cyclosporin A. *J. Biol. Chem.* **267**, 13768–13771.

Mouginot, D., Feltz, P., and Schlichter, R. (1991). Modulation of GABA-gated chloride currents by intracellular Ca^{2+} in cultured porcine melanotrophs. *J. Physiol.* (*London*) **437**, 109–132.
Muegge, K., and Durum, S. K. (1990). Cytokines and transcription factors. *Cytokine* **2**, 1–8.
Mulkey, R. M., Herron, C. E., and Malenka, R. C. (1993). An essential role for protein phosphatases in hippocampal long-term depression. *Science* **261**, 1051–1055.
Mulkey, R. M., Endo, S., Shenolikar, S., and Malenka, R. C. (1994). Involvement of a calcineurin/inhibitor-1 phosphatase cascade in long-term depression. *Nature* (*London*) **369**, 486–488.
Mulle, C., Choquet, D., Korn, H., and Changeux, J. P. (1992). Calcium influx through nicotinic receptors in rat central neurons: Its relevance to cellular regulation. *Neuron* **8**, 135–143.
Mumby, M. C., and Walter, G. (1993). Protein serine/threonine phosphatases: Structure, regulation, and functions in cell growth. *Physiol. Rev.* **73**, 673–699.
Muramatsu, T., and Kincaid, R. L. (1992). Molecular cloning and chromosomal mapping of the human gene for the testis-specific catalytic subunit of calmodulin-dependent protein phosphatase (calcineurin A). *Biochem. Biophys. Res. Commun.* **188**, 265–271.
Muramatsu, T., Rathna Giri, P., Higuchi, S., and Kincaid, R. L. (1992). Molecular cloning of a calmodulin-dependent phosphatase from murine testis: Identification of a developmentally expressed nonneuronal isoenzyme. *Proc. Natl. Acad. Sci. U.S.A.* **89**, 529–533.
Murphy, B. J., Rossie, S., De Jongh, K. S., and Catterwall, W. A. (1993). Identification of the sites of selective phosphorylation and dephosphorylation of the rat brain Na^+ channel α subunit by cAMP-dependent protein kinases and phosphoprotein phosphatases. *J. Biol. Chem.* **268**, 27355–27362.
Murthy, A. S. N., and Flavin, M. (1983). Microtubule assembly using the microtubule-associated protein MAP-2 prepared in defined states of phosphorylation with protein kinase and phosphatase. *Eur. J. Biochem.* **137**, 37–46.
Nakamura, T., Liu, Y., Hirata, D., Namba, H., Harada, S.-I., Hirokawa, T., and Miyakawa, T. (1993). Protein phosphatase type 2B (calcineurin)-mediated, FK506-sensitive regulation of intracellular ions in yeast is an important determinant for adaptation to high salt stress conditions *EMBO J.* **12**, 4063–4071.
Nast, C. C., Adler, S., Artishevsky, A., Kresser, C. T., Ahmed, K., and Anderson, P. S. (1991). Cyclosporine induces elevated procollagen alpha 1 (I) mRNA levels in the rat renal cortex. *Kidney Int.* **39**, 631–638.
Nghiem, P., Ollick, T., Gardner, P., and Schulman, H. (1996). Interleukin-2 transcriptional block by multifunctional Ca^{2+}/calmodulin kinase. *Nature* (*London*) **371**, 347–350.
Nichols, R. A., Suplick, G. R., and Brown, J. M. (1994). Calcineurin-mediated protein dephosphorylation in brain nerve terminals regulates the release of glutamate. *J. Biol. Chem.* **269**, 23817–23823.
Numann, R., Catterwall, W. A., and Scheuer, T. (1991). Functional modulation of brain sodium channels by protein kinase C phosphorylation. *Science* **254**, 115–118.
O'Keefe, S. J., Tamura, J., Kincaid, R. L., Tocci, M. J., and O'Neill, E. A. (1992). FK-506- and CsA-sensitive activation of the interleukin-2 promoter by calcineurin. *Nature* (*London*) **357**, 692–694.
Paliogianni, F., Kincaid, R. L., and Boumpas, D. T. (1993). Prostaglandin E2 and other cyclic AMP elevating agents inhibit interleukin 2 gene transcription by counteracting calcineurin-dependent pathways. *J. Exp. Med.* **178**, 1813–1817.

Pallen, C. J. and Wang, J. H. (1984). Regulation of calcineurin by metal ions. *J. Biol. Chem.* **259,** 6134–6141.

Pallen, C. J., Valentine, K. A., Wang, J. H., and Hollenberg, M. D. (1985). Calcineurin-mediated dephosphorylation of the human placental membrane receptor for epidermal growth factor urogastrone. *Biochemistry* **24,** 4727–4730.

Parsons, J. N., Wiederrecht, G. J., Salowe, S., Burbaum, J. J., Rokosz, L. L., Kincaid, R. L., and O'Keefe, S. J. (1994). Regulation of calcineurin phosphatase activity and interaction with the FKBP12-FK506 binding protein complex. *J. Biol. Chem.* **269,** 19610–19616.

Perrino, B. A., Hashimoto, Y., and Soderling, T. R. (1990). Analysis of the autoinhibitory domain of calcineurin. *FASEB J.* **4,** A2237.

Perrino, B. A., Fong, Y. L., Brickey, D. A., Saitoh, Y., Ushio, Y., Fukunaga, K., Miyamoto, E., and Soderling, T. R. (1992). Characterization of the phosphatase activity of a baculovirus-expressed calcineurin A isoform. *J. Biol. Chem.* **267,** 15965–15969.

Perrino, B. A., Ng, L. Y., and Soderling, T. R. (1995). Calcium regulation of calcineurin phosphatase activity by its B subunit and calmodulin: Role of the autoinhibitory domain. *J. Biol. Chem.* **270,** 340–346.

Pinna, L. A., and Donella-Deana, A. (1994). Phosphorylated synthetic peptides as tools for studying protein phosphatases. *Biochim. Biophys. Acta* **1222,** 415–431.

Politino, M., and King, M. M. (1987). Calcium- and calmodulin-sensitive interactions of calcineurin with phospholipids. *J. Biol. Chem.* **262,** 10109–10113.

Politino, M., and King, M. M. (1990). Calcineurin–phospholipid interactions. *J. Biol. Chem.* **265,** 7619–7622.

Polli, J. W., Billingsley, M. L., and Kincaid, R. L. (1991). Expression of the calmodulin-dependent protein phosphatase, calcineurin, in the rat brain: Developmental patterns and the role of nigrostriatal innervation. *Dev. Brain Res.* **63,** 105–119.

Price, E. R., Zydowski, L. D., Jin, M., Baker, C. H., McKeon, F. D., and Walsh, C. T. (1991). Human cyclophilin B: A second cyclophilin gene encodes a peptidyl-prolyl isomerase with a signal sequence. *Proc. Natl. Acad. Sci. U.S.A.* **88,** 1903–1907.

Pujol, M. J., Bosser, R., Vendrell, M., Serratosa, J., and Bachs, O. (1993). Nuclear calmodulin-binding proteins in rat neurons. *J. Neurochem.* **60,** 1422–1428.

Rao, A. (1995). NFATp, a cyclosporin-sensitive transcription factor implicated in cytokine gene induction. *J. Leukocyte Biol.* **57,** 536–542.

Rao, A., Luo, C., and Hogan, P. G. (1997). Transcription factors of the NFAT family: Regulation and function. *Annu. Rev. Immunol.* **15,** 707–747.

Rasmussen, C., and Garen, C. (1993). Activation of calmodulin-dependent enzymes can be selectively inhibited by Histone H1. *J. Biol. Chem.* **268,** 23788–23791.

Rathna Giri, P., Marietta, C. A., Higuchi, S., and Kincaid, R. L. (1992). Molecular and phylogenetic analysis of calmodulin-dependent protein phosphatase (calcineurin) catalytic subunit genes. *DNA Cell Biol.* 11, 415–424.

Renoir, J., Mercier-Bodard, C., Hoffmann, K., Le Bihan, S., Ning, Y., Sanchez, E. R., Handschumacher, R. E., and Baulieu, E. (1995). Cyclosporin A potentiates the dexamethasone-induced mouse mammary tumor virus-chloramphenicol acetyltransferase activity in LMCAT cells: A possible role for different heat shock protein-binding immunophilins in glucocorticosteroid receptor-mediated gene expression. *Proc. Natl. Acad. Sci. U.S.A.* **92,** 4977–4981.

Ringe, D. (1991). Immunosuppression. Binding by design. *Nature (London)* **351,** 185–186.

Robinson, P. J., Liu, J. P., Powell, K. A., Fyske, E. M., and Sudhof, T. C. (1994). Phosphorylation of dynamin 1 and synaptic-vesicle recycling. *Trends Neurosci.* **17**, 348–353.

Rosenmund, C., Carr, D. W., Bergeson, S. E., Nilaver, G., Scott, J. D., and Westbrook, G. L. (1994). Anchoring of protein kinase A is required for modulation of AMPA/kainate receptors on hippocampal neurons. *Nature (London)* **368**, 853–856.

Rubin, C. S. (1994). A kinase anchor proteins and the intracellular targeting of signals carried by cyclic AMP. *Biochim. Biophys. Acta* **1224**, 467–479.

Ruff, V. A., and Leach, K. L. (1995). Direct demonstration of NFATp dephosphorylation and nuclear localization in activated HT-2 cells using a specific NFATp polyclonal antibody. *J. Biol. Chem.* **270**, 22602–22607.

Schaefer, A., Magocsi, M., Stocker, U., Fandrich, A., and Marquardt, H. (1996). Ca^{2+}/calmodulin-dependent and -independent down-regulation of *c-myb* mRNA levels in erythropoietin-responsive murine erythroleukemia cells. *J. Biol. Chem.* **271**, 13484–13490.

Schreiber, S. L. (1991). Chemistry and biology of the immunophilins and their immunosuppressive ligands. *Science* **251**, 283–287.

Schroeder, J. I., and Hagiwara, S. (1989). Cytosolic calcium regulates ion channels in the plasma membrane of Vicia faba guard cells. *Nature (London)* **338**, 427–430.

Schwaninger, M., Blume, R., Oetjen, E., Lux, G., and Knepel, W. (1993). Inhibition of cAMP-responsive element-mediated gene transcription by cyclosporin A and FK506 after membrane depolarization. *J. Biol. Chem.* **26**, 23111–23115.

Schwaninger, M., Blume, R., Kruger, M., Lux, G., Oetjen, E., and Knepel, W. (1995). Involvement of the Ca^{2+}-dependent phosphatase calcineurin in gene transcription that is stimulated by cAMP through cAMP response elements. *J. Biol. Chem.* **270**, 8860–8866.

Shaw, K.T. Y., Ho, A. M., Raghavan, A., Kim, J., Jain, J., Park, J., Sharma, S., Rao, A., and Hogan, P. G. (1995). Immunosuppressive drugs prevent a rapid dephosphorylation of the transcription factor NFAT1 in stimulated immune cells. *Proc. Natl. Acad. Sci. U.S.A.* **92**, 11205–11209.

Sheldon, P. W., and Aghajanian, G. K. (1991). Excitatory responses to serotonin 5-HT in neurons of the rat piriform cortex: evidence for mediation by 5-HT_{1C} receptors in pyramidal cells and 5-HT_2 receptors in interneurons. *Synapse* **9**, 208–218.

Shi, Y., Sahai, B. M., and Green, D. R. (1989). Cyclosporin A inhibits activation-induced cell death in T-cell hybridomas and thymocytes. *Nature (London)* **339**, 625–626.

Shibasaki, F., and McKeon, F. D. (1995). Calcineurin functions in Ca^{2+}-activated cell death in mammalian cells. *J. Cell Biol.* **131**, 735–743.

Shibasaki, F., Price, E. R., Milan, D., and McKeon, F. D. (1996). Role of kinases and the phosphatase calcineurin in the nuclear shuttling of transcription factor NF-AT4. *Nature (London)* **382**, 370–373.

Shipston, M. J., Hernando, F., Barnard, R. J. O., and Antoni, F. A. (1994). Glucocorticoid negative feedback in pituitary corticotropes. Pivotal role for calcineurin inhibition of adenylyl cyclase. *Ann. N. Y. Acad. Sci.* **746**, 453–456.

Siekierka, J. J., Hung, S. H. Y., Poe, M., Lin, C. S., and Sigal, N. H. (1989). A cytosolic binding protein for the immunosuppressant FK506 has peptidyl-prolyl isomerase activity but is distinct from cyclophilin. *Nature (London)* **341**, 755–757.

Sigal, N. H., Dumont, F., Durette, P., Siekierka, J. J., Peterson, L., Rich, D. H., Dunlap, B. E., Staruch, M. J., Melino, M. R., Koprak, S. L., Williams, D., Witzel, B., and

Pisano, J. M. (1991). Is cyclophilin involved in the immunosuppressive and nephrotoxic mechanism of action of cyclosporin A? *J. Exp. Med.* **173**, 619–628.
Sikkink, R., Haddy, A., MacKelvie, S., Mertz, P., Litwiller, R., and Rusnak, F. (1995). Calcineurin subunit interactions: Mapping the calcineurin B binding domain on calcineurin A. *Biochemistry* **34**, 8348–8356.
Singer, G. G., and Abbas, A. K. (1994). The Fas antigen is involved in peripheral but not thymic deletion of T lymphocytes in T-cell receptor transgenic mice. *Immunity* **1**, 365–371.
Smith, C. A., Williams, G. T., Kingston, R. E., Jenkinson, J., and Owen, J. J. T. (1989). Antibodies to CD3/T-cell receptor complex induce cell death by apoptosis in immature T cells in thymic culture. *Nature (London)* **337**, 181–184.
Soderling, T. R. (1990). Protein kinases. Regulation by autoinhibitory domains. *J. Biol. Chem.* **265**, 1823–1826.
Stemmer, P. M., and Klee, C. B. (1991). Serine/threonine phosphatases in the nervous system. *Curr. Opin. Neurobiol.* **1**, 53–64.
Stemmer, P. M., and Klee, C. B. (1994). Dual calcium regulation of calcineurin by calmodulin and calcineurin B. *Biochemistry* **33**, 6859–6866.
Stemmer, P. M., Wang, S., Krinks, M. H., and Klee, C. B. (1995). Factors responsible for the Ca^{2+}-dependent inactivation of calcineurin in brain. *FEBS Lett.* **374**, 237–240.
Stewart, A. A., Ingebritsen, T. S., Manalan, A., Klee, C. B., and Cohen, P. (1982). Discovery of a Ca^{2+}-and calmodulin-dependent protein phosphatase. *FEBS Lett.* **137**, 80–84.
Takahashi, N., Hayano, T., and Suzuki, M. (1989). Peptidyl-prolyl *cis-trans* isomerase is the cyclosporin A-binding protein cyclophilin. *Nature* **337**, 473–475.
Takaishi, T., Saito, N., Kuno, T., and Tanaka, C. (1991). Differential distribution of the mRNA encoding two isoforms of the catalytic subunit of calcineurin in the rat brain. *Biochem. Biophys. Res. Commun.* **174**, 393–398.
Tanaka, T., Ames, J. B., Harvey, T. S., Stryer, L., and Ikura, M. (1995). Sequestration of the membrane-targeting myristoyl group of recoverin in the calcium-free state. *Nature (London)* **376**, 444–447.
Tash, J. S., Krinks, M. H., Patel, J., Means, R. L., Klee, C. B., and Means, A. R. (1988). Identification, characterization, and functional correlation of calmodulin-dependent protein phosphatase in sperm. *J. Cell Biol.* **106**, 1625–1633.
Tocci, M. J., Matkovich, D. A., Collier, K. A., Kwok, P., Dumont, F., Lin, S., Degudicubus, S., Siekierka, J. J., Chin, J., and Hutchinson, N. I. (1989). The immunosuppressant FK506 selectively inhibits expression of early T-cell activation genes. *J. Immunol.* **143**, 718–726.
Tong, G., Shepherd, D., and Jahr, C. E. (1995). Synaptic desensitization of NMDA receptors by calcineurin. *Science* **267**, 1510–1512.
Ucker, D. S., Ashwell, J. D., and Nichas, G. (1989). Activation-driven T cell death: Requirements for *de novo* transactivation and translation and association with genome fragmentation. *J. Immunol.* **143**, 3461–3469.
Ueki, K., Muramatsu, T., and Kincaid, R. L. (1992). Structure and expression of two isoforms of the murine calmodulin-dependent protein phosphatase regulatory subunit (calcineurin B). *Biochem. Biophys. Res. Commun.* **187**, 537–543.
Usuda, N., Arai, H., Sasaki, H., Hanai, T., Nagata, T., Kincaid, R. L., and Higuchi, S. (1996). Differential subcellular localization of neural isoforms of the catalytic subunit of calmodulin-dependent protein phosphatase (calcineurin) in central nervous system

neurons: Immunohistochemistry on formalin-fixed paraffin sections employing antigen retrieval by microwave irradiation. *J. Histochem. Cytochem.* **44,** 13–18.
Victor, R. G., Thomas, G. D., Marban, E., and O'Rourke, B. (1995). Presynaptic modulation of cortical synaptic activity by calcineurin. *Proc. Natl. Acad. Sci. U.S.A.* **92,** 6269–6273.
Villafranca, J. E., Kissinger, C. R., and Parge, H. E. (1996). Protein serine/threonine phosphatases. *Curr. Opin. Biotechnol.* **7,** 397–402.
Walaas, S. I., and Greengard, P. (1991). Protein phosphorylation and neuronal function. *Pharmacol. Rev.* **43,** 299–349.
Walsh, M. P., Busaan, J. L., Fraser, E. D., Fu, S. Y., Pato, M. D., and Michalak, M. (1995). Characterization of the recombinant C-terminal domain of dystrophin: Phosphorylation by calmodulin-dependent kinase II and dephosphorylation by Type 2B protein phosphatase. *Biochemistry* **34,** 5561–5568.
Wang, J. H., and Desai, R. (1977). Modulator binding protein. *J. Biol. Chem.* **252,** 4175–4184.
Wang, X., Culotta, V. C., and Klee, C. B. (1996). Superoxide dismutase protects calcineurin from inactivation. *Nature (London)* **383,** 434–437.
Watanabe, Y., Perrino, B. A., Chang, B. H., and Soderling, T. R. (1995). Identification in the calcineurin A subunit of the domain that binds the regulatory B subunit. *J. Biol. Chem.* **270,** 456–460.
Watanabe, Y., Perrino, B. A., and Soderling, T. R. (1996). Activation of calcineurin A subunit phosphatase activity by its calcium-binding B subunit. *Biochemistry* **35,** 562–566.
Wesselborg, S., Fruman, D. A., Sagoo, J. K., Bierer, B. E., and Burakoff, S. J. (1996). Identification of a physical interaction between calcineurin and nuclear factor of activated T cells (NFATp). *J. Biol. Chem.* **271,** 1274–1277.
Williams, K. R., Hemmings, H. C. J., LoPresti, M. B., Konigsber, W. H., and Greengard, P. (1986). DARPP-32, a dopamine-and cyclic AMP-regulated neuronal phosphoprotein. *J. Biol. Chem.* **261,** 1890–1903.
Wolff, D. J., and Sved, D. W. (1985). The divalent cation dependence of bovine brain calmodulin-dependent phosphatase. *J. Biol. Chem.* **260,** 4195–4202.
Wood, J. G., Wallace, R. W., Whitaker, J. N., and Cheung, W. Y. (1980). Immunocytochemical localization of calmodulin and a heat-labile calmodulin-binding protein (CaM-BP$_{80}$) in basal ganglia of mouse brain. *J. Cell Biol.* **84,** 66–76.
Yamamoto, H., Fukunaga, K., Tanaka, E., and Miyamoto, E. (1983). Ca^{2+}- and calmodulin-dependent phosphorylation of microtubule-associated protein 2 and t factor, and inhibition of microtubule assembly. *J. Neurochem.* **41,** 1119–1125.
Yamauchi, T., and Fujisawa, H. (1983). Disassembly of microtubules by the actions of calmodulin-dependent protein kinase (kinaseII) which occurs only in the brain tissues. *Biochem. Biophys. Res. Commun.* **110,** 287–291.
Yang, S. D., Tallant, E. A., and Cheung, W. Y. (1982). Calcineurin is a calmodulin-dependent phosphatase. *Biochem. Biophys. Res. Commun.* **106,** 1419–1425.
Yang, S. D., Rosen, M. K., and Schreiber, S. L. (1993). A composite FKBP12–FK506 surface that contacts calcineurin. *J. Am. Chem. Soc.* **115,** 819–820.
Yu, S. P., O'Malley, D. M., and Adams, P. R. (1994). Regulation of M current by intracellular calcium in bullfrog sympathetic ganglion cells. *J. Neurosci.* **14,** 3487–3499.
Zhang, W., Zimmer, G., Chen, J., Ladd, D., Li, E., Alt, F. W., Wiederrecht, G. J., Cryan, J., O'Neill, E. A., Seidman, C. E., Abbas, A. F., and Seidman, J. G. (1996). T cell responses in calcineurin Aa-deficient mice. *J. Exp. Med.* **183,** 413–420.

Zhu, D., Cardenas, M. E., and Heitman, J. (1995). Myristoylation of calcineurin B is not required for function or interaction with immunophilin–immunosuppressant complexes in the yeast *Saccharomyces cerevisiae*. *J. Biol. Chem.* **270,** 24831–24838.

Zhu, Y., and Yakel, J. L. (1997). Calcineurin modulates G-protein-mediated inhibition of N-type calcium currents in rat sympathetic neurons. *J. Neurophysiol.* **78,** 1161–1165.

5

Cyclic Nucleotide Regulation by Calmodulin

WILLIAM K. SONNENBURG, GARY A. WAYMAN, DANIEL R. STORM, AND JOSEPH A. BEAVO

Department of Pharmacology
University of Washington
Seattle, Washington 98195

I. Introduction 237
II. Calmodulin-Regulated Adenylyl Cyclase 238
   A. Type I Adenylyl Cyclase 239
   B. Type VIII Adenylyl Cyclase 248
   C. Type III Adenylyl Cyclase 249
III. Calmodulin-Stimulated Phosphodiesterase Isoforms: Genetic Diversity 251
   A. Structure and Domain Organization of Calmodulin-Stimulated
      Phosphodiesterase 257
   B. Kinetic Properties and Activation of Calmodulin-Stimulated
      Phosphodiesterase Isoforms by Calmodulin 262
   C. Selective Calmodulin-Stimulated Phosphodiesterase Inhibitors 266
   D. Regulation of Calmodulin-Stimulated Phosphodiesterase Isoforms by
      Phosphorylation 266
   E. Expression and Role of Calmodulin-Stimulated Phosphodiesterase Isoforms
      in Different Tissues and Cell Types 267
IV. Concluding Remarks 277
   References 278

I. INTRODUCTION

Since its discovery (1) as a subunit and activator of cyclic nucleotide phosphodiesterase (PDE), the calcium signal transducer, calmodulin

(CaM), has been known to mediate the interaction of signal transduction pathways involving the elevation of intracellular calcium with cyclic nucleotide signaling. It is now becoming more evident that not only does CaM regulate the degradation of cyclic nucleotides by activating CaM-stimulated PDEs, but it also modulates in both positive and negative ways the synthesis of cAMP and cGMP via inhibition or activation of adenylyl cyclase (AC) or activation of guanylyl cyclase via a CaM-sensitive nitric oxide synthase (2, 3). This chapter focuses on the role that CaM plays in regulating (a) adenylyl cyclase (adenylate cyclases) and (b) cyclic nucleotide phosphodiesterases, the identity and manner by which CaM regulates these different isozymes, the mechanisms by which CaM elicits its regulatory function, the distribution of these enzymes in the different cell types and organs of mammals, and the physiological contexts in which these enzymes are important for the "normal" signal transduction and functioning of cells, organs, and systems.

II. CALMODULIN-REGULATED ADENYLYL CYCLASES

The adenylyl cyclases catalyze the formation of cAMP, an important intracellular message in almost all animal cells. They are regulated by stimulatory and inhibitory receptors coupled to their catalytic subunits through stimulatory (G_s) and inhibitory (G_i) GTP-binding, regulatory proteins. Intracellular Ca^{2+} also modulates adenylyl cyclase activity in several tissues, and the Ca^{2+}-sensitive adenylyl cyclases provide mechanisms for "cross-talk" between Ca^{2+} and cAMP signal transduction systems. In some cases, adenylyl cyclases function as signal integrators and respond synergistically to multiple extracellular and intracellular signals.

Characterization of adenylyl cyclases was advanced greatly by purification of the enzymes and isolation of cDNA clones encoding mammalian adenylyl cyclases. There are at least nine distinct adenylyl cyclases, each having its own unique regulatory properties. Although adenylyl cyclases are regulated by multiple effector molecules, including Ca^{2+} and hormone receptors, each enzyme has unique regulatory properties. The diversity of this enzyme family reflects the need for different mechanisms for regulating cAMP levels in animal cells and the variety of physiological processes that are regulated by intracellular cAMP. This section focuses on the CaM-regulated enzymes, type I (1AC) and type VIII (8AC), as well as on type III (3AC), an enzyme that is regulated by CaM kinase II.

A. Type I Adenylyl Cyclase

The existence of distinct CaM-stimulated and CaM-insensitive adenylyl cyclases in brain was first demonstrated by the separation of two isoforms of the enzyme using CaM–Sepharose affinity chromatography (4). In addition, polyclonal antibodies were isolated that distinguished between CaM-sensitive and CaM-insensitive adenylyl cyclases in brain (5). The isolation of cDNA clones encoding 1AC (CaM-sensitive) and type II adenylyl cyclase (2AC; CaM-insensitive) confirmed the existence of at least two classes of adenylyl cyclases (6, 7). 1AC has been of particular interest to neurobiologists because it is neurospecific, and data from invertebrates and mammals suggest that it is important for synaptic plasticity as well as for learning and memory.

1. Regulation of 1AC by Ca^{2+} and G_s-Coupled Receptors

1AC is directly stimulated by Ca^{2+} and CaM *in vitro* (8, 9) and *in vivo* (10, 11). Half-maximal stimulation is at 150 nM free Ca^{2+} and 20 nM CaM. To characterize Ca^{2+} stimulation of the enzyme *in vivo*, the enzyme was stably expressed in HEK-293 cells and its sensitivity to Ca^{2+} was examined using the Ca^{2+} ionophore A23187 to increase intracellular Ca^{2+} (10). Although Ca^{2+} does not affect intracellular cAMP in control cells transfected with empty expression vector alone, A23187 causes a significant cAMP increase in cells expressing 1AC. The increase in cAMP depends on the concentration of extracellular Ca^{2+} applied and is detectable within a few minutes after addition of the Ca^{2+} ionophore.

HEK-293 cells express muscarinic acetylcholine receptors that are coupled to the mobilization of intracellular Ca^{2+} pools. Treatment of these cells with carbachol increases intracellular-free Ca^{2+}. Carbachol also increases cAMP levels approximately three-fold in HEK-293 cells stably expressing 1AC, but is without effect on cAMP in control cells lacking 1AC. Half-maximal stimulation is at 30 μM carbachol, and cAMP increases are inhibited by the muscarinic acetylcholine receptor antagonist, atropine. Carbachol-stimulated increases in intracellular Ca^{2+} and cAMP are completely blocked by BAPTA, an intracellular Ca^{2+} chelator. Furthermore, a CaM-insensitive mutant of 1AC, containing a point mutation within its CaM-binding domain, is not stimulated by an increase in intracellular-free Ca^{2+} (12). This indicates that Ca^{2+} stimulation of 1AC *in vivo* is mediated by direct interactions of a Ca^{2+}/CaM complex with the enzyme.

2. Synergistic Stimulation of 1AC by Ca^{2+} and Receptor Activation

To examine the effect of intracellular Ca^{2+} on the coupling of G_s-activated receptors to 1AC *in vivo*, stably transfected HEK-293 cell lines expressing the glucagon receptor with 1AC or 3AC were prepared (13). HEK-293 cells express endogenous β-adrenergic receptors that couple to stimulation of 3AC but not 1AC. Activation of glucagon receptors also stimulates 3AC but not 1AC. These data indicate that G_s, as well as β-adrenergic and glucagon receptors, are functionally expressed in HEK-293 cells. Since $G_s\alpha$ subunit stimulation of 1AC *in vitro* is inhibited by β/γ subunits (14, 15) and receptor activation of G_s releases β/γ subunits, the insensitivity of 1AC to G_s-coupled receptor activation may be due to β/γ subunit inhibition. To address this issue, the sensitivity of 1AC to activation by the G_s-coupled 5HT-7 receptor *in vivo* was examined when β/γ subunit-binding proteins were coexpressed (16). Coexpression of transducin α subunit with 1AC elicits a substantial ($\sim$4-fold) stimulation of 1AC by serotonin. Experiments in which the β-adrenergic receptor kinase 1-carboxyl terminus was used as the β/γ subunit scavenger give similar results. These data suggest that β/γ subunits release from dissociating G_s heterotrimers inhibits 1AC and prevents receptor activation through $G_s\alpha$.

Although 1AC is not stimulated by isoproterenol alone, it is synergistically activated by combinations of A23187 with increasing concentrations of isoproterenol (13). To test the generality of this phenomena and to ensure that the insensitivity to isoproterenol alone is not due to low levels of endogenous β-adrenergic receptors in HEK-293 cells, stable transformants coexpressing glucagon receptors and 1AC were also analyzed. In the absence of Ca^{2+} increases, the enzyme is not stimulated by glucagon. However, it is stimulated by glucagon when it is also activated by Ca^{2+}. Intracellular Ca^{2+} signals generated by muscarinic agonists also increase the sensitivity of 1AC to G_s-coupled receptors.

Synergism between intracellular Ca^{2+} and hormones might be attributable to activation of Ca^{2+} or cAMP-stimulated protein kinases that phosphorylate a component of the adenylyl cyclase system. However, synergism between Ca^{2+} and isoproterenol is not blocked by KN 62, an inhibitor of CaM kinases, calphostin C, a protein kinase C inhibitor, or the PKA inhibitors H89 or Rp-cAMP. Similarly, carbachol and isoproterenol synergism is not blocked by these inhibitors, suggesting that activation of a protein kinase, in response to cAMP or Ca^{2+}, is not required for these phenomena.

Synergistic activation of 1AC by Ca^{2+} and G_s-coupled receptors is dependent on binding of CaM to 1AC and activation of the enzyme by Ca^{2+}. A CaM-insensitive mutant of 1AC containing an Arg in place of Phe-503 within the CaM-binding domain (FR-1AC) was used to demonstrate this requirement. This mutant enzyme is insensitive to CaM *in vitro*, but its basal and forskolin-stimulated activities are indistinguishable from wild-type 1AC (11). Because FR-1AC is not synergistically stimulated by isoproterenol and A23187, the increased sensitivity to β-adrenergic receptors caused by elevated intracellular Ca^{2+} requires CaM binding to 1AC. The authors hypothesize that CaM binding to the catalytic subunit of 1AC enhances its affinity for activated $G_s\alpha$ and prevents β/γ subunit binding to the enzyme.

3. Stimulation of 1AC by Protein Kinase C

Activation of protein kinase C (PKC) by phorbol esters stimulates the activity of several adenylyl cyclases *in vivo* (17). To determine if PKC modulates the activity of 1AC, the effects of phorbol esters were examined *in vivo* (18). The phorbol ester, TPA, stimulates 1AC 200% within 15 min after treatment. To evaluate the specificity of the cAMP increase caused by TPA, the effects of two other phorbol esters were examined. TPA and PDBu are potent activators of PKC and both stimulate 1AC. However, the inactive phorbol ester, 4-a-PMA, does not stimulate 1AC. at concentrations up to 100 nM. Stimulation of 1AC and other adenylyl cyclases by PKC provides an interesting mechanism for cross-talk between phosphoinositide and cAMP signal transduction systems. This may play an important role in various physiological processes, including modulation of synaptic plasticity in the nervous system. For example, PKC activity is stimulated during some forms of long-term potentiation (LTP) in the hippocampus, which may contribute to increased cAMP associated with LTP.

4. Mechanisms for Inhibition of 1AC

Although mechanisms for inhibition of 1AC *in vivo* have not been characterized completely, CaM stimulation of the enzyme is inhibited by M4-muscarinic, somatostatin, and dopamine D2 receptor agonists (16, 19). Inhibition is blocked by pertussis toxin, indicating that receptor coupling is most probably mediated by G_i. This inhibition is due primarily to $G_i\alpha$ and not β/γ subunits as coexpression of β/γ subunit-binding proteins with 1AC does not affect somatostatin inhibition (16). These data are consistent with studies showing that CaM stimulation of 1AC is directly inhibited by GTP-activated $G_i\alpha$ *in vitro* (20).

1AC is also inhibited by CaM kinase IV *in vivo* (21). Expression of constitutively active or wild-type CaM kinase IV inhibits Ca^{2+} stimulation of 1AC without affecting basal or forskolin-stimulated activities. 1AC has two CaM kinase IV consensus phosphorylation sequences near its CaM-binding domain at Ser-545 and Ser-552. Conversion of either serine to alanine by mutagenesis destroys CaM kinase IV inhibition of the adenylyl cyclase, suggesting that the activity of this enzyme may be directly inhibited by CaM kinase IV phosphorylation *in vivo*. CaM kinase IV (22) and 1AC (23) are both expressed in the neocortex, hippocampus, and cerebellar cortex with high levels of mRNA expression in granule cells. In neurons expressing both CaM kinase IV and 1AC, elevations in intracellular Ca^{2+} may produce only transient or submaximal concentrations of cAMP.

5. Tissue and Cellular Distribution of 1AC

An analysis of the distribution of 1AC mRNA in bovine and rodent tissues indicates that this enzyme is neurospecific (24). The only bovine tissues showing a positive signal for 1AC mRNA are brain, retina, and adrenal medulla. Several cultured cell lines, including neuroblastoma cell N1E-115, neurogliohybridoma cell NG-108, rat glioma 36B-10 cell, and PC-12 cells, also do not express mRNA for 1AC. The restricted expression of 1AC to neural tissues contrasts sharply with some of the other mammalian enzymes that show fairly broad distribution in both neural and nonneuronal tissues.

Within the brain the distribution of mRNA encoding the 1AC was also examined by *in situ* hybridization (24). *In situ* hybridizations in adult mouse and rat brain indicate that 1AC is highly expressed in specific areas of brain, including the hippocampal formation, the neocortex, entorhinal cortex, cerebellum cortex, and parts of the olfactory system (23). The dentate gyrus in the hippocampal formation showed very intense labeling that was associated with the granule cell layer. Moderately strong labeling is also evident in association with the pyramidal cells in CA1, CA2, and CA3 layers of the hippocampus. 1AC mRNA is not detected in the brain stem. Because 1AC is not expressed throughout the brain, suggesting that it probably does not play a general "housekeeping" role (e.g., regulation of cell metabolism) and may be important for specific neuronal functions. The neurospecific expression of the 1AC and its limited distribution in brain are consistent with the proposal that this enzyme may be important for some forms of synaptic plasticity.

6. Disruption of 1AC Gene Leads to Deficiencies in LTP and Spatial Learning

Although the physiological function of 1AC in brain is not known, studies have demonstrated a possible role for the Ca^{2+}-sensitive adenylyl cyclases in memory and learning. The first evidence that Ca^{2+}-sensitive adenylyl cyclases may be important for synaptic plasticity came from studies of the *Drosophila* learning mutant, rutabaga. Rutabaga is an X-linked recessive mutant that is deficient in associative learning (25, 26). In contrast to wild-type *Drosophila*, the rutabaga fly lacks Ca^{2+} and CaM-sensitive adenylyl cyclase activity. Furthermore, the gene for an adenylyl cyclase similar to the 1AC maps to the rutabaga locus on the X chromosome. A single point mutation in this gene is sufficient to destroy all Ca^{2+}-stimulated adenylyl cyclase activity (27). Analyses of neuromuscular transmission in rutabaga larvae have revealed impaired synaptic transmission and posttetanic potentiation as well as abnormal responses to direct application of dibutyryl-cAMP (28).

Using targeted gene disruption of 1AC, Wu *et al.* (29) have shown that mice deficient in 1AC have altered synaptic plasticity as well as learning and memory. 1AC mutant mice have normal growth, motor coordination, and longevity. They show no detectable anatomical differences in the hippocampus, neocortex, or cerebellum. Compared to wild type mice, Ca^{2+}-sensitive adenylyl cyclase activities in the cerebellum, neocortex, and hippocampus of mutant mice are decreased 62, 38, and 46%, respectively. Furthermore, the Ca^{2+} sensitivity of the residual adenylyl cyclase activity in mutant mice is lower than 1AC and is consistent with the Ca^{2+} sensitivity of 8AC. There was only a minor reduction in Ca^{2+}-sensitive adenylyl cyclase activity in the brain stem.

To determine if 1AC is crucial for synaptic plasticity, several forms of hippocampal LTP were compared in wild-type and mutant mice. Both wild-type and mutant mice exhibit long-lasting CA1 LTP (L-LTP) that persists beyond 3 hr; however, there are several quantitative differences in the response of the mutant mice that are evident during the first hour after stimulation. The rate of increase of the excitatory postsynaptic potential (EPSP) slope for the mutant mice after tetanic stimulation (from 1 to 30 min) is half that of wild-type mice (1.3 ± 0.05 vs 2.6 ±

0.1; $p < 0.001$). The maximum field EPSP slope above baseline is also reduced approximately 40% in mutant mice. Facilitation of the synaptic response that occurred after tetanic stimulation developed more slowly and reached a lower level in mutants lacking 1AC. Mossy fiber/CA3 LTP in mutant and wild-type mice was also analyzed. This form of LTP is presynaptic in origin and maintenance and is independent of NMDA receptors. 1AC mutant mice show greatly depressed mossy fiber/CA3 LTP, indicating that the 1AC is important for this type of LTP (Villacres and Storm, unpublished observations). What is the molecular role of 1AC for LTP? The coupling of the Ca^{2+} and cAMP systems may result in simultaneous or sequentially ordered activation of Ca^{2+} and cAMP-stimulated protein kinases or may provide positive feedback regulation of Ca^{2+} channels by PKA. For example, Nicoll and colleagues (30) have proposed that mossy fiber/CA3 LTP may depend on Ca^{2+} stimulation of 1AC. Entry of Ca^{2+} into the presynaptic terminal is hypothesized to activate 1AC and causes a persistent increase of glutamate release through PKA. The fact that 1AC mutant mice show depressed mossy fiber/CA3 LTP is consistent with this hypothesis.

Mutant and wild-type mice were analyzed for spatial learning by the Morris water task. In this test, both sets of animals show decreased escape latencies with training, and there are no statistically significant differences in the ability of the mutant and wild-type mice to find the visible or hidden platform. However, escape latencies in the hidden platform task are a poor indicator of spatial learning and even rodents with hippocampal lesions that affect other forms of spatial learning can learn to find the hidden platform in the Morris water task (31). A better indicator of spatial learning is the transfer test in which the animal is trained to find the hidden platform at a specific site in the pool. The platform is then removed, and the number of times that a mouse swims across the target area or the time in the target quadrant is quantitated. There were significant and reproducible differences in transfer test behavior between the mutant and wild-type mice. Wild-type mice crossed the target area 6 ± 0.4 times during a 60-sec trial whereas mutant mice cross only 4 ± 0.4 times ($p < 0.002$). The difference in transfer ability was also evident when the time in various quadrants was analyzed. Only wild-type mice showed a bias for the target quadrant. They spent 46% ± 3.0 of their time in the target quadrant searching for the platform. Mutant mice show no significant preference for the target quadrant A (27% ± 2.0), indicating an impaired ability in this specific task ($p < 0.001$). These data illustrate that mutant mice have a significant and lasting place navigational impairment that is dissociated from visual, motivational, or motor requirements of the test. The relationships between the defects in spatial memory and LTP have not been established.

7. Developmental Expression of 1AC

Synaptogenesis and the expression of LTP in rodents occur during the first 3 weeks following birth with a gradual decline at later stages of development (32–34). Because 1AC may be important for the regulation of synaptic plasticity, basal and Ca^{2+}-stimulated adenylyl cyclase activities were measured in rodent brains during the first 3 weeks of postnatal development (35). Ca^{2+}-stimulated adenylyl cyclase activity in membranes from the rat hippocampus increased 5.5-fold between PD1 and PD16 and declined after PD16. Although basal activity also increased during this period, the relative increase in Ca^{2+}-stimulated adenylyl cyclase activity was greater.

The developmental increase in Ca^{2+}-stimulated adenylyl cyclase in rat brain could be due to changes in gene expression or modulation of adenylyl cyclase activity by other proteins. Consequently, 1AC mRNA was quantitated in brains from PD2, PD8, and PD16 rats by Northern analysis. In the hippocampus, cerebellum, or whole brain, 1AC mRNA increased during the period from P2 to P16. In the hippocampus it increased sevenfold between PD2 and PD16. This is consistent with the increase in Ca^{2+}-stimulated adenylyl cyclase activity in this tissue. In contrast, 8AC mRNA increased only twofold between PD2 and PD16, suggesting that the developmental increase in Ca^{2+}-stimulated adenylyl cyclase activity in brain is due primarily to the expression of 1AC.

To evaluate the contribution of 1AC to the developmental increase in Ca^{2+}-stimulated activity, adenylyl cyclase activity was measured in the cerebellum from developing wild-type and 1AC mutant mice. Like rats, wild-type mice showed a significant increase in basal and Ca^{2+}-stimulated adenylyl cyclase activity during the first 2 weeks of postnatal development. The developmental increase in Ca^{2+}-stimulated activity was greater than basal activity (6.5- vs 3.0-fold). Mutant mice showed only a 2-fold increase in Ca^{2+}-stimulated adenylyl cyclase activity between P2 and P16 and no significant increase in basal activity. Changes in expression of the 1AC during the period of long-term potentiation development are consistent with the hypothesis that this enzyme is important for neuroplasticity in vertebrates.

8. Role of 1AC in Melatonin Synthesis

The development, maturation, and reproduction of mammals undergo seasonal or circadian fluctuations. One of the endogenous entrainable clocks controlling circadian rhythm is in the suprachiasmatic nucleus (SCN) of the hypothalamus. It is regulated by several factors, including light and neurohumoral agents (36, 37). The pineal gland also plays an important role in the circadian cycle by functioning as a neuroendocrine transducer. During the dark/light cycle, input from the retina is relayed to the pineal gland via the SCN as rhythmic increases in nocturnal norepinephrine (NE). This results in the synthesis of melatonin, which is important for reproduction, aging, and sleep.

Melatonin biosynthesis in the pineal gland and retina is regulated by cAMP, which stimulates transcription of genes encoding enzymes important for circadian expression of melatonin, including serotonin N-acetyltransferase (NAT). Nocturnal release of NE at the pineal increases cAMP by activation of β- and α_1-adrenergic receptors. 1AC is expressed in the pineal and retina and may play a major role in regulating melatonin biosynthesis because of its unique regulatory properties. It is hypothesized that costimulation of 1AC by Ca^{2+} and β-adrenergic receptors may generate cAMP signals of sufficient strength and duration to regulate the transcription of specific genes important for melatonin synthesis (Fig. 1).

It was discovered that Ca^{2+}-sensitive adenylyl cyclase activity in rat (38) and mouse (M. D. Nielsen and D. R. Storm, 1997) pineal undergoes a diurnal variation with highest levels expressed in the dark. However, Ca^{2+}-stimulated adenylyl cyclase activity in the neocortex, cerebellum, and hippocampus does not show circadian variation. The variation in expression of Ca^{2+}-sensitive adenylyl cyclase activity in rat pineal is mirrored by a circadian pattern of expression of 1AC mRNA. 1AC mRNA shows maximum expression at midday and the minimum at night (0:00–3:00). In addition, 1AC mRNA expression at midday is reduced 30% by the administration of isoproterenol, a β-adrenergic agonist. Administration of the β-adrenergic antagonist, propranolol, 2 hr before the dark phase increased 1AC mRNA at midnight.

5. Cyclic Nucleotide Regulation by Calmodulin

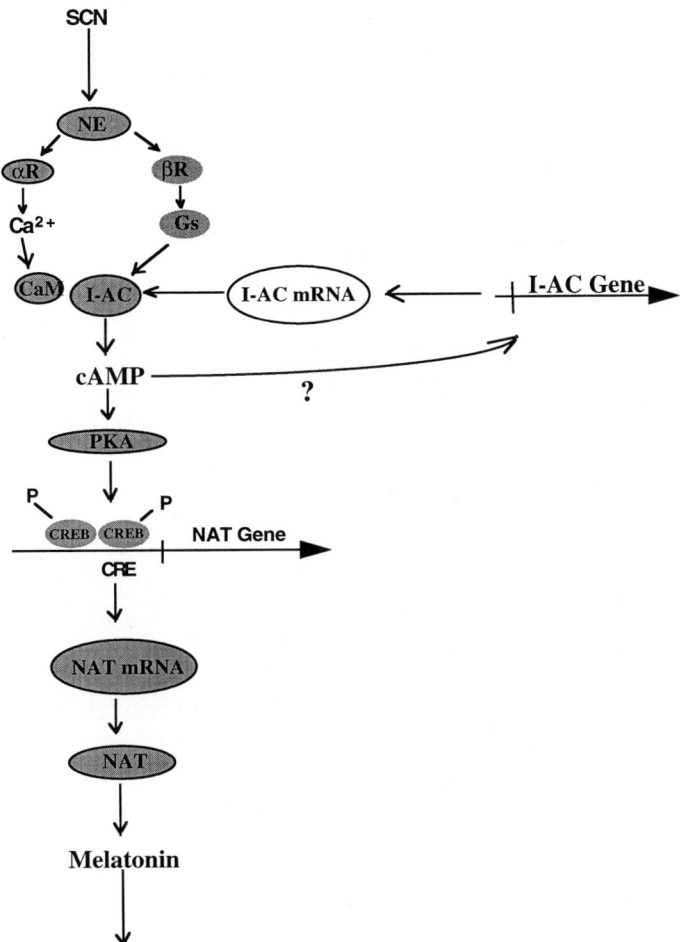

Fig. 1. 1AC functions as a regulatory gatekeeper for pineal melatonin synthesis. It is hypothesized that nocturnal NE stimulates β-adrenergic receptors coupled to 1AC and α1-adrenergic receptors coupled to the liberation of intracellular Ca^{2+}. Synergistic activation of 1AC by Ca^{2+} and activated G_s increases cAMP levels and increases the transcription of the NAT, as well as NAT activity and melatonin synthesis. The expression of 1AC mRNA may be negatively controlled by cAMP. 1AC enzyme activity is maximal at night whereas its mRNA is maximally expressed during the day. This time lapse is not unusual for a membrane enzyme.

These data indicate that the generation of 1AC mRNA may be directly regulated by NE and indirectly by the light cycle.

B. Type VIII Adenylyl Cyclase

1. Regulatory Properties of 8AC

The only other Ca^{2+} and CaM-sensitive adenylyl cyclase is 8AC. This enzyme was first cloned in 1994 by Cali *et al.,* (38a) and not much is known about its regulatory properties and physiological functions. 8AC is stimulated by Ca^{2+} and CaM *in vitro.* However, its Ca^{2+} sensitivity is approximately four to five times lower than 1AC (35). It is also stimulated by increases in intracellular Ca^{2+}, presumably through CaM, although the mechanism for Ca^{2+} stimulation *in vivo* has not been yet defined. Like 1AC, 8AC is not stimulated by G_s-coupled receptors *in vivo,* even though it is stimulated by GTP-activated $G_s\alpha$ *in vitro.* In contrast to 1AC, 8AC is not synergistically stimulated by Ca^{2+} and G_s-coupled receptors *in vivo.* The activity of 8AC is not attenuated by either G_i-coupled receptors or CaM kinase IV *in vivo.* 8AC apparently functions as a pure Ca^{2+} detector and responds to relatively high concentrations of intracellular Ca^{2+} in the micromolar range.

2. Physiological Role of 8AC

Although the physiological functions of 8AC are not known, it may contribute to some forms of synaptic plasticity. For example, 8AC may respond to high-level Ca^{2+} signals generated by the stimulus paradigm used to generate L-LTP in the hippocampus. In this respect, it is interesting that 1AC mutant mice showed quantitative defects only in the early stage of CA1 L-LTP but not in the later stages of L-LTP, which depend on transcription. The two Ca^{2+}-stimulated adenylyl cyclases may both contribute to CA1 LTP as complementary activities responding to low and high Ca^{2+} signals. 1AC may be particularly important for amplifying initial Ca^{2+} signals whereas 8AC may respond to higher Ca^{2+} signals and couple Ca^{2+} to cAMP-stimulated transcription. The relative contribution of the two Ca^{2+}-sensitive adenyly cyclases for various types of synaptic plasticity will be defined more clearly when 8AC mutant mice become available.

C. Type III Adenylyl Cyclase

1. Regulatory Properties of 3AC

In contrast to 1AC and 8AC, which are stimulated by Ca^{2+} and CaM *in vitro*, 3AC is not stimulated by Ca^{2+} and CaM alone. To determine if Ca^{2+} and receptor-activated G_s synergistically stimulate 3AC in membranes, the sensitivity of the enzyme to CaM and Ca^{2+} was analyzed in the presence of glucagon. In membrane preparations, 3AC is stimulated by glucagon with an EC_{50} of 7 nM (9). Glucagon-stimulated 3AC activity is enhanced only 45% by CaM and Ca^{2+}, and the EC_{50} for glucagon was not significantly affected by CaM. Nevertheless, these data suggest that Ca^{2+} and hormones might synergistically activate 3AC *in vivo*. Therefore, the Ca^{2+} sensitivity of hormone-stimulated 3AC was evaluated in intact HEK-293 cells (39). Glucagon stimulates 322-fold *in vivo*. However, increases in intracellular Ca^{2+}, generated by A23187 and extracellular Ca^{2+}, inhibited glucagon-stimulated 3AC activity 60% and had no effect on basal activity. Isoproterenol and forskolin-stimulated 3AC activities are also inhibited by increases in intracellular Ca^{2+}. The Ca^{2+} dependencies for inhibition of 3AC and stimulation of 1AC *in vivo* were compared using A23187 and various amounts of extracellular Ca^{2+}. Glucagon-stimulated 3AC activity was inhibited by Ca^{2+} concentrations that stimulated 1AC. The concentration of free intracellular Ca^{2+} for half-maximal inhibition of glucagon-stimulated 3AC activity is estimated at 150 to 200 nM.

Inhibition of 3AC by intracellular Ca^{2+} is not mediated by G_i, cAMP-dependent protein kinase, or protein kinase C. However, Ca^{2+} inhibition of 3AC is antagonized by KN-62, a CaM kinase inhibitor. In addition, constitutively activated CaM kinase II inhibits the enzyme (39). These data suggest that CaM kinase II regulates the activity of 3AC by direct phosphorylation or by an indirect mechanism involving phosphorylation of a protein that inhibits 3AC. However, 3AC is phosphorylated *in vivo* when intracellular Ca^{2+} is increased and that phosphorylation is prevented by CaM kinase inhibitors (40). Site-directed mutagenesis of a CaM kinase II consensus site (Ser-1076 to Ala-1076) in 3AC greatly reduced Ca^{2+}-stimulated phosphorylation and inhibition of 3AC *in vivo*. These data support the hypothesis that Ca^{2+} inhibition of 3AC is due to direct phosphorylation of the enzyme by CaM kinase II *in vivo*.

2. Hormone Stimulation of 3AC Induces Ca^{2+} Oscillations

Although intracellular-free Ca^{2+} can affect cAMP levels by modulation of adenyly cyclase or phosphodiesterase activities, cAMP can also affect intracellular Ca^{2+} by regulating Ca^{2+} ion channel activity. Because activation of PKA can increase intracellular-free Ca^{2+} and hormone stimulation of 3AC is inhibited by Ca^{2+}, one might expect hormone stimulation of 3AC to generate Ca^{2+} oscillations. This question was addressed using HEK-293 cells expressing the glucagon receptor and 3AC (41). HEK-293 cells stably expressing the glucagon receptor (293-G), the glucagon receptor with 1AC (1AC-G), or 3AC (3AC-G) were treated with glucagon and individual cells were Ca^{2+} imaged using Fura-2. Treatment of 293-G cells with glucagon causes a single spike of intracellular Ca^{2+}. Cells expressing 1AC and the glucagon receptor give a similar response; a single peak of Ca^{2+} with no additional increase with subsequent exposures to glucagon. In contrast, Ca^{2+} oscillations are generated when cells expressing the glucagon receptor and 3AC were treated with glucagon. These oscillations are dependent on the continued presence of glucagon and are not generated by transient exposure to the hormone. Exposure of 3AC expressing cells to isoproterenol or forskolin also causes Ca^{2+} oscillations that strongly resemble those induced by glucagon. Ca^{2+} oscillations were dependent on the activity of PKA and CaM kinases but were not solely due to cAMP increase as dibutyryl-cAMP or Sp-cAMP does not stimulate Ca^{2+} oscillations. Although Ca^{2+} oscillations are not dependent on extracellular Ca^{2+}, they were blocked when IP3-sensitive Ca^{2+} pools were depleted.

What is the mechanism for hormone-stimulated Ca^{2+} oscillations in 3AC expressing cells? Data are most consistent with the model schematically depicted in Fig. 2. When 3AC is activated by hormones, cAMP accumulation stimulates PKA, which phosphorylates and activates IP3 receptors localized on an intracellular Ca^{2+} store. This phosphorylation promotes a release of stored intracellular Ca^{2+}. As intracellular Ca^{2+} rises, 3AC activity is attenuated by CaM kinase II and intracellular cAMP levels decrease due to the actions of cAMP phosphodiesterases. When cAMP levels drops below a threshold point and the IP3 receptor is dephosphorylated, Ca^{2+} is resequestered. CaM kinase II activity decreases and 3AC is dephosphorylated, removing the inhibitory signal. This allows the cycle to repeat, if 3AC is chronically exposed to an activator such as forskolin or hormone. The periodicity of the oscillations may depend on the concentrations of adenylyl cyclase, phosphodiesterases, and protein kinases found in a particular cell. This model also predicts that cells expressing 3AC will also show cAMP oscillations.

3. Role of 3AC in Olfactory Signal Transduction

Although 3AC is expressed in several tissues, including heart and brain (42), it is most abundant in olfactory tissue (43). Because 3AC is stimulated by G_s-coupled receptors *in vivo* and inhibited by Ca^{2+}, it seems likely that its regulatory properties may explain at least part of the kinetics for odorant-induced cAMP changes in olfactory cilia. Various odorants stimulate rapid cAMP increases in olfactory cilia, which rise and fall within milliseconds to seconds, depending on the preparation examined and the technique used to measure the kinetics for cAMP changes (44). These increases in cAMP are likely due to stimulation of the 3AC and other adenylyl cyclases through G_s-coupled olfactory receptors. There are several possible mechanisms for the subsequent decreases in cAMP, including the actions of CaM-dependent cyclic nucleotide phosphodiesterases. Because intracellular Ca^{2+} is elevated during odorant exposure (45, 46), Ca^{2+} inhibition of 3AC activity and stimulation of CaM-sensitive phosphodiesterases may both contribute to the biphasic cAMP response.

III. CALMODULIN-STIMULATED PHOSPHODIESTERASE ISOFORMS: GENETIC DIVERSITY

Calmodulin-stimulated phosphodiesterase (PDE) isozymes catalyze the hydrolysis of cAMP and cGMP (47). In the absence of CaM, cyclic nucleotide hydrolytic activity is low but detectable. In the presence of calcium-bound CaM, the activity of these isozymes increases four- to sevenfold. The large growth in newly identified PDE isozymes has inspired a change in PDE nomenclature (for a detailed review, see Ref. 48) that is based on the primary structure predicted from cloned cDNAs or genes, direct amino acid sequencing of purified enzymes, or other

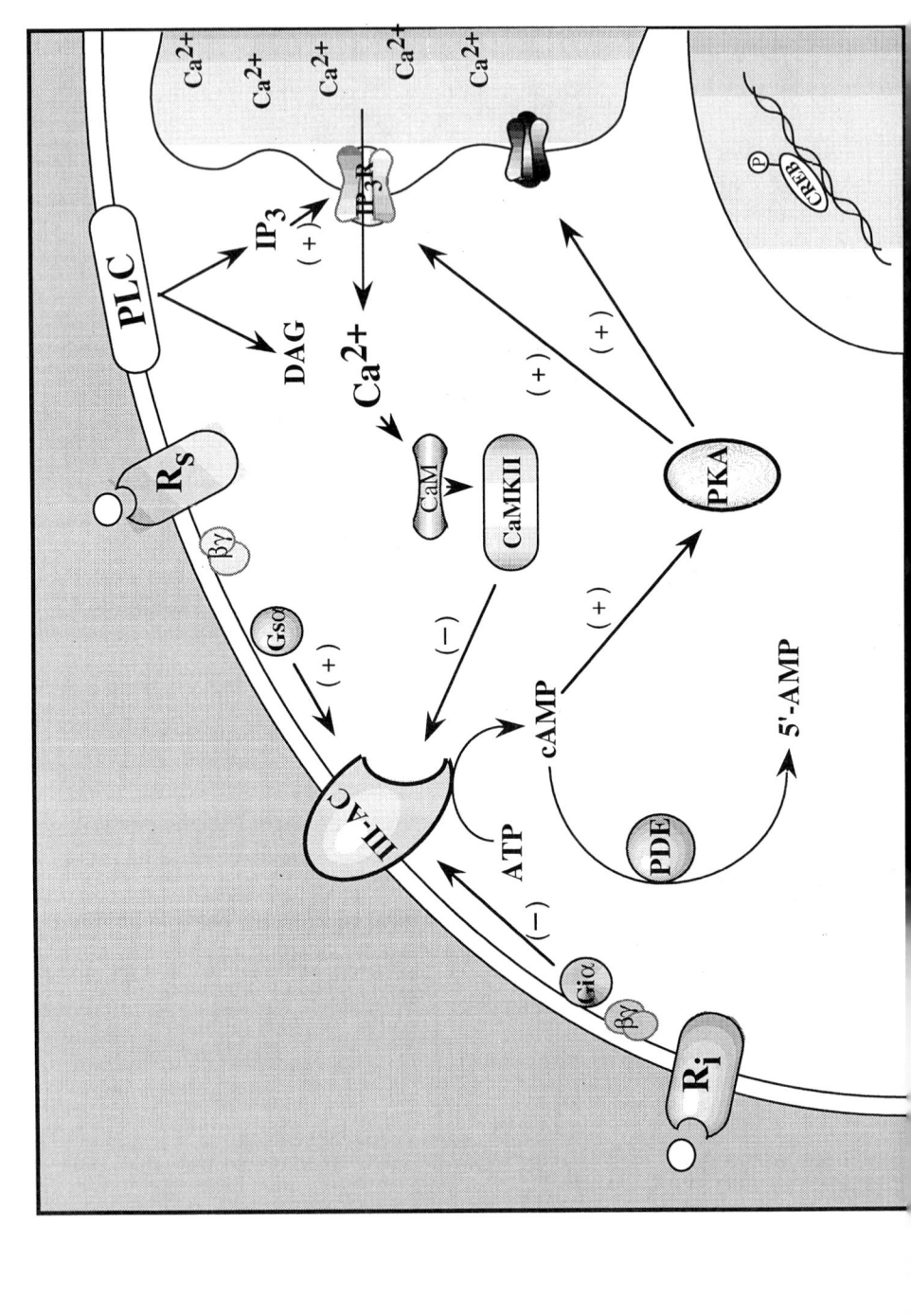

characteristics that can clearly place it in a new or existing gene family. Prior to this change, PDE isoforms were referred to by hallmark biophysical and kinetic attributes (49). For example, two CaM-stimulated PDE isoforms abundant in bovine brain tissue were designated as the 61-kDa CaM–PDE and the 63-kDa CaM–PDE in reference to the mobility of the large subunit of these two different isozymes on SDS–polyacrylamide gels. In accordance with the new nomenclature, *CaM-stimulated* PDEs are now designated as PDE1s, followed by alphabet and numeral suffixes indicating the gene designation. The genus and species from which a particular PDE gene/polypeptide is isolated precede the PDE designation. Using the two isozymes described earlier as an example, the bovine 61-kDa CaM–PDE is now designated BTPDE1A2 (BT; *Bos taurus,* 1; CaM-stimulated PDE, A; gene A, 2; alternative splice variant number 2 of gene A) and the bovine 63-kDa CaM–PDE is designated BTPDE1B1 (B; gene B, 1; alternative splice variant number 1 of gene B). This systematic method of naming PDE isozymes is desirable in that it is based on the nucleotide/amino acid sequence of the gene/polypeptide and is therefore less ambiguous than past methods.

Currently, three different PDE1 genes have been identified (50) and are thought to give rise to nine mRNAs encoding eight unambiguously unique PDE1 polypeptides. The structures of these cDNAs (51–59) are shown schematically in Fig. 3. A pairwise comparison of a limited stretch of relatively conserved nucleotides within the coding regions of PDE1 cDNAs from four different species were made to determine the uncorrected similarity and corrected evolutionary distances (60, 61). The results of these analyses are summarized in Table I, and the corresponding phylogenetic tree (62) is displayed in Fig. 4. It is clear from this analysis that three

---

Fig. 2. Mechanism for hormone-stimulated Ca^{2+} oscillations in cells expressing 3AC. It is hypothesized that stimulation of 3AC by hormones or forskolin leads to activation of PKA, stimulation of IP3 receptors, and increases in intracellular Ca^{2+}. As intracellular Ca^{2+} increases, 3AC activity is inhibited and cAMP levels are decreased by cAMP phosphodiesterases. When cAMP drops below a threshold level, Ca^{2+} is resequestered and the cycle is repeated as long as activators of 3AC are present. Rs, adenylyl cyclase stimulatory receptor; 3AC, type III adenylyl cyclase; CaM, calmodulin; PKC, protein kinase C; PKA, cAMP-dependent protein kinase; PLC, phospholipase C; DAG, diacylglycerol; IP3, inositol triphosphate; IP3R, IP3 receptor/channel; CaMK II/IV, CaM kinase type II or IV; PDE, cAMP phosphodiesterase.

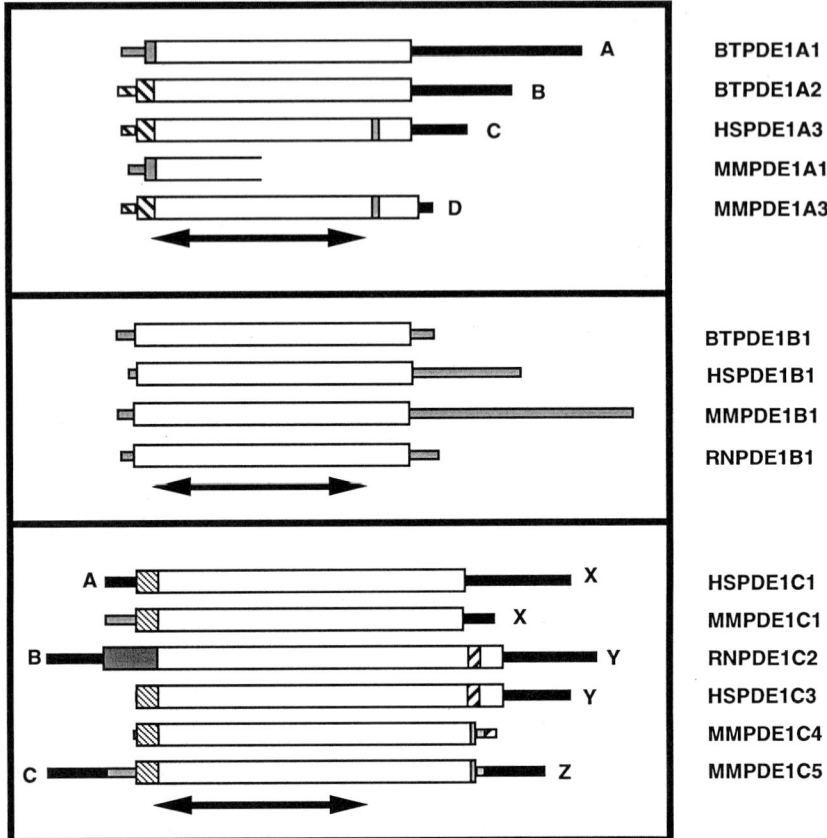

Fig. 3. Schematic representations of PDE1 cDNAs isolated from different species. For each clone, the large box represents the coding sequence and the thinner lines denote untranslated regions. Except for untranslated regions shaded black, regions within a clone representation sharing a common pattern have similar sequences. Untranslated regions shaded black sharing a common letter also have similar sequences. The double-ended arrow indicates the sequence used to calculate evolutionary distances (Table I) for constructing the phylogenetic tree shown in Fig. 4. The accession numbers corresponding to each of the clones shown here are as follows: BTPDE1A1, L34069; BTPDE1A2, M90358; HSPDE1A3, U40370; MMPDE1A1, PCR product (not submitted); MMPDE1A3, U56649; BTPDE1B1, M94867; HSPDE1B1, U56976; MMPDE1B1, L01695; RNPDE1B1, M94537; HSPDE1C1, U40371; MMPDE1C1, L76944; RNPDE1C2, L41045; HSPDE1C3, U40372; MMPDE1C4, L76947; MMPDE1C5, L76946.

TABLE I
Nucleotide Sequence Similarities and Evolutionary Distance Matrices between cDNA Pairs Encoding PDE1 Isoforms[a]

		Percentage similarity										
		1	**2**	**3**	**4**	**5**	**6**	**7**	**8**	**9**	**10**	
Percentage divergence	**1**		94	88	65	65	66	66	73	74	72	BTPDE1A2
	2	7		89	65	66	67	67	73	73	72	HSPDE1A3
	3	15	13		67	67	68	68	72	73	72	MMPDE1A3
	4	71	70	63		92	89	89	67	69	68	BTPDE1B1
	5	69	65	61	9		89	90	68	68	68	HSPDE1B1
	6	67	64	60	12	11		96	68	69	69	MMPDE1B1
	7	67	63	60	13	11	4		68	69	69	RNPDE1B1
	8	42	43	46	59	58	56	57		89	90	HSPDE1C1
	9	41	43	44	55	58	54	54	13		96	MMPDE1C1
	10	45	46	46	56	59	55	55	12	5		RNPDE1C2
		1	**2**	**3**	**4**	**5**	**6**	**7**	**8**	**9**	**10**	

[a] Isolated from bovine, human, mouse, and rat species. Numbers 1 through 10 (in bold type) demarking each row and column of the matrix correspond to the sequences listed in the far right column. The range of nucleotides used in this comparison were chosen by initially performing a multiple sequence alignment to identify a contiguous region that was homologous among all of the cDNAs. The sequence was also edited to ensure maintenance of codon boundaries in cases where small gaps (three to six nucleotides) were introduced. This region constitutes 1251 nucleotides within the open reading frame of the cDNAs and is indicated in Fig. 3 by double arrows. Both scores were determined using DISTANCES software (University of Wisconsin Genetics Computer Group) set as either uncorrected (percentage similarity) or corrected (percentage divergence) by the Jin-Nei γ distance method. Similarity is defined as the percentage of identical matches between sequence pairs. The evolutionary distance (divergence) is expressed as the percentage of substitutions after corrections are made for transitions and transversions.

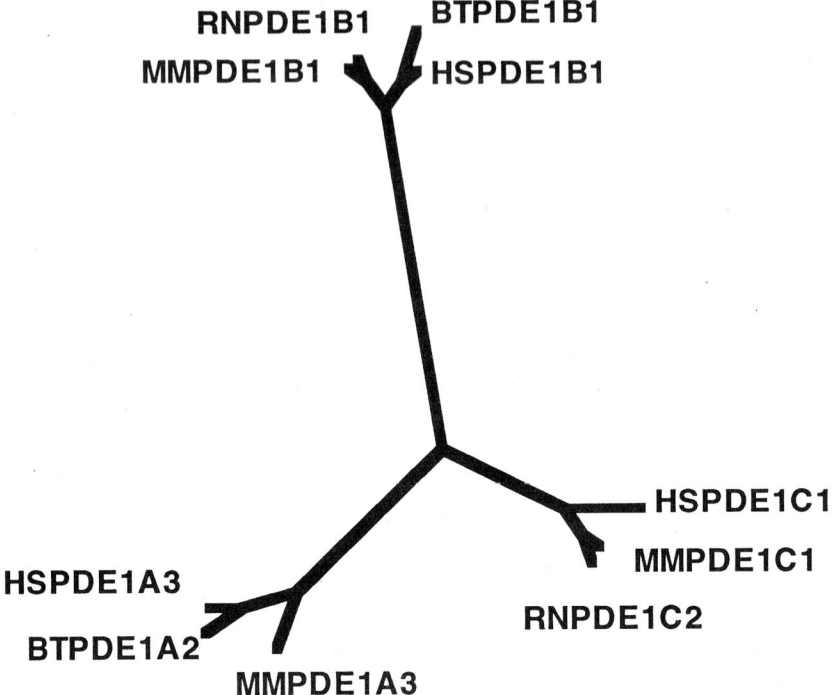

Fig. 4. Phylogenetic tree showing the evolutionary distances among PDE1 genes from different species. The evolutionary distances described in Table I were used to construct an unrooted tree using the neighbor-joining method and TreeView software.

PDE1 subfamilies have evolved, presumably from a common progenitor, and are designated families A, B, and C, as discussed earlier. Moreover, PDE1B isozymes constitute the most evolutionarily distant subfamily, yet appear more closely related to PDE1C than PDE1A isoforms.

Based on the current PDE1 cDNA sequences, it is also apparent that PDE1A and PDE1C genes can undergo alternative splicing to give rise to multiple variants (Fig. 3). Divergence in the amino-terminal coding sequences of PDE1A and PDE1C isoforms and in the carboxyl-terminal coding sequences of PDE1C isoforms has been noted in at least two species (59), suggesting that the exons that give rise to these variants have been conserved throughout evolution and probably encode a physiologically important functionality. Some evidence suggests that the

PDE1B gene may also undergo alternative splicing. RNase protection analysis of bovine kidney total RNA with a riboprobe corresponding to the 5'-untranslated region (UTR) and amino-terminal coding sequence of BTPDE1B1 is only partially protected (54). The simplest interpretation of this result would suggest that an alternatively spliced variant of PDE1B1 having a divergent amino-terminal coding sequence is expressed in bovine kidney, yet there is currently no other experimental evidence supporting this observation.

The divergent amino-terminal regions are likely to influence the sensitivity to activation of PDE1 isozymes by CaM (63), whereas little is known about the divergent carboxyl-terminal regions of PDE1C isoforms (discussed in detail later). Relative to bovine species, PDE1A cDNAs from both mouse and human species (53) have been shown to contain a short insertion near the carboxyl-terminal end of the coding sequence (Fig. 3). As a result of this observation, these cDNAs have been given a new designation in the PDE1A family as PDE1A3 isoforms. Assignment of these cDNAs as new PDE1A variants needs to be interpreted with caution, however, as it is presently not clear whether this insertion is the result of a newly identified exon or merely divergence of an existing exon. In the latter case, it would be concluded that these isoforms are actually homologs of PDE1A2 isoforms rather than a new alternatively spliced variant. An unambiguous assignment of PDE1 isoforms may become more clear in the future as the genomic sequences are determined from multiple species. Although the genes for these different subclasses of PDE1 isozymes have not been cloned or characterized at this writing, it is clear from the sequence homology of the cDNAs that three genes exist, at least two of which undergo alternative splicing, resulting in greater structural diversity of the encoded polypeptides.

A. Structure and Domain Organization of Calmodulin-Stimulated Phosphodiesterase

Like all PDEs identified to date, PDE1 isozymes are multimeric macromolecules composed of two identical subunits containing a catalytic core and regulatory elements that are required for activation by calcium-bound calmodulin. It was recognized early after the discovery of PDE1 that these domains were discrete functional elements (64). When purified bovine brain PDE1 (most likely PDE1A2) was subjected to partial proteolysis by trypsin in the presence of EGTA (apoenzyme), a catalytic activity was produced having a specific activity equal to the CaM-activated, nonproteolyzed PDE1, yet the enzyme was insensitive to activation by calcium and CaM (64, 65). The loss of the ability of calcium-bound CaM to activate the enzyme suggested that a regulatory domain containing a CaM-binding region must have been degraded, leaving an intact catalytic domain having a subunit relative mobility of 36 kDa. Moreover, activation of the enzyme by partial proteolysis also suggested that an inhibitory domain may also have been removed from the catalytic core. In another study (66), partial proteolysis of bovine PDE1 with chymotrypsin resulted in a fully activated, CaM-insensitive activity having a subunit relative mobility of 45 kDa. Interestingly, the proteolytic fragment could also associate with CaM, yet no further activation was observed with this preparation. This result suggested that the 45-kDa chymotryptic PDE1 fragment possessed both catalytic and CaM-binding functions. However, the lack of any activation by CaM further suggested that a discrete inhibitory domain within the PDE1 subunit, which serves to maintain the enzyme in a less active conformation in the absence of calcium-bound CaM, was degraded by chymotrypsin. Therefore, based on these early studies, PDE1 isozymes appeared to be chimeric, possessing as many as three discrete functional domains: catalytic, CaM binding, and inhibitory.

Prodigious advancement in our understanding of the domain organization of all PDE enzymes, as well as PDE1 isoforms, was made on determination of the primary structure of these isozymes either by direct amino acid sequencing of purified enzyme preparations or by deducing the amino acid sequence of cloned cDNAs. Conserved among all PDEs (63) is a 250-residue segment located near the carboxyl-terminal end of the molecule (Fig. 5). Because the function common to all PDE isoforms is to catalyze cyclic nucleotide hydrolysis, it was deduced that this segment probably defines the catalytic core of the enzyme. Experimental evidence in support of this hypothesis was forthcoming when the 36-kDa tryptic fragment of BTPDE1A2 was shown to contain this conserved region (67). Since this report, studies with other PDE families have yielded similar results (68–71).

Like the catalytic domain, the location of a CaM-binding domain was deduced by comparing the primary structures of two closely related PDE1A isoforms; BTPDE1A2 and BTPDE1A1 (67, 72). These isoforms have identical primary structures except for the divergent amino-terminal residues (Fig. 3). As might be expected, they also have nearly identical kinetic properties (73). An important functional difference, however, is

5. Cyclic Nucleotide Regulation by Calmodulin

A

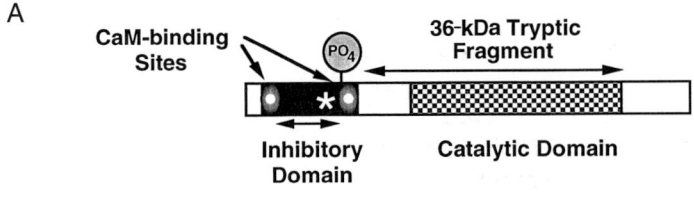

B

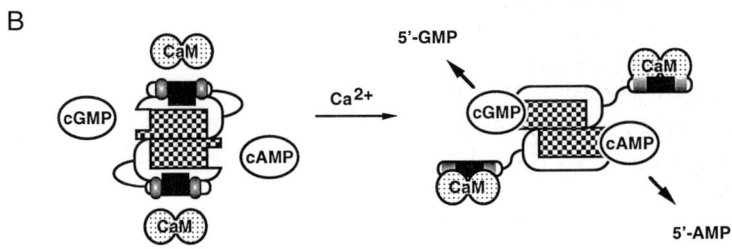

Fig. 5. (A) Domain organization of PDE1A1/2 isozymes. The asterisk indicates the location of residues that are required for the enzyme to remain in a relatively inactive conformation in the absence of calcium-bound CaM. (B) Hypothetical mechanism whereby calcium-bound CaM activates PDE1A isozymes. In the absence of calcium, the inhibitory domain interacts with a complementary region resulting in relatively inactive conformation. In the presence of calcium, the fully liganded CaM interacts with the amino-terminal, carboxyl-terminal, or both CaM-binding domains, changing the conformation of the enzyme and relieving inhibition, allowing the catalysis of cyclic nucleotide hydrolysis. In this depiction, the inhibitory domain is presumed to interact with a noncatalytic region that changes the kinetic properties by inducing a conformational change in the catalytic site. Although it is known that the enzyme is a homodimer, it has not been established whether the subunits are aligned in a parallel or antiparallel configuration.

that the concentration of CaM required to elicit a half-maximal stimulation of BTPDE1A1 is 10-fold lower than that for BTPDE1A2 (52, 73). Based on these observations, it was hypothesized that the divergent amino-terminal regions of these two PDE1 isoforms probably encoded CaM-binding domains having different affinities for CaM (63). Sequences corresponding to the most likely CaM-binding region within BTPDE1A1 (residues 5–28) and BTPDE1A2 (residues 21–44) were found using an algorithm designed to identify potential regions that may form a basic, amphipathic helix, which is a common structural motif among CaM-binding peptides (67, 72). Approximately half of the sequence encompassing a likely CaM-binding site included residues within the divergent aminoterminus of BTPDE1A1 and BTPDE1A2. Peptides corresponding to these putative CaM-binding regions were synthesized and assayed for association with CaM. These peptides bound CaM–agarose in a calcium-dependent manner and competitively inhibited activation of BTPDE1A2 by CaM. A similar CaM-binding site was also identified in BTPDE1B1 in the same relative location as for both BTPDE1A isoforms (72). Thus, the divergent amino-terminal residues of BTPDE1A1 and BTPDE1A2 contribute residues that encode two functionally different CaM-binding domains and probably define a CaM-binding domain for all PDE1 isoforms.

In an effort to identify and map the hypothetical inhibitory domain of PDE1 isozymes, the BTPDE1A2 cDNA was subjected to deletion mutagenesis, and the cognate polypeptides were expressed using the baculovirus system and characterized (52). If an inhibitory domain does in fact exist, conservative deletion of the CaM-binding domain described earlier would be predicted to result in a CaM-insensitive activity, having a specific activity equal to the nonactivated wild-type enzyme. Surprisingly, the resultant activity of this deletion mutant was nearly identical to wild-type BTPDE1A2. However, equally important is the observation that the CaM sensitivity of this activity was not the same as wild-type BTPDE1A1, which requires 10-fold less CaM than BTPDE1A2 to achieve half-maximal activation. Therefore, these results indicated that, in addition to the CaM-binding domain that must exist in the divergent amino-terminal region, a second CaM-binding domain must exist within the primary structure of the deletion mutant.

Further deletion mutagenesis of the aminoterminus of BTPDE1A2 eventually revealed the existence of a component of a putative inhibitory domain (52). Deletion of BTPDE1A2 residues 1 through 98 resulted in a polypeptide that was completely insensitive to CaM and had a specific activity equal to that of CaM-activated, wild-type BTPDE1A2. A decrease in CaM sensitivity and an increase in specific activity were first detected after deleting residues 1 through 89. Therefore, it appeared that one component of an inhibitory domain that functions to hold the enzyme in a less active conformation was located between residues 89 through 98 of BTPDE1A2 (Fig. 5A).

A second CaM-binding domain was identified using similar methods for identification of the amino-terminal CaM-binding domains described earlier (52). Because calculations used for predicting basic amphipathic helices (74) indicated that no other potential CaM-binding domains existed in the BTPDE1A2 sequence, seven overlapping 24 amino acid peptides corresponding to residues 76 through 149 were synthesized and tested for their ability to interact with CaM. Three of these peptides collectively spanning residues 108 through 149 could potently titrate CaM, as evidenced by a loss in activation of BTPDE1A2 by 3–5 nM CaM. Moreover, the deletion mutant encompassing residues 109–530 of BTPDE1A2 could be immunoprecipitated by a monoclonal antibody reactive with only the CaM-bound form of the enzyme. These experiments indicated that a functional CaM-binding domain is located between residues 108 and 149 of BTPDE1A2 and appears to be in close proximity to, but separate from, the inhibitory domain (Fig. 5A).

A summary of the domain organization of BTPDE1A2 based on the experiments described earlier and a model depicting a possible mechanism by which CaM activates PDE1 isozymes is shown in Fig. 5. The identity and location of three types of functionally distinct domains reside in the primary structure of the polypeptide: (1) a catalytic domain constituting 250 residues near the carboxyl-terminal end of the molecule; (2) two CaM-binding domains separated by approximately 90 amino acids, with one domain lying within the divergent amino-terminal end of the molecule; and (3) an inhibitory domain located near the internal CaM-binding region. The discovery that each monomer of BTPDE1A isozymes possesses two CaM-binding domains again raises questions regarding the stoichiometry of CaM binding. It is tempting to speculate that two molecules of CaM bind each domain on a single PDE1 monomer. However, numerous reports show that the binding stoichiometry of CaM with the PDE1 monomer is unity (65, 75–77). It is certainly possible that one molecule of CaM associates with only one of the two binding sites within the PDE1A monomer to activate the enzyme. For example, in the case of PDE1A1 and PDE1A2 isozymes where the apparent affinity for CaM is approximately 10-fold greater for the former versus the latter, CaM may bind to the second CaM-binding site to activate PDE1A2, whereas in the case of PDE1A1, the amino-terminal CaM-binding domain may dominate as it presumably has a higher affinity for CaM. However, it is also likely that CaM associates with both binding domains to activate PDE1A, as phosphorylation (see later) of the second CaM-binding site lowers the activation potency of CaM for both PDE1A1 and PDE1A2 (77, 78). In any case, these are hypothetical possibilities that need to be proven experimentally.

It is not known at this writing if, in fact, the inhibitory domain constitutes more than residues 89 through 98 as indicated in Fig. 5A. Moreover, it is likely that this inhibitory domain may have a complementary site somewhere within the primary structure of BTPDE1A2 with which it interacts physically to exert its function (Fig. 5B). Some protein kinases and phosphatases possess inhibitory domains that interact with the catalytic site to inhibit the activity of these enzymes (79). This mechanism may also apply to PDE1 isozymes, whereby the inhibitory domain interacts with the catalytic site, effectively blocking access of the cyclic nucleotide substrate. However, it is equally likely that the inhibitory domain interacts with a noncatalytic site of BTPDE1A2 and, through this interaction, possibly alters the conformation of the catalytic domain, rendering it less active. Further studies will be needed to determine more precisely how the inhibitory domain regulates the activity of this enzyme.

Regarding the quaternary structure of the molecule, biophysical studies seem to indicate that PDE1As and perhaps other PDE1 isozymes are composed of two identical subunits and that the catalytic core, if not the entire enzyme, is fairly ellipsoidal (47, 52, 80). The domain involved in subunit association, or dimerization, is currently not known, but it is not located between residues 1 and 109 of BTPDE1A2. Moreover, it is not known whether the two subunits associate in a parallel or an antiparallel fashion.

Since the discovery of the PDE1C isozymes, additional structure differences with as yet no known functions exist (53, 59). The divergent carboxyl-terminal portions of PDE1C1, PDE1C2, and PDE1C4 isozymes may either be unimportant functionally or, more likely, play other roles, such as docking motifs to target the enzymes to specific subcellular sites or signal transduction microenvironments. Clarification of these speculations awaits further experimentation.

B. Kinetic Properties and Activation of Calmodulin-Stimulated Phosphodiesterase Isoforms by Calmodulin

The kinetic properties of recombinant isozymes from PDE1A, PDE1B, and PDE1C subfamilies (59) are summarized in Fig. 6. Several reports (81–84) seem to indicate that binding of CaM to PDE1 isozymes results in an increase in maximum velocity (V_{max}), with little change in the Michaelis constant (K_m). In the presence of calcium-bound CaM, all PDE isozymes display reasonably typical Michaelis–Menton kinetics (47, 52, 59). PDE1A and PDE1B isoforms are both similar in that they

5. Cyclic Nucleotide Regulation by Calmodulin

ISOFORM	STRUCTURE	K_m, cAMP (μM)	K_m, cGMP (μM)	V_{max} (cAMP/cGMP)	EC_{50}, Ca^{2+} (μM)
PDE1A1		113	5.1	2–3	0.3
PDE1A2		113	5.1	2–3	2.0
PDE1B1		24.3	2.7	0.9	1.3
PDE1C1		3.5	2.2	1.3	3.0
PDE1C2		1.2	1.1	1.2	0.8
PDE1C3		0.3	0.6	≈1	N.D.
PDE1C4/5		1.1	1.0	1.0	2.4

Fig. 6. Summary of differences in the primary structure and kinetic properties of PDE1A, PDE1B, and PDE1C subfamilies. Lines represent the primary structure of each isozyme. Divergent regions arising from alternatively spliced genes are indicated by boxes (common patterns indicate similar sequences). Michaelis constants (K_m) for either cAMP or cGMP, maximum velocity cAMP/cGMP ratios [V_{max} (cAMP/cGMP)], and the concentration of calcium to achieve half-maximal activation (EC_{50}, Ca^{2+}) of each PDE1 isozyme (in the presence of excess CaM) are shown in the tabulation on the right-hand side (N.D., not determined).

have relatively low K_m values for cGMP and somewhat higher constants for cAMP, suggesting that these isozymes may be more important for regulating cGMP metabolism. Whether this is true may depend on the cell type and subcellular compartment in which these enzymes are localized, as well as the concentration of these enzymes within a particular compartment. The major difference between these two different isozyme families is in V_{max} ratios: PDE1A isoforms have a higher V_{max} for cAMP than for cGMP, whereas for PDE1B enzymes, the V_{max} ratios is nearly unity. In contrast to PDE1A and PDE1B isoforms, members of the PDE1C gene family have similar K_m values for both cGMP and cAMP, and the K_m values are in fact lower than any of the values reported for either PDE1A or PDE1B isozymes, typically calculated to be 1 μM or lower.

Finally, it is worth noting that the specific activity of these enzymes is also quite different. Values between 20 and 30 μmol cGMP hydrolyzed per minute per milligram PDE protein have been reported for PDE1B enzyme preparations, whereas the specific activity typically ranges between 200 and 300 μmol cGMP hydrolyzed per minute per milligram PDE protein for PDE1A isoforms (52, 73). Currently, the only value for the specific activity of PDE1C enzymes has been obtained from extracts of Sf9 (*Spodoptera frugiperda* fall armyworm ovary) cells infected with a recombinant baculovirus, having a specific activity value of approximately 3 μmol cGMP hydrolyzed per minute per milligram of PDE protein (A. A. Zhao, K. S. Kwak, W. K. Sonnenburg, and J. A. Beavo, 1996). Because PDE1 enzymes have similar molecular masses, the cGMP-hydrolyzing "capacity" of PDE1A isoforms may be considered about 10-fold greater than for either PDE1B or PDE1C isoforms. In contrast, PDE1C enzymes may have greater cAMP-hydrolyzing "capacity" than either PDE1A or PDE1B isoforms under physiological conditions.

The property that varies the most among PDE1 isozymes within a gene family is the potency of activation by calcium-bound CaM. Fully calcium-liganded CaM appears to be the form that interacts most strongly with PDE1 enzymes (47), therefore it is likely that the diversity in the activation potency of CaM is mainly due to structural differences in the CaM-binding sites of each PDE1 variant. Figure 6 summarizes the relative potencies of calcium-bound CaM used to activate a number of recombinant PDE1 isoforms by varying the concentration of calcium in the assay buffer while keeping the concentration of CaM constant (59). The CaM activation potency is nearly 10-fold greater for PDE1A1 than for PDE1A2 (85) and can be attributed to the divergence in the amino-terminal domains discussed earlier (52, 67, 72). Similarly for PDE1C isoforms, where two amino-terminal divergent isoforms exist (PDE1C2 versus PDE1C1, 3, and 4/5; Fig. 6), the difference in CaM activation potency is approximately 3-fold. Even though there is considerable divergence of the amino-terminal nucleotides of all PDE1s between isozyme subfamilies, there is enough similarity in the CaM activation potencies to suggest that there may be some sequence similarity on the amino acid level. For example, in Fig. 6, PDE1A2 and PDE1C4/5 have nearly identical CaM activation potencies, whereas for PDE1A1 and PDE1C2, the CaM activation potencies are quite dissimilar from all other PDE1s. Based on these functional similarities and differences, it can be predicted that the divergent amino-terminal residues of PDE1A2 and PDE1C4/5 might in fact display some sequence similarity. A comparison of the alignment quality of the amino-terminal divergent sequences for PDE1

TABLE II

Pairwise Comparison of Alignment Quality of Nonrandomized versus Randomized Divergent Amino-Terminal Amino Acid Sequences among PDE1 Isoforms[a]

		Alignment quality						
		1	2	3	4	5		
Randomized	1		6.1	6.0	10.9	8.8	1	MMPDE1A1
alignment	2	6.6 ± 1.2		14.9	31.1	14.4	2	MMPDE1A3
quality	3	7.2 ± 1.1	10.4 ± 1.1		20.2	12.6	3	MMPDE1B1
	4	8.2 ± 0.9	12.2 ± 1.6	12.1 ± 1.8		14.1	4	MMPDE1C4/5
	5	8.1 ± 0.9	14.0 ± 1.2	15.5 ± 1.1	14.4 ± 1.2		5	RNPDE1C2
		1	2	3	4	5		

[a] Divergent amino-terminal sequences were analyzed using the GAP software (University of Wisconsin Genetics Computer Group). The second sequence in the comparison was randomized 10 times and compared to the first sequence to calculate a mean quality score ± the standard error. Quality is determined using a scoring matrix that is dependent on the evolutionary distance of the mismatched residues, gap length, and gap number. An alignment quality greater than the randomized quality score, and outside of the standard error, is interpreted as significant in similarity.

isoforms is shown in Table II. Based on these data, PDE1C4/5 displays sequence similarity with PDE1A2, followed by PDE1B1 (Fig. 7). No significant sequence similarity could be detected among any of the other isoforms (Table II). This result is consistent with the rank order of

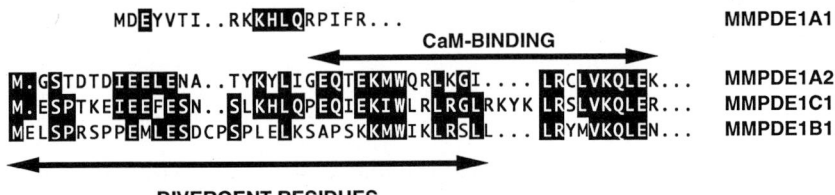

Fig. 7. Alignment of divergent, amino-terminal putative CaM-binding regions from different PDE1 isozymes having significant alignment scores (Table II). Residues in black boxes indicate amino acids that are identical. Residues that constitute a putative CaM-binding region are overlined by a double-headed arrow and are based on findings for PDE1A and PDE1B isozymes. Residues underlined with a double-headed arrow indicate regions of sequence divergence within isozyme subfamilies, probably arising from an alternatively spliced gene.

similarity in CaM activation potencies among these isozymes. This correlation between functional CaM activation potencies and structural similarity, albeit loose similarity, adds strength to the argument that the amino-terminal domains of these enzymes participate in CaM binding and, ultimately, in the regulation of PDE1 activation by calcium-bound CaM.

C. Selective Calmodulin-Stimulated Phosphodiesterase Inhibitors

Currently, agents that selectively inhibit only PDE1 isozymes are lacking. It has been reported frequently that vinpocetine is a PDE1-selective inhibitor; however, there is now fairly good evidence to refute this claim (86). Two compounds initially identified as natural products of microbial fermentation have been reported as fairly potent and selective PDE1 inhibitors. The agent KS-505a (87) apparently inhibits CaM activation of PDE1 by interacting with a CaM-binding domain on PDE1A2. The divergent aminoterminal CaM-binding domain is likely to be the site of action, as KS-505a is a less potent inhibitor of PDE1A1 and the 36-kDa CaM-independent, catalytic trypsin fragment of PDE1A2. The fact that the catalytic domain does not appear to be a primary site of action for this compound also provides a rational basis for its isozyme selectivity. However, KS-505a inhibits PDE1B1, PDE1C1, PDE1C2, and PDE1C4/5 recombinant enzymes with potencies nearly equal to that for PDE1A2 (59). A second agent, SCH45752 (88), has been shown to be a 100-fold more selective inhibitor of PDE1A isoforms than PDE3. However, little is known about its inhibitory potency for PDE isozymes from other families.

D. Regulation of Calmodulin-Stimulated Phosphodiesterase Isoforms by Phosphorylation

Phosphorylation may be a mechanism for modulating the CaM sensitivity of certain PDE1 isozymes *in vivo* (47). PDE1A1 (77) and PDE1A2 (47, 89) have been shown to be phosphorylated *in vitro* by cAMP-dependent protein kinase at concentrations that are probably physiologically relevant, and phosphorylation results in a very dramatic rightward shift in the potency of CaM activation (Fig. 8A). A similar phenomenon is observed when PDE1B is phosphorylated *in vitro* by the autophosphorylated form of CaM-dependent protein kinase II (47). Phosphorylation has been shown to be rapid, quantitative, and reversible as calcineurin can catalyze the dephosphorylation of these enzymes. The sites of phosphorylation have been determined for BTPDE1A2 by mass spectrum analysis of proteolytic peptides from pure, phosphorylated BTPDE1A2 and have been confirmed by mutagenesis of the cDNA encoding BTPDE1A2 (89). Phosphate is nearly quantitatively incorporated into BTPDE1A2 serine residue 120 and, using longer incubation conditions, serine 138 also becomes phosphorylated (Fig. 8B). The residue that is most important for the decrease in CaM activation potency is residue 120. Phosphorylation of serine residue 138 decreases the CaM activation potency only slightly. Because the potential cAMP-dependent protein kinase phosphorylation site is conserved in PDE1A isoforms across different species (Fig. 8B), it seems likely that phosphorylation would occur *in vivo* under appropriate physiological conditions.

E. Expression and Role of Calmodulin-Stimulated Phosphodiesterase Isoforms in Different Tissues and Cell Types

In cells that express PDE1 isozymes, hormones or neurotransmitters causing an elevation in intracellular calcium would be predicted to have opposing actions on hormones that stimulate adenylyl or guanylyl cyclases. An elevation in intracellular calcium would increase the concentration of calcium-bound CaM, which would associate with PDE1, thereby lowering cAMP and cGMP concentrations by increasing cyclic nucleotide hydrolysis. In order to understand the physiological role that this enzyme family might play, it is important to identify the cellular sites of expression. Much progress has been made toward this goal and it may be possible to predict the role that PDE1 isozymes may play in different physiological contexts.

1. Central Nervous System

The brain is one of the richest sources of PDE1 activity, and all three PDE1 genes have been localized in brain (90–93). It is noteworthy that expression, whether by *in situ* hybridization or by immunocytochemistry,

A

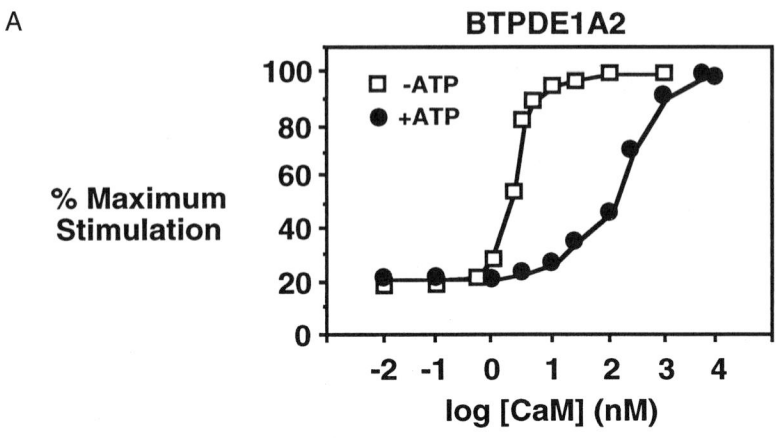

B

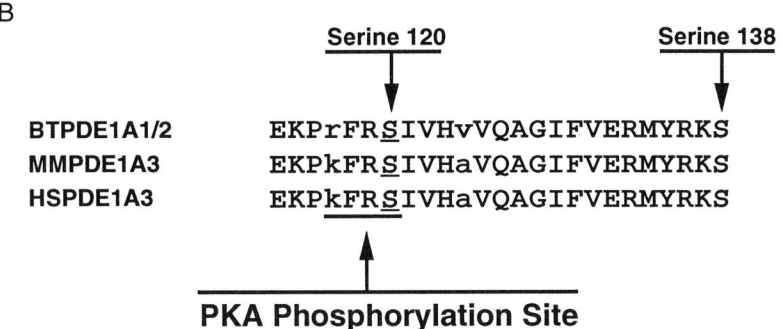

Fig. 8. (A) Consequence of BTPDE1A2 phosphorylation. Recombinant BTPDE1A2 was incubated with a PKA catalytic subunit in the absence (□) or presence (●) of ATP and was subsequently assayed for cGMP hydrolytic activity in the presence of increasing concentrations of calcium-bound CaM. The activity is expressed as a percentage of the maximal stimulation of nonphosphorylated (-ATP) BTPDE1A2 by CaM. (B) Phosphorylation sites of PDE1A isozymes. The phosphorylation site of BTPDE1A2 was determined experimentally. The phosphorylation site consensus sequence and the serine residue that is phosphorylated by the catalytic subunit of cAMP-dependent protein kinase (PKA) are conserved among bovine (BTPDE1A1/2), mouse (MMPDE1A3), and human (HSPDE1A3) species.

has been detected only in neurons. This should not be interpreted to indicate that these isozymes are not expressed in glia, however. This observation may merely reflect the possibility that cyclic nucleotide signaling is less robust in nonneuronal brain cell types (90).

The PDE1A2 isoform appears to be specifically expressed in brain (51) and is most concentrated in hippocampal layers CA1 through CA4 and in layer five and six of the cerebral cortex (93). Somewhat less expression has also been detected in the basal ganglia, particularly in the nucleus accumbens and in some scattered neurons in the caudate putament (93). The PDE1B1 isoform also appears to be strongly expressed in brain (54–56). PDE1B1 mRNA can be detected readily in many different nuclei throughout the mouse brain (92, 93) but is most concentrated in the basal ganglia, dentate gyrus of the hippocampus, olfactory bulb and tubercle, and cerebellar Purkinje cells. Fairly strong hybridization has also been observed in all layers of the cerebral cortex.

PDE1C isozymes display a pattern of expression that is again different than that observed for either PDE1A or PDE1B (59). The strongest hybridization has been detected in cerebellar granule cells and Purkinje cells, the interpolar spinal trigeminal nucleus, and the central amygdaloid nucleus. Somewhat less PDE1C hybridization has been detected in the caudate putamen and olfactory bulb glomerular and external plexiform layers. Among the members constituting the PDE1C gene subfamily, PDE1C1 and PDE1C5 mRNAs appear to be the most abundant in mouse brain, and some differential expression of these two mRNAs has been observed. For example, only PDE1C1 mRNA was detected in the caudate putamen and interpolar spinal trigeminal nucleus. In the cerebellum, PDE1C1 mRNA was detected in granule cells and Purkinje cells. However, only PDE1C5 mRNA was detected in cerebellar granule cells.

At this time, one can only speculate on the role that these different PDE1 isozymes may subserve in the various functions of the brain. Experimental evidence in this regard is lacking due in part to the fact that no truly selective PDE1 inhibitors exist and also because more than one PDE gene is expressed in most cells. However, one interesting study suggests that a PDE1 isozyme may be involved in the regulation of a potassium current in certain neurons of the sea slug, *Aplysia californica* (94, 95), and therefore may serve as a useful starting point for the possible function of these isozymes in mammalian neurons. In *Aplysia* neuron R15, agents that raise cAMP formation in these cells increase an inwardly rectifying potassium current (I_r). Conversely, repeated depolarization that results in a rise in intracellular calcium inactivates I_r. Because (a) PDE1 activity is present in these cells and (b) the nonselective PDE inhibitor 3-isobutyl-1-methylxanthine blocks this inactivation, it is likely that a PDE1 isozyme mediates this response. Regulation of I_r is one important component for determining the amplitude and frequency of the bursting pattern that occurs spontaneously in these sensory neurons. These studies are an excellent example demonstrating how PDE1 activation may mediate a calcium-signaling pathway involved in neuroplasticity.

In mammalian neurons it is possible that PDE1 isozymes also regulate ionic currents by decreasing cyclic nucleotide concentrations within various subcellular compartments. Conversely, inhibition of PDE1 activation by phosphorylation may also play a role in a feedforward manner, causing a greater accumulation of cAMP or cGMP for a longer duration under certain conditions (47, 96). A model depicting how this may occur in a neuron is shown in Fig. 9. A neurotransmitter released from an interneuron would cause an elevation in cAMP in a postsynaptic neuron expressing PDE1A2, resulting in the phosphorylation of PDE1A2, inactivating the enzyme. Subsequent neurotransmission along this pathway would be predicted to be amplified as rises in intracellular calcium via voltage-sensitive calcium channels (97, 98) or ligand-gated ion channels (97, 99) would be predicted to activate 1AC but not stimulate cyclic nucleotide hydrolysis via activation of PDE1A2. Moreover, PDE1A2 would not be able to hydrolyze cGMP formed in response to the generation of nitric oxide. The result might be a larger amplitude in the pulse of cyclic nucleotide for a longer duration. This hypothetical mechanism for regulating cyclic nucleotide formation may contribute to certain forms of neuronal plasticity (30, 100–102).

Very little data indicate which subcellular compartments within the neuron express the various PDE1 isozymes. Some histochemical evidence suggests that a CaM-sensitive cGMP hydrolytic activity is concentrated in the postsynaptic densities within the rat striatum, which is likely to be PDE1B1 (90, 103). However, some evidence also shows that cGMP hydrolytic activity is also present in synaptosomes isolated from rat brain, suggesting that a PDE1 activity may also be present in some presynaptic neurons (92, 104). Therefore, signal transduction pathways mediated by PDE1 activation may also play a role in neurotransmitter release.

2. *Olfactory Neurons*

In olfactory neurons, cAMP plays a pivotal role in the signaling events involved in odorant detection (105). The rapid rise and fall of the cAMP

5. Cyclic Nucleotide Regulation by Calmodulin

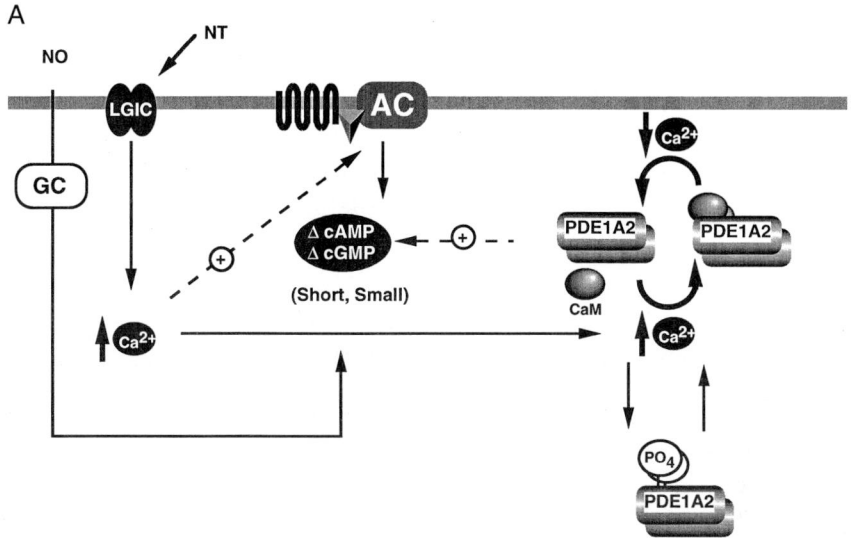

Fig. 9. Hypothetical role of PDE1A2 phosphorylation in neuroplasticity. (A) Occupation of ligand-gated ion channels (LGIC) with a neurotransmitter (NT) would be expected to elevate intracellular calcium, which would sequentially activate type I adenylyl cyclase (AC) and PDE1A2, resulting in a short, small change in cAMP formation. A similar scenario is possible for modulating cGMP formation by nitric oxide (NO) and soluble guanylyl cyclase (GC). (B) Activation of adenylyl cyclase either via occupation of a G_s-linked receptor with a neurotransmitter or possibly repetitive stimulation (not shown) might be predicted to raise intracellular cAMP high enough to activate cAMP-dependent protein kinase (PKA), catalyzing the phosphorylation of PDE1A2. Consequently, *subsequent* activation of adenylyl cyclase or guanylyl cyclase by neurotransmitters or nitric oxide (not shown in order to emphasize PDE1A2 phosphorylation pathway) would result in a larger and longer rise in intracellular cyclic nucleotides, as the phosphorylated form of PDE1A2 is less likely to be activated by calcium and CaM (feedforward). This change in cAMP and/or cGMP metabolism might also increase the phosphorylation state of other signaling macromolecules, such as voltage- and ligand-gated ion channels, possibly increasing the efficiency of neurotransmission and causing even larger changes in intracellular calcium. Depending on the rate of cAMP synthesis and PKA activation state, activation of the protein phosphatase, calcineurin, by calcium-bound CaM would be predicted to catalyze the dephosphorylation of PDE1A2, thereby allowing PDE1A2 to be activated during subsequent stimulation of the neuron with a neurotransmitter (feedback).

second messenger and the electrical signal generated to propagate odorant perception indicate that this response must involve tightly regulated synthesis and degradation of this cyclic nucleotide (44, 106, 107).

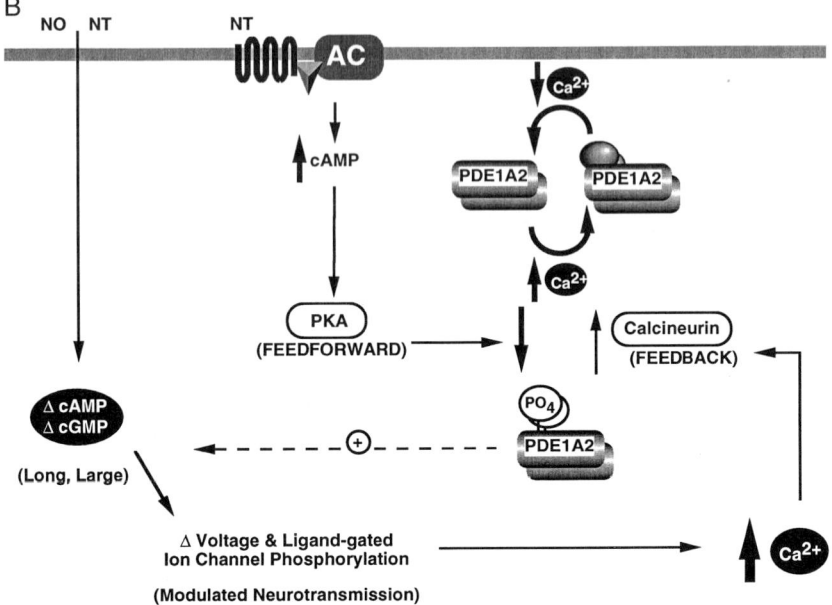

Fig. 9. Continued

PDE1C2 is well suited for the rapid termination of the cAMP signal as it is very concentrated in the cilia of olfactory epithelia, has a very low K_m for cAMP, and is the PDE1C isoform that is most sensitive to activation by calcium-bound CaM (58, 59, 108). A model depicting the role of PDE1C2 in olfactory signal transduction is shown in Fig. 10. On binding of an odorant to a receptor on the surface of the cilia, 3AC activation mediated by the olfactory-specific GTP-binding stimulatory protein, Golf, occurs, leading to a rapid rise in cAMP (105). A cAMP-gated ion channel subsequently binds the cyclic nucleotide, causing the channel to open, and a rise in intraciliary calcium occurs. This rise in calcium is not only a component of the propagation of odorant receptor activation, but a messenger for modulating the signal. A likely sequence of events subsequent to the rise in intraciliary calcium is the activation of CaM, which (a) may indirectly inhibit 3AC [(39), but see Ref. (44)], (b) activate PDE1C2 causing a reduction in cAMP, and (c) inhibit channel activation by cAMP (109). Rapid termination of an odorant signal

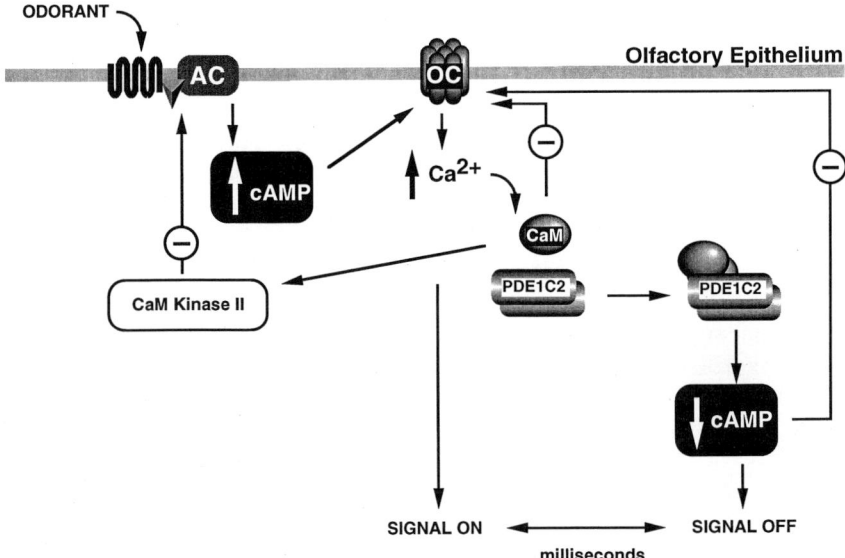

Fig. 10. Role of PDE1C2 in the signal transduction pathway involved in sensing odorants. In the cilia of olfactory epithelium, an odorant binds to a G-protein (inverted pyramid icon)-linked odorant receptor (seven-transmembrane-spanning icon), resulting in the activation of type III adenylyl cyclase (AC). The cAMP formed in response to this stimulus binds to an olfactory cyclic nucleotide-activated channel (OC), resulting in a rise in intracellular calcium and neurotransmission from the epithelium to the olfactory bulb. The signal is transient due to (a) hydrolysis of cAMP by CaM-activated PDE1C2, (b) a decrease in the affinity of cAMP for the OC due to binding of CaM, and (c) phosphorylation and inhibition of type III adenylyl cyclase by CaM kinase.

may be important as it would allow an animal to sample the environment repeatedly over a short period of time, yielding information such as distance and direction of movement of a predator, prey, or food source, which can then be integrated with information from other sensory input.

3. Testis and Sperm

PDE1 activity is also very concentrated in testis and sperm from mouse and rat species. Early kinetic data suggested that two different PDE1 isoforms may be expressed in testis (96), and this observation has been substantiated more recently by RNase protection analysis of testis total

RNA surveyed with PDE1 isoform-specific riboprobes (53, 59). In mouse testis, several PDE1C isoforms have been detected; PDE1C5, PDE1C1, and PDE1C4 mRNA have relative amounts approximately equal to 11, 3, and 1 arbitrary units, respectively (59). However, little PDE1C mRNA was detected in rat testis (58). PDE1A mRNA and protein have also been detected in mouse testis and epididymal sperm, respectively (C. Yan and J. A. Beavo, 1996). The precise PDE1A isoform is not currently known, but RNase protection and Northern analysis of testis RNA suggest that this isoform may be different than PDE1A1, PDE1A2, or PDE1A3. *In situ* hybridization studies in mouse testis have revealed that high concentrations of both PDE1A and PDE1C mRNAs are differentially expressed in germ cells at specific stages of development (C. Yan and J. A. Beavo, 1996). PDE1C mRNA expression occurs earlier than PDE1A; maximal hybridization signals are detected in pachytene spermatocytes and gradually decline throughout round spermatid stages of development. In contrast, PDE1A gene expression peaks at a transition from round spermatid to elongated spermatid development and gradually declines to undetectable levels in spermatogonia.

Currently, the signaling events that PDE1C plays in the testis have no clear physiological context. However, there is growing evidence that mature sperm undergo chemoattraction in the female reproductive tract (110, 111) and that the signaling events involved closely parallel those of olfaction. For example, sperm have been shown to express cyclic nucleotide-gated ion channels (112, 113) and olfactory receptor homologs (114), which, when occupied, presumably activate adenylyl cyclase (115). A major difference between olfaction and chemoattraction, however, is that in sperm, cGMP is probably the second messenger that activates the cyclic nucleotide-gated channel (113). Therefore, homologs of the guanylyl cyclase/atrial natriuretic peptide receptors (111) and a PDE1A isoform, molecules that are more likely to synthesize and degrade cGMP, are also likely to be involved in this pathway. So like olfactory signal transduction, the rapid termination of a chemoattractant signal may be necessary in order for sperm to sample the environment and respond appropriately to these sensory-like cues to maximize successful fertilization.

4. Cardiovascular System

Expression of all three PDE1 genes has been detected in the cardiovascular system, although the level of expression for each individual gene

5. Cyclic Nucleotide Regulation by Calmodulin

seems to vary among species. In bovine species, the heart appears to be one of the richest sources of the PDE1A1 variant, typically yielding enzyme protein approximately one-twenty-fifth that of PDE1A2 purified from brain tissue (73). In rat heart, PDE1 isoforms constitute about 60% of the total cGMP hydrolytic activity; PDE1A1 accounts for roughly 85% of this activity, and the remaining 15% appears to be contributed by a PDE1C variant (W. K. Sonnenburg, K. S. Kwak, and J. A. Beavo, 1996). Relative to PDE1A1 mRNA expression, mRNA-encoding PDE1C isoforms appear to be more abundant in mouse (58) and human (53) species when compared to rat. PDE1B mRNA has also been detected in total RNA isolated from rat heart, but is barely detectable and therefore may be expressed in a small subpopulation of the cells (W. K. Sonnenburg, K. S. Kwak, and J. A. Beavo, 1996). Based on immunocytochemistry using a polyclonal antibody having preferential reactivity to PDE1A isoforms, the major cardiac cell type expressing PDE1A1 is the smooth muscle of muscular arteries. Much less, if any, immunoreactivity was detected in veinous smooth muscle beds. Interestingly, PDE1A immunoreactivity was also detected in nerve fibers innervating the rat heart. It is not yet known, however, what types of nerve fibers display this immunoreactivity (e.g., adrenergic versus cholinergic) and which PDE1A isoform is expressed in this particular neuron. Although undetectable by immunocytochemistry, PDE1A activity is also present in cardiac interstitial fibroblasts cultured from rat heart and constitutes nearly all of the cGMP hydrolytic activity in these cells (W. K. Sonnenberg, K. S. Kwak, and J. A. Beavo, 1996). It is not yet known if this isoform is expressed in intact tissue. It is now apparent that at least in rat heart (116), very little PDE1 activity is expressed in cardiomyocytes and therefore may not contribute significantly to the regulation of cyclic nucleotide hydrolysis in this cell type.

A high proportion of the cGMP hydrolytic activity (70–80%) in vascular tissue has been reported to be PDE1 (117, 118), although the precise isoforms have not been identified. More recent studies (119) seem to indicate that the expression of PDE1 isozymes in vascular smooth muscle varies considerably depending on both the species and the phenotypic form of smooth muscle (contractile versus proliferative). For example, in human aortic smooth muscle extracts, both PDE1A and PDE1B activities have been detected in a ratio of 2:1. Macaque aortic smooth muscle extracts also contain both PDE1A and PDE1B activities, but at a much different ratio (1 : 10). When tissue from the medial layer of a human aorta is explanted and cultured, an induction of PDE1C occurs and PDE1A expression is reduced, resulting in 10-fold greater PDE1C activity than PDE1A activity. However, for smooth muscle cells derived from macaque aortic medial layer explants, PDE1B remains the predominant CaM-dependent activity and PDE1A expression is no longer detectable. Finally, PDE1A is the only detectable PDE1 activity in rat aortic smooth muscle cells.

One of the possible physiological roles for PDE1 is probably best understood in smooth muscle. Hormone-induced smooth muscle contraction is mediated by the mobilization of intracellular calcium (120), which would be predicted to also activate PDE1. Conversely, hormones that raise intracellular cGMP or cAMP promote the relaxation of smooth muscle. Therefore, the putative role for PDE1 isozymes might be to ensure robust constriction by hydrolyzing cyclic nucleotides that would otherwise oppose the increase in smooth muscle tone. This role is based on the following observations: (a) the stimulation of smooth muscle constriction correlates with the degree of activation of PDE1 (121, 122); (b) PDE inhibitors thought to be PDE1 selective raise intracellular cAMP and cGMP in smooth muscle treated with constricting agents and, conversely, PDE1-selective inhibitors do not raise intracellular cAMP and cGMP in noncontracted smooth muscle (118, 123–125); and finally (c) treatment of precontracted smooth muscle with PDE1-selective inhibitors correlates positively with smooth muscle relaxation (88, 124). Some caution needs to be exercised, however, when interpreting these pharmacological studies, as the selectivity of the PDE1 inhibitors overlap somewhat with other PDEs (86) expressed in smooth muscle. Nonetheless, data seem to support the idea that activation of PDE1 serves to promote smooth muscle contraction by hydrolyzing cyclic nucleotides under certain physiological conditions.

The effect of PDE1 inhibition on the cardiovascular system in the whole animal has also been investigated using the PDE1 inhibitor vinpocetine (126). In the anesthetized dog, administration of vinpocetine intravenously resulted in a decrease in blood pressure and total peripheral resistance, a modest increase in cardiac output, and no effect on heart rate. The decrease in blood pressure and peripheral resistance is consistent with vinpocetine selectively inhibiting PDE1 activity, but may also be due to the inhibition of other PDE activities in blood vessels (86). The increase in cardiac output may be attributed indirectly to the inhibition of vascular smooth muscle or nonmyocyte PDE1 activity in heart (116). However, the cardiac output increase may also be due to the inhibition of other PDE isozymes, particularly PDE3A, which would likely result in positive inotropy (127). Thus, the effect of PDE1 inhibition on these broad physiological parameters remains an open question.

A new role for PDE1 in the regulation of vascular smooth muscle proliferation is currently emerging. Proliferation of smooth muscle is antagonized by agents that elevate intracellular cAMP. Conversely, agents that stimulate proliferation of smooth muscle in many cases involve elevation of intracellular calcium (128). In human smooth muscle cells in culture (a potential model for smooth muscle cells displaying the proliferative phenotype), a PDE1C isoform accounts for most of the cAMP hydrolytic activity (119). Therefore, activation of PDE1C may be a mechanism for ensuring that a robust proliferative response is attained with certain types of growth factor signaling involving elevation in the intracellular concentration of calcium. Moreover, because agents that elevate cAMP in smooth muscle cells, including phosphodiesterase inhibitors (129, 130), have been shown to inhibit proliferation, PDE1C may be one of several important targets for therapeutic intervention of atherosclerosis or restenosis following balloon angioplasty.

5. Other Tissues

PDE1 gene expression has also been observed in other tissues. Most notably, a PDE1A isoform, presumably PDE1A1, appears to be concentrated in the outer and inner medulla of kidney (51). PDE1B gene expression has also been detected in only the kidney inner medulla (54). Speculation on the role that these isozymes might play awaits determination of the cell types that express these isozymes.

PDE1B1 gene expression has been detected in a human lymphoblastoid B-cell line (57). PDE1B1 mRNA and activity could be inhibited by transfection with antisense oligonucleotides, and inhibition of PDE1B1 expression or activity induced apoptosis. PDE1B1 is not expressed in peripheral blood lymphocytes, but, like bovine mononuclear cells (131), is induced in response to mitogenic stimulation by phytohemagglutinin and therefore may play a role in the maintenance of mitogenically activated lymphocytes.

IV. CONCLUDING REMARKS

Both adenylyl cyclases and cyclic nucleotide phosphodiesterases can be regulated by calmodulin, providing a logical mechanism for the interaction of calcium-signaling pathways with hormonal responses regulated by cyclic nucleotides. In the case of adenylyl cyclases, calcium-bound CaM can activate specific isoforms, providing a mechanism by which calcium can potentiate cyclic nucleotide synthesis. Calmodulin-stimulated phosphodiesterases are activated by CaM, catalyzing the hydrolysis of cyclic nucleotides. Based on the estimated activation constants for these isozymes, it is likely that Ca^{2+} would sequentially activate AC and PDE1, and the duration and amplitude of the cAMP formed would be dependent on which isozymes of PDE1 and AC are expressed in any given cell type. For some AC and PDE1 isoforms, phosphorylation catalyzed by different kinases can inhibit activity, thereby providing an additional layer of regulation. These properties, therefore, may be important for feedforward or feedback modulation of cyclic nucleotide formation in some cell types. Many investigators have determined the gene structure of these enzymes, which has proven essential for developing probes to determine the tissue distribution of each of the adenylyl cyclase and CaM-dependent phosphodiesterase isoforms. From this newly gained knowledge, the possible physiological roles that these CaM-regulated isoforms of adenylyl cyclase and phosphodiesterase play can be deduced. In the case of one adenylyl cyclase isoform, gene disruption techniques have uncovered the role of this enzyme in learning and memory. This methodology will undoubtedly be useful for exploring the function of other adenylyl cyclases and phosphodiesterases in the future. It is also clear that the genetic diversity of CaM-regulated adenylyl cyclases and phosphodiesterases is vast. From the structure of the encoded polypeptides, the domain organization of these enzymes has been advanced considerably, and with this newly gained knowledge agents can be rationally designed for therapeutic intervention.

REFERENCES

1. Cheung, W. Y. (1971). Cyclic 3′,5′-nucleotide phosphodiesterase. Evidence for and properties of a protein activator. *J. Biol. Chem.* **246,** 2859–2869.
2. Masters, B. S., McMillan, K., Sheta, E. A., Nishimura, J. S., Roman, L. J., and Martasek, P. (1996). Neuronal nitric oxide synthase, a modular enzyme formed by convergent evolution: Structure studies of a cysteine thiolate-liganded heme protein that hydroxylates L-arginine to produce NO as a cellular signal. *FASEB J.* **10,** 552–558.
3. McDonald, L. J., and Murad, F. (1996). Nitric oxide and cyclic GMP signaling. *Proc. Soc. Exp. Biol. Med.* **211,** 1–6.

4. Westcott, K. R., LaPorte, D. C., and Storm, D. R. (1979). Resolution of adenylyl cyclase sensitive and insensitive to Ca^{2+} and CDR by CDR-Sepharose affinity chromatography. *Proc. Natl. Acad. Sci. U.S.A.* **76,** 204–228.
5. Rosenberg, G. B., and Storm, D. R. (1987). Immunological distinction between calmodulin-sensitive and calmodulin-insensitive adenylyl cyclases. *J. Biol. Chem.* **262,** 7623–7628.
6. Krupinski, J., Coussen, F., Bakalyar, H. A., Tang, W. J., Feinstein, P. G., Orth, K., Slaughter, C., Reed, R. R., and Gilman, A. G. (1989). Adenylyl cyclase amino acid sequence: Possible channel- or transporter-like structure. *Science* **244,** 1558–1564.
7. Feinstein, P. G., Schrader, A., Bakalyar, H. A., Tang, W.-J., Krupinski, J., and Gilman, A. G. (1991). Molecular cloning and characterization of a calcium calmodulin-insensitive adenylyl cyclase from rat brain. *Proc. Natl. Acad. Sci. U.S.A.* **88,** 10173–10177.
8. Tang, W. J., Krupinski, J., and Gilman, A. G. (1991). Expression and characterization of calmodulin activated (type I) adenylyl cyclase. *J. Biol. Chem.* **266,** 8595–8603.
9. Choi, E. J., Xia, Z., and Storm, D. R. (1992). Stimulation of the type III olfactory adenylyl cyclase by calcium and calmodulin. *Biochemistry* **31,** 6492–6498.
10. Choi, E. J., Wong, S. T., Hinds, T. R., and Storm, D. R. (1992). Calcium and muscarinic agonist stimulation of type I adenylyl cyclase in whole cells. *J. Biol. Chem.* **267,** 12440–12442.
11. Wu, Z., Wong, S. T., and Storm, D. R. (1995). Modification of the calcium and calmodulin sensitivity of the type I adenylyl cyclase by mutagenesis of its calmodulin binding domain. *J. Biol. Chem.* **268,** 23766–23768.
12. Wu, Z., Wong, S. T., and Storm, D. R. (1993). Modification of the calcium and calmodulin sensitivity of the type I adenylyl cyclase by mutagenesis of its calmodulin binding domain. *J. Biol. Chem.* **268,** 23766–23768.
13. Wayman, G. A., Impey, S., Wu, Z., Kindsvogel, W., Prichard, L., and Storm, D. R. (1994). Synergistic activation of the type I adenylyl cyclase by Ca^{2+} and Gs-coupled receptors *in vivo*. *J. Biol. Chem.* **269,** 25400–25405.
14. Tota, M. R., Xia, Z., Storm, D. R., and Schimerlik, M. I. (1990). Reconstitution of muscarinic receptor mediated inhibition of adenylyl cyclase. *Mol. Pharmacol.* **37,** 950–957.
15. Tang, W. J., and Gilman, A. G. (1991). Type specific regulation of adenylyl cyclase by G protein beta/gamma subunits. *Science* **254,** 1500–1503.
16. Nielsen, M. D., Chan, G. C. K., Poser, S. W., and Storm, D. R. (1996). Differential regulation of type I and type VIII Ca^{2+}-stimulated adenylyl cyclases by Gi-coupled receptors *in vivo*. *J. Biol. Chem.* **271,** 33308–33316.
17. Jacobowitz, O., Chen, J., and Premont, R. T. (1993). Stimulation of specific types of Gs-stimulated adenylyl cyclases by phorbol ester treatment. *J. Biol. Chem.* **268,** 3829–3832.
18. Choi, E. J., Wong, S. T., Dittman, A. H., and Storm, D. R. (1993). Phorbol ester stimulation of the type I and type III adenylyl cyclases in whole cells. *Biochemistry* **32,** 1891–1894.
19. Dittman, A. H., Weber, J. P., Hinds, T. R., Choi, E. J., Migeon, J. C., Nathanson, N. M., and Storm, D. R. (1994). A novel mechanism for coupling of m4 muscarinic acetylcholine receptors to calmodulin-sensitive adenylyl cyclases: Crossover from G protein-coupled inhibition to stimulation. *Biochemistry* **33,** 943–951.

20. Taussig, R., Iniguez-Lluhi, J. A., and Gilman, A. G. (1993). Inhibition of adenylyl cyclase by Gi-alpha. *Science* **261,** 218–221.
21. Wayman, G. A., Wei, J., Wong, S., and Storm, D. R. (1996). Regulation of type I adenylyl cyclase by CaM-kinase IV *in vivo. Mol. Cell. Biol.* **16,** 6075–6082.
22. Nakamura, Y., Okuno, S., and Sato, F. (1995). An immunohistochemical study of Ca^{2+}/calmodulin-dependent protein kinase IV in the rat central nervous system: Light and electron microscopic observations. *J. Neurosci.* **68,** 181–194.
23. Xia, Z. G., Refsdal, C. D., Merchant, K. M., Dorsa, D. M., and Storm, D. R. (1991). Distribution of mRNA for the calmodulin-sensitive adenylate cyclase in rat brain: Expression in areas associated with learning and memory. *Neuron* **6,** 431–443.
24. Xia, Z., Choi, E. J., Wang, F., Blazynski, C., and Storm, D. R. (1993). Type I calmodulin-sensitive adenylyl cyclase is neural specific. *J. Neurochem.* **60,** 305–311.
25. Livingston, M. S., Sziber, P. P., and Quinn, W. G. (1984). Loss of calcium calmodulin responsiveness in adenylyl cyclase of rutabaga, a *Drosophila* learning mutant. *Cell (Cambridge, Mass.)* **37,** 205–215.
26. Dudai, Y., and Zvi, S. (1984). Adenylate cyclase in the *Drosophila* memory mutant rutabaga displays an altered Ca^{2+} sensitivity. *Neurosci. Lett.* **47,** 19–24.
27. Levin, L. R., Han, P. L., Hwang, P. M., Feinstein, P. G., Davis, R. L., and Reed, R. R. (1992). The *Drosophila* learning and memory gene rutabaga encodes a Ca^{2+}/calmodulin-responsive adenylyl cyclase. *Cell (Cambridge, Mass.)* **68,** 479–489.
28. Zhong, Y., and Wu, C. F. (1991) Altered synaptic plasticity in *Drosophila* memory mutants with a defective cyclic AMP cascade. *Science* **251,** 198–201.
29. Wu, Z. L., Thomas, S. A., Villacres, E. C., Xia, Z., Simmons, M. L., Chavkin, C., Palmiter, R. D., and Storm, D. R. (1995). Altered behavior and long-term potentiation in type I adenylyl cyclase mutant mice. *Proc. Natl. Acad. Sci. U.S.A.* **92,** 220–224.
30. Weisskopf, M. G., Castillo, P. E., Zalutsky, R. A., and Nicoll, R. A. (1994). Mediation of hippocampal mossy fiber long-term potentiation by cyclic AMP. *Science* **265,** 1878–1882.
31. Davis, S., Butcher, S. P., and Morris, R. G. (1992). The NMDA receptor antagonist D-2-amino-5-phosphonopentanoate (D-AP5) impairs spatial learning and LTP *in vivo* at intracerebral concentrations comparable to those that block LTP *in vitro. J. Neurosci.* **12,** 21–34.
32. Pokorny, J., and Yamamoto, T. (1981). Postnatal ontogenesis of hippocampal CA1 area in rats. II. Development of ultrastructure in stratum lacunosum and moleculare. *Brain Res. Bull.* **7,** 113–120.
33. Harris, K. M., and Teyler, T. J. (1984). Developmental onset of long-term potentiation in area CA1 of the rat hippocampus. *J. Physiol. (London)* **346,** 27–48.
34. Teyler, T. J., Perkins, A. T., and Harris, K. M. (1989). The development of long-term potentiation in hippocampus and neocortex. *Neuropsychologia* **27,** 31–39.
35. Villacres, E. C., Wu, Z., Hua, W., Nielsen, M. D., Watters, J. J., Yan, C., Beavo, J., and Storm, D. R. (1995). Developmentally expressed $Ca^{(2+)}$-sensitive adenylyl cyclase activity is disrupted in the brains of type I adenylyl cyclase mutant mice. *J. Biol. Chem.* **270,** 14352–14357.
36. van den Pol, A. N., and Dudek, F. E. (1993). Cellular communication in the circadian clock, the suprachiasmatic nucleus. *Neuroscience (Oxford)* **56,** 793–811.
37. Miller, J. D. (1993). On the nature of the circadian clock in mammals. *Am. J. Physiol.* **264,** R821–R832.

38. Tzavara, E. T., Pouille, Y., Defer, N., and Hanoune, J. (1996). Diurnal variation of the adenylyl cyclase type 1 in the rat pineal gland. *Proc. Natl. Acad. Sci. U.S.A.* **93,** 11208–11212.
38a. Cali, J. J., Zwaagstra, J. C., Mons, N., Cooper, D. M., Krupinski, J. (1994). Type VIII Adenylyl cyclase. A Ca^{2+}/calmodulin-stimulated enzyme expressed in discrete regions of rat brain. *J. Biol. Chem.* **269,** 12190–12195.
39. Wayman, G. A., Impey, S., and Storm, D. R. (1995). Ca^{2+} inhibition of type III adenylyl cyclase *in vivo*. *J. Biol. Chem.* **270,** 21480–21486.
40. Wei, J., Wayman, G., and Storm, D. R. (1996). Phosphorylation and inhibition of type III adenylyl cyclase by calmodulin-dependent protein kinase II *in vivo*. *J. Biol. Chem.* **271,** 24231–24235.
41. Wayman, G. A., Hinds, T. R., and Storm, D. R. (1995). Hormone stimulation of type III adenylyl cyclase causes Ca^{2+} oscillations in HEK-293 cells. *J. Biol. Chem.* **270,** 24108–24115.
42. Xia, Z., Choi, E. J., Wang, F., and Storm, D. R. (1992). The type III calcium/calmodulin-sensitive adenylyl cyclase is not specific to olfactory sensory neurons. *Neurosci. Lett.* **144,** 169–173.
43. Bakalyar, H. A., and Reed, R. R. (1990). Identification of a specialized adenylyl cyclase that may mediate odorant detection. *Science* **250,** 1403–1406.
44. Jaworsky, D. E., Matsuzaki, O., Borisy, F. F., and Ronnett, G. V. (1995). Calcium modulates the rapid kinetics of the odorant-induced cyclic AMP signal in rat olfactory cilia. *J. Neurosci.* **15,** 310–318.
45. Restrepo, D., Miyamoto, T., and Bryant, B. P. (1990). Odor stimuli trigger influx of Ca^{2+} into olfactory neurons of the channel catfish. *Science* **249,** 1166–1168.
46. Hirono, J., Sato, T., and Tonoike, M. (1992). Simultaneous recording of $[Ca^{2+}]I$ increases in isolated olfactory receptor neurons retaining their original spatial relationship in intact tissue. *J. Neurosci. Methods* **42,** 185–194.
47. Wang, J. H., Sharma, R. K., and Mooibroik, M. J. (1990). Calmodulin-stimulated cyclic nucleotide phosphodiesterases. *In* "Cyclic Nucleotide Phosphodiesterases: Structure, Regulation and Drug Action" (J. Beavo and M. D. Houslay, eds.), pp. 19–60. Wiley, Chichester.
48. Beavo, J. A., Conti, M., and Heaslip, R. (1994). Multiple cyclic nucleotide phosphodiesterases. *Mol. Pharmacol.* **46,** 399–405.
49. Beavo, J. A., and Reifsnyder, D. H. (1990). Primary sequence of cyclic nucleotide phosphodiesterase isozymes and the design of selective inhibitors. *Trends Pharmacol. Sci.* **11,** 150–155.
50. Zhao, A. Z., Yan, C., Sonnenburg, W. K., and Beavo, J. A. (1997). Recent advances in the study of Ca^{2+}/CaM activated phosphodiesterases: Expression and physiological functions. *In* "Signal Transduction in Health and Disease," (J. D. Corbin and S. H. Francis, eds.), Vol. 31, pp. 237–251, Lippincott-Raven Publishers, Philadelphia.
51. Sonnenburg, W. K., Seger, D., and Beavo, J. A. (1993). Molecular Cloning of a cDNA Encoding the "61 kDa" calmodulin-stimulated cyclic nucleotide phosphodiesterase: Tissue-specific expression of structurally related isoforms. *J. Biol. Chem.* **268,** 645–652.
52. Sonnenburg, W. K., Seger, D., Kwak, K. S., Huang, J., Charbonneau, H., and Beavo, J. A. (1995). Identification of inhibitory and calmodulin-binding domains of the PDE1A1 and PDE1A2 calmodulin-stimulated cyclic nucleotide phosphodiesterases. *J. Biol. Chem.* **270,** 30989–31000.

53. Loughney, K., Martins, T. J., Harris, E. A., Sadhu, K., Hicks, J. B., Sonnenburg, W. K., Beavo, J. A., and Ferguson, K. (1996). Isolation and characterization of cDNAs corresponding to two human calcium, calmodulin-regulated, 3',5'-cyclic nucleotide phosphodiesterases. *J. Biol. Chem.* **271,** 796–806.
54. Bentley, J. K., Kadlecek, A., Sherbert, C. H., Seger, D., Sonnenburg, W. K., Charbonneau, H., Novack, J. P., and Beavo, J. A. (1992). Molecular cloning of cDNA encoding a "63"-kDa calmodulin-stimulated phosphodiesterase from bovine brain. *J. Biol. Chem.* **267,** 18676–18682.
55. Polli, J. W., and Kincaid, R. L. (1992). Molecular cloning of DNA encoding a calmodulin-dependent phosphodiesterase enriched in striatum. *Proc. Natl. Acad. Sci. U.S.A.* **89,** 11079–11083.
56. Repaske, D. R., Swinnen, J. V., Jin, S. L., Van, W. J. J., and Conti, M. (1992). A polymerase chain reaction strategy to identify and clone cyclic nucleotide phosphodiesterase cDNAs. Molecular cloning of the cDNA encoding the 63-kDa calmodulin-dependent phosphodiesterase. *J. Biol. Chem.* **267,** 18683–18688.
57. Jiang, X., Li, J., Paskind, M., and Epstein, P. M. (1996). Inhibition of calmodulin-dependent phosphodiesterase induces apoptosis in human leukemic cells. *Proc. Natl. Acad. Sci. U.S.A.* **93,** 11236–11241.
58. Yan, C., Zhao, A. Z., Bentley, J. K., Loughney, K., Ferguson, K., and Beavo, J. A. (1995). Molecular cloning and characterization of a calmodulin-dependent phosphodiesterase enriched in olfactory sensory neurons. *Proc. Natl. Acad. Sci. U.S.A.* **92,** 9677–9681.
59. Yan, C., Zhao, A., Bentley, J. K., and Beavo, J. A. (1996). The calmodulin-dependent phosphodiesterase gene PDE1C encodes several functionally different splice variants in a tissue-specific manner. *J. Biol. Chem.* **271,** 25699–25706.
60. The Genetics Computer Group, (1991). Program Manual for the GCC Package. 575 Science Drive, Madison, Wisconsin 53711.
61. Jin, L., and Nei, M. (1990). Limitations of the evolutionary parsimony method of phylogenetic analysis. *Mol. Biol. Evol.* **7,** 82–102.
62. Page, R. D. M. (1996). An application to display phylogenetic trees on personal computers. *Comput. Appl. Biosci.* **12,** 357–358.
63. Charbonneau, H. (1990). Structure–function relationships among cyclic nucleotide phosphodiesterase. *In* "Cyclic Nucleotide Phosphodiesterases: Structure, Regulation and Drug Action" (J. Beavo and M. D. Houslay, eds.), pp. 267–298. Wiley, Chichester.
64. Klee, C. B. (1980). Calmodulin: The coupling factor of the two second messengers Ca^{2+} and cAMP. *In* "Protein Phosphorylation and Bioregulation" (T. G. Podesta and E. J. Gordon, eds.), pp. 61–69. Karger, Basel.
65. Tucker, M. M., Robinson, J. B. J., and Stellwagen, E. (1981). The effect of proteolysis on the calmodulin activation of cyclic nucleotide phosphodiesterase. *J. Biol. Chem.* **256,** 9051–9058.
66. Kincaid, R. L., Stith, C. I. E., and Vaughan, M. (1985). Proteolytic activation of calmodulin-dependent cyclic nucleotide phosphodiesterase. *J. Biol. Chem.* **260,** 9009–9015.
67. Charbonneau, H., Kumar, S., Novack, J. P., Blumenthal, D. K., Griffin, P. R., Shabanowitz, J., Hunt, D. F., Beavo, J. A., and Walsh, K. A. (1991). Evidence for domain organization within the 61-kDa calmodulin-dependent cyclic nucleotide phosphodiesterase from bovine brain. *Biochemistry* **30,** 7931–7940.

68. Jin, S. L., Swinnen, J. V., and Conti, M. (1992). Characterization of the structure of a low Km, rolipram-sensitive cAMP phosphodiesterase. Mapping of the catalytic domain. *J. Biol. Chem.* **267,** 18929–18939.
69. Pillai, R., Kytle, K., Reyes, A., and Colicelli, J. (1993). Use of a yeast expression system for the isolation and analysis of drug-resistant mutants of a mammalian phosphodiesterase. *Proc. Natl. Acad. Sci. U.S.A.* **90,** 11970–11974.
70. Pillai, R. Staub, S. F., and Colicelli, J. (1994). Mutational mapping of kinetic and pharmacological properties of a human cardiac cAMP phosphodiesterase. *J. Biol. Chem.* **269,** 30676–30681.
71. Cheung, P. P., Xu, H., McLaughlin, M. M., Ghazaleh, F. A., Livi, G. P., and Colman, R. W. (1996). Human platelet cGI-PDE: Expression in yeast and localization of the catalytic domain by deletion mutagenesis. *Blood* **88,** 1321–1329.
72. Novack, J. P., Charbonneau, H., Bentley, J. K., Walsh, K. A., and Beavo, J. A. (1991). Sequence comparison of the 63-, 61-, and 59-kDa calmodulin-dependent cyclic nucleotide phosphodiesterases. *Biochemistry* **30,** 7940–7947.
73. Hansen, R. S., Charbonneau, H., and Beavo, J. A. (1988). Purification of calmodulin-stimulated cyclic nucleotide phosphodiesterase by monoclonal antibody affinity chromatography. *Methods Enzymol.* **159,** 543–557.
74. Erickson, V. S., and DeGrado, W. F. (1987). Recognition and characterization of calmodulin-binding sequences in peptides and proteins. *Methods Enzymol.* **139,** 455–478.
75. LaPorte, D. C., Toscano, W. A. J., and Storm, D. R. (1979). Cross-linking of iodine-125-labeled, calcium-dependent regulatory protein to the Ca^{2+}-sensitive phosphodiesterase purified from bovine heart. *Biochemistry* **18,** 2820–2825.
76. Sharma, R. K., Wang, T. H., Wirch, E., and Wang, J. H. (1980). Purification and properties of bovine brain calmodulin-dependent cyclic nucleotide phosphodiesterase. *J. Biol. Chem.* **255,** 5916–5923.
77. Sharma, R. K. (1991). Phosphorylation and characterization of bovine heart calmodulin-dependent phosphodiesterase. *Biochemistry* **30,** 5963–5968.
78. Sharma, R. K., and Wang, J. H. (1985). Differential regulation of bovine brain calmodulin-dependent cyclic nucleotide phosphodiesterase isoenzymes by cyclic AMP-dependent protein kinase and calmodulin-dependent phosphatase. *Proc. Natl. Acad. Sci. U.S.A.* **82,** 2603–2607.
79. Soderling, T. R. (1993). Protein kinases and phosphatases: Regulation by autoinhibitory domains. *Biotechnol. Appl. Biochem.* **18,** 185–200.
80. Martins, T. J. (1984). Cyclic GMP-Stimulated Cyclic Nucleotide Phosphodiesterase: Purification from Bovine Tissues and Characterization. Ph.D. Thesis, University of Washington, St. Louis, Missouri.
81. Kincaid, R. L., Manganiello, V. C., Odya, C. E., Osborne, J. J., Stith, C. I. E., Danello, M. A., and Vaughan, M. (1984). Purification and properties of calmodulin-stimulated phosphodiesterase from mammalian brain. *J. Biol. Chem.* **259,** 5158–5166.
82. Sharma, R. K., and Wang, J. H. (1986). Purification and characterization of bovine lung calmodulin-dependent cyclic nucleotide phosphodiesterase. An enzyme containing calmodulin as a subunit. *J. Biol. Chem.* **261,** 14160–14166.
83. Sharma, R. K., and Wang, J. H. (1986). Calmodulin and Ca^{2+}-dependent phosphorylation and dephosphorylation of 63-kDa subunit-containing bovine brain calmodulin-stimulated cyclic nucleotide phosphodiesterase isozyme. *J. Biol. Chem.* **261,** 1322–1328.

84. Pagani, E. D., Buchholz, R. A., and Silver, P. J. (1992). Cardiovascular cyclic nucleotide phosphodiesterases and their role in regulating cardiovascular function. *Basic Res. Cardiol.* **1,** 73–86.
85. Hansen, R. S., and Beavo, J. A. (1982). Purification of two calcium/calmodulin-dependent forms of cyclic nucleotide phosphodiesterase by using conformation-specific monoclonal antibody chromatography. *Proc. Natl. Acad. Sci. U.S.A.* **79,** 2788–2792.
86. Saeki, T., and Saito, I. (1993). Isolation of cyclic nucleotide phosphodiesterase isozymes from pig aorta. *Biochem. Pharmacol.* **46,** 833–839.
87. Ichimura, M., Eiki, R., Osawa, K., Nakanishi, S., and Kase, H. (1996). KS-505a, an isoform-selective inhibitor of calmodulin-dependent cyclic nucleotide phosphodiesterase. *Biochem. J.* **316,** 311–316.
88. Hegde, V. R., Miller, J. R., Patel, M. G., King, A. H., Puar, M. S., Horan, A., Hart, R., Yarborough, R., and Gullo, V. (1993). SCH 45752—an inhibitor of calmodulin-sensitive cyclic nucleotide phosphodiesterase activity. *J. Antibiot.* **46,** 207–213.
89. Florio, V. A., Sonnenburg, W. K., Johnson, R., Kwak, K. S., Jensen, G. S., Walsh, K. A., and Beavo, J. A. (1994). Phosphorylation of the 61-kDa calmodulin-stimulated cyclic nucleotide phosphodiesterase at serine 120 reduces its affinity for calmodulin. *Biochemistry* **33,** 8948–8954.
90. Ariano, M. A. (1983). Distribution of components of the guanosine 3′,5′-phosphate system in rat caudate-putamen. *Neuroscience (Oxford)* **10,** 707–723.
91. Billingsley, M. L., Polli, J. W., Balaban, C. D., and Kincaid, R. L. (1990). Developmental expression of calmodulin-dependent cyclic nucleotide phosphodiesterase in rat brain. *Brain Res. Dev. Brain Res.* **53,** 253–263.
92. Polli, J. W., and Kincaid, R. L. (1994). Expression of a calmodulin-dependent phosphodiesterase isoform (PDE1B1) correlates with brain regions having extensive dopaminergic innervation. *J. Neurosci.* **14,** 1251–1261.
93. Yan, C., Bentley, J. K., Sonnenburg, W. K., and Beavo, J. A. (1994). Differential expression of the 61 kDa and 63 kDa calmodulin-dependent phosphodiesterases in the mouse brain. *J. Neurosci.* **14,** 973–984.
94. Kramer, R. H., and Levitan, I. B. (1988). Calcium-dependent inactivation of a potassium current in the Aplysia neuron R15. *J. Neurosci.* **8,** 1796–1803.
95. Kramer, R. H., Levitan, E. S., Wilson, M. P., and Levitan, I. B. (1988). Mechanism of calcium-dependent inactivation of a potassium current in Aplysia neuron R15: Interaction between calcium and cyclic AMP. *J. Neurosci.* **8,** 1804–1813.
96. Sonnenburg, W. K., and Beavo, J. A. (1994). Cyclic GMP and regulation of cyclic nucleotide hydrolysis. *In* "Advances in Pharmacology" (F. Murad, ed.), pp. 87–114. Academic Press, San Diego.
97. Levitan, I. B. (1994). Modulation of ion channels by protein phosphorylation and dephosphorylation. *Annu. Rev. Physiol.* **56,** 193–212.
98. Catterall, W. A. (1995). Structure and function of voltage-gated ion channels. *Annu. Rev. Biochem.* **64,** 493–531.
99. Swope, S. L., Moss, S. J., Blackstone, C. D., and Huganir, R. L. (1992). Phosphorylation of ligand-gated ion channels: A possible mode of synaptic plasticity. *FASEB J.* **6,** 2514–2523.
100. Frey, U., Huang, Y. Y., and Kandel, E. R. (1993). Effects of cAMP simulate a late stage of LTP in hippocampal CA1 neurons. *Science* **260,** 1661–1664.

101. Huang, Y. Y., Li, X. C., and Kandel, E. R. (1994). cAMP contributes to mossy fiber LTP by initiating both a covalently mediated early phase and macromolecular synthesis-dependent late phase. *Cell (Cambridge, Mass.)* **79,** 69–79.
102. Zhuo, M., and Hawkins, R. D. (1995). Long-term depression: A learning-related type of synaptic plasticity in the mammalian central nervous system. *Rev. Neurosci.* **6,** 259–277.
103. Ludvig, N., Burmeister, V., Jobe, P. C., and Kincaid, R. L. (1991). Electron microscopic immunocytochemical evidence that the calmodulin-dependent cyclic nucleotide phosphodiesterase is localized predominantly at postsynaptic sites in the rat brain. *Neuroscience (Oxford)* **44,** 491–500.
104. Mayer, B., Klatt, P., Bohme, E., and Schmidt, K. (1992). Regulation of neuronal nitric oxide and cyclic GMP formation by Ca^{2+}. *J. Neurochem.* **59,** 2024–2029.
105. Breer, H., Wanner, I., and Strotmann, J. (1996). Molecular genetics of mammalian olfaction. *Behav. Genet.* **26,** 209–219.
106. Firestein, S., Darrow, B., and Shepherd, G. M. (1991). Activation of the sensory current in salamander olfactory receptor neurons depends on a G protein-mediated cAMP second messenger system. *Neuron* **6,** 825–835.
107. Boekhoff, I., and Breer, H. (1992). Termination of second messenger signaling in olfaction. *Proc. Natl. Acad. Sci. U.S.A.* **89,** 471–474.
108. Borisy, F. F., Ronnett, G. V., Cunningham, A. M., Juilfs, D., Beavo, J., and Snyder, S. H. (1992). Calcium/calmodulin-activated phosphodiesterase expressed in olfactory receptor neurons. *J. Neurosci.* **12,** 915–923.
109. Chen, T. Y., and Yau, K. W. (1994). Direct modulation by $Ca^{(2+)}$-calmodulin of cyclic nucleotide-activated channel of rat olfactory receptor neurons. *Nature (London)* **368,** 545–548.
110. Ralt, D., Goldenberg, M., Fetterolf, P., Thompson, D., Dor, J., Mashiach, S., Garbers, D. L., and Eisenbach, M. (1991). Sperm attraction to a follicular factor(s) correlates with human egg fertilizability. *Proc. Natl. Acad. Sci. U.S.A.* **88,** 2840–2844.
111. Anderson, R. A. J., Feathergill, K. A., Rawlins, R. G., Mack, S. R., and Zaneveld, L. J. (1995). Atrial natriuretic peptide: A chemoattractant of human spermatozoa by a guanylate cyclase-dependent pathway. *Mol. Reprod. Dev.* **40,** 371–378.
112. Weyand, I., Godde, M., Frings, S., Weiner, J., Muller, F., Altenhofen, W., Hatt, H., and Kaupp, U. B. (1994). Cloning and functional expression of a cyclic-nucleotide-gated channel from mammalian sperm. *Nature (London)* **368,** 859–863.
113. Frings, S., Seifert, R., Godde, M., and Kaupp, U. B. (1995). Profoundly different calcium permeation and blockage determine the specific function of distinct cyclic nucleotide-gated channels. *Neuron* **15,** 169–179.
114. Vanderhaeghen, P., Schurmans, S., Vassart, G., and Parmentier, M. (1993). Olfactory receptors are displayed on dog mature sperm cells. *J. Cell Biol.* **123,** 1441–1452.
115. Rojas, F. J., Patrizio, P., Do, J., Silber, S., Asch, R. H., and Moretti, R. I. (1993). Evidence for a novel adenylyl cyclase in human epididymal sperm. *Endocrinology (Baltimore)* **133,** 3030–3033.
116. Bode, D. C., Kanter, J. R., and Brunton, L. L. (1991). Cellular distribution of phosphodiesterase isoforms in rat cardiac tissue. *Circ. Res.* **68,** 1070–1079.
117. Ahn, H. S., Foster, M., Cable, M., Pitts, B. J., and Sybertz, E. J. (1991). Ca/CaM-stimulated and cGMP-specific phosphodiesterases in vascular and nonvascular tissues. *Adv. Exp. Med. Biol.* **308,** 191–197.

118. Ahn, H. S., Crim, W., Pitts, B., and Sybertz, E. J. (1992). Calcium-calmodulin-stimulated and cyclic-GMP-specific phosphodiesterases. Tissue distribution, drug sensitivity, and regulation of cyclic GMP levels. *Adv. Second Messenger Phosphoprotein Res.* **25,** 271–88.
119. Rybalkin, S. D., Bornfeldt, K. E., Sonnenburg, W. K., Rybalkina, I. G., Kwak, K. S., Hanson, K., Krebs, E. G., and Beavo, J. A. (1997). Calmodulin-stimulated cyclic nucleotide phosphodiesterase (PDE1C) is induced in human arterial smooth muscle cells of the synthetic, proliferative phenotype. *J. Clin. Invest.* **100,** 2611–2621.
120. Walsh, M. P., Kargacin, G. J., Kendrick, J. J., and Lincoln, T. M. (1995). Intracellular mechanisms involved in the regulation of vascular smooth muscle tone. *Can. J. Physiol. Pharmacol.* **73,** 565–573.
121. Saitoh, Y., Hardman, J. G., and Wells, J. N. (1985). Differences in the association of calmodulin with cyclic nucleotide phosphodiesterase in relaxed and contracted arterial strips. *Biochemistry* **24,** 1613–1618.
122. Miller, H. W. C., Miller, J. R., Wells, J. N., Stull, J. T., and Kamm, K. E. (1988). Biochemical events associated with activation of smooth muscle contraction. *J. Biol. Chem.* **263,** 13979–13982.
123. Miller, J. R., and Wells, J. N. (1987). Effects of isoproterenol on active force and Ca^{2+} X calmodulin-sensitive phosphodiesterase activity in porcine coronary artery. *Biochem. Pharmacol.* **36,** 1819–1824.
124. Souness, J. E., Brazdil, R., Diocee, B. K., and Jordan, R. (1989). Role of selective cyclic GMP phosphodiesterase inhibition in the myorelaxant actions of M & B 22,948, MY-5445, vinpocetine, and 1-methyl-3-isobutyl-8-(methylamino)xanthine. *Br. J. Pharmacol.* **98,** 725–734.
125. Ahn, H. S., Crim, W., Romano, M., Sybertz, E., and Pitts, B. (1989). Effects of selective inhibitors on cyclic nucleotide phosphodiesterases of rabbit aorta. *Biochem. Pharmacol.* **38,** 3331–3339.
126. Weishaar, R. E., Kobylarz, S. D. C., Keiser, J., Haleen, S. J., Major, T. C., Rapundalo, S., Peterson, J. T., and Panek, R. (1990). Subclasses of cyclic GMP-specific phosphodiesterase and their role in regulating the effects of atrial natriuretic factor. *Hypertension (Dallas)* **15,** 528–540.
127. Beavo, J. A. (1995). Cyclic nucleotide phosphodiesterases: Functional implications of multiple isoforms. *Physiol. Rev.* **75,** 725–748.
128. Bornfeldt, K. E., Raines, E. W., Graves, L. M., Skinner, M. P., Krebs, E. G., and Ross, R. (1995). Platelet-derived growth factor. Distinct signal transduction pathways associated with migration versus proliferation. *Ann. N.Y. Acad. Sci.* **766,** 416–430.
129. Souness, J. E., Hassall, G. A., and Parrott, D. P. (1992). Inhibition of pig aortic smooth muscle cell DNA synthesis by selective type III and type IV cyclic AMP phosphodiesterase inhibitors. *Biochem. Pharmacol.* **44,** 857–866.
130. Pan, X., Arauz, E., Krzanowski, J. J., Fitzpatrick, D. F., and Polson, J. B. (1994). Synergistic interactions between selective pharmacological inhibitors of phosphodiesterase isozyme families PDE III and PDE IV to attenuate proliferation of rat vascular smooth muscle cells. *Biochem. Pharmacol.* **48,** 827–835.
131. Hurwitz, R. L., Hirsch, K. M., Clark, D. J., Holcombe, V. N., and Hurwitz, M. Y. (1990). Induction of a calcium/calmodulin-dependent phosphodiesterase during phytohemagglutinin-stimulated lymphocyte mitogenesis. *J. Biol. Chem.* **265,** 8901–8907.

6

Regulation of Nitric Oxide Synthase by Calmodulin

JINGRU HU* AND LINDA J. VAN ELDIK*,†,‡

*Department of Cell and Molecular Biology
Northwestern University Medical School
†Northwestern Drug Discovery Program
‡Northwestern University Institute for Neuroscience
Chicago, Illinois 60611

I. Introduction 288
II. Nitric Oxide Synthase.. 288
   A. Discovery of Nitric Oxide as a Messenger 288
   B. Nitric Oxide Synthase Classification, Domain Organization, and Catalytic Mechanism 289
   C. Purification and Biochemical Properties and Localization of Nitric Oxide Synthase 294
   D. Nitric Oxide Synthase cDNAs and Genes and Regulation of Expression .. 296
III. Calmodulin and Nitric Oxide Synthase Activation 300
   A. Discovery of Nitric Oxide Synthase as a Calmodulin-Requiring Enzyme .. 300
   B. Characterization of Calmodulin Binding Sites on Nitric Oxide Synthase.... 302
   C. Characterization of Nitric Oxide Synthase-Binding Sites on Calmodulin.... 307
   D. Significance of Calmodulin Binding to Nitric Oxide Synthase 312
IV. Regulation of Nitric Oxide Synthase Activity by Phosphorylation/Dephosphorylation 319
   A. Regulation of Nitric Oxide Synthase by Ca^{2+}/Calmodulin-Dependent Kinases and Phosphatases 319
   B. Regulation of Nitric Oxide Synthase by Protein Kinase C 320
   C. Regulation of Nitric Oxide Synthase by cAMP/cGMP-Dependent Kinases 321
   D. Regulation of Nitric Oxide Synthase by Tyrosine Kinases 322
   E. Autophosphorylation 323
   F. Phosphorylation as a Control Point/Cross-Talk among Signaling Pathways 323
V. Physiological and Pathophysiological Functions of Nitric Oxide 325
   A. Biochemical Reactivity of Nitric Oxide................................ 325

B. Roles of Nitric Oxide in Central Nervous System 328
  C. Roles of Nitric Oxide in Cardiovascular System...................... 331
  D. Roles of Nitric Oxide in Immune and Inflammatory Systems 332
  E. Pharmacological Implications 332
VI. Conclusion 333
  References 334

I. INTRODUCTION

Nitric oxide synthase (NOS; EC 1.14.13.39) is an enzyme responsible for the synthesis of nitric oxide (NO), an important signaling molecule in most cell types of mammalian tissues. Nitric oxide synthase catalyzes the conversion of the amino acid L-arginine into L-citrulline and generates NO stoichiometrically. Three forms of NOS have been classified based on genetic cloning, origin of the tissue from which the enzyme was isolated, and the constitutive versus inducible nature of the enzyme. All of the three isoforms of NOS contain a calmodulin (CaM)-binding domain in their amino acid sequences, and their enzymatic activity is regulated by CaM. This chapter focuses on the interactions of CaM with NOS, the significance of CaM in NOS function, and the multiple modes of regulation of the enzyme. As background, an introduction to NOS is also presented: its history, structure, and functions. Finally, the chapter concludes with a brief discussion of some of the physiological and pathological functions of NO. Numerous aspects of NOS research cannot be covered adequately here, nor can every primary literature reference be cited. Throughout this chapter, therefore, the reader will be referred to recent reviews for more in-depth coverage of specific aspects discussed.

II. NITRIC OXIDE SYNTHASE

A. Discovery of Nitric Oxide as a Messenger

The ability of the free radical gas NO to function as a messenger molecule in a variety of cell types was demonstrated conclusively in the late 1980s, but these finding were based on a number of earlier observations. One of the first reports of biological activity of NO was the observation in 1977 that NO could stimulate guanylate cyclase activity and elevate tissue levels of cyclic GMP (cGMP), resulting in smooth muscle relaxation (Arnold et al., 1977). The finding by Furchgott and Zawadzki (1980) that acetylcholine-induced blood vessel relaxation required the presence of an intact endothelium led to the postulation of an endothelium-derived relaxing factor (EDRF), which appeared to act by increasing cGMP production in the vascular smooth muscle. Experiments *in vitro* using macrophage cell lines showed that cytotoxicity and formation of nitrate and nitrite in response to lipopolysaccharide (LPS) and interferon-γ (IFN-γ) were dependent on the presence of L-arginine in the culture medium (Iyenger et al., 1987). In the early 1980s, Deguchi and Yoshioka (1982) showed that L-arginine could stimulate guanylate cyclase in a cytosolic fraction from neuroblastoma cells, suggesting that L-arginine or a metabolite could activate guanylate cyclase in the brain. These apparently unrelated observations coalesced in the late 1980s with the demonstrations that EDRF was NO (Ignarro et al., 1988; Furchgott, 1988; Palmer et al., 1987), that NO could be produced enzymatically from L-arginine in vascular endothelial cells (Palmer et al., 1988), that glutamate could stimulate brain cGMP via formation of an EDRF-like factor (Garthwaite et al., 1988), and that NO-forming enzymatic activity was present in brain tissue extracts (Knowles et al., 1989; Schmidt et al., 1989). As discussed later, these findings stimulated an explosion of research to identify the enzyme responsible for NO synthesis, i.e., the nitric oxide synthase. It is clear now that NO, as a signaling molecule, is a major regulator in the cardiovascular, immune, and nervous systems. As discussed in more detail at the end of this chapter, NO has also been implicated as causal or contributing to pathophysiological conditions, including many disease states such as vascular shock, stroke, and diabetes.

B. Nitric Oxide Synthase Classification, Domain Organization, and Catalytic Mechanism

Nitric oxide synthase enzymatic activity was quickly purified by a number of investigators and shown to exist in a variety of tissues (Bredt and Snyder, 1990; Mayer et al., 1990; Schmidt et al., 1991; Schmidt and

Murad, 1991; Ohshima et al., 1992; Hiki et al., 1992; Hope et al., 1991). Three NOS isoforms have now been identified and classified, and there is a confusing array of nomenclatures for the enzymes based on the historical order of their purification, the cell or tissue of origin, or the constitutive or inducible expression of the enzyme. Table I lists NOS nomenclatures and some examples of their tissue distribution. The first NOS isoform (NOS I, bNOS, nNOS, ncNOS) was purified from brain, was shown to be constitutively expressed, and was found to require Ca^{2+} and CaM for activity. A distinct Ca^{2+}/CaM-dependent, constitutive NOS isoform (NOS III, eNOS, ecNOS) was isolated from endothelial cells and found to be associated with EDRF formation. An inducible NOS isoform (NOS II, iNOS, macNOS), originally isolated from macrophages, contains CaM as a tightly bound subunit, and its NOS activity is independent of elevations in intracellular Ca^{2+} levels. A nitric oxide research subcommittee of the International Union of Pharmacology Committee on Receptor Nomenclature and Drug Classification has recommended a uniform nomenclature for the NOS isoforms (Moncada et al., 1997). This chapter follows these recommendations and uses the terms nNOS, eNOS, and iNOS for the three isoforms.

All three NOS isoforms contain two distinct structural domains (Fig. 1): a reductase (C-terminal) domain, which is similar to cytochrome P450 and contains binding sites for NADPH and flavins (FMN and FAD); and an oxygenase (N-terminal) domain, which contains binding regions for heme, L-arginine, and tetrahydrobiopterin (BH_4). The CaM-binding site is located between the two domains. The reaction catalyzed by NOS is illustrated in Fig. 2 and is an oxidative reaction utilizing L-arginine as a substrate to produce L-citrulline and NO. The reaction requires molecular oxygen (O_2) and reducing equivalents from NADPH. Flavins (FMN and FAD), heme, and BH_4 are cofactors of the enzyme. The specific activities of the purified NOS isoforms are in the range of 900–1600 nmol/min/mg protein, suggesting comparable catalytic efficiency.

The NOS enzymatic reaction consists of two successive monooxygenase reactions: oxidation of the guanidino nitrogen of the L-arginine substrate to a N-hydroxyarginine (N-OH-L-arginine) intermediate, and a subsequent oxidation of N-OH-L-arginine to form L-citrulline and NO (Fig. 2). NADPH is required as an electron donor. The NOS flavins are thought to serve an electron storage and transfer function, accepting electrons from NADPH and transferring them to the heme moiety, leading to oxidation of the substrate L-arginine. Calmodulin binding to NOS facilitates the electron transfer flow between reductase and

TABLE I

Nitric Oxide Synthase Isoforms

Recommended nomenclature[a]	Previous designations	Description	Tissue distribution[b]
Neuronal NOS (nNOS)	Type I NOS NOS-I bNOS ncNOS	Constitutive NOS First identified in neurons Ca^{2+}/CaM regulated Cytosol > membrane Low (pM) NO output	Central and peripheral neurons, skeletal muscle, certain epithelial cells, renal macula densa cells, pancreatic islet cells
Inducible NOS (iNOS)	Type II NOS NOS-II macNOS hepNOS	Inducible NOS First identified in macrophages CaM as a tightly bound subunit Cytosol > membrane High (μM) NO output	Macrophages, Kupffer cells, hepatocytes, fibroblasts, glioma cells, microglia, astrocytes, myocardium, vascular smooth muscle cells, liver, mesangial cells, endothelial cells, articular chondrocytes, polymorphonuclear neutrophils
Endothelial NOS (eNOS)	Type III NOS NOS-III ecNOS	Constitutive NOS First identified in endothelium Ca^{2+}/CaM regulated Membrane > cytosol Low (pM) NO output	Endothelial cells, CA1 neurons, neutrophils, mast cell, platelets, adrenal medullary cells, astrocytes, kidney tubular epithelial cells, B and T lymphocytes

[a] From Moncada et al. (1997)
[b] Examples only.

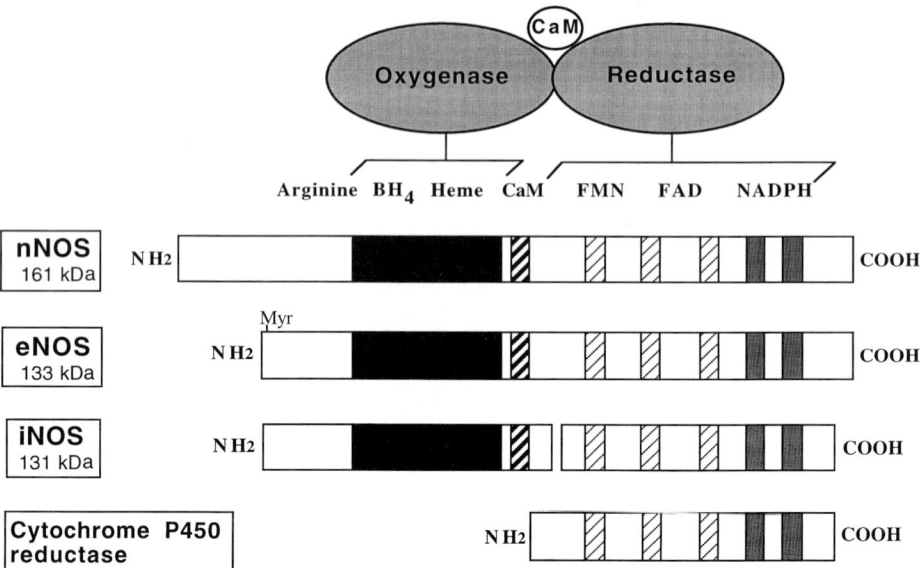

Fig. 1. Nitric oxide synthase domain organization. Nitric oxide synthase contains oxygenase and reductase domains. The oxygenase domain contains binding regions for heme, L-arginine, and tetrahydrobiopterin (BH$_4$) and is responsible for oxidation of the L-arginine substrate. The reductase domain has binding sites for NADPH and flavins (FMN and FAD). The CaM-binding domain is located between oxygenase and reductase domains. All three isoforms of NOS have a similar reductase domain structure to cytochrome P450 reductase. Endothelial NOS has a myristoylated glycine (Myr) at residue 2.

oxygenase domains of NOS; more discussion of this mechanism is included in Section IIID. It has been proposed (Abu-Soud and Stuehr, 1993) that the binding of Ca^{2+}/CaM leads to proper alignment of the two domains for efficient electron transfer and enzyme activity. This model would explain the nNOS and eNOS activation by increased intracellular Ca^{2+}, which would lead to CaM binding to nNOS and eNOS, and subsequent enzyme activity. The model would also be consistent with the insensitivity of iNOS to increases in intracellular Ca^{2+} concentrations, as CaM is tightly bound to iNOS and thus domains would already be properly aligned and the enzyme would be active.

6. Regulation of NO Synthase by Calmodulin

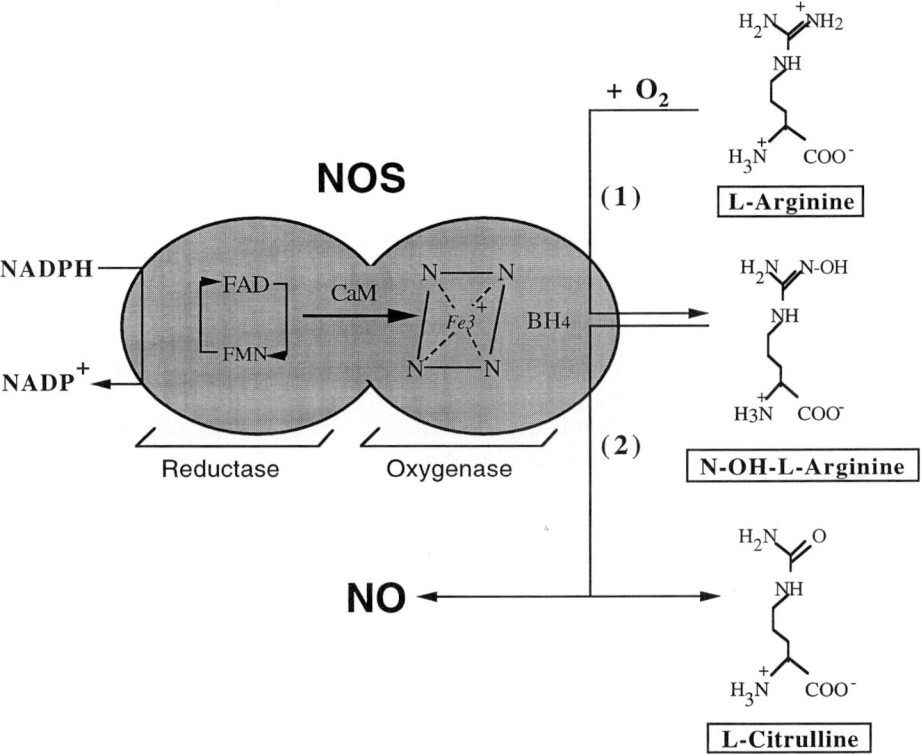

Fig. 2. Catalytic mechanism of NOS activity. The reaction catalyzed by NOS consists of two successive monooxygenase reactions: (1) oxidation of the guanidino nitrogen of the L-arginine substrate to produce a N-hydroxy-L-arginine (N-OH-L-arginine) intermediate, and (2) oxidation of N-OH-L-arginine to form L-citrulline and NO. Electrons donated by NADPH are transferred into the flavins (FAD and FMN), then passed to the heme iron, leading to oxidation of the substrate L-arginine. Calmodulin facilitates the flavin-to-heme electron transfer flow between reductase and oxygenase domains of NOS. Modified from Marletta (1993).

Although this model is attractive, it should be kept in mind that CaM may not actually exist "free-floating" in the cell, waiting to bind a target enzyme when the free intracellular Ca^{2+} levels increase. Calmodulin may already be bound to nNOS and eNOS under resting cell Ca^{2+} concentrations, but in a conformation that does not lead to enzyme activation. When the intracellular Ca^{2+} concentrations increase in response to a stimulus, the binding of Ca^{2+} to CaM may shift the equilibrium to allow CaM to assume the appropriate set of conformations required for stimulation of the NOS catalytic activity. This would be similar to other enzymes where CaM is a subunit of the enzyme (see the chapter by Lukas *et al.*). Calmodulin that is more tightly bound to the iNOS isoform may assume appropriate conformations that favor NOS catalytic activity even under resting cell Ca^{2+} concentrations. Thus, in the cell, CaM may be associated with its NOS target enzyme, and enzyme activation is triggered by increases in Ca^{2+} (for nNOS and eNOS) or the availability of L-arginine substrate (for iNOS).

An interesting finding is that nNOS from rat cerebellum is also able to use arginine-containing oligopeptides such as bradykinin as substrate (Chen and Rosazza, 1996) and can oxidize both N- and C-terminal arginines of bradykinin to citrullines. Different from using arginine as substrate, however, nNOS-catalyzed oxidation of bradykinin requires the presence of Ca^{2+} but appears to be CaM independent. For a more detailed discussion of the catalytic mechanism of NOS, the reader is referred to several reviews (Griffith and Stuehr, 1995; Marletta, 1993, 1994).

C. Purification and Biochemical Properties and Localization of Nitric Oxide Synthase

The appreciation that nNOS requires CaM for activity enabled the development of a purification strategy utilizing CaM-based affinity chromatography and the successful purification of a NOS isoform from cerebellum (Bredt and Snyder, 1990). Since then, a number of convenient purification protocols for the three NOS isoforms have been developed (for reviews, see Förstermann and Gath, 1996; Garvey *et al.,* 1996; Stuehr, 1996). A typical purification strategy utilizes solubilization and chromatography on 2′,5′-ADP agarose, followed by anion-exchange or CaM affinity chromatography. Specific protocol modifications are introduced based on the nature and tissue sources of the different isoforms. For example, natural sources of nNOS such as cerebellum express relatively low levels of the protein. To concentrate nNOS prior to chromatography, an ammonium sulfate precipitation step can be carried out on the soluble fraction of the tissue homogenate. In the case of eNOS, which is found mainly in the particulate fraction of endothelial cells, solubilization of the particulate fraction with a detergent such as CHAPS is required.

The NOS isoforms are widely distributed in many types of cells in cardiovascular, immune, and central nervous systems. Examples of the variety of tissues that contain NOS are shown in Table I. Different NOS isoforms can also be expressed in the same cell types. For example, both iNOS and eNOS are found in endothelial cells and astrocytes. All three NOS isoforms can be recovered from cells in both soluble and particulate fractions. However, nNOS and iNOS are found mainly in the soluble fraction of cells or tissue homogenates. The nNOS purified from cerebellum is a primarily cytosolic protein with a monomeric M_r of 150,000–160,000 M_r on sodium dodecyl sulfate–polyacrylamide gels (SDS-PAGE) and is a homodimer under native conditions (Bredt and Snyder, 1990; Schmidt et al., 1991; Mayer et al., 1990). The iNOS purified from murine macrophages is a cytosolic protein with M_r of 130,000 and is a dimer under native conditions (Hevel et al., 1991; Stuehr et al., 1991). However, the eNOS isolated from aortic endothelial cells (M_r of 135,000) is found largely associated with cell membranes (Pollock et al., 1991; Förstermann et al., 1991). The eNOS contains an N-terminal myristoylation site, which has been shown to be important for the membrane association of this isoform (Lamas et al., 1992; Sessa et al., 1992, 1993; Liu and Sessa, 1994; Busconi and Michel, 1993). Another study (Robinson et al., 1995) has suggested that palmitoylation may also contribute to the association of eNOS with membranes. As discussed further in Section III D, electrostatic interactions between negatively charged membrane phospholipids and basic residues in the eNOS CaM-binding site are also important for membrane association. Studies (Shaul et al., 1996; Feron et al., 1996) have demonstrated that eNOS is localized primarily to plasmalemmal caveolae, specialized domains of the plasma membrane where a number of signaling molecules are compartmentalized, and that this subcellular targeting requires acylation of the eNOS. In the caveolae, eNOS interacts with an integral membrane protein caveolin, and this interaction is blocked by Ca^{2+}/CaM binding to eNOS (Michel et al., 1997). Based on these data, Michel et al. (1997) have proposed a model of regulation of eNOS in endothelial cells wherein the inhibitory eNOS–caveolin complex is disrupted by CaM binding to eNOS, with a subsequent activation of the enzyme.

Although nNOS and iNOS are generally cytosolic and eNOS is primarily membrane associated, the situation is somewhat more complex. For example, nNOS can localize to plasma membrane and endoplasmic reticulum in cerebellum (Hiki *et al.*, 1992; Hecker *et al.*, 1994), nNOS can associate with the postsynaptic density (PSD) by interactions with PSD proteins through a PDZ domain in the amino-terminal region of nNOS (Brenman *et al.*, 1996), a large percentage of nNOS in skeletal muscle is membrane bound (Kobzik *et al.*, 1994; Nakane *et al.*, 1993; Brenman *et al.*, 1995), and iNOS in macrophages has been found associated with the particulate fraction (Förstermann *et al.*, 1992). The subcellular distribution of NOS can be regulated under physiological conditions, e.g., the eNOS can be translocated from membrane to cytosol upon phosphorylation by protein kinases (Matsubara *et al.*, 1996; Michel *et al.*, 1993). There also appear to be developmental changes in the subcellular localization of nNOS in cerebellum and cerebrum (Matsumoto *et al.*, 1993). For example, particulate nNOS activity in the cerebrum increased during the first week of development, then decreased, and became almost undetectable in adult rats. In contrast, the cytosolic nNOS in cerebellum showed low activity in newborn rats, then constantly increased until adulthood (Matsumoto *et al.*, 1993).

The detailed mechanisms by which nNOS and iNOS associate with membranes and the physiological significance of the subcellular localization of the NOS isoforms are just beginning to be elucidated. However, as will be discussed later in this chapter, the CaM-binding domain of NOS appears to be important in the membrane association, and the association with membranes can be regulated by both Ca^{2+}/CaM and phosphorylation.

D. Nitric Oxide Synthase cDNAs and Genes and Regulation of Expression

The first NOS cDNA was isolated from rat cerebellum (Bredt *et al.*, 1991) and contained an open reading frame encoding a protein of 160,458 kDa, in good agreement with protein purification studies. Since that time, cDNAs encoding the three NOS isoforms have been isolated from a number of species and tissues (for review, see Förstermann and Kleinert, 1995). The deduced amino acid sequences predict proteins of the appropriate molecular weight based on protein purification results (160–161 kDa for nNOS, 130–131 kDa for iNOS, and 133 kDa for eNOS). The similarity among the three human genes is 50–60%. There is a high degree of amino acid sequence conservation of each NOS

6. Regulation of NO Synthase by Calmodulin

isoform from different species (80–90%) and from different cell types of the same species (~99%). However, the overall sequence conservation between any two NOS isoforms is somewhat lower (50–60% identity). The regions of highest homology among the three NOS isoforms are associated with binding sites for substrate, cofactors, and CaM (see Fig. 1). For example, the C-terminal region of NOS shows homology to NADPH cytochrome P450 reductase, with similar consensus binding sites for NADPH, FAD, and FMN. A region in the N-terminal portion of the enzyme also shows higher sequence identity among the three NOS isoforms; this region contains the arginine-binding region and probably the BH_4 and heme-binding sites.

Genes for the three NOS isoforms have been isolated and their chromosomal localizations determined (Table II). Each NOS isoform appears to be encoded by a single gene. The human gene for nNOS is very large (>150 kb), the mRNA is encoded by 29 exons, and the gene is located on chromosome 12 in the 12q24.2–24.31 region (Hall et al., 1994; Xu et al., 1993). The human gene for iNOS is 37 kb, with 26 exons, and is located on chromosome 17 in the 17cen–q12 region (Chartrain et al., 1994; W. M. Xu et al., 1994; Marsden et al., 1994). The human gene for eNOS spans 21–22 kb, with 26 exons, and is located on chromosome 7 in the 7q35–7q36 region (Marsden et al., 1993; Nadaud et al., 1994; Robinson et al., 1994; W. M. Xu et al., 1994).

The expression of the three NOS isoforms has been found to be regulated at a number of different levels, including transcriptional, posttranscriptional and translational, and posttranslational controls. The posttranslational regulation of the enzyme will be discussed further in

TABLE II

Molecular Biology and Biochemistry of Human Nitric Oxide Synthase Isoforms

Characteristic[a]	nNOS	iNOS	eNOS
Chromosomal location	12q24.2–24.31	17cen–17q12	7q35–7q36
Gene size (kb)	160	37	21–22
Gene structure (exons)	29	26	26
Deduced molecular mass (kDa)	161	131	133
Protein length (amino acids)	1433	1153	1203

[a] See Förstermann and Kleinert (1995) for references.

Section IV. This section focuses on the regulation of expression at transcriptional and translational levels.

1. nNOS

An interesting feature of the nNOS gene is that a number of different NOS transcripts exist. This developmental stage- and tissue-dependent transcriptional heterogeneity arises from alternative promoter usage and from 5' structural diversity through the use of eight different first exons followed by splicing to a common second exon that contains the beginning of the coding region (Xie *et al.*, 1995; Fujisawa *et al.*, 1994). Thus, the same amino acid sequence should be produced, but this 5' transcriptional heterogeneity provides a potential mechanism for differential regulation of expression of nNOS in specific tissues or during specific developmental periods. There is also a report (Wang *et al.*, 1997) of an amino-terminal truncated nNOS expressed from a testis-specific mRNA transcript; this truncated nNOS had calcium-dependent catalytic activity comparable to the full-length nNOS.

Although nNOS is considered a constitutive NOS isoform, there have been some reports that nNOS expression can be regulated. For example, upregulation of nNOS mRNA and/or protein has been reported after estrogen treatment (Weiner *et al.*, 1994), after chronic salt loading (Kadowaki *et al.*, 1994), and following axotomy (Herdegen *et al.*, 1993), middle cerebral artery occlusion (Z. G. Zhang *et al.*, 1994), or ventral spinal root avulsion (Wu *et al.*, 1994).

2. iNOS

In resting, unstimulated cells, the iNOS expression is very low or absent. A large and growing number of agents induce iNOS expression, and this iNOS induction is generally regulated at the transcriptional level. The prototypical iNOS-stimulating agents in macrophages are LPS and cytokines such as interleukin-1 (IL-1), IFN-γ, and tumor necrosis factor-α (TNF-α). However, there is an increasing literature that a large number of different cell types are capable of expressing iNOS in response to a variety of stimuli and that stimuli that induce iNOS vary among species and cell types. In addition, a number of different compounds can reduce or block the ability of stimulating factors to induce iNOS expression or can themselves act as either inducers or inhibitors of iNOS expression, depending on the cell type. For a partial listing of agents that affect iNOS expression, see Nathan and Xie (1994a) and Förstermann and Kleinert (1995).

The molecular mechanisms by which compounds stimulate or inhibit iNOS expression are beginning to be defined through analysis of iNOS promoter regions of the gene. Consistent with the ability of a variety of diverse agents to induce expression, the iNOS promoter region has been found to be complex, with a number of potentially important regulatory regions. For example, the murine iNOS promoter contains a number of consensus-binding sites for known transcription factors, such as an IFN-γ response element site, nuclear factor-κB binding sites, nuclear factor IL-6-binding sites, and AP-1 sites (Xie *et al.*, 1993). The transcriptional regulation of the iNOS gene has been reported to vary among species and tissues (see Spitsin *et al.*, 1996; de Vera *et al.*, 1996; Förstermann and Kleinert, 1995; Kolyada *et al.*, 1996). In addition to transcriptional regulation of the iNOS gene, some stimulatory agents can act at the posttranscriptional level by stabilizing iNOS mRNA (Weisz *et al.*, 1994), and some inhibitory agents have been reported to decrease iNOS mRNA stability, reduce mRNA translation, or increase degradation of iNOS protein (Vodovotz *et al.*, 1993; Bogdan *et al.*, 1994; Cetkovic-Cvrlje *et al.*, 1993). A full discussion of iNOS expressional regulation is beyond the scope of this review, but the reader is referred to other reviews for more details (Nathan and Xie, 1994b; Förstermann and Kleinert, 1995).

3. eNOS

The eNOS promoter lacks a typical TATA box but contains proximal Sp1 and GATA motifs that may be utilized, as in some other constitutively expressed genes (Marsden *et al.*, 1993; Robinson *et al.*, 1994; Venema *et al.*, 1994; R. Zhang *et al.*, 1995). There are also a number of putative sites for regulation by AP-1, AP-2, nuclear factor-1, NF-κB, acute-phase reactants, shear stress, and sterols (for a review see Harrison *et al.*, 1996). Although eNOS is characterized as a constitutive isoform, a number of agents have been reported to affect its expression. Some examples are shear stress (Nishida *et al.*, 1992), estradiol (Förstermann and Kleinert, 1995), glucose (Suschek *et al.*, 1994), hypoxia (Mcquillan *et al.*, 1994), and TNF-α (Yoshizumi *et al.*, 1993). However, the molecular mechanisms by which most of these agents regulate eNOS expression have not been defined.

III. CALMODULIN AND NITRIC OXIDE SYNTHASE

A. Discovery of Nitric Oxide Synthase as a Calmodulin-Requiring Enzyme

The first demonstration that NOS is a CaM-requiring enzyme came from the purification of nNOS from rat cerebellum. Bredt and Snyder (1990) found that a crude supernatant fraction from cerebellar extracts exhibited NOS enzyme activity, as assessed by conversion of L-arginine into L-citrulline. Efforts to purify NOS by DEAE anion-exchange chromatography showed that the enzyme activity could be fully recovered on elution with 1 M NaCl. However, on elution with a gradient of NaCl concentration, NOS activity was not recovered from the DEAE column, suggesting that an important cofactor of the NOS was being lost during the purification protocol. Calmodulin was postulated as a possible cofactor because of the requirement of Ca^{2+} for NO formation. This postulation was supported by the demonstration that the addition of CaM to inactive DEAE eluate fractions restored the NOS activity, and the restored enzyme activity was CaM concentration dependent. The half-maximal stimulation (EC_{50}) for Ca^{2+} in stimulation of NOS was ~200 nM in the presence of 1 mM NADPH (Bredt and Snyder, 1990).

Similar findings of a CaM requirement for NOS activity were reported when assaying the NOS activity in neuroblastoma N1E-115 cells by the ability of these cells to activate guanylate cyclase, one of the major intracellular targets of NO, and form cGMP in fibroblast RFL-6 cells (Förstermann et al., 1990a). The principle of the assay is that NO generated from the N1E-115 cells on stimulation diffuses into neighboring target RFL-6 cells and leads to an activation of guanylate cyclase in target RFL-6 cells. Förstermann et al. (1990a) found that stimulation of intact N1E-115 cells with neurotensin released a NO-like factor that was capable of enhancing cGMP levels in fibroblast RFL-6 cells. The response was dependent on the presence of extracellular Ca^{2+}. When using a cytosolic extract of N1E-115 cells, the NOS activity was also highly sensitive to Ca^{2+} with the major increase in activity occurring between 100 and 500 nM Ca^{2+}. Calmodulin-binding drugs (calmidazolium, trifluoperazine, and fendiline) inhibited the enzyme activity (Förstermann et al., 1990a). Similar to the results of Bredt and Snyder (1990) for the rat cerebellar enzyme, Förstermann et al. (1990a) found that when cytosolic extracts from N1E-115 cells were eluted from DE52 anion-exchange

6. Regulation of NO Synthase by Calmodulin

columns in 0.1 M KCl, there was little if any NOS activity, as assayed by the stimulation of cGMP formation in RLF-6 cells. However, NOS activity could be restored if the eluate was assayed in the presence of CaM or if the cytosolic extract was eluted from the DE52 column with 0.3 M KCl, a salt concentration that elutes both NOS activity and CaM. Altogether, these data support the idea that nNOS is regulated by free Ca^{2+} and that its activity requires the presence of CaM. A number of subsequent studies have confirmed the requirement of Ca^{2+}/CaM for nNOS activity and convincingly demonstrated that nNOS can be activated by a number of agonists that lead to elevation of intracellular Ca^{2+} concentration and resultant activation by Ca^{2+}/CaM.

The eNOS isoform, even though largely associated with particulate rather than soluble fractions, has also been found to be a CaM-dependent enzyme. Purified eNOS from solubilized particulate fractions of bovine aortic endothelial cells (Pollock *et al.*, 1991) requires Ca^{2+} and CaM for its full enzyme activity, as determined by various activity assays, such as conversion of L-arginine into L-citrulline, stimulation of guanylate cyclase activity in RFL-6 cells, an increase in levels of nitrite/nitrate, and induction of relaxation of endothelium-denuded vascular smooth muscle strips (Pollock *et al.*, 1991). The eNOS activity can be inhibited by CaM-binding drugs such as calmidazolium (Busse and Mülsch, 1990). Thus, both nNOS and eNOS require Ca^{2+}/CaM, and intracellular Ca^{2+} levels can therefore regulate the constitutive NOS activity in cells containing these NOS isoforms. The absolute requirement of CaM for nNOS and eNOS activity explains crucial roles for Ca^{2+} in endothelium-dependent, NO-mediated smooth muscle relaxation and in neurotransmitter receptor-stimulated NO signaling pathways.

In contrast to the constitutive nNOS and eNOS isoforms, the iNOS isoform was originally thought to be independent of Ca^{2+}/CaM because the iNOS purified from rat macrophage and polymorphonuclear neutrophils (Yui *et al.*, 1991a,b) was insensitive to the addition of EGTA, CaM antagonists, or exogenous CaM. However, it was later shown (Cho *et al.*, 1992) that the iNOS isoform contains a CaM-binding domain and has CaM as a tightly bound subunit even in the absence of elevated intracellular calcium. The requirement of this bound CaM for iNOS activity was suggested by studies demonstrating that expression in *Escherichia coli* of recombinant iNOS encoding the mouse iNOS sequence required coexpression of CaM in order to generate an active form of the enzyme (Wu *et al.*, 1996). The CaM-replete iNOS was dimeric, contained normal quantities of heme, flavins, and tightly bound CaM, and exhibited high iNOS activity, whereas CaM-deficient iNOS was monomeric, devoid of flavins and heme, and had no iNOS activity. Similar results were reported for expression of human iNOS in *E. coli*, which also showed a CaM dependence for activity (Fossetta *et al.*, 1996).

B. Characterization of Calmodulin-Binding Sites on Nitric Oxide Synthase

As depicted in Fig. 1, all three NOS isoforms share a similar domain structure, with a C-terminal reductase domain containing binding sites for NADPH and flavins, and an N-terminal oxygenase domain containing binding sites for heme, BH_4, and the substrate L-arginine. Between the N- and C-terminal domains is a region rich in basic and hydrophobic amino acid residues, which has been found to contain the CaM-binding domain for each NOS isoform. The CaM-binding domains span 20–30 amino acids and have been identified by sequence analysis and synthetic peptide studies. Figure 3 illustrates the CaM-binding domains in the three NOS isoforms from human, mouse, rat, and bovine tissue. Although the CaM-binding sites of each NOS isoform have distinct amino acid sequences, the homology of this region between nNOS and eNOS is higher than that between iNOS and either of the constitutive isoforms.

```
hu-nNOS    730    KRRAIGFKKL  AEAVKFSAKL  MGQ           752
ra-nNOS    725    KRRAIGFKKL  AEAVKFSAKL  MGQ           747
mo-nNOS    725    KRRAIGFKKL  AEAVKFSAKL  MGQ           747

hu-eNOS    491            TRKKTFKEVA  NAVKISASLM        510
bo-eNOS    493            TRKKTFKEVA  NAVKISASLM        512
mo-eNOS    490            TRKKTFKEVA  NAVKISASLM        509

hu-iNOS    509    KRREIPLKVL  VKAVLFACML  MRKTMASRVR    538
ra-iNOS    506    RRREIRFTVL  VKAVFFASVL  MRKVMASRVR    535
mo-iNOS    503    RRREIRFRVL  VKVVFFASML  MRKVMASRVR    532
```

Fig. 3. Calmodulin-binding domains of NOS isoforms from different species. The numbers indicate the amino acid residues encompassing the CaM-binding domain in the respective NOS protein. Nonidentical residues among iNOS from different species are shown in bold face type. hu, human; ra, rat; mo, mouse; bo, bovine.

Characterization of the CaM-binding regions of NOS has been facilitated by utilizing synthetic peptides encompassing the putative CaM-binding domain, an approach similar to that originally used by Lukas *et al.* (1986) for the CaM-binding enzyme, myosin light-chain kinase (MLCK) from smooth muscle. The experimental approaches utilized and the conclusions reached concerning the CaM-binding region of each NOS isoform will be reviewed briefly. It should be noted that there is some disagreement in the literature over the exact amino acid residues that delimit the CaM-binding regions of the NOS isoforms. Nevertheless, there is general consensus on the core residues of the CaM-binding domain from each isozyme required for efficient CaM binding.

1. nNOS

In a study of CaM binding to nNOS, Zhang and Vogel (1994) found that a 23-residue synthetic peptide (KRRAIGFKKLAEAVKF-SAKLMGQ) corresponding to amino acids 725–747 of rat cerebellar nNOS formed a 1:1 stoichiometric complex with CaM in a Ca^{2+}-dependent manner. The binding of the nNOS peptide to CaM was determined by mixing the peptide with CaM at different ratios and analyzing the mobility shift of CaM on a urea–polyacrylamide gel. At a 1:2 molar ratio of the peptide to CaM, approximately half of the CaM formed a CaM–peptide complex, whereas the other half remained in the unbound form. At a 1:1 peptide:CaM ratio, however, all of the CaM appeared as a complex. No further change in the gel profile was seen when the ratio of the peptide to CaM exceeded 1, showing that the peptide binds to CaM in a 1:1 stoichiometry. ^1H nuclear magnetic resonance titration of CaM with the nNOS peptide also showed a 1:1 stoichiometry (Zhang and Vogel, 1994). The binding affinity of the nNOS peptide to CaM was high, with a relative dissociation constant (K_d) of ~2.2 nM, as obtained from a competition experiment where the nNOS peptide inhibited the ability of CaM to stimulate CaM-dependent phosphodiesterase (PDE) activity. A similar affinity constant (K_d 1.8 nM) was determined with a 30-residue nNOS peptide corresponding to amino acids 725–754 (Vorherr *et al.*, 1993). These K_d values determined with synthetic nNOS peptides are close to that determined for the binding of the intact nNOS enzyme to CaM (~3.5 nM) (Schmidt *et al.*, 1991), suggesting that amino acid residues in the 725–747 region probably represent the sole CaM-binding domain in the nNOS enzyme.

Nuclear magnetic resonance and circular dichroism analyses (Zhang and Vogel, 1994) revealed that the nNOS peptide can adopt a nascent α-helical conformation in aqueous solution and bind to CaM in an α-helical conformation. This α helix has the typical amphiphilic structure found in many other CaM-binding peptides. Studies using a specific nitroxide spin-labeled nNOS peptide derivative showed that the C-terminal part of the nNOS peptide binds to the N-terminal part of CaM (M. Zhang et al., 1995). This orientation of binding resembles complexes of CaM with the CaM-binding domains of MLCK (Ikura et al., 1992; Meador et al., 1992), CaM kinase II (Meador et al., 1993), and adenylate cyclase (Craescu et al., 1995). There are also similarities in the amino acid characteristics and spacing in the CaM-binding domain between nNOS and the smooth muscle and skeletal muscle MLCKs. As shown in Fig. 4, the CaM-binding domains of both enzymes contain two anchoring hydrophobic residues (Phe-731 and Leu-744 in nNOS) separated by 12 residues, a hydrophobic residue (Val-738 in nNOS) 7 residues after the first anchoring hydrophobic residue, and a basic residue (Lys-743 in nNOS) preceding the second anchoring hydrophobic residue (Meador et al., 1992; M. Zhang et al., 1995). The secondary structure of CaM is retained during binding of the nNOS and MLCK peptides (M. Zhang et al., 1994). Furthermore, binding of the nNOS peptide increases the affinity of CaM for metal ions and induces interdomain cooperativity in metal ion binding (M. Zhang et al., 1995). Such increases in metal ion binding affinity of CaM on binding of target peptides are also observed

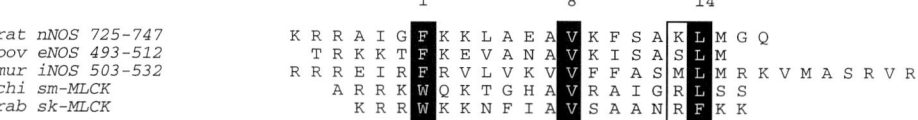

Fig. 4. Alignment of CaM-binding domains from several CaM-regulated enzymes. Proposed CaM-binding domains are from rat nNOS, bovine eNOS, murine iNOS, chicken smooth muscle MLCK (Lukas et al., 1986), and rabbit skeletal muscle MLCK (Blumenthal et al., 1985). Conserved hydrophobic residues, which are marked in black, have an important role in interaction with CaM. The hydrophobic residues at positions 1 and 14 have been proposed to be important in anchoring the peptide in a complex with CaM. The hydrophobic residue at position 8 is conserved among NOS and MLCK isozymes. A basic residue at position 13 (clear box) is conserved between MLCKs and nNOS, but is a Ser or Met in eNOS and iNOS, respectively.

2. eNOS

Examination of the amino acid sequence of eNOS for regions of appropriate length, charge, hydrophobicity, and α-helical potential to qualify as potential CaM-binding sites led to the early identification of a region around amino acid residues 489–519 as the CaM-binding domain. For example, the CaM-binding domain of bovine eNOS was identified as residues 493–512 (Lamas et al., 1992), residues 496–512 (Nishida et al., 1992), or residues 496–519 (Sessa et al., 1992), and the CaM-binding domain of human eNOS was identified as residues 489–512 (Marsden et al., 1992). A more recent study (Venema et al., 1996) with synthetic peptides and chimeric proteins supports the identification of residues 493–512 of bovine eNOS as the CaM-binding domain. A synthetic peptide corresponding to residues 493–512 of eNOS forms a 1:1 complex with CaM in a Ca^{2+}-dependent manner (Venema et al., 1996), and a mutant eNOS with a deletion of this region no longer exhibits CaM-binding activity (Venema et al., 1995). The dissociation constant of the eNOS peptide for CaM binding is 4 nM, as determined by the ability of the peptide to inhibit CaM-mediated activation of nNOS. Replacement of the CaM-binding region of iNOS with eNOS residues 493–512 led to a chimeric iNOS that was both Ca^{2+} and CaM dependent (Venema et al., 1996).

A common feature of the CaM-binding domains of many CaM-regulated enzymes is the presence of hydrophobic and basic residues that form an amphiphilic α helix with positively charged residues and hydrophobic residues on opposite sides of the helix. Figure 4 illustrates this principle for the CaM-binding domains of NOS and MLCK. The eNOS 493–512 region contains a number of hydrophobic and basic residues with the appropriate spacing. For example, there are hydrophobic residues Phe-498, Val-505, and Leu-511 whose relative positions are conserved in nNOS, iNOS, smooth muscle MLCK, and skeletal muscle MLCK, and basic residues at Arg-494, Lys-495, Lys-496, Lys-499, and Lys-506. Mutagenesis of these amino acid residues revealed that Phe-498, Leu-511, and Lys-499 in bovine eNOS are important for CaM binding and enzyme activation (Venema et al., 1996). Phe-498 appears to be a critical determinant of activation because mutation at this residue leads to a

150-fold decrease in the potency of the mutant peptide in inhibiting the CaM-mediated activation of nNOS.

3. iNOS

Based on the identification of the CaM-binding site in bovine eNOS as residues 493–512, the homologous region in iNOS was postulated to represent the CaM-binding domain. However, a synthetic peptide corresponding to those residues in the murine iNOS sequence (501–523) that are aligned with the CaM-binding domain of bovine eNOS (493–512) does not bind to CaM (Venema *et al.*, 1996). Longer iNOS peptides (residues 501–532, Venema *et al.*, 1996; residues 504–532 or 499–532, Anagli *et al.*, 1995; residues 503–532, Xie *et al.*, 1992) do form a 1:1 complex with CaM in the presence of Ca^{2+} (Venema *et al.*, 1996), suggesting that the additional residues from 524 to 532 in iNOS are necessary for high-affinity binding of CaM by iNOS. The K_d values of those longer iNOS peptides are ~1 nM (Anagli *et al.*, 1995; Venema *et al.*, 1996).

It should be noted that the native intact iNOS would be expected to have a much higher affinity for CaM binding because CaM can stay associated with iNOS even in the presence of chelators or denaturants (Cho *et al.*, 1992). Although CaM binds tightly to iNOS, the binding is reversible. The murine iNOS 503–532 peptide can displace endogenous CaM from iNOS enzyme to form a peptide:CaM complex, and the iNOS synthetic peptide can inhibit iNOS enzyme activity in a concentration-dependent manner (Stevens-Truss and Marletta, 1995). The maximal inhibition requires about a 12-fold excess concentration of peptide compared to iNOS, and the inhibition can be reversed by the addition of exogenous CaM.

Interestingly, even though the iNOS isoform is generally considered to exhibit Ca^{2+}-independent activity, the iNOS 501–532 peptide did not form a complex with CaM in the presence of EGTA (Venema *et al.*, 1996) and the iNOS 504–532 and 499–532 peptides showed much lower binding affinity in the presence of EGTA (Anagli *et al.*, 1995). These results indicate that CaM binding by the iNOS peptides was Ca^{2+} dependent. In addition, purified iNOS can be partially inactivated by EGTA (Venema *et al.*, 1996; Stevens-Truss and Marletta, 1995). The issue of the calcium dependence of iNOS binding to CaM has not been completely resolved. It is possible that the binding of iNOS to CaM is not entirely calcium independent but requires very low calcium concentrations. It is also possible that other regions of the iNOS protein besides residues

501–532 might contribute to CaM binding. This possibility has been supported by a study (Ruan *et al.*, 1996) utilizing chimeric NOS proteins in which iNOS CaM-binding region 503–532 was exchanged with the corresponding residues 725–754 of nNOS. Data showed that both the canonical CaM-binding site of mouse iNOS (residues 503–532) and additional residues within the region composed of amino acids 484–726 appear to be necessary for iNOS to bind CaM when free Ca^{2+} concentrations approach zero.

The questions of the reversibility and the calcium requirement for iNOS binding to CaM were also addressed in some detail by Venema *et al.* (1996). They created a series of eNOS/iNOS chimeric proteins where the CaM-binding sequence in eNOS was replaced by the iNOS CaM-binding sequence or the CaM-binding sequence in iNOS was replaced by the eNOS CaM-binding sequence. Replacement of CaM-binding region 501–523 in iNOS by residues 493–512 from eNOS resulted in a chimeric enzyme that was completely dependent on both Ca^{2+} and CaM for enzyme activity, suggesting that iNOS region 501–523 is necessary for the Ca^{2+}/CaM independence of iNOS. Replacement of CaM-binding sequence 493–512 in eNOS by residues 501–532 from iNOS resulted in a chimeric enzyme that was active in the absence of exogenous CaM. However, the enzyme activity required the presence of calcium. These data were interpreted to mean that bound CaM must be in the Ca^{2+}/CaM conformation in order to interact with eNOS in the appropriate way for activation.

Altogether, available data support the region around residues 503–532 of murine iNOS as being the primary CaM-binding domain (and additional residues in other regions of iNOS contributing to CaM binding when Ca^{2+} concentrations approach zero), with key hydrophobic and basic residues being critical to interaction with CaM.

C. Characterization of Nitric Oxide Synthase-Binding Sites on Calmodulin

As discussed in the preceding section, a number of studies have led to the identification of the CaM-binding domain for each of the three NOS isoforms. Much less information is available concerning the regions on CaM that are important for its interaction with NOS. No three-dimensional structure is available for CaM complexed with any NOS-

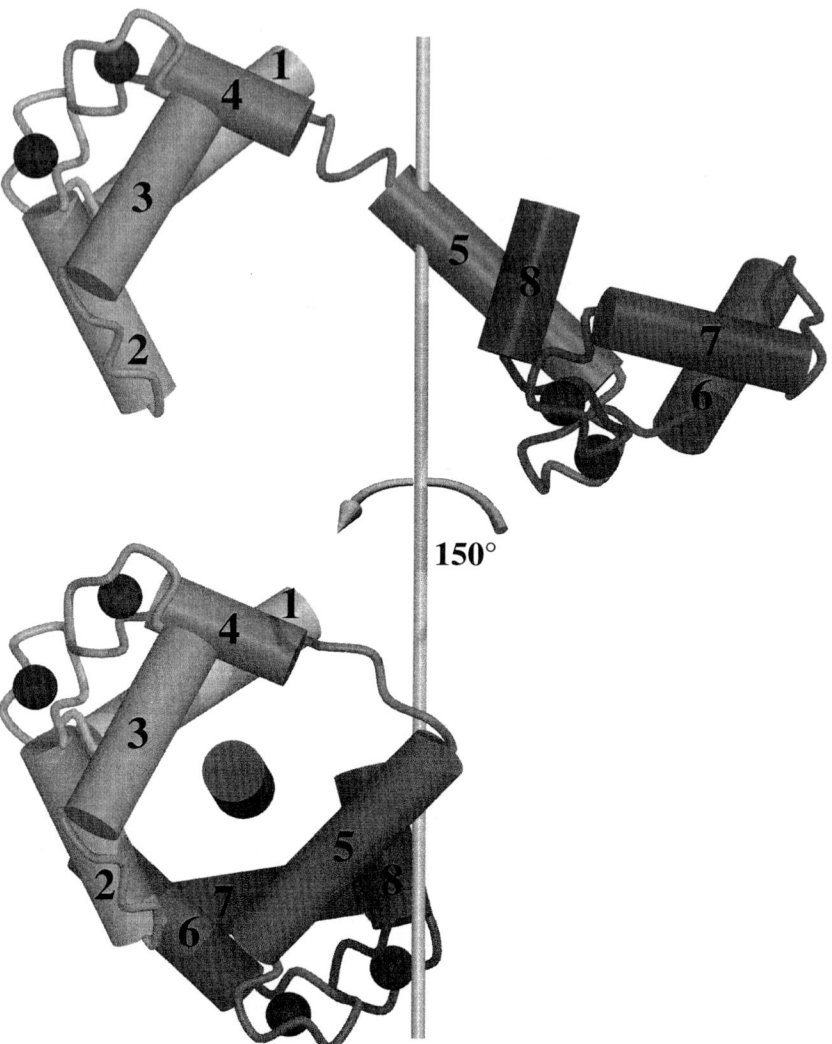

Fig. 5. Structural reorientation of CaM on binding a target peptide. A diagram of the three-dimensional structure of residues 6–146 of Ca^{2+}-saturated CaM (Chattopadhyaya *et al.*, 1992) is shown both in the absence of a target peptide (top) and complexed to the CaM-binding peptide from smooth muscle MLCK (bottom; Meador *et al.*, 1992). The four Ca^{2+} ions are depicted as black balls, and the eight helices in CaM are depicted as rods labeled 1–8 from the N-terminal to C-terminal domains of CaM. The N-terminal domains of each CaM are shown in a similar orientation. The most striking change in CaM when

6. Regulation of NO Synthase by Calmodulin

binding peptide. However, based on precedents with other CaM-binding proteins, the interaction of CaM with NOS is likely to involve both hydrophobic and electrostatic interactions. Because two methionine-rich hydrophobic surfaces of CaM have been postulated to be involved in the binding of CaM to the CaM-binding domains of target proteins (see O'Neil and DeGrado, 1990), the involvement of methionine residues of CaM in binding of the nNOS CaM-binding peptide was studied (M. Zhang et al., 1995). Nuclear magnetic resonance spectroscopic techniques utilizing [^{13}C]methylmethionine and selenomethionine-substituted CaM demonstrated that all eight methionine residues in the two hydrophobic surfaces, but not Met-76 in the central helix of CaM, undergo significant chemical shift changes on binding the nNOS 725–747 peptide. There is some evidence of conformational changes in the central helix region of CaM on binding the nNOS peptide, as pH titration experiments showed that Lys-75 exhibits a large increase in pK_a on peptide binding (M. Zhang et al., 1995).

In the three-dimensional structures of CaM complexed to the CaM-binding peptides from MLCK or CaM-dependent protein kinase II (CaMPKII) (Meador et al., 1992, 1993; Ikura et al., 1992), the two lobes of CaM wrap around and envelop the peptide in a hydrophobic channel diagonal to its long axis, and the top of the channel (referred to as the latch domain) is formed by a junction between CaM helices 2 and 6 in domains 1 and 3. The hydrophobic channel in the CaM : peptide complex is anchored at both ends by hydrophobic residues near each end of the peptide, with 8 (CaMPKII) or 12 (MLCK) amino acid residues separating the anchoring hydrophobic residues. The central helix of Ca^{2+}/CaM unravels and acts as an "expansion joint" to accommodate different relative positionings and sizes of CaM-binding peptides. Electrostatic interactions with basic residues at the NH$_2$-terminal end of CaM-binding peptides also appear to be important in the CaM recognition process. These features are illustrated in Fig. 5 in a diagram comparing the structure of Ca^{2+}/CaM in the absence of peptide (Chattopadhyaya et al.,

---

complexed to a target peptide can be described primarily as a rigid body rotation of 150° around the region near residue 80 in the central helix. The two lobes of CaM (each containing four helices) thus are wrapped around the MLCK peptide, which resides in a hydrophobic channel diagonal to the long axis of CaM. The latch region is formed by a close apposition of helices 2 and 6.

1992) or in a complex with the RS20 peptide analog of the CaM-binding domain from smooth muscle MLCK (Meador et al., 1992). A more comprehensive discussion of how the structure of CaM reorients to accommodate the presence of a target peptide can be found in the chapter by Lukas et al.

Based on available biochemical data, the interaction of NOS with CaM may involve a similar mechanism involving both hydrophobic and electrostatic interactions. A detailed study using a mutagenesis strategy and a series of chimeras between CaM and troponin C (a calcium-binding protein that does not activate nNOS) demonstrated the importance of the helix 2–helix 6 region of CaM in activation of nNOS (Su et al., 1995). Mutation of certain amino acids in this region of CaM (Thr-34, Ser-38, Ala-46, Val-108, Asn-111, and Leu-112) leads to an impairment of nNOS activation. All of these residues are in or near the latch domain of CaM (see Fig. 6), and all of these residues except Asn-111 are phylogenetically conserved in CaMs from *Dictyostelium* to vertebrates (see chapter by Lukas et al. for sequences). In addition, Val-108 and Leu-112 are interior hydrophobic residues that interact directly with CaM-binding peptides from smooth muscle MLCK (Val-108; Meador et al., 1992) or skeletal muscle MLCK (Val-108 and Leu-112; Ikura et al., 1992). Analysis of chimeric proteins where either N-or C-terminal domains of CaM are replaced with the corresponding domains of troponin C showed a loss of the ability of the chimeric proteins to activate nNOS (Su et al., 1995; George et al., 1993, 1996). Furthermore, studies using engineered CaMs where the two pairs of calcium-binding sites are duplicated or exchanged showed that the N-terminal pair (residues 9–75) are essential for activation of nNOS (Persechini et al., 1996a). The N-terminal CaM residues 1–8 do not appear to be involved in activation of nNOS because deletion of these residues from CaM does not affect nNOS activation (Persechini et al., 1996b). Detailed analysis of a number of CaM–troponin C chimeras (Su et al., 1995) demonstrated that domains 1 and 3 of CaM contain amino acid residues essential for nNOS activation. Domains 1 and 3 contain helices 2 and 6.

Thus, although no three-dimensional structure of a CaM:NOS complex is available, data to date suggest that the CaM-binding domain of NOS may interact with CaM in a manner similar to other CaM-binding peptides for which structural information is available. An example is shown in Fig. 6, where the structure of the complex between CaM and the CaM-binding peptide from smooth muscle MLCK is shown. If CaM binds to nNOS in this way, the CaM-binding domain of nNOS may

6. Regulation of NO Synthase by Calmodulin

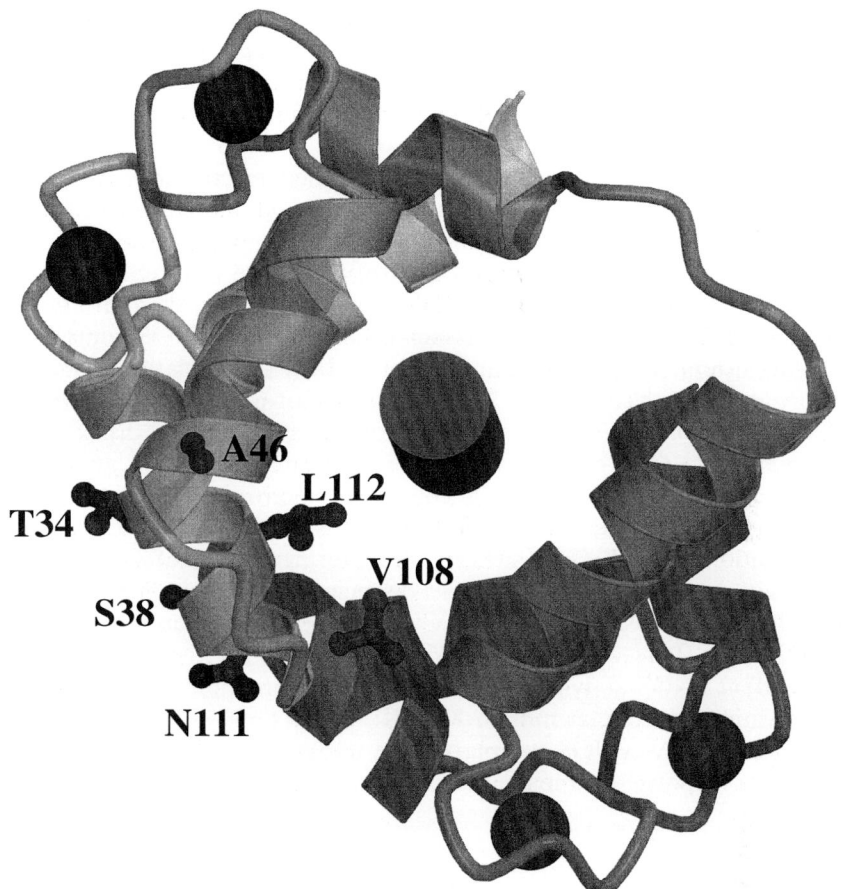

Fig. 6. Proposed importance of latch domain residues in CaM:nNOS interactions. A ribbon diagram of the three-dimensional structure of the CaM:smooth muscle MLCK peptide complex (Meador *et al.,* 1992) is shown. The CaM:MLCK peptide complex is shown in the same orientation as in Fig. 5. The ball-and-stick side chains indicate the location of six residues in CaM (Thr-34, Ser-38, Ala-46, Val-108, Asn-111, Leu-112) that have been postulated based on site-directed mutagenesis studies (Su *et al.,* 1995) to be important in CaM:nNOS interactions. Note that these six residues are near the latch domain in the CaM:peptide complex.

reside in a hydrophobic channel formed by the bending of the lobes of CaM around the nNOS-binding site, with important hydrophobic and electrostatic contact points that stabiize the complex. The validity of this idea will, of course, remain to be confirmed by more extensive structural studies of CaM:NOS peptide complexes.

Although little is known about the detailed mechanisms by which Ca^{2+} dissociates from the complex of CaM:nNOS and leads to enzyme inactivation, Persechini *et al.* (1996c) have shown that the rates of Ca^{2+} dissociation from CaM complexed with nNOS are different from the rates of Ca^{2+} dissociation from CaM complexed with skeletal muscle MLCK. The kinetics of Ca^{2+} dissociation from CaM:nNOS complexes are consistent with a model in which rapid Ca^{2+} dissociation from the two Ca^{2+}-binding sites of the N-terminal lobe of CaM is responsible for the inactivation of nNOS enzyme activity, and a slower Ca^{2+} dissociation from the two Ca^{2+}-binding sites of the C-terminal lobe is coupled to dissociation of the CaM:nNOS complex. Inactivation of CaM:nNOS catalytic activity occurs at least 10-fold faster than the rate of Ca^{2+} dissociation from the two sites in the C-terminal lobe of CaM in the complex. In contrast, three slowly dissociating Ca^{2+}-binding sites are coupled to both dissociation of the CaM:MLCK complex and the enzyme inactivation (Persechini *et al.,* 1996c). Therefore, inactivation of CaM-dependent nNOS activity on removal of Ca^{2+} occurs much faster than inactivation of skeletal muscle MLCK. These rapid Ca^{2+} dissociation kinetics for CaM : nNOS complexes may explain the rapid regulation of nNOS in response to transient changes in intracellular Ca^{2+} concentrations.

D. Significance of Calmodulin Binding to Nitric Oxide Synthase

1. Regulation of NOS Activity by Ca^{2+}/CaM

Even though the three NOS isoforms are structurally related, the mechanisms by which they are regulated are distinct. Figure 7 is a simplified schematic illustration comparing the major regulatory mechanisms for constitutive and inducible NOS isoforms. Levels of intracellular free Ca^{2+} are a crucial switching point in the regulation of constitutive NOS (nNOS and eNOS) activity. Studies using cell extracts or purified enzyme show that activation of nNOS by Ca^{2+} occurs in the nanomolar range

6. Regulation of NO Synthase by Calmodulin

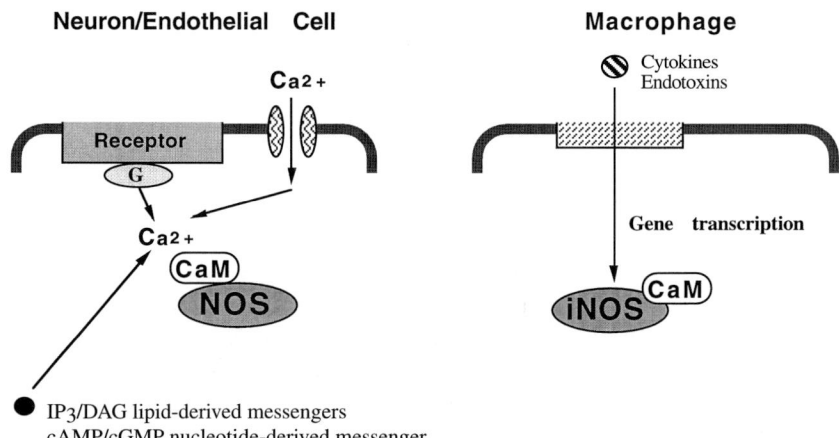

Fig. 7. Modes of regulation of NOS isoforms. The constitutive NOS isoforms nNOS and eNOS present in neurons and endothelial cells are highly Ca^{2+}/CaM dependent. Stimuli that lead to increases in intracellular Ca^{2+} levels can trigger Ca^{2+}/CaM activation of nNOS and eNOS enzyme activity. In contrast, there is little if any iNOS activity in macrophages under normal circumstances. However, cytokines and endotoxins released during certain pathological conditions can stimulate iNOS activity by stimulating transcription of the iNOS gene. The enzyme activity of iNOS is Ca^{2+} independent because of the very tightly bound CaM associated with iNOS.

of Ca^{2+} concentrations (EC_{50} is ~100–400 nM). Any stimuli that are linked to appropriate elevations in intracellular-free Ca^{2+} levels can potentially stimulate the enzyme because the elevated Ca^{2+} levels promote CaM activation of NOS (for a review see Hu and El-Fakahany, 1996). A good example is the muscarinic receptor-coupled activation of nNOS and generation of NO. Stimulation of muscarinic receptors in neurons activates phospholipase C (PLC), which increases the rate of phosphoinositide hydrolysis. A major product of this signaling pathway is inositol trisphosphate (IP_3), which leads to Ca^{2+} release from the endoplasmic reticulum (Berridge, 1993). An excellent correlation exists among the time course of muscarinic receptor-mediated formation of IP_3, elevation of intracellular Ca^{2+} concentration, and activation of nNOS (Surichamorn *et al.*, 1990). There is also good correlation in the rank order of the relative efficacy of muscarinic receptor agonists at increasing phosphoinositide hydrolysis and activating nNOS (Hu and El-Fakahany, 1993). Another example is with ion channel-coupled receptors, where

NO has been linked to the stimulation of glutamate receptors, especially the NMDA subtype. Stimulation of NMDA receptors results in an elevation of intracellular Ca^{2+} levels through both Ca^{2+} influx and Ca^{2+} release from internal stores and subsequent activation of nNOS. Many intracellular Ca^{2+} elevating reagents also exhibit an ability to activate NOS, e.g., Ca^{2+} ionophore A23187, caffeine, and thapsigargin (X. Xu et al., 1994).

As expected based on their dependence on CaM, nNOS and eNOS can be inhibited by a variety of CaM antagonists (Bredt and Snyder, 1990; Hu et al., 1994). The potencies of a range of CaM-binding drugs in inhibition of NOS activity are similar to those in inhibition of CaM-dependent phosphodiesterase activity (Hu et al., 1994). The inhibition of NOS by CaM antagonists can be competitively reversed by the addition of exogenous CaM (Bredt and Snyder, 1990). In addition, other CaM-binding proteins, such as neuromodulin and neurogranin, also regulate NOS activity through competition for CaM binding (Slemmon and Martzen, 1994; Martzen and Slemmon, 1995) and can be used as NOS inhibitors.

Because the constitutive NOS isoforms are poised for rapid activation in response to elevations in intracellular Ca^{2+} concentration and subsequent enzyme activation by CaM, nNOS and eNOS play a role in a variety of physiological functions. For example, nNOS activity in the central nervous system (CNS) and periphery is important in the regulation of cerebral blood flow, modulation of pain, control of long-term potentiation (LTP) and memory, and in neurogenic vasodilation and regulation of smooth muscle tone in the gastrointestinal, respiratory, and urogenital tracts (for a review, see Moncada et al., 1997). Nitric oxide produced in vascular endothelial cells on activation of eNOS is important in the regulation of blood pressure, relaxation of vascular smooth muscle, and inhibition of platelet adhesion and aggregation. These physiological functions of NO generated by nNOS and eNOS occur primarily by the activation of soluble guanylate cyclase in target cells with a resultant increase in cGMP and initiation of downstream signaling events in the cell. Some of these physiological roles are discussed in more detail in Section V of this chapter.

In contrast to the rapid and transient activation of the constitutive NOS isoforms in response to various stimuli, there is little if any iNOS activity under normal conditions. However, NO can be generated in high amounts by iNOS during pathophysiological situations, such as host defense against infection, inflammatory responses, and disease. For example, stimulation of iNOS takes place after a few hours of exposure

6. Regulation of NO Synthase by Calmodulin

to stimuli such as bacterial endotoxins, inflammatory cytokines such as IL-1β or TNF-α, or some glial-derived proteins such as S100β (Hu et al., 1996), leading to sustained high levels of NO production over days. The continuous occupancy by CaM of iNOS explains the steady-state catalysis and high-output NO by this isoform. As depicted schematically in Fig. 7, the regulation of iNOS occurs most generally at the level of gene transcription and mRNA stability (see Förstermann and Kleinert, 1995). There has been increasing attention to the roles of iNOS in the brain because activated microglia and astrocytes express high iNOS activity. The generation of NO in the brain may have significant implications in brain pathophysiology (Moncada and Higgs, 1993; Hu and Van Eldik, 1996). The reader is referred to reviews of iNOS for a further discussion of iNOS regulation and potential significance to pathophysiology (Förstermann and Kleinert, 1995; Gross and Wolin, 1995; Xie and Nathan, 1994).

2. Calmodulin Stimulation of Electron Transfer in NOS

A well-established role of CaM is to facilitate the electron transfer process in NOS. Electrons provided by NADPH are transferred in a linear sequence from the reductase domain to the oxygenase domain, first entering the NOS flavins and passing across the domains to the heme iron. Reduction of the heme iron enables it to bind oxygen and catalyze NO synthesis in the presence of L-arginine substrate. Calmodulin activates both the electron transfer from NADPH into flavins and the interdomain electron transfer between the flavins and heme (Abu-Soud and Stuehr, 1993, Abu-Soud et al., 1994). For example, although electrons from NADPH can load into the flavins of nNOS in the absence of CaM, the flavin-to-heme electron transfer requires CaM. The CaM-stimulated electron transfer process is independent of NOS substrate binding and results in rapid NADPH oxidation in the presence of O_2. Calmodulin has been shown not to affect the affinity of the enzyme for its L-arginine substrate or the ability of the NOS heme iron to bind cyanide or CO ligands (Matsuoka et al., 1994). However, CaM has been found to stimulate four catalytic activities of nNOS: NO synthesis, NADPH oxidation, cytochrome c reduction, and ferricyanide reduction. Studies of apo-nNOS devoid of its bound heme and BH_4 (and thus not able to participate in interdomain electron transfer) showed that cytochrome c and ferricyanide reductions occurred through a mechanism not involving flavin-to-heme electron transfer, i.e., the apo-nNOS appeared to contain a fully functional reductase domain. This finding is consistent with observations of independent NOS reductase and oxygenase domains following proteolytic cleavage (Sheta et al., 1994) and with iNOS exhibiting a functional reductase domain in the absence of bound heme or BH_4 (Baek et al., 1993). Further analysis demonstrated that CaM increases the rate of electron transfer from NADPH into the flavins by a factor of 20 (Abu-Soud et al., 1994), indicating a direct CaM activation of the nNOS reductase domain. In contrast to the regulation of intradomain electron transfer, CaM-dependent stimulation of NOS synthesis and substrate-independent NADPH oxidation appear to require the interdomain transfer of electrons from flavins to heme because these reactions were not activated in apo-nNOS and were inhibited in intact NOS by agents that prevent heme iron reduction (Abu-Soud et al., 1994).

Thus, CaM appears to stimulate two points in the electron transfer process: the rate of electron transfer into flavins and the interdomain electron transfer from flavins to heme. Activation at the first point leads to enhanced catalysis by the reductase domain, such as cytochrome c and ferricyanide reduction. Activation at the second point leads to reduction of the NOS heme iron and initiation of NO synthesis in the presence of L-arginine substrate or reduction of O_2 to form superoxide (O_2^-) in the absence of substrate. A model consistent with this dual mechanism of control of nNOS by CaM is shown in Fig. 8.

In contrast to the mechanisms by which CaM controls electron transfer in nNOS (and probably eNOS), there are some differences in the electron transfer process in iNOS with its tightly bound CaM subunit. All of the NOS isoforms exhibit comparable specific activities for NO synthesis in the presence of L-arginine substrate. However, unlike the CaM stimulation of electron transfer from NADPH to flavins in nNOS, Rafferty and Malech (1996) showed that electron transfer within the reductase domain of iNOS is not regulated by CaM. They found that the cytochrome c reductase activity of a recombinant iNOS reductase domain that lacks the CaM-binding site was the same as that of intact iNOS or a tryptic reductase domain fragment containing the CaM-binding site, suggesting that CaM binding does not play a role in mediating electron transfer within the reductase domain of iNOS. In addition, the basal rate of NADPH oxidation by iNOS is 16-fold greater than that of CaM-free nNOS, suggesting that the tightly bound CaM on iNOS facilitates the transfer of NADPH-derived electrons to its heme. Ca^{2+}/CaM binding to nNOS dramatically increases NADPH oxidation in the presence or absence of L-arginine substrate, whereas the maximal rate of uncoupled

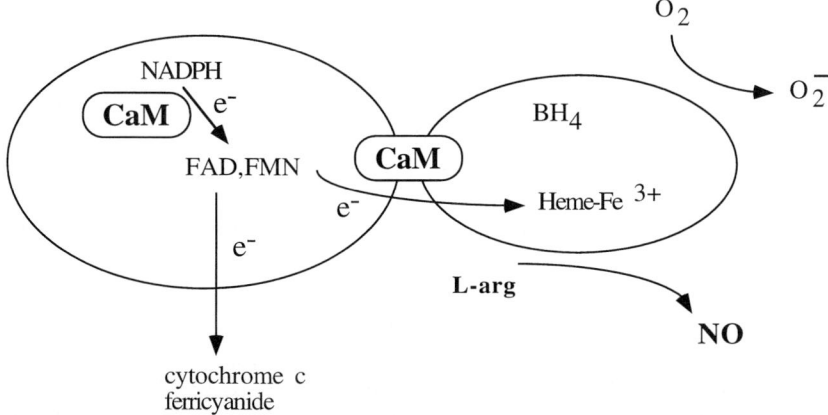

Fig. 8. Model of dual control of NOS by CaM. The reductase domain (left oval) and the oxygenase domain (right oval) of nNOS are depicted. Calmodulin binding to nNOS stimulates two points in the electron transfer process: the rate of electron transfer from NADPH into the flavins (FAD, FMN) and the interdomain electron transfer from flavins to heme. Activation at the first point leads to enhanced catalysis by the reductase domain, such as cytochrome c and ferricyanide reduction. Activation at the second point leads to NO synthesis in the presence of L-arginine substrate or reduction of O_2 to form superoxide (O_2^-) in the absence of substrate. Modified from Abu-Soud et al. (1994).

NADPH oxidation in the absence of L-arginine for iNOS is only one-sixth of that obtained with CaM-bound, L-arginine-free nNOS. On substrate binding, NADPH oxidation by iNOS increases 6-fold, whereas NADPH oxidation by nNOS decreases by 37% (Abu-Soud and Stuehr, 1993). Thus, for nNOS, CaM binding alone triggers a high rate of electron transfer to the heme and allows rapid NO production in response to Ca^{2+}, but can also result in the generation of superoxide in the absence of substrate. For iNOS, the presence of tightly bound CaM likely results in a continuous but suboptimal level of heme iron reduction (and a slow level of uncoupled NADPH oxidation in the absence of L-arginine), which then increases on substrate binding.

Although the reasons for these differences between the constitutive NOS isoforms (nNOS and eNOS) and the inducible iNOS are not known, Griffith and Stuehr (1995) have speculated on why the enzymes have evolved these different mechanisms of control. Under physiological conditions, cells probably maintain sufficient L-arginine levels for nNOS and eNOS to synthesize NO in response to stimuli that increase intracellular Ca^{2+}. Only under certain pathologic conditions of prolonged Ca^{2+} influx would the nNOS or eNOS deplete the L-arginine substrate and generate superoxide. In contrast, iNOS appears capable of continuous heme reduction because of its tight association with the CaM subunit, but this electron transfer is slowed unless the L-arginine substrate is available. This ability to limit superoxide generation in the absence of substrate is an advantage for iNOS because it can function at sites where L-arginine may become limiting, such as in wounds (Albina *et al.*, 1990).

3. Calmodulin Regulation of NOS Subcellular Localization

Another role of CaM in NOS activation is to promote the translocation of eNOS from a membrane-bound form to the cytosolic active form. As mentioned previously, unlike nNOS and iNOS, which are mainly cytosolic, eNOS is largely associated with plasma membranes. Although the N-terminal myristoyl group in eNOS plays a role in the membrane association, studies have shown that the CaM-binding domain of eNOS is also involved in membrane association (Venema *et al.*, 1995; Matsubara *et al.*, 1996). The eNOS is capable of binding to membrane phospholipids through electrostatic interactions between anionic membrane phospholipids and basic residues in the CaM-binding domain of eNOS. Binding of eNOS to phospholipids leads to an inhibition of the catalytic activity of the enzyme. Deletional mutation of the CaM-binding domain of eNOS results in a loss of ability of eNOS to bind to phospholipids and converts eNOS from a membrane to a cytosolic protein when the enzyme is expressed in baculovirus/Sf9 insect cells (Venema *et al.*, 1995). Phosphorylation of the CaM-binding domain of eNOS also results in the translocation of eNOS from membrane to cytosol (Matsubara *et al.*, 1996). A direct binding study using a synthetic eNOS peptide (residues 492–511) encompassing the CaM-binding domain showed that the eNOS peptide binds to CaM and various membrane acidic phospholipids, but not to neutral phospholipids (Matsubara *et al.*, 1996). The addition of $Ca^{2+}/$CaM (but not apo-CaM) to the eNOS peptide:membrane complex released the peptide from the membranes, suggesting that CaM can affect the eNOS–membrane interaction in a calcium-dependent manner. These data indicate that the CaM-binding domain in the eNOS is important for enzyme membrane association and that membrane association can function to inhibit eNOS catalytic activity by interfering with the interaction of eNOS with CaM. A corollary arising from these data is that the subcellular localization and translocation of eNOS from membrane-associated to cytosolic on Ca^{2+}/CaM binding releases the enzyme from the inhibitory effects of phospholipids and leads to enzyme activation.

4. Calmodulin Regulation of NOS Dimerization

All three NOS isoforms are active as homodimers (for references, see Hellermann and Solomonson, 1997; Venema *et al.*, 1997). Thus, regulation of NOS dimer formation can provide an additional mechanism of regulation of NOS enzyme activity. The domains of NOS that are involved in dimerization and the influence of cofactors and prosthetic groups on dimer formation vary for the individual NOS isoforms. For example, iNOS subunit association occurs through the oxygenase domain, whereas nNOS and eNOS dimer formation involves both oxygenase and reductase domains (Venema *et al.*, 1997). Another example is that dimerization of nNOS and iNOS requires BH_4 and heme, whereas eNOS dimerization is not influenced by BH_4 (Venema *et al.*, 1997). The possible involvement of Ca^{2+}/CaM in the modulation of dimer formation has been examined. Calmodulin was found to promote dimerization of the oxygenase domain of human eNOS (Hellermann and Solomonson, 1997), and the authors suggest that CaM-induced dimerization may be a primary event in the activation of eNOS by Ca^{2+}/CaM. However, Venema *et al.* (1997) concluded that CaM is not required for eNOS dimerization. Thus, the question of the role of CaM, if any, in eNOS oligomerization is still unresolved.

IV. REGULATION OF NITRIC OXIDE SYNTHASE ACTIVITY BY PHOSPHORYLATION/DEPHOSPHORYLATION

A. Regulation of Nitric Oxide Synthase by Ca^{2+}/Calmodulin-Dependent Kinases and Phosphatases

Nitric oxide synthase is subject to phosphorylation by Ca^{2+}/CaM-dependent protein kinase II *in vitro*, resulting in a rapid loss of NOS enzyme activity (Nakane *et al.*, 1991; Bredt *et al.*, 1992). The phosphorylation occurs at serine residues on nNOS (Bredt *et al.*, 1992). Nitric oxide synthase is also a substrate for the Ca^{2+}/CaM-activated phosphatase calcineurin (Steiner et al., 1992; Dawson et al., 1993). Activation of calcineurin dephosphorylates NOS and leads to an increase in NOS catalytic activity (Dawson et al., 1993). Accordingly, calcineurin inhibitors, such as FK506 and cyclosporin A, are able to prevent the calcineurin-induced dephosphorylation of NOS, retaining NOS in a phosphorylated inactive form and thereby preventing increases in NOS activity (Dawson et al., 1993; Fast et al., 1993). This mechanism has been implicated in the ability of FK506 to protect cells against glutamate-induced neurotoxicity (Dawson et al., 1993).

B. Regulation of Nitric Oxide Synthase by Protein Kinase C

Controversy exists regarding the role of protein kinase C (PKC) in regulation of NOS activity. Bredt et al. (1992) found that phosphorylation of nNOS following activation of PKC in intact cells leads to a dramatic decrease in NOS catalytic activity. Phosphoamino acid analysis revealed that phosphorylation of the purified nNOS by PKC occurs on serine residues of nNOS (Bredt et al., 1992). In contrast, Nakane et al. (1991) observed that there is increased NOS activity on phosphorylation by PKC. Brüne and Lapetina (1991) reported that PKC does not phosphorylate NOS. The reason for these apparently contradictory results is not clear, but it is possible that there might be different phosphorylation sites under various experimental conditions or variable stability of the NOS due to different purification protocols (Okada, 1995). However, by employing pharmacological tools, many studies have suggested that activation of PKC is involved in receptor-stimulated nNOS activity. For example, in mouse striatal primary neuronal cultures, application of a PKC inhibitor staurosporine or desensitization of PKC by long-term exposure to a phorbol ester selectively reduces NMDA-stimulated NO production, suggesting that there is a specific PKC-mediated activation of NOS on stimulation of NMDA receptors. The activation of NOS by NMDA may involve two mechanisms: an increase in intracellular Ca^{2+} levels and an activation of PKC (Marin et al., 1992). It has been reported that on stimulation of glutamate receptors, PKC activation can be longer lasting than the intracellular Ca^{2+} rise (Martinson et al., 1990; Nishizuka, 1992). Okada (1995) has shown that PKC modulates the Ca^{2+} sensitivity of NOS in rat cerebellar slices and therefore postulates that PKC might serve as a NOS activator after the receptor-stimulated transient increase in intracellular Ca^{2+} concentration has subsided.

Phosphorylation of human eNOS by PKC leads to the membrane-bound enzyme being translocated into the cytoplasm. Phosphorylation of the CaM-binding domain peptide from human eNOS reduced its binding to membrane phospholipids, and one of the PKC phosphorylation sites was identified as Thr-495 in the CaM-binding domain (Matsubara et al., 1996). The CaM-binding domains of nNOS and iNOS were also shown to bind to membrane phospholipids, and CaM could affect the interactions of the NOS peptides with the membranes, depending on calcium concentration (Matsubara et al., 1996). These findings are similar to another CaM-binding PKC substrate, the myristoylated alanine-rich protein kinase C substrate (MARCKS) protein, which is a major cellular substrate of PKC in many cells and contains a phosphorylation site domain composed of 20–25 amino acids rich in basic residues (Graff et al., 1989). Calmodulin is able to bind to the domain and reverse the association of MARCKS with cytoplasmic membranes (Kim et al., 1994). Phosphorylation of the CaM-binding domain of MARCKs also abolishes the phospholipid-binding activity of MARCKs and results in translocation of the protein from membrane to cytosol. Based on this analogy, it is possible that phosphorylation of eNOS by PKC is a mechanism by which eNOS interaction with membrane phospholipids is regulated.

C. Regulation of Nitric Oxide Synthase by cAMP/cGMP-Dependent Kinases

The influence of cyclic nucleotide-dependent kinases on NOS activity has been evaluated by employing in vitro phosphorylation, immunoprecipitation, and enzyme catalytic activity analyses. Nitric oxide synthase can be phosphorylated stoichiometrically by both cAMP- and cGMP-dependent kinases, and the same serine on NOS is phosphorylated by both kinases. This phosphorylation site differs, however, from the sites phosphorylated by PKC and Ca^{2+}/CaM-dependent kinases (Brüne and Lapetina, 1991; Bredt et al., 1992; Dinerman et al., 1994). Immunoprecipitation of phosphorylated NOS in the human kidney 293 cell line transfected with nNOS reveals that treatment of cells with the cGMP analog 8-bromo-cGMP stimulates NOS phosphorylation almost fivefold (Dinerman et al., 1994). Moreover, the phosphorylation of NOS results in a decrease in enzyme catalytic activity as assayed both with purified NOS and in intact cells stimulated with Ca^{2+} ionophore A23187 (Bredt et al., 1992; Dinerman et al., 1994). The cyclic nucleotide-dependent phosphorylation may serve as a potential feedback mechanism of NOS regulation. For example, activation of NOS generates NO, which triggers the activation of guanylate cyclase and elevation in intracellular cyclic GMP levels. The increased cyclic GMP stimulates cGMP-dependent kinase, leading to phosphorylation of NOS and a decrease in NOS activity.

D. Regulation of Nitric Oxide Synthase by Tyrosine Kinases

To date, there has been no direct evidence that tyrosine kinases phosphorylate constitutive NOS isoforms. However, a potential involvement of tyrosine kinases in the regulation of NOS has been suggested by studies using tyrosine kinase inhibitors. In mouse striatal neuronal cultures, for example, activation of nNOS by ionomycin, NMDA, and kainate is blocked by the tyrosine kinase inhibitors lavendustin A and genistein, suggesting that a tyrosine phosphorylation step is needed to obtain full nNOS activity (Rodriguez et al., 1994). In the case of iNOS, many studies have shown that inhibitors of tyrosine kinases, but not inhibitors of PKC or cAMP-dependent kinase, block the ability of LPS and cytokines to stimulate iNOS activity and mRNA levels in a variety of cell types, including human articular chondrocytes, macrophages, cardiac myocytes, endothelial cells, and glial cells (Feinstein et al., 1994; Geng et al., 1995; Paul et al., 1995; Pan et al., 1996; Singh et al., 1996). Moreover, in cultured murine macrophages treated with LPS and IFN-γ, iNOS was shown to be phosphorylated on tyrosine residues by Western blot analysis using antiphosphotyrosine antibodies, and the phosphorylation is an early event coinciding with the appearance of newly synthesized iNOS (Pan et al., 1996). The tyrosine phosphatase inhibitor vanadate increases the levels of iNOS tyrosine phosphorylation, whereas the serine/threonine phosphatase inhibitor okadaic acid or the PKC activator PMA does not. Vanadate treatment of cells also results in a rapid (within 5 min) increase in iNOS enzyme activity (Pan et al., 1996). Data suggest that tyrosine kinases and phosphatases are involved in the posttranslational regulation of iNOS. The tyrosine kinases and phosphatases involved in iNOS regulation have not been identified. However, Singh et al. (1996) suggest that induction of iNOS by IL-1β and IFN-γ in cardiac myocytes and microvascular endothelial cells may require activation of ERK1/ERK2 mitogen-activated protein kinases (MAPKs).

E. Autophosphorylation

Nitric oxide synthase is also subject to autophosphorylation. Watanabe et al. (1996) reported that incubation of purified rat nNOS in the presence of Mg^{2+} and ATP results in phosphorylation of the nNOS. The phosphorylation seems to be an intramolecular autophosphorylation reaction because the ratio of phosphorylation is independent of the nNOS concentration. Rat nNOS contains a nucleotide-binding consensus GQGAGS in its amino acid sequence (residues 166–171), and deletional mutation of the nNOS amino terminus containing the GQGAGS sequence results in loss of phosphorylation (Watanabe et al., 1996). Although the site of autophosphorylation on nNOS has not been defined, eNOS and iNOS do not contain this nucleotide-binding sequence and do not exhibit autophosphorylation ability. The significance of the autophosphorylation is unclear, however, as no studies have been reported concerning functional effects, if any, on enzyme activity or whether autophosphorylation occurs in vivo.

F. Phosphorylation as a Control Point/Cross-Talk among Signaling Pathways

The regulation of NOS by phosphorylation/dephosphorylation is particularly intriguing as a control point for NOS activity because it could represent an intracellular counterbalance mechanism, as depicted in Fig. 9. For instance, Ca^{2+}/CaM can directly activate NOS activity on the one hand and, on the other hand, stimulate Ca^{2+}/CaM-dependent protein kinases, resulting in phosphorylation of NOS and diminishing of the enzyme activity. Ca^{2+}/CaM can also trigger calcineurin activity, leading to dephosphorylation of phosphorylated NOS and an increase in NOS catalytic activity. Phosphorylation of iNOS by tyrosine kinases, but not by PKC, further suggests that constitutive and inducible NOS isoforms

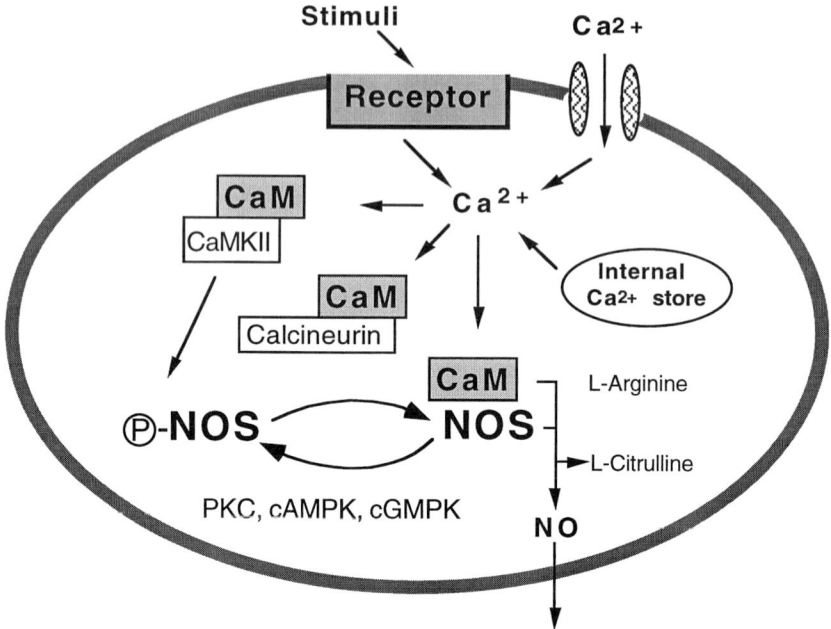

Fig. 9. Phosphorylation as a control point of NOS activity. Stimuli that lead to increases in intracellular Ca^{2+} levels promote CaM activation of constitutive NOS isoforms (nNOS and eNOS). Nitric oxide synthase converts L-arginine into L-citrulline and generates NO. Calmodulin also activates CaM-dependent kinase II (CaMKII), which phosphorylates NOS, resulting in a reduction of NOS activity. Calmodulin also activates calcineurin, which dephosphorylates NOS, resulting in an increase in NOS catalytic activity. Protein kinase C and cyclic nucleotide-dependent kinases also phosphorylate NOS, leading to decreases in NOS activity.

are not only regulated differently by Ca^{2+}/CaM, but are also controlled differentially by downstream intracellular signaling mechanisms.

Another level of complexity in the regulation of NOS is that the intracellular messengers generated by various stimuli can be amplified, selectively cross-talk with other signaling pathways, and subsequently modify NOS activity. For example, it has been reported (Reiser, 1992; Toms and Roberts, 1994) that NO formation induced by Ca^{2+} release from internal Ca^{2+} stores or by stimulation of NMDA receptors can be enhanced by cAMP. Increases in intracellular Ca^{2+} levels can also activate adenylate cyclase and synthesis of cAMP. Multiple intracellular cGMP targets can interact with other second messenger systems, e.g., cGMP-gated ion channels that regulate cation flow, a cGMP-regulated cAMP phosphodiesterase that controls cAMP and Ca^{2+} levels, and cGMP cross-activation of cAMP-dependent protein kinases (for review, see Schmidt et al., 1993). An additional level of complexity in the regulation of NOS is that the enzyme is subject to inhibition by its end product, NO. This may occur through the ability of NO to bind to the iron of the heme cofactor on the enzyme (see Griffith and Stuehr, 1995). Whether this regulatory mechanism plays an important role *in vivo* has not been clearly elucidated (for further discussion, see Hu and El-Fakahany, 1996).

V. PHYSIOLOGICAL AND PATHOPHYSIOLOGICAL FUNCTIONS OF NITRIC OXIDE

It should be evident from the previous discussion that the regulation of NOS is complex and exquisitely modulated at a variety of levels to allow for a coordinated and controlled generation of NO. Nitric oxide generated on activation of NOS functions as a signaling molecule to mediate a diverse range of biological activities. In this section and as illustrated in Fig. 10, the pivotal role of NO on physiology and pathology is summarized by selected examples. The section concludes by discussing the pharmacological implications of NO function.

A. Biochemical Reactivity of Nitric Oxide

The nature of NO, being a free radical, makes it an unstable and highly chemically reactive molecule. Nitric oxide undergoes a number of chemical reactions, many of which occur under physiological conditions (for review, see Ignarro, 1990). Some examples of these chemical reactions are

1. Reactivity with oxygen to yield nitrogen dioxide gas or nitrite in solution: $2NO + O_2 \rightarrow 2NO_2^-$

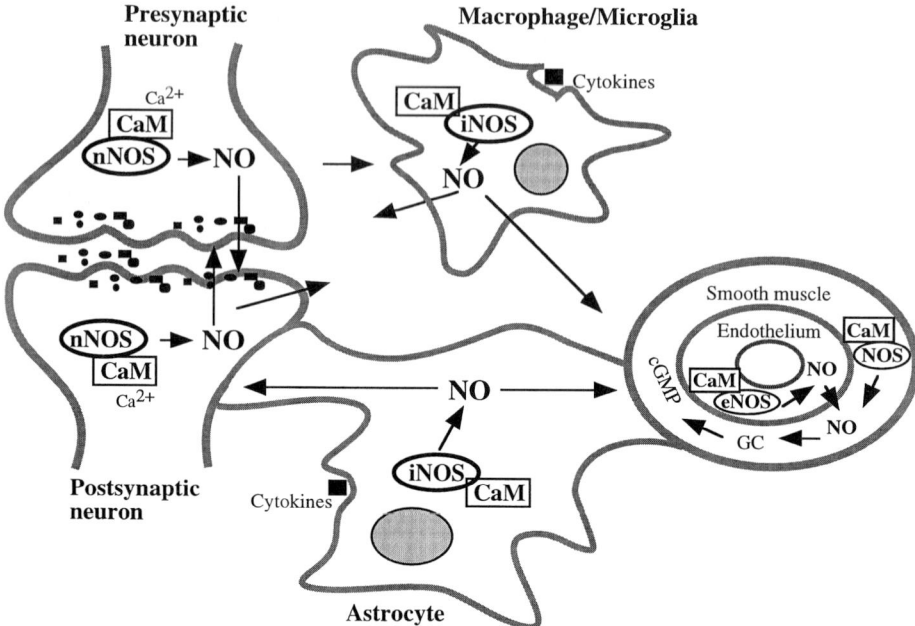

Fig. 10. Nitric oxide signaling in the nervous system. In response to stimuli that increase intracellular Ca^{2+} concentrations, CaM activation of nNOS or eNOS can lead to generation of NO. In neurons, NO can be produced in either presynaptic or postsynaptic neurons, can act as a neurotransmitter, or can modulate the release of other neurotransmitters. Nitric oxide produced by endothelial cells can lead to the relaxation of neighboring smooth muscle cells, through stimulation of guanylate cyclase (GC) and generation of cGMP and downstream signaling events. In response to a variety of stimuli, including cytokines, there is a stimulation of iNOS production and subsequent generation of NO. Nitric oxide can act on multiple cell types, including neurons, astrocytes, macrophages/microglia, endothelial, and smooth muscle cells, to mediate a diverse array of biological activities.

2. Reactivity with superoxide anion to yield the unstable intermediate peroxynitrite anion, which rearranges to form nitrate: $NO + O_2^- \rightarrow ONOO^- \rightarrow NO_3^-$
3. Reactivity with oxyhemoglobin to yield methemoglobin and nitrate: $HbO_2 + NO \rightarrow metHb + NO_3^-$
4. Reactivity with thiols (R-SH) to yield S-nitrosothiols (R-SNO): $R\text{-}SH + NO \rightarrow R\text{-}SNO + H^+$. An example is the reaction between cysteine and NO to yield S-nitrosocysteine.

5. Reactivity with heme iron to yield nitrosyl-heme adducts.
6. Other nitrosation reactions.

As is evident from these examples, NO has the ability to undergo a number of chemical reactions and exist in a variety of redox states. It must be kept in mind, therefore, that the biological actions of NO depend on the chemical reactions it undergoes, particularly its reactions with another free radical such as superoxide anion (O_2^-), a transition metal such as heme iron, or O_2. In addition, these NO adducts and reaction products are also capable of further reactions to produce products that may have biological activity (for review, see Gross and Wolin, 1995; Stamler et al., 1992). Thus, the overall biological consequences of NO generation depend on a complex interplay of variables, including the chemical reactions of NO itself and the variety of redox forms in which NO can exist, the availability of reactive oxygen species and antioxidant defenses, the chemical environment in which NO is being produced, and the array of NO target molecules present.

Because of the chemical nature of NO, NO interacts with a variety of intracellular target proteins (for example, see Zhang and Snyder, 1995). The NO action occurs at cell membranes, cytosolic compartments, and nuclear locations. For example, NO can interact with membrane NMDA receptors by nitrosylating the thiol groups of the receptor, resulting in a decrease in receptor activity and ion-channel sensitivity (Lipton et al., 1993). Nitric oxide interacts with a number of intracellular proteins and with protein kinases, leading to activation of downstream biochemical events. In cytosol, a good example of NO interaction is the activation of soluble guanylate cyclase, a heme-containing protein, activation of which leads to formation of cyclic GMP. This mechanistic pathway is thought to be involved in a variety of NO-mediated responses, such as smooth muscle relaxation, synaptic transmission, and regulation of blood pressure. Nitric oxide also interacts with hemoglobin and myoglobin, reactions which form the basis for several assays for NO formation. Furthermore, NO is able to nitrosylate iron–sulfur proteins such as NADH:ubiquinone oxidoreductase, NADH:succinate oxidoreductase, and cis-aconitase, a mechanism that has been implicated in the cytotoxicity of macrophage-derived NO. Nitric oxide also binds to the nonheme iron of ribonucleotide reductase to inhibit DNA synthesis (Lepoivre et al., 1990) and binds to the iron in ferritin, an iron storage protein, liberating the iron that can cause lipid peroxidation. Nitric oxide also interacts with some DNA-binding proteins such as AP-1 and NF-κB through interaction with their cysteine residues. Additionally, the highly oxidative species peroxynitrite and hydroxyl radicals generated through reaction of NO with superoxide anion can induce cell membrane damage. Moreover, NOS itself is a heme-containing protein, interaction of which with its product NO results in reduction of the enzyme activity. This is implicated in feedback regulation of NOS (for review, see Stamler, 1994; Hu and El-Fakahany, 1996).

B. Roles of Nitric Oxide in Central Nervous System

Nitric oxide function in CNS was first demonstrated by Garthwaite *et al.* (1988), who showed that cerebellar granule cells produce NO in response to stimulation of their NMDA subtype of glutamate receptors. Nitric oxide can be produced presynaptically, to modulate neurotransmitter release, and postsynaptically in response to stimulation of neurotransmitter receptors. Nitric oxide produced diffuses out to act on neighboring cellular elements, such as neighboring neurons, astrocytes, and endothelial cells (see Fig. 10). There is extensive literature on the plethora of NO roles in the nervous system. Only some selected examples of NO function are provided here; the reader is referred to other reviews for more in-depth discussion (Zhang and Snyder, 1995; Garthwaite and Boulton, 1995; Vincent, 1994; Murphy and Grzybicki, 1996; Rand and Li, 1995).

1. Synaptic Plasticity

The NO production in response to stimulation of NMDA receptors is particularly intriguing because NMDA receptors are importantly involved in neuronal plasticity, including roles in neurite outgrowth and synaptic transmission. The most well-studied models of synaptic plasticity are long-term potentiation (LTP) and long-term depression (LTD), which refer to a persisting experience-dependent synaptic enhancement (LTP) or decline (LTD) that typically follows either the brief high-frequency stimulation of certain synapses or the pairing of postsynaptic depolarization with lower frequency synaptic activation (Bliss and Collingridge, 1993; Zorumski and Izumi, 1993). The LTP and LTD are considered to represent the cellular correlates of learning and memory. It has been shown that NO is involved in modulation of NMDA receptor-dependent LTP and LTD. For example, postsynaptic intracellular injection of nitroarginine or extracellular application of hemoglobin, a cell-impermeant NO scavenger, completely blocks LTP in the CA1 region of the hippocampus, an effect reversed by L-arginine, whereas NO or NO donor compounds produce a long-lasting enhancement in synaptic efficacy (O'Dell et al., 1991; Schuman and Madison, 1991). Blockade of LTD by nitroarginine and hemoglobin is also observed in cerebellum where the stimulation of climbing and parallel fibers evokes LTD of parallel fiber–Purkinje cell transmission (Shibuki and Okada, 1991). This system has been implicated as the cellular mechanism for cerebellar motor learning (Ito, 1989). Taken together, the observations suggest that NO can act as a retrograde messenger, modulating pre- and postsynaptic activities. For example, stimulus-evoked glutamate release from certain presynaptic neurons stimulates postsynaptic NMDA receptors, eliciting a Ca^{2+} increase in the postsynaptic neurons and Ca^{2+}/CaM activation of their nNOS. Nitric oxide produced can diffuse from the postsynaptic neurons across the synaptic cleft to act in a retrograde fashion to affect the presynaptic neurons. This could be a mechanism of self-regulation of NOS activity at synapses (for discussion, see Hu and El-Fakahany, 1996).

2. Neurotransmitter Release

Nitric oxide generated after various stimuli not only exerts its effects on postsynaptic neurons, but also diffuses across the synaptic cleft and causes release of neurotransmitters at presynaptic nerve terminals. Nitric oxide has been invoked as important in NMDA receptor-mediated neurotransmitter release in several brain regions *in vitro* (see Montague et al., 1994; Jones et al., 1995). It has been shown through the use of *in vivo* microdialysis sampling techniques that NO modulates release of a variety of neurotransmitters, including aspartate, glutamate, GABA, dopamine, taurine, acetylcholine, and serotonin (Guevara-Guzman et al., 1994). Nitric oxide also regulates the release of neuropeptides and hormones (Rettori et al., 1992; Karanth et al., 1993; Garry et al., 1994). The primary mechanism by which NO regulates neurotransmitters and neuropeptides release is thought to be through activation of cGMP-dependent pathways in presynaptic neurons.

3. Neurodestruction

Another major role of NO in the CNS is to mediate glutamate-induced neurotoxicity. The cell death associated with cerebral ischemia is thought to involve glutamatergic stimulation via NMDA receptors. Excessive stimulation of NMDA receptors induces prolonged increases in intracellular Ca^{2+} levels, resulting in the release of large amounts of NO. NMDA receptor antagonists such as MK-801 and NOS inhibitors such as nitroarginine and the NO scavenger hemoglobin block the neurotoxicity elicited by NMDA and related excitatory amino acids (Dawson et al., 1991).

4. Neuroprotection

In addition to mediating glutamate-induced neurotoxicity under some circumstances, the situation is more complex as NO has also been found to be able to play a neuroprotective role in NMDA-induced neurotoxicity. For example, in rat cortical neurons, only NOS-positive neurons are able to inhibit the glutamate-mediated increase in intracellular Ca^{2+} levels, whereas in the presence of a NOS inhibitor these neurons lose their ability to regulate Ca^{2+} responses (Ikeda et al., 1993). A similar finding is seen in striatal neurons where the NMDA-mediated Ca^{2+} responses are slowly but strongly diminished after the incubation of cells with L-arginine (Manzoni and Bockaert, 1993). Data suggest that activation of NOS in neurons plays an essential role in feedback regulation of the glutamate receptors. Application of NO donor compounds into cortical neuronal cultures protects against NMDA neurotoxicity through inhibition of the NMDA-enhanced Ca^{2+} increase (Lei et al., 1992). Nitric oxide is thought to interact with sulfhydryl groups on the redox modulatory site of the NMDA receptor–channel complex and lead to S-nitrosylation of the NMDA receptor, producing a persistent inhibition of the NMDA-evoked responses. For example, sodium nitroprusside (SNP), a nitroso compound with strong reductive NO^+ character, can react with the thiol of the NMDA receptors. The S-nitrosylation of the NMDA receptors decreases excessive Ca^{2+} influx and prevents NMDA receptor-mediated neurotoxicity. Thus, SNP can serve as a neuroprotective agent in this aspect (Lipton et al., 1993). Taken together, NO may mediate both neurodestruction and neuroprotection, depending on its redox status, with NO free radical (NO) being neurodestructive and nitrosonium (NO^+) being neuroprotective (for review, see Lipton and Stamler, 1994).

Nitric oxide is also produced by glial cells in CNS. Activation of iNOS in response to neuronal injury or inflammatory stimuli results in the generation of large amounts of NO, which can diffuse into neighboring neurons, eliciting neuroprotective or neurodestructive responses, and into smooth muscle cells, regulating cerebral blood flow. Microglial cells and astrocytes have been implicated in many pathological states such as gliosis, ischemia, and the pathogenesis of diseases such as multiple sclerosis, Alzheimer's disease, Parkinson's disease, and the dementia of acquired immunodeficiency syndrome (for review, see Moncada and Higgs, 1993; Gross and Wolin, 1995).

C. Roles of Nitric Oxide in Cardiovascular System

The identification of NO as an EDRF typifies its role in the cardiovascular system. Activation of eNOS in endothelial cells on stimulation of neurotransmitter receptors, such as acetylcholine and bradykinin, causes vasodilatory responses. Endogenous NO-dependent vasodilation is essential for the regulation of blood flow and blood pressure. Nitric oxide is also implicated in various vascular reflexes. For instance, systemic vasodilation evoked by hypoxia involves endogenous NO formation and is prevented by nitroarginine. Nitric oxide also regulates the vascular system through inhibition of platelet aggregation and adhesion. The molecular mechanism of NO action is mainly through activation of guanylate cyclase and generation of cyclic GMP (for review, see Bredt and Snyder, 1994; Umans and Levi, 1995).

However, activation of iNOS in both endothelial and vascular smooth muscle cells in response to inflammatory stimuli contributes to inflammation-induced vasodilation. The amount of NO produced under these conditions is far greater than that produced by neurotransmitter receptor agonist-triggered eNOS activation. For example, in endotoxin shock in animals, the increase in the generation of NO is directly related to the degree of hypotension. Such vascular relaxation is resistant to vasoconstrictors and can be prevented by treatment with glucocorticoids and NOS inhibitors. Furthermore, endotoxin also induces iNOS activity in venous smooth muscle, and enhanced NO synthesis in the myocardium and endocardium may account for the venous pooling and cardiac dysfunction associated with endotoxemia (for review, see Moncada and Higgs, 1993).

Dysregulation of NO production has been implicated in the pathogenesis of cardiovascular disease. Impaired NO-related, endothelium-dependent vasorelaxation has been observed in essential hypertension, the endothelial dysfunction of hypercholesterolemia, and pulmonary arterial hypertension of chronic obstructive lung disease (Moncada and Higgs, 1993). In fact, nitrovasodilators have been used clinically for about 100 years and are still widely used in the treatment of cardiovascular diseases.

D. Roles of Nitric Oxide in Immune and Inflammatory Systems

The major source of NO in the immune system is from activated macrophages, although NO in inflammatory reactions can also originate from blood vessels, neutrophils, and lymphocytes on activation of iNOS (for review, see Moncada and Higgs, 1993). Activation of iNOS in macrophages contributes to the cytotoxic activities of macrophages against bacteria or other microorganisms. The cytotoxic activity of NO can result from its inhibition of enzymes involved in mitochondrial respiration and DNA synthesis. Nitric oxide toxicity may also result from its ability to interact with oxygen-derived species such as superoxide anion to form toxic substances such as peroxynitrite. These substances can be cytotoxic through disruption of cell membranes by peroxidation of membrane lipids, dysfunction of mitochondrial respiration by inhibition of cytochrome oxidase and electron transport, and damage of DNA by base deamination. Numerous studies have been published concerning the effects of NO generated from iNOS on immune system function and dysfunction. The reader is referred to reviews for further discussions of iNOS regulation and potential significance to pathophysiology (Förstermann and Kleinert, 1995; Gross and Wolin, 1995; Xie and Nathan, 1994).

E. Pharmacological Implications

The first inhibitor of NOS to be identified was N^{ω}-monomethyl-L-arginine (L-NMMA), an analog of L-arginine (Hibbs *et al.*, 1987). Since then, over 100 different NOS inhibitors have been developed (for review and references, see Moore and Handy, 1997; Fukuto and Chaudhuri, 1995). The complexity of the NOS enzyme with its multiple binding domains (e.g., for L-arginine substrate, CaM, NADPH, flavins, heme, and BH_4) has resulted in the identification of inhibitors that target each of these domains. However, although some of the known inhibitors show a modest degree of selectivity for one or another NOS isoform, most of the available inhibitors affect both inducible and constitutive NOS isoforms. This can be a problem for the future development of clinically effective NOS-based drugs because of the possibility of unwanted side effects. For example, nNOS inhibitors might be useful in treating cerebral ischemia or postoperative pain, but might lead to unwanted increases in blood pressure if they also inhibit eNOS and endothelial NO formation. Inhibitors of iNOS might be used therapeutically for septic shock or chronic inflammatory diseases, but might also result in impaired memory acquisition, gastrointestinal disturbances, or impotence if nNOS is also inhibited in a chronic manner. Much effort is currently being directed toward the development of inhibitors with better selectivity toward specific NOS isoforms, for potential therapeutic uses. Development of such inhibitors at the very least will provide pharmacological reagents to investigate the precise roles of each NOS isoform in various biological processes. This fundamental knowledge should also further define therapeutic opportunities for intervention through inhibition of specific NO pathways.

Another clinical application that is emerging is the use of NO donors, such as organic nitrates and nitrites, inorganic nitroso compounds, and S-nitrosothiols (see Moncada et al., 1997). These compounds are generally used as vasodilators, but are also being examined for their utility in the treatment of thrombotic disorders, injured arteries, malfunctions of the gastrointestinal and respiratory tracts, and diabetic impotence (see Moncada and Higgs, 1993).

VI. CONCLUSION

Based on a search of Medline databases, almost 15,000 papers have been published related to NO since its identification in 1987 as a biological messenger molecule. It is obviously impossible in one review to adequately do justice to the entire field. This chapter has focused primarily on NOS in terms of its binding to CaM and on the consequences of CaM activation of the enzyme. Although much is known about the interaction of CaM and NOS, there are still several important questions about mechanism and regulation that need to be addressed. What is the three-dimensional structure of the NOS:CaM complex, i.e., how do the two proteins interact at the molecular level? How similar structurally is the interaction of CaM with iNOS vs nNOS and eNOS? What is the mechanistic importance of the different mechanisms of CaM interaction with the constitutive vs inducible NOS isoforms? How are CaM:NOS interactions and regulation altered in pathophysiological conditions? It is hoped that the answers to these questions about this intriguing CaM-regulated enzyme system will emerge in the near future.

ACKNOWLEDGMENTS

Research in the authors' laboratory relevant to nitric oxide was supported by NIH Grant AG13939 and Illinois Department of Health Grant 83880358. We thank Steve Weigand for advice and preparation of Figs. 5 and 6.

REFERENCES

Abu-Soud, H. M., and Stuehr, D. J. (1993). Nitric oxide synthases reveal a role for calmodulin in controlling electron transfer. *J. Biol. Chem.* **269**, 32047–30250.

Abu-Soud, H. M., Yoho, L., and Stuehr, D. J. (1994). Calmodulin controls neuronal nitric-oxide synthase by a dual mechanism. *Proc. Nat. Acad. Sci. U.S.A.* **90**, 10769–10772.

Albina, J. E., Mills, C. D., Henry, W. L., Jr., and Caldwell, M. D. (1990). Temporal expression of different pathways of L-arginine metabolism in healing wounds. *J. Immunol.* **144**, 3877–3880.

Anagli, J., Hofmann, F., Quadroni, M., Vorherr, T., and Carafoli, E. (1995). The calmodulin-binding domain of the inducible (macrophage) nitric oxide synthase. *Eur. J. Biochem.* **233**, 701–708.

Arnold, W. P., Mittal, C. K., Katsuki, S., and Murad, F. (1977). Nitric oxide activates guanylate cyclase and increases guanosine 3',5'-cyclic monophosphate levels in various tissue preparations. *Proc. Natl. Acad. Sci. U.S.A.* **74**, 3203–3207.

Baek, K. J., Thiel, B. A., Lucas, S., and Stuehr, D. J. (1993). Macrophage nitric oxide synthase subunits. *J. Biol. Chem.* **268**, 21120–21129.

Berridge, M. J. (1993). Inositol trisphosphate and calcium signalling. *Nature (London)* **361**, 315–325.

Bliss, T. V. P. and Collingridge, G. L. (1993). A synaptic model of memory: Long-term potentiation in the hippocampus. *Nature (London)* **361**, 31–39.

Blumenthal, D. K., Takio, K., Edelman, A. M., Charbonneau, H., Titani, K., Walsh, K. A., and Krebs, E. G. (1985). Identification of the calmodulin-binding domain of skeletal muscle myosin light chain kinase. *Proc. Natl. Acad. Sci. U.S.A.* **82**, 3187–3191.

Bogdan, C., Vodovotz, Y., Paik, J., Xie, Q. W., and Nathan, C. (1994). Mechanism of suppression of nitric oxide synthase expression by interleukin-4 in primary mouse macrophages. *J. Leukocyte Biol.* **55**, 227–233.

Bredt, D. S., and Snyder, S. H. (1990). Isolation of nitric oxide synthetase, a calmodulin-requiring enzyme. *Proc. Natl. Acad. Sci. U.S.A.* **87,** 682–685.
Bredt, D. S., and Snyder, S. H. (1994). Nitric oxide: A physiologic messenger molecule. *Annu. Rev. Biochem.* **63,** 175–195.
Bredt, D. S., Hwang, P. M., Glatt, C., Lowenstein, C., Reed, R. R., and Snyder, S. H. (1991). Cloned and expressed nitric oxide synthase structurally resembles cytochrome P-450 reductase. *Nature (London)* **351,** 714–718.
Bredt, D. S., Ferris, C. D., and Snyder, S. H. (1992). Nitric oxide synthase regulatory sites: Phosphorylation by cyclic AMP-dependent protein kinase C, and calcium-calmodulin protein kinase; identification of flavin and calmodulin binding sites. *J. Biol. Chem.* **267,** 10976–10981.
Brenman, J. E., Chao, D. S., Xia, H., Aldape, K., and Bredt, D. S. (1995). Nitric oxide synthase complexed with dystrophin and absent from skeletal muscle sarcolemma in Duchenne muscular dystrophy. *Cell (Cambridge, Mass.)* **82,** 743–752.
Brenman, J. E., Chao, D. S., Gee, S. H., McGee, A. W., Craven, S. E., Santillano, D. R., Wu, Z., Wang, F., Xia, H., Peters, M. F., Froehner, S. C., and Bredt, D. S. (1996). Interaction of nitric oxide synthase with the postsynaptic density protein PSD-95 and α-1 syntrophin mediated by PDZ motifs. *Cell (Cambridge, Mass.)* **84,** 757–767.
Brüne, B., and Lapetina, E. G. (1991). Phosphorylation of nitric oxide synthase by protein kinase A. *Biochem. Biophys. Res. Commun.* **181,** 921–926.
Busconi, L. and Michel, T. (1993). Endothelial nitric oxide synthase. N-terminal myristoylation determines subcellular localization. *J. Biol. Chem.* **268,** 8410–8413.
Busse, R., and Mülsch, A. (1990). Calcium-dependent nitric oxide synthesis in endothelial cytosol is mediated by calmodulin. *FEBS Lett.* **265,** 133–136.
Cetkovic-Cvrlje, M., Sandler, S., and Eizirik, D. L. (1993). Nicotinamide and dexamethasone inhibit interleukin-1-induced nitric oxide production by RINm5F cells without decreasing messenger ribonucleic acid expression for nitric oxide synthase. *Endocrinology (Baltimore)* **133,** 1739–1743.
Chartrain, N. A., Geller, D. A., Koty, P. P., Sitrin, N. F., Nussler, A. K., Hoffman, E. P., Billiars, T. R., Hutchinson, N. I., and Mudgett, J. S. (1994). Molecular cloning, structure and chromosomal localization of the human inducible nitric oxide synthase gene. *J. Biol. Chem.* **269,** 6765–6772.
Chattopadhyaya, R., Meador, W. E., Means, A. R., and Quiocho, F. A. (1992). Calmodulin structure refined at 1.7 Å resolution. *J. Mol. Biol.* **228,** 1177–1192.
Chen, Y., and Rosazza, J. P. N. (1996). Oligopeptides as substrates and inhibitors for a new constitutive nitric oxide synthase from rat cerebellum. *Biochem. Biophys. Res. Commun.* **224,** 303–308.
Cho, H. J., Xie, Q.-W., Calaycay, J., Mumford, R. A., Swiderek, K. M., Lee, T. D., and Nathan, C. (1992). Calmodulin is a subunit of nitric oxide synthase from macrophages. *J. Exp. Med.* **176,** 599–604.
Craescu, C. T., Bouha, A., Mispelter, J., Diesis, E., Popescu, A., Chiriae, M., and Bârzu, O. (1995). Calmodulin binding of a peptide derived from the regulatory domain of *Bordetella pertussis* adenylate cyclase. *J. Biol. Chem.* **270,** 7088–7096.
Dawson, V. L., Dawson, T. M., London, E. D., Bredt, D. S., and Snyder, S. H. (1991). Nitric oxide mediates glutamate neurotoxicity in primary cortical culture. *Proc. Natl. Acad. Sci. U.S.A.* **88,** 6368–6371.
Dawson, T. M., Steiner, J. P., Dawson, V. L., Dinerman, J. L., Uhl, G. R., and Snyder, S. H. (1993). Immunosuppressant FK506 enhances phosphorylation of nitric oxide

synthase and protects against glutamate neurotoxicity. *Proc. Natl. Acad. Sci. U.S.A.* **90**, 9808-9812.

Deguchi, T., and Yoshioka, M. (1982). L-arginine identified as an endogenous activator for soluble guanylate cyclase from neuroblastoma cells. *J. Biol. Chem.* **257**, 10147-10152.

De Vera, M. E., Shapiro, R. A., Nussler, A. K., Mudgett, J. S., Simmons, R. L., Morris, S. M., Billiar, T. R., and Geller, D. A., (1996). Transcriptional regulation of human inducible nitric oxide synthase (NOS2) gene by cytokines: Initial analysis of the human NOS2 promoter. *Proc. Natl. Acad. Sci. U.S.A.* **93**, 1054-1059.

Dinerman, J. L., Steiner, J. P., Dawson, T. M., Dawson, V., and Snyder, S. H. (1994). Cyclic nucleotide dependent phosphorylation of neuronal nitric oxide synthase inhibits catalytic activity. *Neuropharmacology* **33**, 1245-1251.

Fast, D. J., Lynch, R. C., and Leu, R. W. (1993). Cyclosporin A inhibits nitric oxide production by L929 cells in response to tumor necrosis factor and interferon-γ. *J. Interferon Res.* **13**, 235-240.

Feinstein, D. L., Galea, E., Cermak, J., Chugh, P., Lyandvert, L. and Reis, D. J. (1994). Nitric oxide synthase expression in glial cells: Suppression by tyrosine kinase inhibitors. *J. Neurochem.* **62**, 811-814.

Feron, O., Belhassen, B. L., Kobzik, L., Smith, T. W., Kleely, R. A., and Michel, T. (1996). Endothelial nitric oxide synthase targeting to caveolae. Specific interactions with caveolin isoforms in cardiac myocytes and endothelial cells. *J. Biol. Chem.* **271**, 22810-22814.

Förstermann, U., and Gath, I. (1996). Purification of nitric oxide synthase. *Methods Enzymol.* **268**, 334-339.

Förstermann, U., and Kleinert, H. (1995). Nitric oxide synthase: Expression and expressional control of the three isoforms. *Arch. Pharmacol.* **352**, 351-364.

Förstermann, U., Gorsky, L. E., Pollock, J. S., Ishii, K., Schmidt, H. H. H. W., Heller, M., and Murad, F. (1990a). Hormone-induced biosynthesis of endothelium-derived relaxing factor/nitric oxide-like material in N1E-115 neuroblastoma cells requires calcium and calmodulin. *Mol. Pharmacol.* **38**, 7-13.

Förstermann, U., Gorsky, L. E., Pollock, J. S., Schmidt, H. H. H. W., Heller, M., and Murad, F. (1990b). Regional distribution of EDRF/NO-synthesizing enzyme(s) in rat brain. *Biochem. Biophys. Res. Commun.* **168**, 727-732.

Förstermann, U., Pollock, J. S., Schmidt, H. H. H. W., Heller, M., and Murad, F. (1991). Calmodulin-dependent endothelium derived relaxing factor/nitric oxide synthase activity is present in the particulate and cytosolic fraction of bovine aortic endothelial cells. *Proc. Natl. Acad. Sci. U.S.A.* **88**, 1788-1792.

Förstermann, U., Schmidt, H. H. H. W., Kohlhaas, K. L., and Murad, F. (1992). Induced RAW264.7 macrophages express soluble and particulate nitric oxide synthase: Inhibition by transforming growth factor-beta. *Eur. J. Pharmacol.* **225**, 161-165.

Fossetta, J. D., Niu, X. D., Lunn, C. A., Zavodny, P. J., Narula, S. K., and Lundell, D. (1996). Expression of human inducible nitric oxide synthase in *Escherichia coli. FEBS Lett.* **379**, 135-138.

Fugisawa, H., Ogura, T., Jurashima, Y., Yokoyama, T., Yamashita, J., and Esumi, H. (1994). Expression of two types of nitric oxide synthase mRNA in human neuroblastoma cell lines. *J. Neurochem.* **63**, 140-145.

Fukuto, J. M., and Chaudhuri, G. (1995). Inhibition of constitutive and inducible nitric oxide synthase: Potential selective inhibition. *Annu. Rev. Pharmacol. Toxicol.* **35**, 165-194.

6. Regulation of NO Synthase by Calmodulin

Furchgott, R. F. (1988). Studies on relaxation of rabbit aorta by sodium nitrite: The basis for the proposal that the acid-activatable inhibitory factor from retracto penis is inorganic nitrite and the endothelium-derived relaxing factor is nitric oxide. *In* "Vasodilation: Vascular Smooth Muscle, Peptides, Autonomic Nerves and Endothelium" (P. M. Vanhoutte, ed.), pp. 401–414. Raven, New York.

Furchgott, R. F., and Zawadzki, J. V. (1980). The obligatory role of endothelial cells in the relaxation of arterial smooth muscle by acetylcholine. *Nature (London)* **288,** 373–376.

Garry, M. G., Richardson, J. D., and Hargreaves, K. M. (1994). Sodium nitroprusside evokes the release of immunoreactive calcitonin gene-related peptide and substance P from dorsal horn slices via nitric oxide-dependent and nitric-independent mechanisms. *J. Neurosci.* **14,** 4329–4337.

Garthwaite, J., and Boulton, C. L. (1995). Nitric oxide signaling in the central nervous system. *Annu. Rev. Physiol.* **57,** 683–706.

Garthwaite, J., Charles, S. J. and Chess-Williams, R. (1988). Endothelium-derived relaxing factor release on activation of NMDA receptors suggests role as intercellular messenger in the brain. *Nature (London)* **336,** 385–388.

Garvey, E. P., Furfine, E. S., and Sherman, P. A. (1996). Purification of nitric oxide synthase. *Method Enzymol.* **268,** 339–349.

Geng, Y., Maier, R., and Lotz, M. (1995). Tyrosine kinases are involved with the expression of inducible nitric oxide synthase in human articular chondrocytes. *J. Cell. Physiol.* **163,** 545–554.

George, S. E., Su, Z., Fan, D., and Means, A. R. (1993). Calmodulin-cardiac troponin C chimeras. Effects of domain exchange on calcium binding and enzyme activation. *J. Biol. Chem.* **268,** 25213–25220.

George, S. E., Su, Z., Fan, D., Wang, S., and Johnson, J. D. (1996). The fourth EF-hand of calmodulin and its helix-helix components: Impact on calcium binding and enzyme activation. *Biochemistry* **35,** 8307–8313.

Graff, J. M., Stumpo, D. J., and Blackshear, P. J. (1989). Characterization of the phosphorylation sites in the chicken and bovine myristoylated alanine-rich C kinase substrate protein, a prominent cellular substrate for protein kinase C. *J. Biol. Chem.* **264,** 11912–11919.

Griffith, O. W., and Stuehr, D. J. (1995). Nitric oxide synthases: Properties and catalytic mechanism. *Annu. Rev. Physiol.* **57,** 707–736.

Gross, S. S., and Wolin, M. S. (1995). Nitric oxide: Pathophysiological mechanisms. *Annu. Rev. Physiol.* **57,** 737–769.

Guevara-Guzman, R., Emson, P. C., and Kendrick, K. M. (1994). Modulation of *in vivo* striatal transmitter release by nitric oxide and cyclic GMP. *J. Neurochem.* **62,** 807–810.

Haiech, J., Kilhoffer, M. C., Lukas, T. J., Craig, T. A., Roberts, D. M., and Watterson, D. M. (1991). Restoration of the calcium binding activity of mutant calmodulins toward normal by the presence of a calmodulin binding structure. *J. Biol. Chem.* **266,** 3427–3431.

Hall, A. V., Antoniou, H., Wang, Y., Cheung, A. H., Arbus, A. M., Olson, S. L., Lu, W. C., Kau, C. L., and Marsden, P. A. (1994). Structural organization of the human neuronal nitric oxide synthase gene. *J. Biol. Chem.* **269,** 33082–33090.

Harrison, D. G., Sayegh, H., Ohara, Y., Inoue, N., and Venema, R. C. (1996). Regulation of expression of the endothelial cell nitric oxide synthase. *Clin. Exp. Pharmacol. Physiol.* **23,** 251–255.

Hecker, M., Mülsch, A., and Busse, R. (1994). Subcellular localization and characterization of neuronal nitric oxide synthase. *J. Neurochem.* **62,** 1524–1529.

Hellerman, G. R., and Solomonson, L. P. (1997). Calmodulin promotes dimerization of the oxygenase domain of human endothelial nitric-oxide synthase. *J. Biol. Chem.* **272,** 12030–12034.

Herdegen, T., Brecht, S., Mayer, B., Leah, J., Kummer, W., Vravo, R., and Zimmermann, M. (1993). Long-lasting expression of JUN and KROX transcription factors and nitric oxide synthase in intrinsic neurons of the rat brain following axotomy. *J. Neurosci.* **13,** 4130–4145.

Hevel, J. M., White, K. A., and Marletta, M. A. (1991). Purification of the inducible macrophage nitric oxide synthase. *J. Biol. Chem.* **266,** 22789–22791.

Hibbs, J. B., Jr., Tainer, R. R., and Vavrin, Z. (1987). Macrophage cytotoxicity: Role for L-arginine deaminase and imino nitrogen oxidation to nitrite. *Science* **235,** 473–476.

Hiki, K., Hattori, R., Kawai, C., and Yui, Y. (1992). Purification of insoluble nitric oxide synthase from rat cerebellum. *J. Biochem. (Tokyo)* **111,** 556–558.

Hope, B. T., Michael, G. J., Knigge, K. M. and Vincent, S. R. (1991). Neuronal NADPH-diaphorase is a nitric oxide synthase. *Proc. Natl. Acad. Sci. U.S.A.* **88,** 2811–2814.

Hu, J., and El-Fakahany, E. E. (1993). Role of intracellular and intercellular communication by nitric oxide in coupling of muscarinic receptors to activation of guanylate cyclase in neuronal cells. *J. Neurochem.* **61,** 578–585.

Hu, J., and El-Fakahany, E. E. (1996). Intricate regulation of nitric oxide synthase in neurons. *Cell. Signal.* **8,** 185–189.

Hu, J., and Van Eldik, L. J. (1996). S100β induces apoptotic cell death in cultured astrocytes via a nitric oxide-dependent pathway. *Biochim. Biophys. Acta* **1313,** 239–245.

Hu, J., Lee, J.-H. and El-Fakahany, E. E. (1994). Inhibition of neuronal nitric oxide synthase by antipsychotic drugs. *Psychopharmacology* **114,** 161–166.

Hu, J., Castets, F., Guevara, J. L., and Van Eldik, L. J., (1996). S100β stimulates inducible nitric oxide synthase and expression of mRNA levels in rat cortical astrocytes. *J. Biol. Chem.* **271,** 2543–2547.

Ignarro, L. J., Byrns, R. E., and Wood, K. S. (1988). Biochemical and pharmacological properties of endothelial-derived relaxing factor and its similarity to nitric oxide radical. *In* "Vasodilation: Vascular Smooth Muscle, Peptides, Autonomic Nerves and Endothelium" (P. M. Vanhoutte, ed.), pp. 427–436. Raven, New York.

Ignarro, L. J. (1990). Biosynthesis and metabolism of endothelium-derived nitric oxide. *Annu. Rev. Pharmacol. Toxicol.* **30,** 535–560.

Ikeda, J., Ochiai, K., Morita, I., and Murota, S.-I. (1993). Endogenous nitric oxide blocks calcium influx induced by glutamate in neurons containing NADPH diaphorase. *Neurosci. Lett.* **158,** 193–196.

Ikura, M., Clore, G. M., Gronenborn, A. M., Zhu, G., Klee, C. B., and Bax, A. (1992). Solution structure of a calmodulin–target peptide complex by multidimensional NMR. *Science* **256,** 632–638.

Ito, M. (1989). Long-term depression. *Annu. Rev. Neurosci.* **12,** 85–102.

Iyengar, R., Stuehr, D. J., and Marletta, M. A. (1987). Macrophage synthesis of nitrite, nitrate and N-nitrosamines: Precursors and roles of the repiratory burst. *Proc. Natl. Acad. Sci. U.S.A.* **84,** 6369–6373.

Jones, N. M., Loiacono, R. E., and Beart, P. M. (1995). Roles for nitric oxide as an intra- and interneuronal messenger at NMDA release-regulating receptors: Evidence from

6. Regulation of NO Synthase by Calmodulin

studies of the NMDA-evoked release of [3H]noradrenaline and D-[3H]aspartate from rat hippocampal slices. *J. Neurochem.* **64,** 2057–2063.

Kadowaki, K., Kishimoto, J., Leng, G., and Emson, P. C. (1994). Upregulation of nitric oxide synthase (NOS) gene expression together with NOS activity in the rat hypothalamo–hypophysial system after chronic salt loading—arginine vasopressin and oxytocin secretion. *Endocrinology (Baltimore)* **134,** 1011–1017.

Karanth, S., Lyson, K., and McCann, S. M. (1993). Role of nitric oxide in interleukin-2 induced corticotropin-releasing factor release from incubated hypothalami. *Proc. Natl. Acad. Sci. U.S.A.* **88,** 3383–3385.

Kim, J., Shishido, T., Jiang, X., Aderem, A., and McLaughlin, S. (1994). Phosphorylation, high ionic strength, and calmodulin reverse the binding of MARCKS to phospholipid vesicles. *J. Biol. Chem.* **269,** 28214–28219.

Knowles, R. G., Palacios, M., Palmer, R. M. J., and Moncada, S. (1989). Formation of nitric oxide from L-arginine in the central nervous system: A transduction mechanism for stimulation of the soluble guanylate cyclase. *Proc. Natl.Acad. Sci. U.S.A.* **86,** 5159–5162.

Kobzik, L., Reid, M. B., Bredt, D. S., and Stamler, J. S. (1994). Nitric oxide in skeletal muscle. *Nature (London)* **372,** 546–548.

Kolyada, A. Y., Savikovsky, N., and Madias, N. E. (1996). Transcriptional regulation of the human iNOS gene in vascular smooth muscle cells and macrophages: Evidence for tissue specificity. *Biochem. Biophys. Res. Commun.* **220,** 600–605.

Lamas, S., Marsden, P. A., Li, G. K., Tempst, P., and Michel, T. (1992). Endothelial nitric oxide synthase: Molecular cloning, characterization of a distinct constitutive enzyme isoform. *Proc. Natl. Acad. Sci. U.S.A.* **89,** 6348–6352.

Lei, S. Z., Pan, Z.-H., Aggarwal, S. K., Chen, H.-S. V., Hartman, J., Sucher, N. J., and Lipton, S. A. (1992). Effects of nitric oxide production on the redox modulatory site of the NMDA receptor–channel complex. *Neuron* **8,** 1087–1099.

Lepoivre, M., Chenais, B., Yapo, A., Lemaire, G., Thelander, L., and Tenu, J. P. (1990). Alterations of ribonucleotide reductase activity following induction of the nitrite-generating pathway in adenocarcinoma cells. *J. Biol. Chem.* **265,** 14143–14149.

Lipton, S. A., and Stamler, J. S. (1994). Actions of redox-related congeners of nitric oxide at the NMDA receptor. *Neuropharmacology* **33,** 1229–1233.

Lipton, S. A., Choi, Y. B., Pan, Z.-H., Lei, S. Z., Chen, H. S. V., Sucher, N. J., Loscalzo, J., Singel, D. J., and Stamler, J. S. (1993). A redox-based mechanism for the neuroprotective and neurodestructive effects of nitric oxide and related nitroso-compounds. *Nature (London)* **364,** 626–632.

Liu, J., and Sessa, W. C. (1994). Identification of covalently bound amino-terminal myristic acid in endothelial nitric oxide synthase. *J. Biol. Chem.* **269,** 11691–11694.

Lukas, T. J., Burgess, W. H., Prendergast, F. G., Lau, W., and Watterson, D. M. (1986). Calmodulin binding domains: Characterization of a phosphorylation and calmodulin binding site from myosin light chain kinase. *Biochemistry* **25,** 1458–1464.

Lukas, T. J., Mirzoeva, S., and Watterson, D. M. (1998). Calmodulin regulation of protein kinases. Calmodulin and Signal Transduction. Academic Press, Orlando, FL (this book).

Mcquillan, L. P., Leung, G. K., Marsden, P. A., Kostyk, S. K., and Kourembanas, S. (1994). Hypoxia inhibits expression of eNOS via transcriptional and posttranscriptional mechanisms. *Am. J. Physiol.* **36,** H1921–H1927.

Manzoni, O., and Bockaert, J. (1993). Nitric oxide synthase activity endogenously modulates NMDA receptors. *J. Neurochem.* **61,** 368-370.

Marin, P., Lafon-Cazal, M., and Bockaert, J. (1992). A nitric oxide synthase activity selectively stimulated by NMDA receptors depends on protein kinase C activation in mouse striatal neurons. *Eur. J. Neurosci.* **4,** 425-432.

Marletta, M. A. (1993). Nitric oxide synthase structure and mechanism. *J. Biol. Chem.* **268,** 12231-12234.

Marletta, M. A. (1994). Nitric oxide synthase: Aspects concerning structure and catalysis. *Cell (Cambridge, Mass.)* **78,** 927-930.

Marsden, P. A., Schappert, K. T., Chen, H. S., Flowers, M., Sundell, C. L., Wilcox, J. N., Lamas, S., and Michel, T. (1992). Molecular cloning and characterization of human endothelial nitric oxide synthase. *FEBS Lett.* **307,** 287-293.

Marsden, P. A., Heng, H. H., Scherer, S. W., Stewart, R. J., Hall, A. V., Shi, X. M., Tsui, L. C., and Schappert, K. T. (1993). Structure and chromosomal localization of the human constitutive endothelial nitric oxide synthase gene. *J. Biol. Chem.* **268,** 17478-17488.

Marsden, P. A., Heng, H. H. Q., Duff, C. L., Shi, X. M., Tsui, L. C., and Hall, A. V. (1994). Localization of the human gene for inducible nitric oxide synthase (NOS2) to chromosome 17Q11.2 -Q12. *Genomics* **19,** 183-185.

Martinson, E. A., Trilivas, I., and Brown, J. H. (1990). Rapid protein kinase C-dependent activation of phospholipid D leads to delayed 1,2-diglyceride accumulation. *J. Biol. Chem.* **265,** 22282-22287.

Martzen, M. R., and Slemmon, J. R. (1995). The dentritic peptide neurogranin can regulate a calmodulin-dependent target. *J. Neurochem.* **64,** 92-100.

Matsubara, M., Titani, K., and Taniguchi, H. (1996). Interaction of calmodulin-binding domain peptides of nitric oxide synthase with membrane phospholipids: Regulation by protein phosphorylation and Ca^{2+}-calmodulin. *Biochemistry* **35,** 14651-14658.

Matsumoto, T., Pollock, J. S., Nakane, M., and Förstermann, U. (1993). Developmental changes of cytosolic and particulate nitric oxide synthase in rat brain. *Dev. Brain Res.* **73,** 199-203.

Matsuoka, A., Stuehr, D. J., Olson, J. S., Clark, P., and Ikeda-Saito, M. (1994). L-Arginine and calmodulin regulation of the heme iron reactivity in neuronal nitric oxide synthase. *J. Biol. Chem.* **269,** 20335-20339.

Mayer, B., John, M., and Böhme, E. (1990). Purification of a Ca^{2+}/calmodulin-dependent nitric oxide synthase from porcine cerebellum. *FEBS Lett.* **277,** 215-219.

Meador, W. E., Means, A. R., and Quiocho, F. A. (1992). Target enzyme recognition by calmodulin: 2.4 Å structure of a calmodulin-peptide complex. *Science* **257,** 1251-1255.

Meador, W. E., Means, A. R., and Quiocho, F. A. (1993). Modulation of calmodulin plasticity in molecular recognition on the basis of x-ray structures. *Science* **262,** 1718-1721.

Michel, J. B., Feron, O., Sacks, D., and Michel, T. (1997). Reciprocal regulation of endothelial nitric oxide synthase by Ca^{2+}-calmodulin and caveolin. *J. Biol. Chem.* **272,** 15583-15586.

Michel, T., Li, G. K., and Busconi, L. (1993). Phosphorylation and subcellular translocation of endothelial nitric oxide synthase. *Proc. Natl. Acad. Sci. U.S.A.* **90,** 6252-6256.

Moncada, S., and Higgs, A. (1993). Mechanisms of disease: The L-arginine-nitric oxide pathway. *N. Engl. J. Med.* **329,** 2002-2012.

Moncada, S., Higgs, A., and Furchgott, R. (1997). XIV. International union of pharmacology nomenclature in nitric oxide research. *Pharmacol. Rev.* **49**, 137–142.

Montague, P. R., Gancayco, C. D., Winn, M. J., Marchase, R. B., and Friedlander, M. J. (1994). Role of NO production in NMDA receptor-mediated neurotransmitter release in cerebral cortex. *Science* **263**, 973–977.

Moore, P. K., and Handy, R. L. C. (1997). Selective inhibitors of neuronal nitric oxide synthase—is no NOS really good NOS for the nervous system? *Trends Pharmacol. Sci.* **18**, 204–211.

Murphy, S., and Grzybicki, D. (1996). Glial NO: Normal and pathological roles. *Neuroscientist* **2**, 90–99.

Nadaud, S., Bonnardeaux, A., Lahrop, M., and Soubrier, F. (1994). Gene structure, polymorphism, and mapping of the human endothelial nitric oxide synthase gene. *Biochem. Biophys. Res. Commun.* **198**, 1027–1033.

Nakane, M., Mitchell, J., Förstermann, U., and Murad, F. (1991). Phosphorylation by calcium calmodulin-dependent protein kinase II and protein kinase C modulates the activity of nitric oxide synthase. *Biochem. Biophys. Res. Commun.* **180**, 1396–1402.

Nakane, M., Schmidt, H. H. H. W., Pollock, J. S., Förstermann, U., and Murad, F. (1993). Cloned human brain nitric oxide synthase is highly expressed in skeletal muscle. *FEBS Lett.* **316**, 175–180.

Nathan, C., and Xie, Q.-W. (1994a) Regulation of biosynthesis of nitric oxide. *J. Biol. Chem.* **269**, 13725–13728.

Nathan, C., and Xie, Q.-W. (1994b). Nitric oxide synthase: Roles, tolls, and controls. *Cell (Cambridge, Mass.)* **78**, 915–918.

Nishida, K., Harrison, D. G., Navas, J. P., Fisher, A. A., Dockery, S. P., Uematsu, M., Nerem, R. M., Alexander, R. W., and Murphy, T. J. (1992). Molecular cloning and characterization of the constitutive bovine aortic endothelial cell nitric oxide synthase. *J. Clin. Invest.* **90**, 2092–2096.

Nishizuka, Y. (1992). Intracellular signaling by hydrolysis of phospholipids and activation of protein kinase C. *Science* **258**, 607–614.

O'Dell, T. J., Hawkins, R. D., Kandel, E. R., and Arancio, O. (1991). Tests of the roles of two diffusible substances in long-term potentiation: Evidence for nitric oxide as a possible early retrograde messenger. *Proc. Natl. Acad. Sci. U.S.A.* **88**, 11285–11289.

O'Neil, K. T., and DeGrado, W. F. (1990). How calmodulin binds its targets: Sequence independent recognition of amphiphilic alpha-helices. *Trends Biochem. Sci.* **15**, 59–64.

Ohshima, H., Oguchi, S., Adachi, H., Iida, S., Suzuki, H., Sugimura, T., and Esumi, H. (1992). Purification of nitric oxide synthase from bovine brain: Immunological characterization and tissue distribution. *Biochem. Biophys. Res. Commun.* **183**, 238–244.

Okada, O. (1995). Protein kinase C modulates calcium sensitivity of nitric oxide synthase in cerebellar slices. *J. Neurochem.* **64**, 1298–1304.

Palmer, R. M. J., Ferrige, A. G., and Moncada, S. (1987). Nitric oxide release accounts for the biological activity of endothelium-derived relaxing factor. *Nature (London)* **327**, 524–526.

Palmer, R. M. J., Ashton, D. S., and Moncada, S. (1988). Vascular endothelial cells synthesize nitric oxide from L-arginine. *Nature (London)* **333**, 664–666.

Pan, J., Burgher, K. L., Szczepanik, A. M., and Ringheim, G. E. (1996). Tyrosine phosphorylation of inducible nitric oxide synthase: Implication for potential post-translational regulation. *Biochem. J.* **314**, 889–894.

Paul, A., Pendreigh, R. H., and Plevin, R. (1995). Protein kinase C and tyrosine kinase pathways regulate lipopolysaccharide-induced nitric oxide synthase activity in RAW 264.7 murine macrophage. *Br. J. Pharmacol.* **114,** 482–488.

Persechini, A., Gansz, K., and Paresi, R. J. (1996a). Activation of myosin light chain kinase and nitric oxide synthase activities by engineered calmodulin with duplicated or exchanged EF hand pairs. *Biochemistry* **35,** 224–228.

Persechini, A., Gansz, K. J., and Paresi, R. J. (1996b). A role in enzyme activation for the N-terminal leader sequence in calmodulin. *J. Biol. Chem.* **271,** 19279–19282.

Persechini, A., White, H. D., and Gansz, K. J. (1996c). Different mechanisms for Ca^{2+} dissociation from complexes of calmodulin with nitric oxide synthase or myosin light chain kinase. *J. Biol. Chem.* **271,** 62–67.

Pollock, J. S., Förstermann, U., Mitchell, J. A., Warner, T. D., Schmidt, H. H. H. W., Nakane, M., and Murad, F. (1991). Purification and characterization of particulate endothelium-derived relaxing factor synthase from cultured and native bovine aortic endothelial cells. *Proc. Natl. Acad. Sci. U.S.A.* **88,** 10480–10484.

Rafferty, S., and Malech, H. L. (1996). High reductase activity of recombinant NOS2 flavoprotein domain lacking the calmodulin binding regulatory sequence. *Biochem. Biophys. Res. Commun.* **220,** 1002–1007.

Rand, M. J., and Li, C. B. (1995). Nitric oxide as a neurotransmitter in peripheral nerves: Nature of transmitter and mechanism of transmission. *Annu. Rev. Physiol.* **57,** 659–682.

Reiser, G. (1992). Nitric oxide formation caused by Ca^{2+} release from internal stores in neuronal cell line is enhanced by cyclic AMP. *Eur. J. Pharmacol.* **227,** 89–93.

Rettori, V., Gimeno, M., Lyson, K., and McCann, S. M. (1992). Nitric oxide mediates norepinephrine-induced prostaglandin E2 release from the hypothalamus. *Proc. Natl. Acad. Sci. U.S.A.* **89,** 11543–11546.

Robinson, L. J., Weremowicz, S., Morton, C. C., and Michel, T. (1994). Isolation and chromosomal localization of the human endothelial nitric oxide synthase (NOS3) gene. *Genomics* **19,** 350–357.

Robinson, L. J., Busconi, L., and Michel, T. (1995). Agonist-modulated palmitoylation of endothelial nitric oxide synthase. *J. Biol. Chem.* **270,** 995–998.

Rodriguez, J., Quignard, J.-F., Fagni, L., Lafon-Cazal, M., and Bockaert, J. (1994). Blockade of nitric oxide synthesis by tyrosine kinase inhibitors in neurons. *Neuropharmacology* **33,** 1267–1274.

Ruan, J., Xie, Q.-W., Hutchinson, N., Cho, H., Wolfe, G. C., and Nathan, C. (1996). Inducible nitric oxide synthase requires both the canonical calmodulin-binding domain and additional sequences in order to bind calmodulin and produce nitric oxide in the absence of free Ca^{2+}. *J. Biol. Chem.* **271,** 22679–22686.

Schmidt, H. H. H. W., and Murad, F. (1991). Purification and characterization of a human NO synthase. *Biochem. Biophys. Res. Commun.* **181,** 1372–1377.

Schmidt, H. H. H. W., Wilke, P., Evers, B., and Böhme, E. (1989). Enzymatic formation of nitrogen oxides from L-arginine in bovine brain cytosol. *Biochem. Biophys. Res. Commun.* **165,** 284–291.

Schmidt, H. H. H. W., Pollack, J. S., Nakane, M., Gorsky, L. D., Förstermann, U., and Murad, F. (1991). Purification of a soluble isoform of guanylyl cyclase-activating factor synthase. *Proc. Natl. Acad. Sci. U.S.A.* **88,** 365–369.

Schmidt, H. H. H. W., Lohmann, S. M., and Walter, U. (1993). The nitric oxide and cGMP signal transduction system: Regulation and mechanism of action. *Biochim. Biophy. Acta* **1178,** 153–175.

Schuman, E. M., and Madison, D. V. (1991). A requirement for the intercellular messenger nitric oxide in long-term potentiation. *Science* **254,** 1503–1506.

Sessa, W. C., Harrison, J. K., Barber, C. M., Zeng, D., Durieux, M. E., D'Angelo, D. D., Lynch, K. R., and Peach, M. J. (1992). Molecular cloning and expression of a cDNA encoding endothelial cell nitric oxide synthase. *J. Biol. Chem.* **27,** 15274–15276.

Sessa, W. C., Barber, C. M., and Lynch, K. R. (1993). Mutation of *N*-myristoylation site converts endothelial cell nitric oxide synthase from a membrane to a cytosolic protein. *Circ. Res.* **72,** 921–924.

Shaul, P. W., Smart, E. J., Robinson, L. J., German, Z., Yuhanna, I. S., Ying, Y, Anderson, R. G. W., and Michel, T. (1996). Acylation targets endothelial nitric-oxide synthase to plamalemmal caveolae. *J. Biol. Chem.* **271,** 6518–6522.

Sheta, E. A., McMillan, K., and Masters, B. S. S. (1994). Evidence for a bidomain structure of constitutive cerebellar nitric oxide synthase. *J. Biol. Chem.* **269,** 15147–15153.

Shibuki, K., and Okada, D. (1991). Endogenous nitric oxide release required for long-term synaptic depression in the cerebellum. *Nature (London)* **349,** 326–328.

Surichamorn, W., Forray, C., and El-Fakahany, E. E. (1990). Role of intracellular Ca^{2+} mobilization in muscarinic and histamine receptor-mediated activation of guanylate cyclase in N1E-115 neuroblastoma cells: Assessment of the arachidonic acid release hypothesis. *Mol. Pharmacol.* **37,** 860–869.

Singh, K., Balligand, J.-L., Fischer, T. A., Smith, T. W., and Kelly, R. A. (1996). Regulation of cytokine-inducible nitric oxide synthase in cardiac myocytes and microvascular endothelial cells. *J. Biol. Chem.* **271,** 1111–1117.

Slemmon, J. R., and Martzen, M. R. (1994). Neuromodulin (GAP-43) can regulate a calmodulin-dependent target *in vitro. Biochemistry* **33,** 5653–5660.

Spitsin, S. V., Koprowski, H., and Michaels, F. H. (1996). Characterization and functional analysis of the human inducible nitric oxide synthase gene promoter. *Mol. Med.* **2,** 226–235.

Stamler, J. S. (1994). Redox signaling: Nitrosylation and related target interactions of nitric oxide. *Cell (Cambridge, Mass.)* **78,** 931–936.

Stamler, J. S., Singel, D. J., and Loscalzo, J. (1992). Biochemistry of nitric oxide and its redox-active forms. *Science* **258,** 1898–1902.

Steiner, J. P., Dawson, T. M., Fotuhi, M., Glatt, C. E., Snowman, A. M., Cohen, N., and Snyder, S. H. (1992). High brain densities of the immunophilin FKBP colocalized with calcineurin. *Nature (London)* **358,** 584–587.

Stevens-Truss, R., and Marletta, M. A. (1995). Interaction of calmodulin with the inducible murine macrophage nitric oxide synthase. *Biochemistry* **34,** 15638–15645.

Stuehr, D. J. (1996). Purification and properties of nitric oxide synthase. *Methods Enzymol.* **268,** 324–333.

Stuehr, D. J., Cho, H. J., Kwon, N. S., Weise, M. F., and Nathan, C. F. (1991). Purification and characterization of the cytokine-induce macrophage nitric oxide synthase: An FAD- and FMN-containing flavoprotein. *Proc. Natl. Acad. Sci. U.S.A.* **88,** 7773–7777.

Su, Z., Blazing, M. A., Fan, D., and Gorge, S. E. (1995). The calmodulin-nitric oxide synthase interaction. *J. Biol. Chem.* **270,** 29117–29122.

Suschek, C., Fehsel, K., Kröncke, K. D., Sommer, A., and Kolb-Bachofen, V. (1994). Primary cultures of rat islet capillary endothelial cells—constitutive and cytokine-inducible macrophage-like nitric oxide synthases are expressed and activities regulated by glucose concentration. *Am. J. Pathol.* **145,** 685–695.

Toms, N. J., and Roberts, P. (1994). NMDA receptor-mediated stimulation of rat cerebellar nitric oxide formation is modulated by cyclic AMP. *Eur. J. Pharmacol.* **266,** 63–66.
Umans, J. G., and Levi, R. (1995). Nitric oxide in the regulation of blood flow and arterial pressure. *Annu. Rev. Physiol.* **57,** 771–790.
Venema, R. C., Nishida, K., Alexander, R. W., Harrison, D. G., and Murphy, T. J. (1994). Organization of the bovine gene encoding the endothelial nitric oxide synthase. *Biochim. Biophys. Acta* **1218,** 413–420.
Venema, R. C., Sayegh, H. S., Arnal, J.-F., and Harrison, D. G. (1995). Role of the enzyme calmodulin-binding domain in membrane association and phospholipid inhibition of endothelial nitric oxide synthase. *J. Biol. Chem.* **270,** 14705–14711.
Venema, R. C., Sayegh, H. S., Kent, J. D., and Harrison, D. G. (1996). Identification, characterization, and comparison of the calmodulin-binding domains of the endothelial and inducible nitric oxide synthases. *J. Biol. Chem.* **271,** 6435–6440.
Venema, R. C., Ju, H., Zou, R., Ryan, J. W., and Venema, V. J. (1997). Subunit interactions of endothelial nitric-oxide synthase. Comparison to the neuronal and inducible nitric-oxide synthase isoforms. *J. Biol. Chem.* **272,** 1276–1282.
Vincent, S. R. (1994). Nitric oxide: a radical neurotransmitter in the central nervous system. *Prog. in Neurobiol.* **42,** 129–160.
Vodovotz, Y., Bogdan, C., Paik, J., Xie, Q.-W., and Nathan, C. (1993). Mechanisms of suppression of macrophage nitric oxide release by transforming growth factor β. *J. Exp. Med.* **178,** 605–613.
Vogel, H. J. (1994). The Merck Frosst Award Lecture 1994. Calmodulin: a versatile calcium mediator protein. *Biochem. Cell Biol.* **72,** 357–376.
Vorherr, T., Knöpfel, L., Hofmann, F., Mollner, S., Pfeuffer, T., and Carafoli, E. (1993). The calmodulin binding domain of nitric oxide synthase and adenylyl cyclase. *Biochemistry* **32,** 6081–6088.
Wang, Y., Goligorsky, M. S., Lin, M., Wilcox, J. N., and Marsden, P. A. (1997). A novel, testis-specific mRNA transcript encoding an NH2-terminal truncated nitric-oxide synthase. *J. Biol. Chem.* **272,** 11392–11401.
Watanabe, Y., Terachima, K., and Hidaka, H. (1996). Neuronal nitric oxide synthase specific autophosphorylation in baculovirus/sf9 insect cell system. *Biochem. Biophys. Res. Commun.* **219,** 638–643.
Weiner, C. P., Lizasoain, I., Baylis, S. A., Knowles, R. G., Charles, I. G., and Moncada, S. (1994). Induction of calcium-dependent nitric oxide synthases by sex hormones. *Proc. Natl. Acad. Sci. U.S.A.* **91,** 5212–5216.
Weisz, A., Oguichi, S., Cicatiello, L., and Esumi, H. (1994). Dual mechanism for the control of inducible-type NO synthase gene expression in macrophages during activation by interferon-gamma and bacterial lipopolysaccharide—transcriptional and post-transcriptional regulation. *J. Biol. Chem.* **269,** 8324–8333.
Wu, W., Liuzzi, F. J., Schinco, F. P., Depto, A. S., Li, Y., Mong, J. A., Dawson, T. M., and Snyder, S. H. (1994). Neuronal nitric oxide synthase is induced in spinal neurons by traumatic injury. *Neuroscience (Oxford)* **61,** 719–726.
Wu, C., Zhang, J., Abu-Soud, H., Ghosh, D., and Stuehr, D. H. (1996). High-level expression of mouse inducible nitric oxide synthase in *Escherichia coli* requires coexpression with calmodulin. *Biochem. Biophys. Res. Commun.* **222,** 439–444.
Xie, Q.-W., and Nathan, C. (1994). The high-output nitric oxide pathway: Role and regulation. *J. Leukocyte Biol.* **56,** 576–582.

Xie, Q.-W., Cho, H. J., Calaycay, J., Mumford, R. A., Swiderek, K. M., Lee, T. D., Ding, A., Troso, T., and Nathan, C. (1992). Cloning and characterization of inducible nitric oxide synthase from mouse macrophages. *Science* **256,** 225–228.

Xie, Q.-W., Whisnant, R., and Nathan, C. (1993). Promoter of the mouse gene encoding calcium-independent nitric oxide synthase confers inducibility by interferon gamma and bacterial lipopolysaccharide. *J. Exp. Med.* **177,** 1779–1784.

Xie, J. L., Roddy, P., Rife, T. K., Murad, F., and Young, A. P. (1995). Two closely linked but separable promoters for human neuronal nitric oxide synthase gene transcription. *Proc. Natl. Acad. Sci. U.S.A.* **92,** 1242–1246.

Xu, W. M., Gorman, P., Sheer, D., Bates, G., Kishimoto, J., Lizhi, L., and Emson, P. (1993). Regional localization of the gene coding for human brain nitric oxide synthase (NOS1) to 12q24.2-24.31 by fluorescent *in situ* hybridization. *Cytogenet. Cell Genet.* **64,** 62–63.

Xu, W. M., Charles, I. G., Moncada, S., Gorman, P., Sheer, D., Liu, L. Z., and Emson, P. (1994). Mapping of the genes encoding human inducible and endothelial nitric oxide synthase (NOS2 and NOS3) to the pericentric region of chromosome 17 and to chromosome 7, respectively. *Genomics* **21,** 419–422.

Xu, X., Star, R., Tortorici, G., and Muallem, S. (1994). Depletion of intracellular Ca^{2+} stores activates nitric oxide synthase to generate cGMP and regulate Ca^{2+} influx. *J. Biol. Chem.* **269,** 12645–12653.

Yoshizumi, M., Perrella, M. A., Burnett, J. C., and Lee, M. E. (1993). Tumor necrosis factor downregulates an endothelial nitric oxide synthase messenger RNA by shortening its half-life. *Circ. Res.* **73,** 205–209.

Yui, Y., Hattori, R., Kosuga, K., Eizawa, H., Hiki, K., Ohkawa, S., Ohnishi, K., Terao, S., and Kawai, C. (1991a) Calmodulin-independent nitric oxide synthase from rat polymorphonuclear neutrophils. *J. Biol. Chem.* **266,** 3369–3371.

Yui, Y., Hattori, R., Kosuga, K., Eizawa, H., Hiki, K., and Kawai, C. (1991b). Purification of nitric oxide synthase from rat macrophages. *J. Biol. Chem.* **266,** 12544–12547.

Zhang, J., and Snyder, S. H. (1995). Nitric oxide in the nervous system. *Annu. Rev. Pharmacol. Toxicol.* **35,** 213–233.

Zhang, M., and Vogel, H. J. (1994). Characterization of the calmodulin-binding domain of rat cerebellar nitric oxide synthase. *J. Biol. Chem.* **269,** 981–985.

Zhang, M., Fabian, H., Mantsch, H. H., and Vogel, H. J. (1994). Isotope-edited Fourier transform infrared spectroscopy studies of calmodulin's interaction with its target peptides. *Biochemistry* **33,** 10883–10888.

Zhang, M., Yuan, T., Aramini, J., and Vogel, H. J. (1995). Interaction of calmodulin with its binding domain of rat cerebellar nitric oxide synthase. *J. Biol. Chem.* **270,** 29901–20907.

Zhang, R., Min, W., and Sessa, W. C. (1995). Functional analysis of the human endothelial nitric oxide synthase promoter. Sp1 and GATA factors are necessary for basal transcription in endothelial cells. *J. Biol. Chem.* **270,** 15320–15326.

Zhang, Z. G., Chopp, M., Gautam, S., Zaloga, C., Zhang, R. L., Schmidt, H. H. H. W., Pollock, J. S., and Förstermann, U. (1994). Upregulation of neuronal nitric oxide synthase and mRNA, and selective sparing of nitric oxide synthase-containing neurons after focal cerebral ischemia in rat. *Brain Res.* **654,** 85–95.

Zorumski, C. F., and Izumi, Y. (1993). Nitric oxide and hippocampal synaptic plasticity. *Biochem. Pharmacol.* **46,** 777–785.

7

Calmodulin-Binding Proteins of the Cytoskeleton

NATHALIE M. BONAFÉ AND JAMES R. SELLERS

Laboratory of Molecular Cardiology
National Heart, Lung, and Blood Institute
National Institutes of Health
Bethesda, Maryland 20892-1872

I. Introduction 348
II. Calmodulin-Binding Motor Proteins 350
    A. Myosin Structure 350
    B. Calmodulin-Binding Properties of Individual Classes of Myosins 357
    C. Kinesins: A Microtubular-Dependent Family of Motors 366
III. Calmodulin-Binding Muscle Proteins 366
    A. Smooth Muscle Proteins 367
    B. Skeletal Muscle Proteins 370
IV. Actin-Binding Proteins of the Membrane Cytoskeleton 375
    A. Membrane Cytoskeleton 375
    B. Spectrins 375
    C. Adducin 377
    D. Interaction between Spectrin and Adducin 378
V. Other Cytoplasmic Calmodulin-Binding Proteins 378
    A. Synapsin 378
    B. Myristoylated Alanine-Rich C Kinase Substrate 379
    C. Desmocalmin 380
VI. Calmodulin-Binding Microtubule-Binding Proteins 380
    A. Microtubule-Associated Proteins 380
    B. Mip-90 381
    C. Stable Tubulin-Only Polypeptides 382
VII. Summary 382
    References 383

I. INTRODUCTION

The cytoplasm contains three types of filament systems, actin filaments, microtubules, and intermediate filaments, that contribute toward many cellular functions, including structural rigidity, shape, adhesion, locomotion of internal vesicles, cell locomotion, and cell division. All three of these filaments systems are built from small protein subunits. Actin is obviously the subunit for actin filaments. Tubulin dimers (α/β) polymerize into microtubules and various intermediate filament proteins such as desmin, lamin, vimentin, or keratin form homopolymers to create different types of intermediate filaments. Many proteins, falling into several family groups, have been identified whose function is to control the state of actin assembly in cells (Pollard and Cooper, 1986; Matsudaira, 1991). Monomeric actin in the cell is typically complexed with actin monomer-binding proteins such as profilin or thymosin $\beta 4$ as the ionic conditions and actin concentration in cells would favor polymerization. Actin filaments may be either isolated single filaments or tethered together into a loose meshwork or aligned into tight bundles such as is found in polarized microvilli structures. The equilibrium between the various states of actin is controlled by different signaling pathways such as those involving the phosphoinositide cycle, two Ras-related proteins, Rac and Rho, and receptor-mediated pathways that operate through Ras (Kimura et al., 1996; Amano et al., 1996; Ridley et al., 1992; Ridley and Hall, 1992; Janmey, 1994; Goldschmidt-Clermont et al., 1990). For example, one set of experiments demonstrated that microinjection of fibroblast with Rac inhibited growth factor-induced reorganization of the cytoskeletal, whereas microinjection of Rho induced rapid assembly of stress fibers and adhesion plaques (Ridley et al., 1992; Ridley and Hall, 1992).

In many cases, actin serves a structural function within the cell, but in some instances it provides a molecular railroad track to support the movement of vesicles and perhaps other cargo by myosin. The polymerization of actin itself provides the motile force for some biological processes such as movement of *Listeria,* a pathogenic bacteria, within infected cells and the extension of the acrosomal process in some invertebrate sperm cells (Theriot et al., 1992; Tilney and Portnoy, 1989; Tilney et al., 1973). In all of these cases, the state of actin assembly is tightly controlled by various actin-binding proteins.

The membrane cytoskeleton is vital for controlling cell shape, rigidity, and cell adhesion. There are specialized proteins that bind to actin and

7. Calmodulin-Binding Proteins of the Cytoskeleton

either directly or indirectly to integral membrane proteins in order to create the membrane cytoskeleton. Some membranes, such as that of the erythrocyte, must be capable of deformation whereas others must be able to perform endocytosis or exocytosis. Some cells are very mobile whereas others are more firmly affixed to the substratum. Thus the membrane cytoskeleton performs different roles in different cells.

A skeletal muscle fiber represents an extremely specialized form of the cytoskeleton. Here, actin- and myosin-containing filaments interdigitate in a precise and ordered fashion. In skeletal muscle the length of both of these filaments is tightly regulated. Ca^{2+}-dependent regulatory proteins are built into the sarcomere to regulate actin–myosin interaction. Two giant proteins, titin and nebulin, contribute toward this order. The sarcolemma or membrane of muscle is subject to great stress and is thought to be stabilized by dystrophin, the genetic target of Duchenne muscular dystrophy. Dystrophin binds to actin and to a set of associated proteins that link the actin cytoskeleton to the extracellular matrix.

Microtubules represent another important structural component to cell shape and polarity. During karyokinesis, they function to segregate chromosomes. In some cells they exist in special, highly ordered structures such as cilia and flagella. In most cell types, microtubules are in a dynamic equilibrium with tubulin dimers, and an individual microtubule can be observed to undergo periods of elongation, followed by periods of dramatic shortening. In contrast, microtubules of neuronal cells are more stable, and the set of proteins thought to be responsible for this stability will be discussed. Comparatively few proteins have been discovered that modulate the assembly state of microtubules (compared to actin), but several of these are calmodulin (CaM)-binding proteins. Other functions for microtubules are as a track for the movement of cytoplasmic dynein and kinesins (a large family of microtubule-dependent motors).

While critically important to cells, intermediate filaments are much more poorly understood. Their role appears to be entirely structural, as no intermediate filament motors have been found. There are many types of intermediate filaments, some of them cell or tissue specific. Unlike actin and tubulin, which are highly conserved, members of the intermediate filament protein family are quite divergent.

Calmodulin plays a crucial role in many cellular functions, as is well documented in other chapters of this volume. The structure and functional properties of the cytoskeleton are also tightly linked to calmodulin. This chapter details the interaction of calmodulin with various cytoskeletal proteins. For convenience of discussion, the types of interacting cytoskeletal proteins have been classified into (1) motor proteins, (2) muscle proteins, (3) membrane cytoskeletal proteins, and (4) microtubule-binding proteins.

The interaction of cytoskeletal proteins with calmodulin can take many forms. It might be a regulatory switch for an enzymatic reaction. The most widely studied interaction of calmodulin with a cytoskeletal protein is with myosin light-chain kinase (MLCK). This enzyme phosphorylates the regulatory light chain of myosin, activating its enzymatic activity, which serves as the trigger for contractile processes. Myosin light-chain kinase is totally inactive in the absence of bound calmodulin. The interaction between MLCK and calmodulin is well studied and is covered extensively elsewhere in this volume. In addition, many cytoskeletal proteins are targets of other calmodulin-dependent protein kinases, as will be discussed later. Calcineurin, a calmodulin-dependent phosphatase, is also present in cells and may participate in cytoskeletal events. In other cases the interaction of calmodulin with a cytoskeletal protein might be a nonenzymatic regulatory switch controlling protein–protein interaction or might merely provide structural stability to a protein. The interaction between calmodulin and the target might be Ca^{2+} independent or be strengthened or weakened by Ca^{2+}.

The first portion of this chapter deals with calmodulin binding to unconventional myosins where calmodulin at least plays a structural role in addition to possible regulatory roles. Later portions deal with calmodulin binding to various other cytoskeletal targets.

II. CALMODULIN-BINDING MOTOR PROTEINS

A. Myosin Structure

It is now recognized that myosins exist as an extended superfamily of molecular motors that are organized into at least 12 distinct classes (Fig. 1) (Mooseker and Cheney, 1995; Cope et al., 1996; Sellers et al., 1996; Sellers and Goodson, 1995). The common features are a relatively conserved motor domain, a light chain-binding neck region, and a tail portion that probably serves to anchor the motor or, in the case of myosin II, to allow for self-association into filaments (Fig. 2). The conventional thick filament-forming myosins (myosin II class) have two light chains

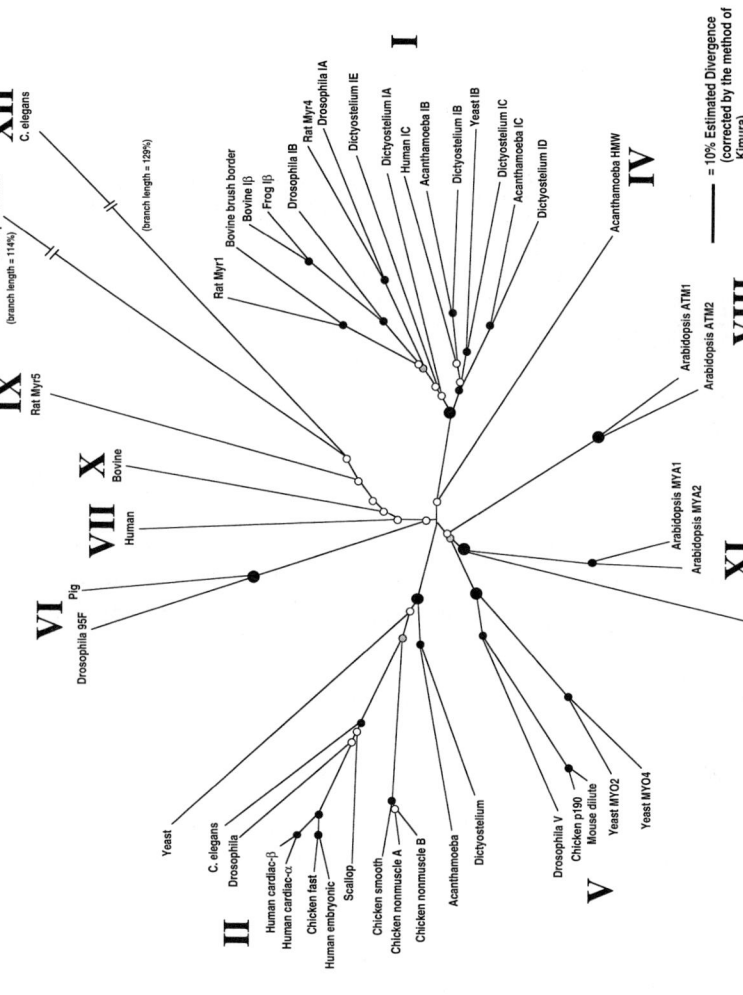

Fig. 1. Phylogenetic analysis of the myosin superfamily. Modified from Sellers *et al.* (1996) and including an entry for *Drosophila* myosin V (N. M. Bonafé and J. R. Sellers, 1998).

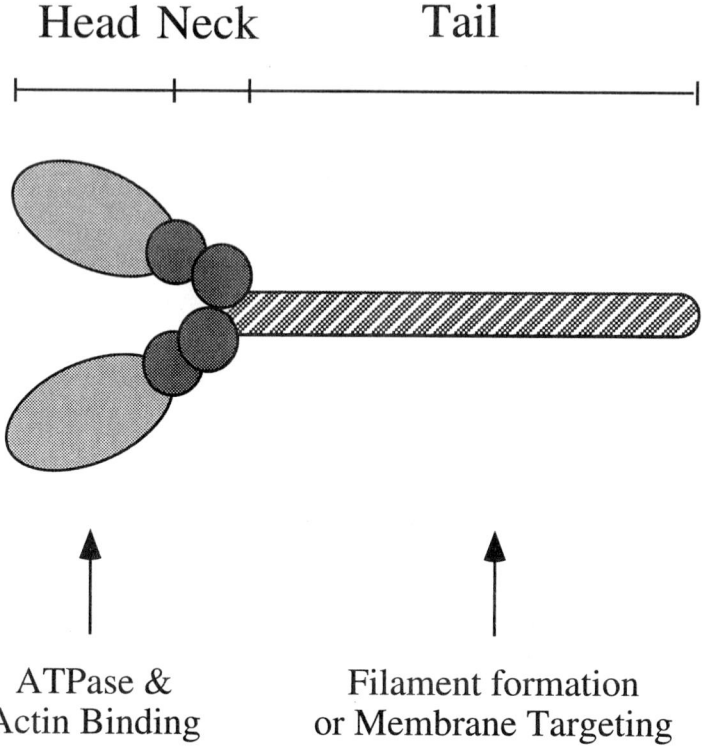

Fig. 2. Schematic diagram of a hypothetical myosin molecule showing the head, neck, and tail domains. This myosin has two light chains bound in the neck region.

termed regulatory (RLC) and essential (ELC) that bind to an extended α-helical stretch of the heavy chain (Rayment et al., 1993; Sellers and Goodson, 1995; Adelstein and Sellers, 1996). In several other classes of myosin there is good evidence that calmodulin instead of distinct light chains is bound to the neck region of the myosin heavy chain. This evidence ranges from partial protein sequence, to Ca^{2+}-induced molecular shifts on SDS–polyacrylamide gels, to interaction with anticalmodulin antibodies, to gel overlay experiments, and to binding of proteins or their fragments to calmodulin affinity columns. In some cases, calmodulin has been shown to bind directly to purified myosins or to fusion proteins containing myosin sequences by chromatography on a calmodulin-affinity column or by coimmunoprecipitation of calmodulin–myosin complexes. In all cases, the light chain (calmodulin)-binding site on the heavy chain is an extended α helix with the consensus sequence of IQXXXRGXXXR (I, isoleucine; Q, glutamine; R, arginine; G, glycine) (Cheney and Mooseker, 1992) (Fig. 3). This "IQ" motif is present in one to six repeats depending on the myosin. Calmodulin binding to the IQ motifs is different from calmodulin binding to many target helices in that it does not aways require Ca^{2+}. As discussed later Ca^{2+} actually weakens the binding of calmodulin to some IQ motifs.

Although there are no three-dimensional structures available for the neck region of a calmodulin-binding myosin, there are two crystal structures of this region of conventional myosin. Rayment *et al.* (1993) solved the structure of chicken fast skeletal muscle subfragment 1 (S-1), which

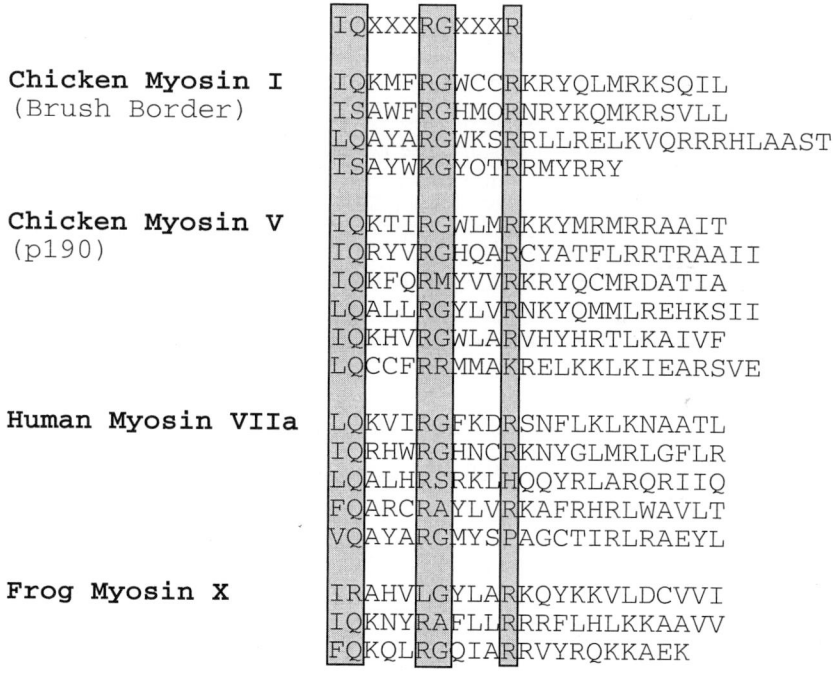

Fig. 3. Alignments of several IQ motifs. IQ motifs of several unconventional myosins were aligned manually. Note that the stagger difference between IQ motifs is constant in human myosin VIIa and frog myosin X, but is variable in chicken myosin V.

consists of the motor domain and the light chain-binding neck region, and Xie et al. (1994) solved the structure of the light chain binding region of scallop myosin. The latter authors termed this structure the "regulatory domain" as the enzymatic activity of scallop myosin is regulated by Ca^{2+} binding to one of the classes of light chains. Both of these studies gave insight into the mechanism of light chain binding to myosin.

Using the structure of the scallop regulatory domain as a model, Houdusse and Cohen (1995) speculated on the relative importance of the various residues of the consensus IQ motif and on the possible mode of interaction of calmodulin with the target helix. Both light chains and calmodulin have amino- and carboxyl-terminal lobes separated by a central helix. Each lobe is formed from a pair of EF hands that comprise the Ca^{2+}-binding domains. They noted that the first portion of the heavy chain IQ motif, IQXXXR, is better conserved than the later residues and is critical for the binding of the carboxyl-terminal lobe of the light chains. They also noted that the fifth position of the IQ motif is usually apolar, often containing an isoleucine (I) residue that contributes toward the stability of the interaction. The second portion of the IQ motif, GXXXR, is less well conserved and only plays a minor role in lobe positioning and conformation.

Each lobe can exist in one of three conformations: open, closed, and semiopen. Houdusse and Cohen (1995) predicted that the conformation of calmodulin (in absence of Ca^{2+}) interacting with the target IQ helix would be different depending on whether the IQ motif is complete. The carboxyl-terminal lobe of calmodulin is expected to be in the semiopen mode, assuming that the first portion of the IQ motif is intact. This is the same conformation adopted by both the light chains of scallop myosin. The amino-terminal lobe is expected to be in the closed conformation if the IQ motif is complete, but in the open conformation if it is incomplete.

Houdusse et al. (1996) also modeled another important aspect of the interaction between calmodulin and myosins. The stagger distance between the two IQ motifs in myosin IIs is always 26 residues, but can vary from 22 to as much as 36 residues in unconventional myosins. Because the IQ motifs within a myosin molecule are present on a continuous α-helix, which has 3.6 residues per turn of the helix, it is clear that changing the stagger from 26 residues to any other number will change the orientation of one calmodulin compared to its neighbors and/or the distance between adjacent calmodulins. Figure 4 shows models of calmodulin heavy-chain interactions for the case where the stagger between IQ motifs is 23 residues (A), 25 residues (B,C), and 26 residues

7. Calmodulin-Binding Proteins of the Cytoskeleton

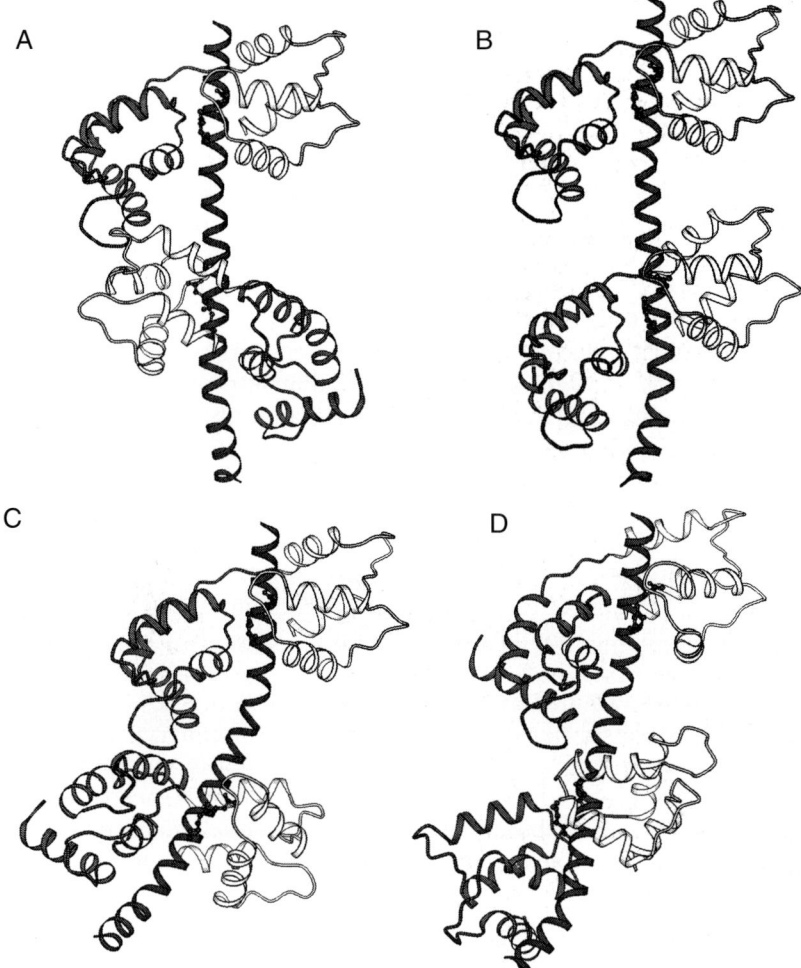

Fig. 4. Models for multiple calmodulins bound to IQ motifs separated by different numbers of residues. Possible interaction between two calmodulins bound to a heavy chain helix with tandem IQ motifs separated by 23 (A), 25 (B,C), or 26 residues (D), respectively. The heavy chain helices in A and B are modeled as straight, whereas it is allowed to bend in models C and D. Amino-terminal lobes of the modeled calmodulins are shaded dark and carboxyl-terminal lobes are white. Adapted from Houdusse *et al.* (1996).

(D). The difference between model B and model C is that in B the target helix is assumed to be straight (as it is in model A) whereas in model C it is assumed to be bent. It should be noted that the bending angle of the heavy-chain helix is different between scallop myosin and chicken skeletal myosin (Rayment et al., 1993; Xie et al., 1994). The bent helix model in Fig. 4C is preferred for the 25-residue stagger over the straight helix model in Fig. 4B as there are no interactions stabilizing the target helix and no interactions between adjacent calmodulin molecules in the latter model. The heavy-chain helix is modeled as bent in Fig. 4D where the stagger is 26 residues as it allows loop 1 of the top calmodulin to interact with linker 3 of the second calmodulin. In all the models the amino-terminal lobe of calmodulin is shaded darkly and the carboxyl-terminal lobes are light. It can be seen that introduction of only two amino acids between the two IQ motifs changes the neck region from a form in which the amino-terminal lobe of the first calmodulin interacts with the carboxyl-terminal lobe of the second (Fig. 4A) to a form in which the amino-terminal lobes of both calmodulins interact (Fig. 4C). It is also likely that the calmodulin bound to the first IQ motif might have interactions with the globular portion of the myosin head, which would not be available for subsequent calmodulins. Thus, it is clear that not all calmodulins bound to the IQ motifs of a given myosin will be equivalent. Taking into account the possible conformations of calmodulin (perhaps determined by the nature of the IQ motif described earlier) and the variable stagger distances between IQ motifs, it is evident that two different myosins, each containing four IQ motifs, might have necks with very different properties, depending on the sequence of the neck region. Similarly, Ca^{2+} binding may have differential effects on different calmodulin–IQ complexes and, in fact, it is known that Ca^{2+} can be activating or inhibitory, depending on the myosin isoform (see later). Taken together, these phenomena could have a tremendous impact on the rigidity of the neck domain and on the regulation of the molecule.

The preservation of the light-chain-binding neck region in all classes of myosin suggests a common function. It is likely that this region operates as a lever arm to amplify smaller ATP-dependent conformational changes in the head region in order to increase the length of the power stroke during actin filament translocation. The speed at which a myosin can translocate actin is a function of the length and frequency of the power stroke. Therefore, a myosin with a longer neck should be able to translocate actin faster than a myosin with a shorter neck if the stroke frequency remains the same. This can be tested using an *in vitro* motility assay in which myosin bound to a surface translocates fluorescently labeled actin filaments (Sellers *et al.*, 1993). Evidence to support the neck/lever arm hypothesis comes from two sources. Uyeda *et al.* (1996) constructed and expressed mutant *Dictyostelium* myosin II molecules in which IQ motifs were either deleted or duplicated to give myosins with either short necks or long necks compared to wild-type myosins, which contain two light chains per neck. Their *in vitro* motility data demonstrated a correlation between speed of movement and length of the lever arm. In another study, the *Dictyostelium* myosin heavy chain was truncated at amino acid 761, which deletes the light-chain binding neck region and the rod (Uyeda *et al.*, 1996). These truncated heads were fused with one or two 120 amino acid α-actinin repeats, which are rigid, compact structures about 6 nm in length. The length of the one α-actinin–fusion motor was a little shorter than the length of the myosin head plus the normal light-chain-bound neck region, whereas that of the two α-actinin–fusion motor was a little longer. The velocity of the two α-actinin–fusion motor was the fastest, followed by wild-type myosin and then the shorter one α-actinin–fusion motor, suggesting that the neck region is indeed a lever arm (Anson *et al.*, 1996).

Some myosin heavy-chain genes can be alternatively spliced to include additional IQ motifs (Ruppert *et al.*, 1993; Kellerman and Miller, 1992; Halsall and Hammer, 1990). If the neck does function as a lever arm, then the alternatively spliced isoform containing more IQ motifs should translocate actin filaments faster than the shorter isoform if the same motor domain is present. In addition to these structural considerations, the neck region is implicated in the regulation of some myosins. Ca^{2+} binding to one of the light chains of scallop myosin acts as a trigger to release the light-chain-dependent inhibition of the myosin enzymatic properties (Jancso and Szent-Györgyi, 1994). Smooth and nonmuscle myosin II are regulated by phosphorylation of one of the light chains by a Ca^{2+}/calmodulin-dependent enzyme termed myosin light-chain kinase (Sellers, 1991).

B. Calmodulin-Binding Properties of Individual Classes of Myosins

1. Myosin I

Many of the unconventional myosins depicted in Fig. 1 have never been studied as proteins and their enzymatic properties, *in vivo* localizations, and cellular functions are unknown. One well-studied example of a calmodulin-binding unconventional myosin is chicken intestinal epithelial brush border myosin I, which was the first of the vertebrate unconventional myosins discovered (Carboni et al., 1988; Collins and Borysenko, 1984; Howe and Mooseker, 1983). It was first described as a 110-kDa calmodulin-binding protein in the intestinal epithelial brush border that was extractable by ATP (Matsudaira and Burgess, 1979) and was subsequently shown to be a myosin on the basis of its actin-activated MgATPase activity, its ability to translocate actin filaments in an *in vitro* motility assay, and by its sequence homology to other myosins (Garcia et al., 1989; Mooseker and Coleman, 1989; Collins et al., 1990). That the bound "light chain" was calmodulin was confirmed by direct amino acid sequencing (P. Matsudaira, personal communication). The light chain from myosin IC from *Acanthamoeba castellani* has been partially sequenced by conventional methods and its cDNA was cloned and fully sequenced (Wang et al., 1997). Interestingly, its sequence is different than that of Acanthamoeba calmodulin, however phylogenetic analysis suggests that it is more similar to calmodulin than to either of the two classes of light chains associated with the myosin II class (Fig. 5).

Purified preparations of brush border myosin I probably contain three molecules of calmodulin, which is consistent with the presence of three IQ motifs in the cDNA sequence (Garcia et al., 1989). However, a partial cDNA fragment of brush border myosin contained a sequence that corresponded to an alternatively spliced isoform containing a fourth calmodulin-binding site (Halsall and Hammer, 1990). Some reports suggest that four calmodulins are bound in tissue-purified brush border myosin I (Swanljung-Collins and Collins, 1991). It is not clear whether the longer alternatively spliced isoform was purified or whether the differences lie in different methods of stoichiometric analysis.

Ca^{2+} inhibited the ability of brush border myosin I to move actin filaments in an *in vitro* motility assay (Wolenski et al., 1993; Collins et al., 1990). At least one of the bound calmodulin molecules dissociates on treatment with Ca^{2+}, contributing to the inhibition (Collins et al., 1990). Once calmodulin was dissociated, the myosin remained inactive, even in the presence of EGTA, whereas activity could be restored by the readdition of calmodulin. Ca^{2+} also inhibited the actin-activated MgATPase activity via an increase in the MgATPase activity of brush border myosin I in the absence of actin such that there was no subsequent increase in activity on the addition of actin (Collins et al., 1990).

The inhibition of the enzymatic activity of myosin I by Ca^{2+} is in direct contrast to the effect of Ca^{2+} on nonmuscle myosin II where Ca^{2+}

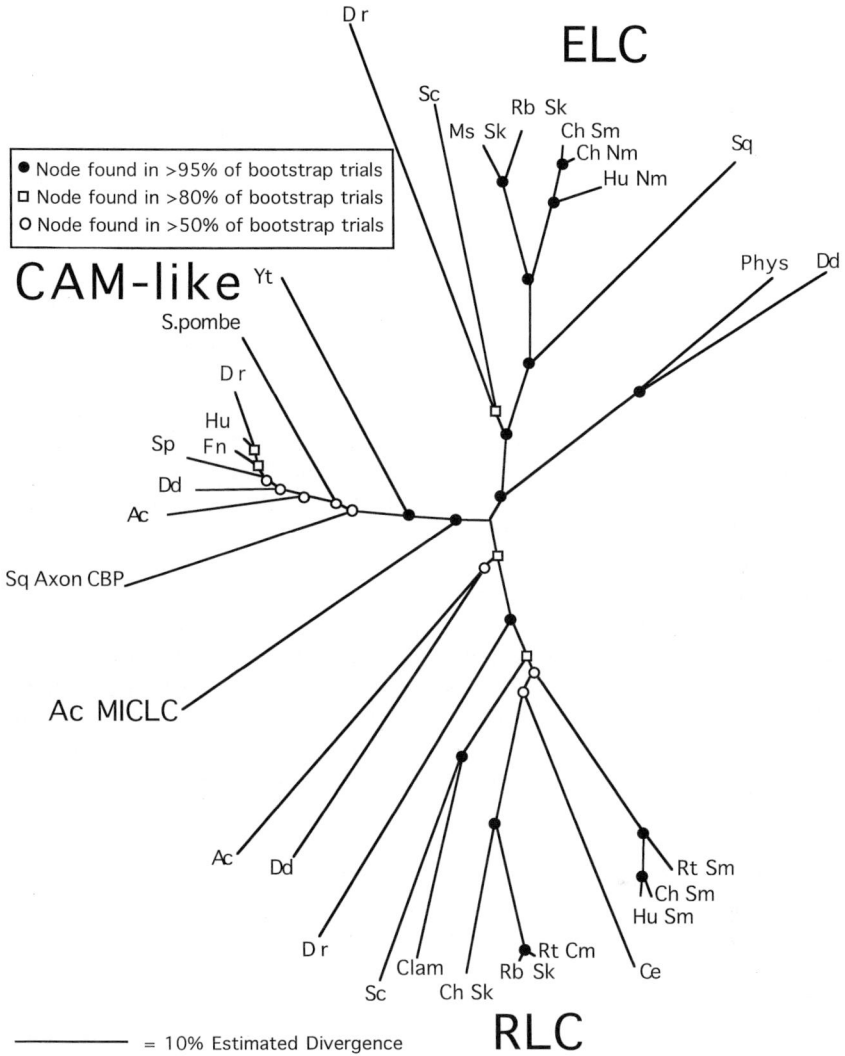

Fig. 5. Phylogenetic analysis of regulatory light chains (RLC), essential light chains (ELC), calmodulin (CAM-like), and the light chain (Ac MICLC) for *Acanthamoeba* myosin IC. Adapted from Wang et al. (1997). Note that the *Acanthamoeba* myosin IC light chain is most similar to calmodulin, but is not identical to the authentic *Acanthamoeba* calmodulin. RLCs: Ac, *Acanthamoeba:* Dd, *Dictyostelium;* Dr, *Drosophila;* Ce, *C. elegans;* ChNM, chicken nonmuscle; ChS, chicken skeletal muscle; ChSm, chicken smooth muscle; Fn, potato late blight fungus; HuNm, human nonmuscle; HuSm, human smooth muscle; MsSk, mouse skeletal muscle; Phys, *Physarum;* RbSk, rabbit skeletal muscle; RtCm, rat cardiac muscle; RtSm, rat smooth muscle; Sc, scallop; Sp, spinach; Sq, squid; Sq Axon CBP, squid axon calcium-binding protein; Yt, yeast, *Saccharomyces cerevisiae.*

indirectly activates the enzymatic activity of nonmuscle myosin II via the calmodulin-dependent phosphorylation of its regulatory light chain by myosin light-chain kinase (Sellers, 1991). This suggests that these two classes of myosin are reciprocally regulated. When myosin I is activated, myosin II is not and vice versa. In general, the two myosins are also differentially localized in the cell (Fukui *et al.*, 1989; Conrad *et al.*, 1993).

Myosin I isoforms fall into several subclasses based on sequence analysis (Mooseker and Cheney, 1995), and seven myosin I gene loci have been identified to date in mouse (Hasson *et al.*, 1996). It is likely that different subclasses may have varying functions and enzymatic properties. The effect of Ca^{2+} on the enzymatic activity of two other subclasses has also been explored. The motility of myosin Iα was inhibited in the presence of Ca^{2+}, and its actin-activated MgATPase activity was reduced but not abolished (Briggs *et al.*, 1994). Two reports examined the properties of myosin Iβ. Using tissue-purified protein, the first report found that actin-activated MgATPase activity was increased in the presence of Ca^{2+} (Barylko *et al.*, 1992), whereas another report using recombinant baculovirus-expressed protein found that Ca^{2+} inhibited both actin-activated MgATPase activity and *in vitro* motility activity (Zhu *et al.*, 1996). Another subclass of myosin I has been proposed to possess two classes of calmodulin-binding sites based on studies with fusion proteins. The first IQ motif binds calmodulin in the absence of free Ca^{2+} whereas the second IQ motif binds calmodulin only in the presence of Ca^{2+} (Bähler *et al.*, 1994).

A given cell may express multiple myosin I isoforms that are differentially distributed within the cell to either the plasma membrane or to different membranous organelles or vesicles (Baines *et al.*, 1992; Baines and Korn, 1990). Baines *et al.* (1992) examined the localization of three myosin I isoforms in *Acanthamoeba* using isoform-specific antibodies. Although there was some overlap, in general, each isoform had its own characteristic distribution. An example of this was the myosin IC isoform, which was the only isoform found associated with the contractile vacuole. In this regard, it is interesting that the contractile vacuole of *Dictyostelium* is highly enriched in calmodulin along with an unidentified myosin that cross-reacts with myosin IC-specific antibodies (Zhu and Clarke, 1992). Brush border myosin I is prominently localized in the apical brush border of enterocytes where it appears to link the plasma membrane with the actin bundle (Matsudaira and Burgess, 1979). No direct motile role has been determined for the myosin in the microvillus, and one speculation is that its function is to target and concentrate calmodulin

7. Calmodulin-Binding Proteins of the Cytoskeleton

into the brush border microvilli. In this regard, it should be noted that the calmodulin concentration of the brush border has been estimated to be about 0.6 mM (Collins and Borysenko, 1984). Another speculation is that its role is structural, to stiffen the microvillus. Fath and Burgess (1993) have shown brush border myosin I to be localized on vesicles of Golgi origin and suggested that it plays a role in the transport of these vesicles to the microvilli. Several studies report that myosin I, in general, does not appear to be in stress fibers (which have a high local concentration of myosin II), but is more concentrated toward the periphery of the cell (Conrad *et al.*, 1993; Wagner *et al.*, 1992). Myosin Iβ is found in the stereocilia of hair cells in the amphibian cochlea (Gillespie *et al.*, 1993). Instead of being located uniformly along the actin bundle of the stereocilia, it is targeted to a patch near the tip. It has been proposed to be involved in the process of adaptation via a motile process involving the "tip" link, a thin strand of material linking one stereocilia to another (Gillespie *et al.*, 1993; Walker and Hudspeth, 1996). The tip link is proposed to control the opening of a Ca^{2+} channel involved in sensory perception.

It is very likely that the tail domains serve as localization targets (Doberstein and Pollard, 1992; Adams and Pollard, 1989; Ruppert *et al.*, 1995). A myosin I tail-binding protein has been found, suggesting that different membranes may have specific receptors for myosins (Xu *et al.*, 1995). The tail domains of the different subclasses of myosin I are divergent and, depending on the subclass, may have interesting motifs such as *src* homology 3 (SH3) motifs, polybasic regions, or a region rich in glutamic acid, proline, and alanine residues (GPA region) (Sellers *et al.*, 1996; Mooseker and Cheney, 1995). Additional sequence motifs are found in some of the other classes of myosin such as pleckstrin homology domains, GTPase-activating (GAP) domains, and talin homology domains (Mooseker and Cheney, 1995; Cope *et al.*, 1996; Sellers *et al.*, 1996; Sellers and Goodson, 1995). Many of these motifs are commonly found in proteins that interact with the cytoskeleton or are involved in signal transduction.

2. Myosin III

The myosin III class is currently represented by a single member that is specifically localized to photoreceptor cells of *Drosophila* where it is referred to as NinaC, based on its genetic locus (Porter *et al.*, 1992). Phylogenetic analysis reveals that myosin III is more divergent in sequence than most classes of myosin, and indeed its molecular structure is novel among myosins (Sellers and Goodson, 1995). A protein kinase domain is found at the amino terminus followed by a conserved myosin motor domain, a neck region, and a tail. Two alternatively spliced isoforms differ in the sequence of their tail and neck regions (Montell and Rubin, 1988). One isoform with a molecular mass of 174 kDa (p174) has two IQ motifs and a longer tail region. It is found exclusively in the polarized actin microvilli of the rhabdomere. The second isoform has a molecular mass of 132 kDa (p132). It has an amino-terminal sequence that is identical with that of p174 up to the first IQ domain, but then its sequence diverges to form a much shorter tail. Thus p132 shares the first IQ sequence with p174, which is the only IQ motif found in p132. This isoform has a cytoplasmic localization within the cell body of the rhabdomere (Porter et al., 1992).

Genetic analysis has shown that deletion of the entire p174 isoform results in an abnormal electrical retinogram (ERG), rapid light-induced degeneration of the retina, and lack of localization of calmodulin to the microvilli (Porter et al., 1995; Porter and Montell, 1993). Deletion of the 132-kDa isoforms has little if any effect on these parameters. A truncated derivative of p174, which was missing 217 amino acids following the motor domain, was produced (Porter et al., 1993). This myosin had the kinase and motor domains intact, but not the calmodulin-binding IQ motifs. When expressed in transgenic flies, the mutant myosin displayed a normal localization to the rhabdomeres, but most of the calmodulin was in the extracellular central matrix rather than in the rhabdomeres and subrhabdomeral cytoplasm. These flies also had a defect in phototransduction, but did not undergo the light-induced retinal degeneration that is characteristic of myosin III null flies. In a more detailed study, separate deletions of the first IQ motif (termed C1) and the second IQ (C2) motif of p174 were performed (Porter et al., 1995). With p174ΔC1 transgenic flies, staining was seen exclusively, although at less intensity compared to wild type, in the rhabdomeres and not in the cell bodies. The ERG of these flies was almost like wild-type flies. In p174ΔC2 transgenic flies (which also expressed wild-type p132) the relative calmodulin staining intensity in the rhabdomere was reduced compared to wild type, but there was normal expression in the cell body. These flies exhibited an ERG similar to myosin III null flies. Using a different electrophysiological paradigm, both transgenic fly lines were found to be defective in the termination of the photoresponse (Porter et al., 1995). The nature of the calmodulin-binding sites was also studied *in vitro*. Interestingly, calmodulin binding to myosin III requires Ca^{2+} (Porter et al., 1995).

Little is known about the enzymatic function of myosin III. It has not yet been purified and studied enzymatically, and certain amino acids found near the active site that are highly conserved in most myosins are not as well conserved in myosin III. The kinase domain has been expressed as a recombinant fragment and shown to possess kinase activity (Ng et al., 1996). Although part of its function may be to maintain the structural integrity of the actin microvillus, these studies strongly suggest that it may also be responsible for calmodulin targeting to the microvilli. Whether this is in the form of a direct, active transport facilitated by the myosin motor domain or passive accumulation is not known. Several other potential targets of calmodulin in the rhabdomere are possible, including calmodulin-sensitive adenyl cyclase (adenylate cyclase), protein kinases, phosphatases, and channels. In addition, the motor domain of myosin III may position its kinase moiety to phosphorylate specific proteins (Porter et al., 1993).

3. Myosin V

Myosin V was first purified as a 190-kDa calmodulin-binding ATPase from brain (Larson et al., 1990; Espindola et al., 1992). It was subsequently cloned from chicken and shown to be homologous to the protein encoded by the *dilute* loci in mouse of which there are many mutant alleles (Mercer et al., 1991; Espreafico et al., 1992). In addition, it has now been found in *Caenorhabditis elegans* (Genbank accession No. U52516) and *Drosophila* (N. M. Bonafé and J. R. Sellers, 1998) and is known to be represented by two genes in mouse (Hasson et al., 1996) and yeast (Johnston et al., 1991; Haarer et al., 1994). It has a conserved myosin motor domain at the amino terminus followed by six IQ domains and a tail that is able to dimerize, but apparently does not form thick filaments. It is proposed that calmodulin associates with the neck region along with two other low molecular weight proteins that probably are identical with the essential light chains of conventional myosins (Espindola et al., 1996). The effect of the six IQ domains is readily apparent in structural studies of myosin V where the head/neck region appears twice as long as that of myosin II (Fig. 6). The function of myosin V is inferred from the phenotypes of the *dilute* mutations and the ablation of the genes in yeast. The term "dilute" comes from the pale coat color of these mutant mice, indicating a role for myosin V in the transport of melanosomes (Mercer et al., 1991; Wei et al., 1997). Mice homozygous for some *dilute* mutations have neurological seizures that also suggest a

Myosin II

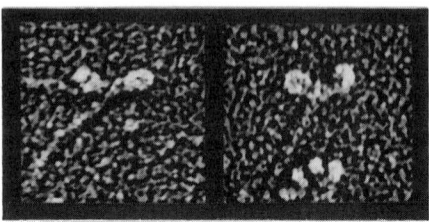

Myosin V

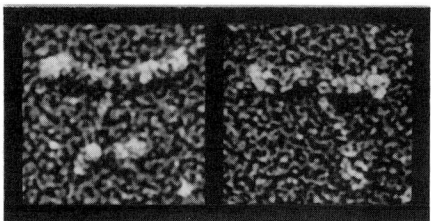

scale bar ——
50nm

Fig. 6. Electron micrographs of rotary-shadowed images of *Acanthamoeba* myosin II and chicken myosin. Images were provided by Dr. Richard Cheney (University of North Carolina). Printed with permission.

role in neurosecretion. The yeast phenotype associated with temperature-sensitive mutations of the MYO2 gene is a block in budding and the accumulation of small vesicles (Govindan *et al.*, 1995; Johnston *et al.*, 1991). The cells arrest as large unbudded cells and have mislocalized actin and chitin (Johnston *et al.*, 1991). Both of these phenotypes are consistent with a role in vesicle translocation from the mother cell into the bud. Interestingly, calmodulin also strongly localizes to the bud site in wild-type yeast, but not in yeast where the actin cytoskeleton is disrupted (Brockerhoff and Davis, 1992).

In vitro analysis of the enzymatic properties of myosin V shows that it has a low actin-activated MgATPase activity in the absence of Ca^{2+}. The addition of Ca^{2+} markedly activates the MgATPase in the presence of actin (Nascimento *et al.*, 1996; Cheney *et al.*, 1993). However, the rate of *in vitro* movement of actin filament slowly declines with time in the presence of Ca^{2+} (Cheney *et al.*, 1993). This decline in the rate of actin filament translocation may be due to dissociation of some of the bound calmodulin in the presence of Ca^{2+} (Nascimento *et al.*, 1996). In support of this idea, the addition of exogenous calmodulin in the presence of Ca^{2+} gives a rate that is about 80% of that of myosin V in the absence of Ca^{2+}.

4. Other Myosin Classes

Myosin from classes VI, VII, and IX has not yet been purified to homogeneity, but there have been some studies of partially purified proteins. Myosin VI has been identified in both *Drosophila* and mammalian tissues (Hasson and Mooseker, 1994; Kellerman and Miller, 1992). In *Drosophila* embryos, it is associated with small motile vesicles that may provide the membrane necessary for cellularization (Mermall *et al.*, 1994). In mammals, it is present in a wide variety of tissues (Hasson and Mooseker, 1994). In LCC-PK cells, it has a punctate staining pattern throughout the cytoplasm that does not colocalize strongly with actin. A mouse deafness model (Snell's waltzer) has mutations in its myosin VI gene (Avraham *et al.*, 1995). In the ear, it is expressed in the cochlea exclusively in hair cells (Avraham *et al.*, 1995). Mammalian myosin VI contains only a single IQ motif that appears to bind calmodulin on the basis of mobility shift assays (Hasson and Mooseker, 1994). The myosin VIIa gene is also associated with deafness in both mice and humans (Weil *et al.*, 1995; Gibson *et al.*, 1995). This gene is defective in *Shaker-1* mice and in human Usher IB syndrome, which is characterized by profound deafness at birth followed by blindness due to retinitis pigmentosis. It is expressed in cochlea, retina, testes, lung, and kidney (Chen *et al.*, 1996). In the cochlea, it is expressed in the inner and outer hair cells where it is found in the apical stereocelia and in the cytoplasm (Hasson *et al.*, 1995). In the eye, it is found in the apical actin-rich domain of the pigmented epithelial cell layer (Hasson *et al.*, 1995). It has five

IQ motifs, but no direct data demonstrate that calmodulin is the associated light chain.

Myosin IX has been identified in rat and human tissue by cDNA cloning (Wirth *et al.*, 1996; Reinhard *et al.*, 1995). It possesses four IQ motifs, but calmodulin has not been directly identified as the light chain. The tail region of this myosin contains a GAP homology domain and a C_2H_6-type, zinc-binding domain. Expression of the GAP domain demonstrated it to be active in stimulating the GTP hydrolysis of members of the ras-related rho subfamily of small G-proteins (Reinhard *et al.*, 1995). Therefore, this putative calmodulin-binding myosin may interact directly with the rho family GTPases, which have been implicated in the regulation of actin filament localization and in the regulation of myosin II activity via effects on the myosin phosphatase. Myosins from classes VIII, X, XI, and XII have not yet been isolated even in crude form. Table I list the number of IQ motifs for each of these.

C. Kinesins: A Microtubular-Dependent Family of Motors

Kinesins represent a large superfamily of microtubule-dependent molecular motors. A novel member of this superfamily has been cloned from the plant *Arabidopsis*. It was termed AtKCBP and is a calmodulin-binding kinesin (Reddy *et al.*, 1996). The calmodulin-binding site was localized to the carboxyl-terminal region of the molecule and binding of calmodulin required Ca^{2+}. A 23 amino acid peptide derived from the putative-binding site was shown to interact in a Ca^{2+}-dependent manner with calmodulin. The sequence of this peptide does not bear homology with the IQ motif and is not found in any other kinesins to date other than newly cloned homologs of AtKCPB from other plants. The binding of microtubules to bacterially expressed AtKCBP was inhibited by the presence of Ca^{2+} and calmodulin (Song *et al.*, 1997).

III. CALMODULIN-BINDING MUSCLE PROTEINS

Several calmodulin-binding proteins are expressed abundantly in smooth and striated muscle tissue, and these proteins are often also expressed in other cells and tissues or have nonmuscle isoforms. In some cases the proteins are thought to be muscle-specific proteins, although

7. Calmodulin-Binding Proteins of the Cytoskeleton

TABLE I

Myosins IQ Domain Composition

Myosin	Source	Number of IQs	Evidence for CaM binding[a]	Reference
I	Rat myr4	1–4	1, 2	Bähler et al. (1994)
	Rat myr3	3–6	3	Stöffler et al. (1995)
	Rat myr1	2	3	Rüppert et al. (1993)
	Chicken BB	3–4	2, 5	Carboni et al. (1988)
				P. T. Matsudaira, (personal communication)
III	Drosophila			
	p174	2	4	Porter et al. (1995)
	p132	1	4	Porter et al. (1995)
IV	Acanthamoeba	1	0	Horowitz and Hammer (1990)
V	Chicken p190	6	2	Cheney et al. (1993)
VI	Porcine	1	1, 2	Hasson and Mooseker (1994)
	Drosophila	1	0	Kellerman and Miller (1992)
VII	Human	5	0	Chen et al. (1996)
VIII	Arabidopsis	4	0	Knight and Kendrick-Jones (1993)
IX	Rat myr5	4	0	Reinhard et al. (1995)
X	Rana catesbeiana	3	0	D. P. Corey (personal communication)
XI	Arabidopsis	6	0	Kinkema et al. (1994)
XII	Caenorhabditis elegans	3	0	Accession No. Z66563

[a] 0, none; 1, immunoprecipitation; 2, gel shift; 3, overlay; 4, affinity chromatography; and 5, sequencing.

with the rapid pace of discovery of protein homologs, this is clearly a dangerous claim to make. This section discusses some of the properties of these proteins.

A. Smooth Muscle Proteins

1. Caldesmon

Caldesmon is an actin-binding and calmodulin-binding protein found in smooth muscle and nonmuscle cells (Marston and Redwood, 1991; Sobue and Sellers, 1991). It has been proposed to be a major regulatory element of smooth muscle contractility based on its ability to inhibit the actin-activated MgATPase activity of smooth muscle myosin *in vitro*. Caldesmon is probably encoded by a single gene, but is present as alternatively spliced isoforms of different molecular weights in smooth muscle vs nonmuscle cell types (Bryan *et al.*, 1989; Hayashi *et al.*, 1989, 1991). The smooth muscle isoform, termed high molecular weight caldesmon (*h*-caldesmon) has a molecular mass of about 87 kDa (although it migrates with an apparent molecular mass of 130 kDa on SDS–polyacrylamide gels). The nonmuscle isoform has a molecular mass of about 59 kDa and is identical to the smooth muscle isoform except for the deletion of an internal segment of 234 amino acids. Caldesmon has a calmodulin-binding site located in the carboxyl-terminal region of the molecule (Riseman *et al.*, 1989; Bartegi *et al.*, 1990). Extensive characterization of the calmodulin-binding site suggests that three portions of the caldesmon heavy chain are involved, including the area around Trp-716, Trp-749, and Trp-779 (Zhuang *et al.*, 1995; Huber *et al.*, 1996) (sequence numbers are for the human high molecular weight isoform; the widely studied avian isoform is somewhat shorter in length due to differences at the amino terminus). The binding affinity is in the range of 10^6–10^7 M^{-1}. Caldesmon also contains a carboxyl-terminal actin-binding site and an amino-terminal myosin-binding site (Velaz *et al.*, 1989; Chalovich *et al.*, 1987). These two domains are separated by an elongated central region to give a length for *h*-caldesmon of 80 μm as measured by rotary shadowing in the electron microscope (Mabuchi and Wang, 1991).

In vitro studies have shown that caldesmon inhibits the actin-activated MgATPase activity of smooth muscle myosin and that this inhibition is reversed in the presence of Ca^{2+} and calmodulin (Marston and Lehman, 1985; Sobue *et al.*, 1985). When tropomyosin is also bound to the thin filaments, caldesmon inhibits the actin-activated MgATPase activity at a molar ratio of one caldesmon per 7–14 actin monomers (Marston and Lehman, 1985). Various models have been proposed to account for the inhibition of the MgATPase activity of myosin at the lower molar ratios of caldesmon to actin (Marston and Redwood, 1991). There are conflicting reports about the mechanism of the Ca^{2+}/calmodulin reversal of the inhibition of myosin activity by caldesmon. It is known that caldesmon binds more weakly to actin in the presence of calmodulin than in its absence. Some reports suggest that the reversal of inhibition is due to the calmodulin-induced dissociation of caldesmon from the actin filaments, whereas other reports suggest that the inhibition can be reversed by calmodulin even under conditions where the Ca^{2+}/calmodulin–

caldesmon complex is still bound to the actin filaments (Chalovich et al., 1987; Smith et al., 1987; Marston et al., 1994). Several models have been proposed to explain the "latch" state of smooth muscle contraction where the muscle is able to maintain tension with little expenditure of energy by invoking a caldesmon-linked tethering of actin and myosin filaments by virtue of its amino-terminal myosin-binding site and carboxyl-terminal actin-binding site, although there are other kinetic models that evoke slowly cycling myosin cross-bridges (Strauss and Murphy, 1996).

The interaction of caldesmon with mutant calmodulins was examined (Medvedeva et al., 1995). Data show that many mutations had little or no effect on the Ca^{2+}-dependent interactions between the two proteins, but the mutations of the first and second Ca^{2+}-binding sites affect the interaction dramatically. Because the affinity of caldesmon for calmodulin is relatively weak, it has been proposed that there are other Ca^{2+}-binding proteins that bind to caldesmon and reverse its inhibition of the MgATPase activity of myosin. Candidates for such a protein are the S100 proteins and caltropin (Pritchard and Marston, 1991; Skripnikova and Gusev, 1989; Mani et al., 1992).

In nonmuscle cells, caldesmon is localized predominantly to stress fibers colocalizing with tropomyosin (Yamashiro-Matusmura and Matusmura, 1988) where it might play a role in stabilizing actin filaments against gelsolin-induced depolymerization in addition to the obvious role of regulating the interaction of myosin with actin (Ishikawa et al., 1989). Nonmuscle tropomyosin isoforms are shorter than their muscle homologs and bind more weakly to actin (Novy et al., 1993). Caldesmon has also been shown to increase the binding affinity of these tropomyosin isoforms to actin (Yamashiro-Matusmura and Matusmura, 1988). Interestingly, caldesmon binds more weakly to actin filaments during mitosis than during interphase (Yamashiro et al., 1990). This effect determined to be due to a mitosis-specific phosphorylation of caldesmon (presumably by cdc2 kinase) that reduces its affinity for actin (Yamakita et al., 1992).

2. Calponin

Calponin is a relatively recently discovered protein whose function is unclear and controversial. It was initially discovered as a calmodulin-dependent actin-binding protein from smooth muscle that could inhibit the actin-activated MgATPase activity of smooth muscle myosin (Winder and Walsh, 1990; Abe et al., 1990; Takahashi et al., 1988). However, nearly stoichiometric (with actin) quantities of calponin were required for this and tropomyosin had no effect on this inhibition.

Localization studies indicate that calponin may have little or nothing to do with the regulation of actomyosin interaction. Many proteins of smooth muscle cells are quite compartmentalized. There is a "contractile" domain that contains α-actin, myosin, tropomyosin, and caldesmon, and a "structural" domain that contains desmin, filamin, and β-actin, but not myosin. Calponin appears to be localized exclusively or mainly to the structural domain in close alignment with desmin, an intermediate filament protein (North et al., 1994; Mabuchi et al., 1996). In vitro evidence shows that it can interact directly with desmin (Wang and Gusev, 1996). Immunoprecipitation studies using various antibodies suggest that calponin is not associated with the same subset of actin filaments that contain caldesmon in smooth muscle (Lehman, 1991).

Other evidence against its possible role as a Ca^{2+}-dependent actomyosin regulatory protein comes from kinetic analysis of the interaction between Ca^{2}/calmodulin and calponin. Although the affinity of calponin for calmodulin ($K_d = 80$ nM) is higher than that of caldesmon, its rate of binding to calmodulin is too slow to be an effective switch (Winder et al., 1993).

B. Skeletal Muscle Proteins

The contractile apparatus of skeletal muscle is a highly ordered array of myosin-containing thick filaments and actin-containing thin filaments that interdigitate to form the characteristic striated appearance observed during light or electron microscopy (Fig. 7). In addition to being capable of active shortening, the muscle fiber is elastic. Resting muscle can be stretched until the thick and thin filaments no longer overlap, but in doing so a resistive force or passive tension is developed (Helmes et al., 1996; Wang et al., 1993; Horowitz, 1992). An underlying contributor to both the order and the inherent elasticity of muscle is a set of very large and long proteins that have been recently discovered.

1. Titin and Related Proteins

Titin (also called connectin) is the third most abundant protein in the skeletal and cardiac muscle sarcomere, comprising about 10% of the

7. Calmodulin-Binding Proteins of the Cytoskeleton 371

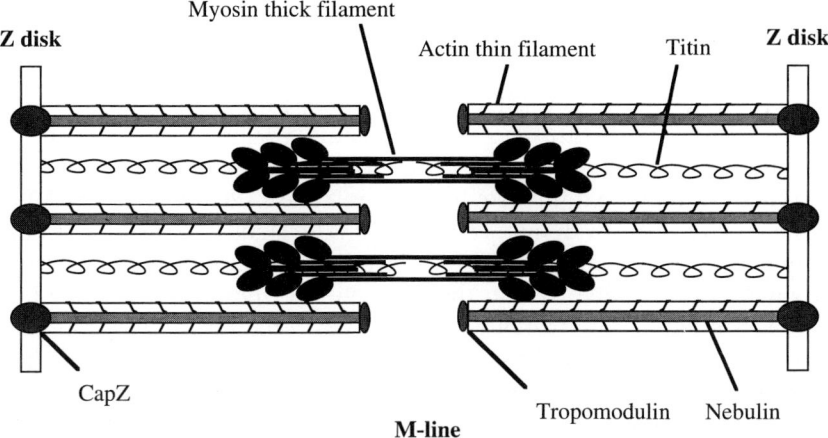

Fig. 7. Schematic diagram of a striated muscle sarcomere showing that a titin molecule is continuous from the Z-line to the center of the thick filament (M-line) and that nebulin is continuous for the length of the thin filament.

myofibrillar protein mass (Trinick, 1991). The molecular mass of titin has been found to be 3000 kDa following cloning and sequencing of the 82-kb cDNA encoding for the complete nucleotide sequence of titin (Labeit and Kolmerer, 1995). Smaller homologs of titin, termed minititin, projectin, and twitchin, have been identified in invertebrate muscle (Ayme-Southgate *et al.*, 1991; Heierhorst *et al.*, 1994; Benian *et al.*, 1989; Nave and Weber, 1990; Nave *et al.*, 1991; Vibert *et al.*, 1993).

Rotary shadowing of isolated titin molecules using electron microscopy shows that titin is about 1 μm long, and immunoelectron microscopy with epitope-mapped antibodies demonstrates that a titin molecule in muscle spans the distance from the M-line to the Z-line with the globular amino-terminal head region binding near the M-line and its carboxyl terminus near the Z-line (Fürst *et al.*, 1988). This molecular arrangement makes titin well suited to serve as an elastic band in keeping the thin filaments centered during contraction (Horowits *et al.*, 1986; Horowits, 1992). The source of the elasticity is found in the molecular structure of titin. It is a multidomain protein, consisting primarily of fibronectin type III and C2 immunoglobulin repeating motifs (Labeit *et al.*, 1990). The concomitant presence of these two motifs is a characteristic feature of myosin-binding proteins as they are found in a growing number of proteins such as C-protein, M-protein, skelemin, and myosin light-chain kinases (Linke *et al.,* 1996; Einheber and Fischman, 1990). The source of the elasticity of titin is found in its carboxyl-terminal region, which is localized in the I-band region (Fürst *et al.,* 1988).

In addition, both titin and twitchin contain, close to their carboxyl termini, a putative kinase domain with extensive sequence homology to the catalytic domain of myosin light-chain kinase (Schneider *et al.,* 1985; Pato and Kerc, 1988; Benian *et al.,* 1989). The significance of the kinase domain of titin is unclear, but twitchin-like molecules in invertebrates might be directly involved in the regulation of contraction–relaxation cycles. Mutants of the twitchin-encoding *unc-22* gene of *Caenorhabditis elegans* are phenotypically characterized by a constant twitch of their body wall muscles and an abnormal sarcomere structure (Williams and Waterston, 1994; Waterston *et al.,* 1980). These abnormalities can be rescued by missense mutations located in the head of unc-54-encoded myosin heavy chains, which suggests that twitchin may somehow exibit its function via its interaction with the myosin heads, possibly through its kinase domain (Moerman *et al.,* 1982). Kinase activity has been reported for the expressed protein kinase domain of *C. elegans* twitchin in the form of autophosphorylation of itself and phosphorylation of model peptides based on the myosin light sequence (Lei *et al.,* 1994) and for molluscan twitchin in the form of autophosphorylation and phosphorylation of the light chains of molluscan myosin (Heierhorst *et al.,* 1995). The expressed kinase domain of both twitchin and titin binds CaM in the presence of Ca^{2+} (Gautel *et al.,* 1995; Heierhorst *et al.,* 1994), but calmodulin does not activate the kinase activity of the enzymes (Heierhorst *et al.,* 1996). S100, another Ca^{2+}-binding protein, has been shown to enhance the activity of molluscan twitchin significantly (Heierhorst *et al.,* 1996). Because S100 is an abundant protein in muscle, it is the likely regulatory element for this family of kinases. However, even in light of a Ca^{2+}-dependent regulatory protein (S100) for both twitchin and titin and the identification of the regulatory light chains of myosin as a potential substrate, the precise function of this kinase activity is unknown as it would likely have access to only a small number of myosin molecules within the confines of the sarcomeric lattice.

2. Nebulin

Nebulin is a family of giant actin-binding protein of 600–900 kDa, present exclusively in skeletal muscle sarcomere and absent from cardiac muscle in verebrates (Zhang *et al.*, 1996; Kruger *et al.*, 1991). Although nebulin is absent from cardiac muscle, there is nevertheless a smaller protein termed nebulette (Moncman and Wang, 1995) with some sequence homology found in this tissue. Nebulin forms long nonelastic filaments that are associated with F-actin (Wright *et al.*, 1993). A single nebulin polypeptide spans about 1 μm and is attached via its C terminus to the Z-line (Wright *et al.*, 1993). Nebulin has been proposed to be a length-regulating template for actin as the characteristic length of thin filaments is proportional to the size of the expressed nebulin molecule (Wright *et al.*, 1993). It has a repeating sequence (module) of 35 residues with a super repeat of seven modules. Root and Wang (1994) proposed a role for nebulin in regulating actomyosin interaction. Recombinant fragments of seven to eight modules bound with high affinity to actin, the myosin head, and calmodulin. The stoichiometry of binding was approximately one actin monomer and one myosin bound/nebulin fragment. The authors showed that nebulin fragments from the amino-terminal half, situated in the actomyosin overlap region of the sarcomere, inhibit actomyosin Mg^{2+}-ATPase activity and the sliding velocity of actin over myosin in *in vitro* motility assays. Calmodulin reversed the ATPase inhibition and accelerated actin sliding in a Ca^{2+} dependent manner. Calmodulin also clearly affected the binding of nebulin to F-actin and to myosin heads, and evidence was presented for the binding of the seven- to eight-module fragment to calmodulin attached to a solid phase. The authors proposed a model where nebulin in skeletal muscle behaves similarly to caldesmon in smooth muscle.

3. Dystrophin and Utrophin

Dystrophin is another large protein, originally found in the skeletal muscle, whose exact role has yet to be elucidated. It is a 427-kDa elongated protein, consisting of four major structural domains (Ervasti and Campbell, 1993). There is an amino-terminal domain, which binds F-actin, an elongated rod-like central domain, a cysteine-rich domain homologous to α-actinin, and a small carboxyl-terminal domain. It is localized in the cytosolic face of the muscle membrane where it probably functions to anchor actin filaments to the membrane via a set of membrane-anchored glycoproteins (Ervasti and Campbell, 1993; Byers *et al.*, 1991). Another protein, termed utrophin, is expressed ubiquitously and has a similar localization to dystrophin (Blake *et al.*, 1996). The two proteins share considerable sequence and domain homologies (Winder et al., 1995).

Mutations in the dystrophin gene give rise to the Duchenne/Becker muscular dystrophies (DMD), which are fatal degenerative sex-linked genetic diseases of muscle (Hoffman and Kunkel, 1989; Hoffman et al., 1987). Two hypotheses are commonly put forth to explain the function of dystrophin in muscle. One is that dystrophin stabilizes the sarcolemmal membrane and protects it against the stress of muscle contraction (Hoffman and Kunkel, 1989). The second possible function is in the regulation of the intracellular Ca^{2+} concentration (Franco and Lansman, 1990). In this regard it is interesting that an elevated level of intracellular Ca^{2+} has been reported in DMD muscle cells leading to a increased protein degradation rate in the cell (Turner et al., 1991).

Calmodulin binding to dystrophin and utrophin has been studied, but the results have been controversial. Madhavan et al. (1992) proposed two calmodulin-binding consensus sequences on dystrophin; one located in the amino-terminal part of dystrophin and the other in the carboxyl-terminal part of the molecule. Using affinity chromatography and ELISA, Bonet-Kerrache et al. (1994) showed that the amino-terminal domain of dystrophin was specifically bound to calmodulin in a Ca^{2+}-dependent manner. Because the amino-terminal region also contains the binding sites to actin, the authors investigated the effect of calmodulin on the dystrophin-actin interaction using various recombinant fragments from the amino-terminal region and observed that the Ca^{2+}-dependent interaction of calmodulin to dystrophin did not regulate the binding of actin of dystrophin. However, when Jarrett and Foster (1995) investigated the same question with slightly different recombinant proteins, they confirmed the existence of a calmodulin-binding site, but demonstrated a partially inhibitory effect of calmodulin on the actodystrophin interaction. An effect of calmodulin on the interaction of an amino-terminal fragment of utrophin with actin was also observed (Winder and Kendrick-Jones, 1995). More recently, the unique carboxyl-terminal domain has been demonstrated to exhibit a high-affinity calmodulin binding (Anderson et al., 1996). Because the carboxyl-terminal domain of dystrophin provides sites of attachment of the protein to membrane proteins, calmodulin binding may also play a role in modulating these interactions with the membrane. Furthermore, the interaction of dystrophin with its various dystrophin-associated proteins may be modulated in other ways. For instance, Madhavan and Jarrett (1994) reported the phosphorylation of dystrophin (sequence 2618–3074) in a $Ca^{2+}/$

calmodulin-dependent manner by a protein kinase activity endogenous to the dystrophin glycoprotein complex.

IV. ACTIN-BINDING PROTEINS OF THE MEMBRANE CYTOSKELETON

A. Membrane Cytoskeleton

The membrane cytoskeleton underlying the cytoplasmic surface of the plasma membrane plays a crucial role for many biological processes, including providing cell rigidity, mobilization of membrane receptors, endocytosis, exocytosis, cell polarity and morphology, cell adhesion, and membrane lipid turnover (Bennett and Gilligan, 1993). The erythrocyte cytoskeleton is very well studied (Lux and Palek, 1995). The major membrane peripheral proteins of mammalian erythrocytes are spectrin and actin, which form the two-dimensional structural lattice of the membrane skeleton (Bennett, 1990a; Ursitti *et al.,* 1991). Other proteins have been isolated from the membrane skeleton and are likely to participate in the cytoskeletal assembly and in the maintenance of the membrane cytoskeleton. These accessory proteins include protein 4.1, protein 4.9, adducin, tropomyosin, and tropomodulin. The cytoskeleton is linked to the bilayer through the association of spectrin with ankyrin and protein 4.1, which are attached to integral membrane proteins (Bennett and Gilligan, 1993; Bennett, 1990b) (Fig. 8). Under certain pathological conditions, erythrocytes undergo dramatic shape changes in response to increasing influxes of Ca^{2+} ion, suggesting that Ca^{2+} plays a major role in modulating the deformability of erythroid shape by affecting interactions between the membrane cytoskeletal elements (Quist and Roufogalis, 1976).

B. Spectrins

Spectrins are a family of acidic, largely α-helical actin-binding proteins of about 200–260 nm in length (Taniguchi and Manenti, 1993). Brain spectrin was initially termed fodrin, and the spectrin isoform in the

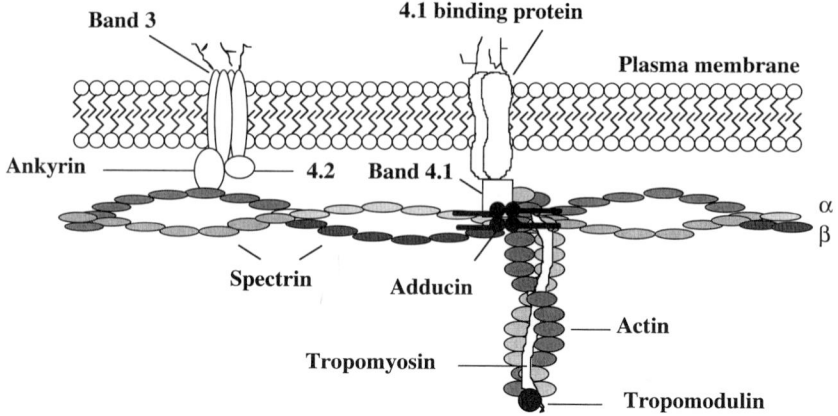

Fig. 8. Schematic diagram of the membrane cytoskeleton of the erythrocyte.

terminal web region of the chicken intestinal epithelial brush border cells was termed TW240/260. Spectrins are composed of rod-shaped α and β subunits laterally associated in an antiparallel dimeric orientation. Two such dimers further associate to form a tetrameric structure (Viel and Branton, 1996). α subunits of spectrin are composed of 22 domains, most of which are modules of a 106-residue repeat that is also found in other actin cross-linking proteins such as α-actinin and dystrophin. No role for actin binding has yet been ascribed to this motif, however, so its function may be to provide a method for elongating the molecule. β-Spectrins contain 19 domains arranged in three distinct segments. Seventeen of these domains are the 106-residue module (Yan et al., 1993). The region of spectrin involved in the association with actin contains the carboxyl terminus of the α subunit and the amino terminus of the β subunit (Karinch et al., 1990). The amino terminus of the β subunit contains a conserved actin-binding motif that is found in a number of other actin-binding proteins (Matsudaira, 1991). Most α-spectrins (but not the erythroid isoform) have a calmodulin-binding site located near the center of the molecule between repeats 11 and 12 (Leto et al., 1989; Harris et al., 1988; Sri Widada et al., 1989). The sequence of amino acids in this region is predicted to form the type of amphipathic helices that are typical of calmodulin targets (Leto et al., 1989).

The Ca^{2+}-dependent proteolysis of nonerythroid α-spectrin (fodrin) has been implicated in the regulation of secretion, neutrophil activation, and long-term potentiation in neurons (Perrin *et al.*, 1987; Jesaitis *et al.*, 1988; Lynch and Baudry, 1984). Calpain I cleaves spectrin in the middle of the α subunit in the calmodulin-binding region and in the COOH-terminal third of the β subunit (Harris *et al.*, 1988; Harris and Morrow, 1990). Cleavage at the β site requires calmodulin, which binds with high affinity to a single site on the α subunit (at the protease-sensitive site). *In vitro*-binding assays, nondenaturing gel electrophoresis, and velocity sedimentation identified a linkage between the Ca^{2+}-dependent calpain proteolysis of spectrin and the ability of calmodulin to regulate the self-association of spectrin and its interaction with actin. Therefore, three functional states appear to exist: (i) intact spectrin, which constitutively forms tetramers and binds F-actin; (ii) α-cleaved spectrin, which does not self-associate and binds F-actin in the presence of calmodulin; and (iii) α,β-cleaved spectrin, which neither forms tetramers or binds actin (Harris and Morrow, 1990).

It has been shown that one of the carboxyl-terminal domains that is not part of the 106-residue repeat motif has homology to calmodulin (Travé *et al.*, 1995a). Sequence analysis predicts the presence of four EF-hands within this segment. Nuclear magnetic resonance studies confirm that Ca^{2+} binding modifies the structure of the helical segments within this region (Travé *et al.*, 1995b). Thus, there are at least three methods by which Ca^{2+} could affect spectrin–protein interactions: via calmodulin binding to spectrin, via Ca^{2+}-dependent proteolysis of spectrin, and via direct Ca^{2+} binding to spectrin.

C. Adducin

Adducin was initially discovered as a calmodulin-binding protein in erythrocytes and is now also known to be expressed in brain as well as most other tissues (Gardner and Bennett, 1986). Adducin has been reported as a candidate for the calmodulin-dependent modulation of the spectrin/actin complex (Gardner and Bennett, 1987). It is a heterodimeric protein, composed of an α subunit (103 kDa) plus either a β (97 kDa) or γ (77 kDa) subunit (Joshi *et al.*, 1991; Joshi and Bennett, 1990). The β and γ subunits are homologous proteins to γ adducin. The domain structure of adducin includes an amino-terminal 39-kDa globular domain connected by a 9-kDa neck domain to an elongated, unstructured carboxyl-terminal protease-sensitive tail domain that has homologies with the MARCKS protein (Joshi and Bennett, 1990). The last 100 amino acids of the tail domain contain the binding sites for the spectrin/actin (Matsuoka *et al.*, 1996; Hughes and Bennett, 1995). The β subunit of adducin binds to calmodulin in a Ca^{2+}-dependent manner with intermediate affinity of about $5 \times 10^6 \, M^{-1}$ (Gardner and Bennett, 1986). A calmodulin-binding region was localized to residues 425–461 (Scaramuzzino and Morrow, 1993); however, a 31 amino acid peptide from the carboxyl-terminal MARCKS homology domain was shown to be the dominant binding site for calmodulin (Matsuoka *et al.*, 1996). This region also contains two putative consensus sequences for phosphorylation by PKA and PKC (Matsuoka *et al.*, 1996). Phosphorylation of β-adducin or the 31-residue peptide by either kinase inhibited calmodulin binding, and the presence of bound calmodulin inhibited the rate of phosphorylation of both subunits by PKA and of the β-adducin by PKC. This is of importance as PKA phosphorylation of adducin reduced its affinity for the spectrin/actin complex (Matsuoka *et al.*, 1996).

D. Interaction between Spectrin and Adducin

Two adducin dimers associate to form a tetrameric structure. The adducin self-association sites are in the amino-terminal globular domains, which allows the carboxyl-terminal tails to project outward. Adducin tetramers interact with the spectrin/actin complex through the tail regions. Using truncated monomeric spectrin fragments, Li and Bennett (1996) showed that spectrin interacts with actin via its first repeat domain and with adducin via its second repeat domain. The juxtaposition of calmodulin-binding sites, Ca^{2+}-dependent proteolytic sites, phosphorylation sites, and spectrin-binding sites on adducin and spectrin opens the possibility for multiple regulatory mechanism of this protein–protein interaction within the cell.

V. OTHER CYTOPLASMIC CALMODULIN-BINDING PROTEINS

A. Synapsin

The cytoskeletal architecture of the presynaptic terminal (neurosynaptic junction, electric organ, cerebellar cortex) involves two major elements: actin filaments, which form a network mostly associated with the presynaptic plasma membranes and active zones, and microtubules. Synapsin I appears to be a critical regulatory protein in the process of neuronal exocytosis (Rosahl et al., 1995). In mature neurons, synapsin I localizes on the cytoplasmic side of synaptic vesicles and appears to be present in virtually all nerve terminals (De Camilli and Jahn, 1990). Many biochemical studies indicate that synapsin I binds to actin, tubulin, spectrin, neurofilaments, and synaptic vesicles (Goold et al., 1995; Huttner et al., 1983; Goldenrign et al., 1986; Baines and Bennett, 1986; Bähler and Greengard, 1987). In particular it is an F-actin bundling protein (Petrucci and Morrow, 1987; Bähler and Greengard, 1987). In addition, it is one of the primary substrates for the Ca^{2+}/calmodulin-dependent protein kinases that are activated on depolarization of nerves endings (Goold et al., 1995). There are three primary sites of phosphorylation by these kinases; serine-9 (site I), serine-568 (site II), and serine-605 (site III) (Czernik et al., 1987). Site I is near the interaction site with actin and tubulin (Bähler et al., 1989). Phosphorylation at this site reduces the extent of F-actin bundling to some extent. Sites II and III are near the interaction site with synaptic vesicles (Benfenati et al., 1992).

Calmodulin interacts directly with synapsin I with a midrange binding affinity of about $2 \times 10^6 \, M^{-1}$, suggesting a second possible mode of regulation in addition to phosphorylation (Goold and Baines, 1994; Okabe and Sobue, 1987; Hayes et al., 1991). Two binding sites for Ca^{2+}/calmodulin were identified in the head region of synapsin I (Goold et al., 1995). Ca^{2+}/calmodulin had an inhibitory effect on F-actin bundling by synapsin I, and this effect was potentiated when synapsin I was phosphorylated at the various sites (Goold et al., 1995).

B. Myristoylated Alanine-Rich C Kinase Substrate

Myristoylated alanine-rich C kinase substrate (MARCKS) is a major acidic actin-binding phosphoprotein with an apparent M_r of 68,000–87,000 that binds calmodulin and its specifically phosphorylated by PKC during diverse cellular processes, including neurosecretion, growth factor-dependent mitogenesis, and leukocyte activation (Aderem et al., 1988; Patel and Kligman, 1987; Qin et al., 1996). This phosphoprotein has been also colocalized with microtubules, vinculin, and talin. Both phosphorylated and dephosphorylated forms interact with actin in a specific and saturable manner, but the protein is redistributed from the plasma membrane to the cytoplasm when phosphorylated (Hartwig et al., 1992). It is particularly of interest because MARCKS appears to mediate interactions between PKC- and calmodulin-mediated signaling pathways. It is composed of two highly conserved domains: an amino-terminal membrane-binding domain that also contains a myristoylation site and a highly basic α-helical domain of about 25 residues in the middle of the polypeptide chain (McIlroy et al., 1991). This helical segment contains all four of the protein kinase C phosphorylation sites, a calmodulin-binding site, an actin-binding site, and can associate electrostatically with membrane surfaces containing acidic lipids (McIlroy et al., 1991). Phosphorylation of the MARCKS-derived central peptide dissociates it from the membrane and also dramatically reduces its affinity for calmodulin (Kim et al., 1994; Thelen et al., 1991; Taniguchi and Manenti, 1993). It has been proposed that MARCKS may act to regulate free calmodulin levels through PKC pathways. The membrane attachment serves to make the calmodulin-binding motif a good substrate of PKC, and phosphorylation of MARCKS then increases cellular levels of calmodulin (Qin and Cafiso, 1996).

C. Desmocalmin

Desmocalmin has been purified from desmosomes based on its calmodulin-binding ability (Tsukita, 1985). It is a high molecular mass (240 kDa) protein that interacts with the keratin family of intermediate filaments in a Mg^{2+}-dependent manner (Tsukita, 1985).

VI. CALMODULIN-BINDING MICROTUBULE-BINDING PROTEINS

A. Microtubule-Associated Proteins

In many cell types, microtubules are dynamically instable, experiencing periods of rapid growth, followed by catastrophic shortening, then recovery to start a new growth phase. In neurons, however, it is important to have stable microtubules to support the generation and maintenance of neuronal morphology (Heidemann et al., 1984). That stable microtubules exist in neurons comes from observations that axonal microtubles remain of relatively constant length and orientation. The integrity and intracellular organization of the microtubule network seem to be modulated by molecular signals involving protein-to-protein interactions at the level of tubulin isoforms and a set of microtubule-associated proteins generically called MAPs (Maccioni and Cambiazo, 1995). The MAP2 and tau proteins belong to the same class of MAPs. They are characterized by a projection domain (amino-terminal) associated with a microtubule-binding domain (carboxyl-terminal) containing three or four repeats of an 18-residue sequence. Tau and MAP2 bind calmodulin in the presence of Ca^{2+} (Lee and Wolff, 1984) with an apparent dissociation constant of calmodulin for MAP2 of 7 μM (Lee and Wolff, 1984).

Weisenberg (1972) noted that the *in vitro* polymerization of tubulin into microtubules was markedly facilitated by reduction of the Ca^{2+} concentration, suggesting that Ca^{2+} may act as a regulator of the state of assembly of microtubules. Ca^{2+} exerts its effect by two principal modes of action. The first is a direct effect on tubulin, whereas the second is a pathway mediated by calmodulin (Berkowitz and Wolff, 1981). In the absence of calmodulin, MAPs promote nucleation and polymerization of microtubules, but the Ca^{2+}/calmodulin–MAP complex contributes toward microtubule depolymerization (Lee and Wolff, 1982).

Tau is a complex family of MAPs, derived from alternative splicing of a tau pre-mRNA, and originally purified based on their promotion of microtubule assembly from purified tubulin (Ennula et al., 1989; Cleveland et al., 1977). Found specifically in neuronal cells along the length of microtubules, tau most likely functions by regulating microtubule dynamics during neurite outgrowth (Drubin et al., 1985). With tau protein synthesis blocked by antisense RNA, neurons were unable to grow an axon (Caceres and Kosik, 1990).

B. Mip-90

More recently, a new microtubule-interacting protein, Mip-90, has been identified (Gonzalez et al., 1995). It binds tightly to a calmodulin–agarose matrix and promotes *in vitro* assembly of tubulin into microtubules. Mip-90 interacts with a β-tubulin peptide βII (422–434) containing the sequence of the MAP2 and tau-binding domains of β-tubulin. No evidence for functional similarities between Mip-90 and tau/MAP2 proteins has been reported. In addition, in contrast to MAPS, Mip-90 is thermosensitive and does not cycle with the microtubular polymer (Gonzalez *et al.*, 1995). The function of this protein remains unknown.

C. Stable Tubulin-Only Polypeptides

Stable tubulin-only polypeptides (STOPs) constitute a family of Ca^{2+}/calmodulin-regulated, microtubule-stabilizing proteins that are responsible for the formation of cold stable microtubules in various tissue extracts (Margolis *et al.*, 1986). These proteins colocalize with stable microtubules *in vivo* and bind to microtubules at a molar ratio of 1:10 with tubulin. They make microtubules extremely resistant to depolymerization by Ca^{2+}, drugs, and cold temperatures (Job *et al.*, 1987). Their effect is abolished by Ca^{2+}/calmodulin concentrations of 1–10 μM (Pirollet *et al.*, 1992; Job *et al.*, 1981). In contrast, MAP2 and tau cannot induce cold stability to microtubules, even when added to assembly systems at very high concentrations (Baas *et al.*, 1994; Job *et al.*, 1985). The primary structure of STOP (100 kDa) has two distinct repeated motifs with no sequence homology with known proteins (Bosc *et al.*, 1996). It also contains a putative SH3-binding motif close to its amino terminus. *In vitro*-translated 100-kDa STOP binds to both microtubules and calmodulin in the presence of Ca^{2+}. When STOP cDNA is expressed in cells that lack cold-stable microtubules, STOP associated with microtubules at 37°C. The microtubules in the transfected cells were cold stable and were resistant to nocodazole-induced depolymerization (Bosc *et al.*, 1996). Early observations showed that STOP activity was controlled by calmodulin kinase-dependent phosphorylation (Job *et al.*, 1983). Indeed, the STOP sequence contains four calmodulin kinase II phosphorylation motifs. The sequence is also present in other motifs that might regulate STOP activity, such as an ATP-binding domain (P-element) and a cdk family consensus phosphorylation site at the carboxyl terminus (Bosc *et al.*, 1996).

VII. SUMMARY

In summary, there is a plethora of calmodulin-binding proteins in the cell, many associated with cytoskeletal proteins. Calmodulin might be functioning to control the movement or association of molecular motors with actin and microtubules. It might also regulate actin and microtubule assembly or association of actin with other proteins or with the plasma membrane. Phosphorylation by various kinases may work synergistically or antagonistically with these calmodulin effects. The interaction of calmodulin with the target protein may be independent of the Ca^{2+} concentration, may require Ca^{2+}, or be inhibited by Ca^{2+}. It is interesting that the freely diffusible concentration of calmodulin in cells is much less than the total calmodulin concentration (Luby-Phelps *et al.,* 1995). Given the extensive association of calmodulin with the cytoskeleton, this fact is not so surprising. In fact, it is possible that certain proteins such as myosin III in *Drosophila* photoreceptors might function either as a calmodulin buffer or, perhaps, even as a calmodulin-dependent Ca^{2+} buffer.

ACKNOWLEDGMENTS

We thank Drs. Robert Adelstein, Harry Meng, and Dennis Buxton for comments on the text and Dr. He Jiang for help in preparing some of the figures. Drs. Richard Cheney, Carolyn Cohen, and Anne Houdusse provide materials used to make some of the figures for which we are grateful. We thank Dr. Richard Valee for helpful discussions.

REFERENCES

Abe, M., Takahashi, K., and Hiwada, K. (1990). Effect of calponin on actin-activated myosin ATPase activity. *J. Biochem. (Tokyo)* **108,** 835–838.

Adams, R. J., and Pollard, T. D. (1989). Binding of myosin I to membrane lipids. *Nature (London)* **340,** 565–568.

Adelstein, R. S., and Sellers, J. R. (1996). Myosin structure and function. *In* "Biochemistry of Smooth Muscle Contraction" (M. Barany, ed.), pp. 3–19. Academic Press, San Diego.

Aderem, A. A., Albert, K. A., Keum, M. M., Wang, J. K., Greengard, P., and Cohn, Z. A. (1988). Stimulus-dependent myristoylation of a major substrate for protein kinase C. *Nature (London)* **332,** 362–364.

Amano, M., Ito, M., Kimura, K., Fukata, Y., Chihara, K., Nakano, T., Matsuura, Y., and Kaibuchi, K. (1996). Phosphorylation and activation of myosin by Rho-associated kinase (Rho-kinase). *J. Biol. Chem.* **271,** 20246–20249.

Anderson, J. T., Rogers, R. P., and Jarrett, H. W. (1996). Ca^{2+}-calmodulin binds to the carboxyl-terminal domain of dystrophin. *J. Biol. Chem.* **271,** 6605–6610.

Anson, M., Geeves, M. A., Kurzawa, S. E., and Manstein, D. J. (1996). Myosin motors with artificial lever arms. *EMBO J* **15,** 6069–6074.

Avraham, K. B., Hasson, T., Steel, K. P., and Kingsley, D. M. (1995). The mouse Snell's *waltzer* deafness gene encodes an unconventional myosin required for structural integrity of inner ear hair cells. *Nat. Genet.* **11,** 369–375.

Ayme-Southgate, A., Vigoreaux, J., Benian, G., and Pardue, M. L. (1991). *Drosophia* has a twitchin/titin-related gene that appears to encode projectin. *Proc. Natl. Acad. Sci. U.S.A.* **88,** 7973–7977.

Baas, P. W., Pienkowski, T. P., Cimbalnik, K. A., Toyama, K. Bakalis, S., Ahmad, F. J., and Kosik, K. S. (1994). Tau confers drug stability but not cold stability to microtubules in living cells. *J. Cell Sci.* **107,** 135–143.

Bähler, M., and Greengard, P. (1987). Synapsin I bundles F-actin in a phosphorylation-dependent manner. *Nature (London)* **326,** 704–707.

Bähler, M., Benfenati, F. Valtorta, F., Czernik, A. J., and Greengard, P. (1989). Characterization of synapsin I fragments produced by cysteine-specific cleavage: A study of their interactions with F-actin. *J. Cell Biol.* **108,** 1841–1850.

Bähler, M., Kroschewski, R., Stöffler, H.-E., and Behrmann, T. (1994). Rat myr 4 defines a novel subclass of myosin I: Identification, distribution, localization, and mapping of calmodulin-binding sites with differential calcium sensitivity. *J. Cell Biol.* **126,** 375–389.

Baines, A. J., and Bennett, V. (1986). Synapsin I is a microtubule-bundling protein. *Nature (London)* **319,** 145–147.

Baines, I. C., and Korn, E. D. (1990). Localization of myosin IC and myosin II in *Acanthamoeba castellanii* by indirect immunofluorescence and immunogold electron microscopy. *J. Cell Biol.* **111,** 1895–1904.

Baines, I. C., Brzeska, H., and Korn, E. D. (1992). Differential localization of *Acanthamoeba* myosin I isoforms. *J. Cell Biol.* **119,** 1193–1203.

Bartegi, A., Fattoum, A., Derancourt, J., and Kassab, R. (1990). Characterization of the carboxyl-terminal 10-kDa cyanogen bromide fragment of caldesmon as an actin-calmodulin-binding region. *J. Biol. Chem.* **265,** 15231–15238.

Barylko, B., Wagner, M. C., Reizes, O., and Albanesi, J. P. (1992). Purification and characterization of a mammalian myosin I. *Proc. Natl. Sci. U.S.A.* **89,** 490–494.

Benfenati, F., Valtorta, F., Rubenstein, J. L., Gorelick, F. S., Greengard, P., and Czernik, A. J. (1992). Synaptic vesicle-associated Ca^{2+}/calmodulin-dependent protein kinase II is a binding protein for synapsin I. *Nature (London)* **359,** 417–420.

Benian, G. M., Kiff, J. E., Neckelmann, N., Moerman, D. G., and Waterston, R. H. (1989). Sequence of an unusually large protein implicated in regulation of myosin activity in *C. elegans. Nature (London)* **342,** 45–50.

Bennett, V. (1990a). Spectrin-based membrane skeleton: A multipotential adaptor between plasma membrane and cytoplasm. *Physiol. Rev.* **70,** 1029–1065.

Bennett, V. (1990b). Spectrin: A structural mediator between diverse plasma membrane proteins and the cytoplasm. *Curr. Opin. Cell Biol.* **2,** 51–56.

Bennett, V., and Gilligan, D. M. (1993). The spectrin-based membrane skeleton and micron-scale organization of the plasma membrane. *Annu. Rev. Cell Biol.* **9,** 27–66.

Berkowitz, S. A., and Wolff, J. (1981). Intrinsic calcium sensitivity of tubulin polymerization. The contributions of temperature, tubulin concentration, and associated proteins. *J. Biol. Chem.* **256,** 11216–11223.

Blake, D. J., Tinsley, J. M., and Davies, K. E. (1996). Utrophin: A structural and functional comparison to dystrophin. *Brain Pathol.* **6,** 37–47.

Bonafé, N. M., and Sellers, J. R. (1998). Molecular characterization of myosin V from *Drosophila melanogaster. J. Muscle Res. Cell Motil.* **19,** 129–141.
Bonet-Kerrache, A., Fabbrizio, E., and Mornet, D. (1994). N-terminal domain of dystrophin. *FEBS Lett.* **355,** 49–53.
Bosc, C., Cronk, J. D., Pirollet, F., Watterson, D. M., Haiech, J., Job, D., and Margolis, R. L. (1996). Cloning, expression, and properties of the microtubule-stabilizing protein STOP. *Proc. Natl. Acad. Sci. U.S.A.* **93,** 2125–2130.
Briggs, M. M., Maready, M., Schmidt, J. M., and Schachat, F. (1994). Identification of a fetal exon in the human fast Troponin T gene. *FEBS Lett* **350,** 37–40.
Brockerhoff, S. E., and Davis, T. N. (1992). Calmodulin concentrates at regions of cell growth in *Saccharomyces cerevisiae. J. Cell Biol.* **118,** 619–629.
Bryan, J., Imai, M., Lee, R., Moore, P., Cook, R. G., and Lin, W.-G. (1989). Cloning and expression of a smooth muscle caldesmon. *J. Biol. Chem.* **264,** 13873–13879.
Byers, T. J., Kunkel, L. M., and Watkins, S. C. (1991). The subcellular distribution of dystrophin in mouse skeletal, cardiac, and smooth muscle. *J. Cell Biol.* **115,** 411–421.
Caceres, A., and Kosik, K. S. (1990). Inhibition of neurite polarity by tau antisense oligonucleotides in primary cerebellar neurons. *Nature (London)* **343,** 461–463.
Carboni, J. M., Conzelman, K. A., Adams, R. A., Kaiser, D. A., Pollard, T. D., and Mooseker, M. S. (1988). Structural and immunological characterization of the myosin-like 110-kD subunit of the intestinal microvillar 110K-calmodulin complex: Evidence for discrete myosin head and calmodulin-binding domains. *J. Cell Biol.* **107,** 1749–1757.
Chalovich, J. M., Cornelius, P., and Benson, C. E. (1987). Caldesmon inhibits skeletal actomyosin subfragment-1 ATPase activity and the binding of myosin subfragment-1 to actin *J. Biol. Chem.* **262,** 5711–5716.
Chen, Z. Y., Hasson, T., Kelley, P. M., Schwender, B. J., Schwartz, M. F., Ramakrishnan, M., Kimberling, W. J., Mooseker, M. S., and Corey, D. P. (1996). Molecular cloning and domain structure of human myosin-VIIa. The gene product defective in usher syndrome 1B. *Genomics* **36,** 440–448.
Cheney, R. E., and Mooseker, M. S. (1992). Unconventional myosins. *Curr. Opin. Cell Biol.* **4,** 27–35.
Cheney, R. E., O'Shea, M. K., Heuser, J. E., Coelho, M. V., Wolenski, J. S., Espreafico, E. M., Forscher, P., Larson, R. E., and Mooseker, M. S. (1993). Brain myosin-V is a two-headed unconventional myosin with motor activity. *Cell (Cambridge, Mass.)* **75,** 13–23.
Cleveland, D. W., Hwo, S. Y., and Kirschner, M. W. (1977). Purification of tau, a microtubule-associated protein that induces assembly of microtubules from purified tubulin. *J. Mol. Biol.* **116,** 207–225.
Collins J. H., and Borysenko, C. W. (1984). The 110,000-Dalton actin-and calmodulin-binding protein from intestinal brush border is a myosin-like ATPase. *J. Biol. Chem.* **259,** 14128–14135.
Collins K., Sellers, J. R., and Matsudaira, P. (1990). Calmodulin dissociation regulates brush border myosin I (110-kD-calmodulin) mechanochemical activity *in vitro. J. Cell Biol.* **110,** 1137–1147.
Conrad, P. A., Giuliano, K. A., Fisher, G., Collins, K., Matsudaira, P. T., and Taylor, D. L. (1993). Relative distribution of actin, myosin I, and myosin II during the wound healing response of fibroblasts. *J. Cell Biol.* **120,** 1381–1391.
Cope, M. J. T. V., Whisstock, J., Rayment, I., and Kendrick-Jones, J. (1996). Conservation within the myosin motor domain implications for structure and function. *Structure* **4,** 969–987.

Czernik, A. J., Pang, D. T., and Greengard, P. (1987). Amino acid sequences surrounding the cAMP-dependent and calcium/calmodulin phosphorylation sites in rat and bovine synapsin I. *Proc. Natl. Acad. Sci. U.S.A.* **84,** 7518–7522.

De Camilli, P., and Jahn, R. (1990). Pathways to regulated exocytosis in neurons. *Annu. Rev. Physiol.* **52,** 625–645.

Doberstein, S. K., and Pollard, T. D. (1992). Localization and specificity of the phospholipid and actin binding sites on the tail of *Acanthamoeba* myosin IC. *J. Cell Biol.* **117,** 1241–1249.

Drubin, D. G., Feinstein, S. C., Shooter, E. M., and Kirschner, M. W. (1985). Nerve growth factor-induced neurite outgrowth in PC12 cells involves the coordinate induction of microtubule assemble and assembly-promoting factors. *J. Cell Biol.* **101,** 1799–1807.

Einheber, S., and Fischman, D. A. (1990). Isolation and characterization of a cDNA clone encoding avian skeletal muscle C-protein: An intracellular member of the immunoglobulin superfamily. *Proc. Natl. Acad. Sci. U.S.A.* **87,** 2157–2161.

Ennulat, D. J., Liem, R. K., Hashim, G. A., and Shelanski, M. L. (1989). Two separate 18-amino acid domains of tau promote the polymerization of tubulin. *J. Biol. Chem.* **264,** 5327–5330.

Ervasti, J. M., and Campbell, K. P. (1993). Dystrophin and the membrane skeleton. *Curr. Opin. Cell Biol.* **5,** 82–87.

Espindola, F. S., Espreafico, E. M., Coelho, M. V., Martins, A. R., Costa, F. R. C., Mooseker, M. S., and Larson, R. E. (1992). Biochemical and immunological characterization of p190-calmodulin complex from vertebrate brain: A novel calmodulin-binding myosin. *J. Cell Biol.* **118,** 359–368.

Espindola, F. S., Cheney, R. E., King, S. M., Suter, D. M., and Mooseker, M. S. (1996). Myosin-V and dynein share a similar light chain. *Mol. Biol. Cell* **7S,** 327a.

Espreafico, F. S., Cheney, R. E., Matteoli, M., Nascimento, A. A. C., De Camilli, P. V., Larson, R. E., and Mooseker, M. S. (1992). Primary structure and cellular localization of chicken brain myosin-V (p190), an unconventional myosin with calmodulin light chains. *J. Cell Biol.* **119,** 1541–1557.

Fath, K. R., and Burgess, D. R. (1993). Golgi-derived vesicles from developing epithelial cells bind actin filaments and possess myosin-I as a cytoplasmically oriented peripheral membrane protein. *J. Cell Biol.* **120,** 117–127.

Franco, R., and Lansman, J. B. (1990). Calcium entry through stretch-inactivated ion channels in mdx myotubes. *Nature (London)* **344,** 670–673.

Fukui, Y., Lynch, T. J., Brzeska, H., and Korn, E. D. (1989). Myosin I is located at the leading edges of locomoting *Dictyostelium* amoebae. *Nature (London)* **341,** 328–331.

Fürst, D. O., Osborn, M., Nave, R., and Weber, K. (1988). The organization of titin filaments in the half-sarcomere revealed by monoclonal antibodies in immunoelectron microscopy: A map of ten nonrepetitive epitopes starting at the Z line extends close to the M line. *J. Cell Biol.* **106,** 1563–1572.

Garcia, A., Coudrier, E., Carboni, J., Anderson, J., Vanderkhove, J., Mooseker, M., Louvard, D., and Arpin, M. (1989). Partial deduced sequence of the 110-kD-calmodulin complex of the avian intestinal microvillus shows that this mechanoenzyme is a member of the myosin I family. *J. Cell Biol.* **109,** 2895–2903.

Gardner, K., and Bennett, V. (1986). A new erythrocyte membrane-associated protein with calmodulin binding activity. Identification and purification. *J. Biol. Chem.* **261,** 1339–1348.

Gardner, K., and Bennett, V. (1987). Modulation of the spectrin-actin assembly by erythrocyte adducin. *Nature (London)* **328**, 359–362.

Gautel, M., Castiglione Morelli, M. A., Pfuhl, M., Motta, A., and Pastore, A. (1995). A calmodulin-binding sequence in the C-terminus of human cardiac titin kinase. *Eur. J. Biochem.* **230**, 752–759.

Gibson, F., Walsh, J., Mburu, P., Varela, A., Brown, K. A., Antonio, M., Beisel, K. W., Steel, K. P., and Brown, S. D. M. (1995). A type VII myosin encoded by the mouse deafness gene *Shaker-1*. *Nature (London)* **374**, 62–64.

Gillespie, P. G., Wagner, M. C., and Hudspeth, A. J. (1993). Identification of a 120 kd hair-bundle myosin located near stereociliary tips. *Neuron* **11**, 581–594.

Goldenrign, J. R., Lasher, R. S., Vallano, M. L., Udea, T., Naito, S., Sternberger, N. H. Sternberger, L. A., and DeLorenzo, R. J. (1986). Association of synapsin I with neuronal cytoskeleton. Identification in cytoskeletal preparations *in vitro* and immunocytochemical localization in brain of synapsin I. *J. Biol. Chem.* **261**, 8495–8504.

Goldschmidt-Clermont, P. J., Machesky, L. M., Baldassare, J. J., and Pollard, T. D. (1990). The actin-binding protein profilin binds to PIP_2 and inhibits its hydrolysis by phospholipase C. *Science* **247**, 1575–1578.

Gonzalez, M., Cambiazo, V., and Maccioni, R. B. (1995). Identification of a new microtubule-interacting protein Mip-90. *Eur. J. Cell Biol.* **67**, 158–169.

Goold, R., and Baines, A. J. (1994). Evidence that two non-overlapping high-affinity calmodulin-binding sites are present in the head region of synapsin I. *Eur. J. Biochem.* **224**, 229–240.

Goold, R., Chan, K. M., and Baines, A. J. (1995). Coordinated regulation of synapsin I interaction with F-actin by Ca^{2+}/calmodulin and phosphorylation: Inhibition of actin binding and bundling. *Biochemistry* **34**, 1912–1920.

Govindan, B., Bowser, R., Novick, R. (1995). The role of Myo2, a yeast class V myosin, in vesicular transport. *J. Cell Biol.* **128**, 1055–1068.

Haarrer, B. K., Petzold, A., Lillie, S. H., and Brown, S. S. (1994). Identification of *MYO4*, a second class V myosin gene in yeast. *J. Cell Sci.* **107**, 1055–1064.

Halsall, D. J., and Hammer III, J. A. (1990). A second isoform of chicken brush border myosin I contains a 29-residue inserted sequence that binds calmodulin. *FEBS Lett.* **267**, 126–130.

Harris, A. S., and Morrow, J. S. (1990). Calmodulin and calcium-dependent protease I coordinately regulate the interaction of fodrin with actin. *Proc. Natl. Acad. Sci. U.S.A.* **87**, 3009–3013.

Harris, A. S., Croall, D. E., and Morrow, J. S. (1988). The calmodulin-binding site in α-fodrin is near the calcium-dependent protease-I cleavage site. *J. Biol. Chem.* **263**, 15754–15761.

Hartwig, J. H., Thelen, M., Rosen, A., Janmey, P. A., Nairn, A. C., and Aderem, A. (1992). MARCKS is an actin filament crosslinking protein regulated by protein kinase C and calcium-calmodulin. *Nature (London)* **356**, 618–622.

Hasson, T., and Mooseker, M. S. (1994). Porcine myosin-VI: Characterization of a new mammalian unconventional myosin. *J. Cell Biol.* **127**, 425–440.

Hasson, T., Heintzelman, M. B., Santos-Sacchi, J., Corey, D. P., and Mooseker, M. S. (1995). Expression in cochlea and retina of myosin VIIa, the gene product defective in Usher syndrome type 1B. *Proc. Natl. Acad. Sci. U.S.A.* **92**, 9815–9819.

Hasson, T., Skowron, J. F., Gilbert, D. J., Avraham, K. B., Perry, W. L., Bement, W. M., Anderson, B. L., Sherr, E. H., Chen, Z. Y., Greene, L. A., Ward, D. C., Corey,

D. P., Mooseker, M. S., Copeland, N. G., and Jenkins, N. A. (1996). Mapping of unconventional myosins in mouse and human. *Genomics* **36,** 431–439.

Hayashi, K., Kanda, K., Kimizuka, F., Kato, I., and Sobue, K. (1989). Primary structure and functional expression of *h*-caldesmon complementary DNA. *Biochem. Biophys. Res. Commun.* **164,** 503–511.

Hayashi, K., Fujio, Y., Kato, I., and Sobue, K. (1991). Structural and functional relationships between *h*- and *l*-caldesomons. *J. Biol. Chem.* **266,** 355–361.

Hayes, N. V. L., Bennett, A. F., and Baines, A. J. (1991). Selective Ca^{2+}-dependent interaction of calmodulin with the head domain of synapsin 1. *Biochem. J.* **275,** 93–97.

Heidemann, S. R., Hamborg, M. A., Thomas, S. J., Song, B., Lindley, S., and Chu, D. (1984). Spatial organization of axonal microtubules. *J. Cell Biol.* **99,** 1289–1295.

Heierhorst, J., Probst, W. C., Wilim, F. S., Buku, A., and Weiss, K. R. (1994). Autophosphorylation of molluscan twitchin and interaction of its kinase domain with calcium/calmodulin. *J. Biol. Chem.* **269,** 21086–21093.

Heierhorst, J., Probst, W. C., Kohanski, R. A., Buku, A., and Weiss, K. R. (1995). Phosphorylation of myosin regulatory light chains by the molluscan twitchin kinase. *Eur. J. Biochem.* **233,** 426–431.

Heierhorst, J., Kobe, B., Feil, S. C., Parker, M. W., Benian, G. M., Weiss, K. R., and Kemp, B. E. (1996). Ca^{2+}/S100 regulation of ginat protein kinases. *Nature (London)* **380,** 636–639.

Helmes, M., Trombitás, K., and Granzier, H. (1996). Titin develops restoring force in rat cardiac myocytes. *Cir. Res.* **79,** 619–626.

Hoffman, E. P., and Kunkel, L. M. (1987). Dystrophin abnormalities in Duchenne/Becker muscular dystrophy. *Neuron* **2,** 1019–1029.

Hoffman, E. P., Brown, R. H., and Kunkel, L. M. (1987). Dystrophin: The protein product of the Duchenne muscular dystrophy locus. *Cell (Cambridge, Mass.)* **51,** 919–928.

Horowits, R. (1992). Passive force generation and titin isoforms in mammalian skeletal muscle. *Biophys. J.* **61,** 392–398.

Horowits, R., Kempner, E. S., Bisher, M. E., and Podolsky, R. J. (1986). A physiological role for titin and nebulin in skeletal muscle. *Nature (London)* **323,** 160–164.

Horowitz, J. A., and Hammer, J. A. 3d (1990). A new Aeanthamoeba myosin heavy chain. Cloning of the gene and immunological identification of the polypeptide. *J. Biol. Chem.* **265,** 20646–20652.

Houdusse, A., and Cohen, C. (1995). Target sequence recognition by the calmodulin superfamily: Implications from light chain binding to the regulatory domain of scallop myosin. *Proc. Natl. Acad. Sci. U.S.A.* **92,** 10644–10647.

Houdusse, A., Silver, M., and Carolyn, C. (1996). A model of Ca^{2+}-free clamodulin binding to unconventional myosins. *Structure* **4,** 1475–1490.

Howe, C. L., and Mooseker, M. S. (1983). Characterization of the 110-kdalton actin, calmodulin-, and membrane-binding protein from microvilli of intestinal epithelial cells. *J. Cell Biol.* **97,** 974–985.

Huber, P. A. J., El-Mezgueldi, M., Grabarek, Z., Slatter, D. A., Levine, B. A., and Marston, S. B. (1996). Multiple-sited interaction of caldesmon with Ca^{2+}-calmodulin. *Biochem. J.* **316,** 413–420.

Hughes, C. A., and Bennett, V. (1995). Adducin: A physical model with implications for function in assembly of spectrin–actin complexes. *J. Biol. Chem.* **270,** 18990–18996.

Huttner, W. B., Schiebler, W., Greengard, P., and De Camilli, P. (1983). Synapsin I (protein I), a nerve terminal-specific phosphoprotein. III. Its association with synaptic vesicles studied in a highly purified synaptic vesicle preparation. *J. Cell Biol.* **96,** 1374–1388.

7. Calmodulin-Binding Proteins of the Cytoskeleton

Ishikawa, R., Yamashiro, S., and Matsumura, F. (1989). Differential modulation of actin-severing activity of gelsolin by multiple isoforms of cultured rat cell tropomyosin. Potentiation of protective ability of tropomyosins by 83-kDa nonmuscle caldesmon. *J. Biol. Chem.* **264,** 7490–7497.

Jancso, A., and Szent-Györgyi, A. G. (1994). Regulation of scallop myosin by the regulatory light chain depends on a single glycine residue. *Proc. Natl. Acad. Sci. U.S.A.* **91,** 8762–8766.

Janmey, P. A. (1994). Phosphoinositides and calcium as regulators of cellular actin assembly and disassembly. *Annu. Rev. Physiol.* **56,** 169–191.

Jarrett, H. W., and Foster, J. L. (1995). Alternate binding of actin and calmodulin to multiple sites on dystrophin. *J. Biol. Chem.* **270,** 5578–5586.

Jesaitis, A. J., Bokoch, G. M., Tolley, J. O., and Allen, R. A. (1988). Lateral segregation of neutrophil chemotactic receptors into actin- and fodrin-rich plasma membrane microdomains depleted in guanyl nucleotide regulatory proteins. *J. Cell Biol.* **107,** 921–928.

Job, D., Fischer, E. H., and Margolis, R. L. (1981). Rapid disassembly of cold-stable microtubules by calmodulin. *Proc. Natl. Acad. Sci. U.S.A.* **78,** 4679–4682.

Job, D., Rauch, C. T., Fischer, E. H., and Margolis, R. L. (1983). Regulation of microtubule cold stability by calmodulin-dependent and -indendent phosphorylation. *Proc. Natl. Acad. Sci. U.S.A.* **80,** 3894–3898.

Job, D., Pabion, M., and Margolis, R. L. (1985). Generation of microtubule stability subclasses by microtubule-associated proteins: Implications for the microtubule "dynamic instability" model. *J. Cell Biol.* **101,** 1680–1689.

Job, D., Rauch, C. T., and Margolis, R. L. (1987). High concentrations of STOP protein induce a microtubule super-stable state. *Biochem. Biopys. Res. Commun.* **148,** 429–434.

Johnston, G. C., Prendergast, J. A., and Singer, R. A. (1991). The *Saccharomyces cerevisiae MYO2* gene encodes an essential myosin for vectorial transport of vesicles. *J. Cell Biol.* **113,** 539–552.

Joshi, R., and Bennett, V. (1990). Mapping the domain structure of human erythrocyte adducin. *J. Biol. Chem.* **265,** 13130–13136.

Joshi, R., Gilligan, D. M., Otto, E., McLaughlin, T., and Bennett, V. (1991). Primary structure and domain organization of human alpha and beta adducin. *J. Cell Biol.* **115,** 665–675.

Karinch, A. M., Zimmer, W. E., and Goodman, S. R. (1990). The identification and sequence of the actin-binding domain of human red blood cell β-spectrin. *J. Biol. Chem.* **265,** 11833–11840.

Kellerman, K. A., and Miller, K. G. (1992). An unconventional myosin heavy chain gene from *Drosophila melanogaster*. *J. Cell Biol.* **119,** 823–834.

Kim, J., Shishido, T., Jiang, X., Aderem, A., and McLaughlin, S. (1994). Phosphorylation, high ionic strength, and calmodulin reverse the binding of MARCKS to phospholipid vesicles. *J. Biol. Chem.* **269,** 28214–28219.

Kimura, K., Ito, M., Amano, M., Chihara, K., Fukata, Y., Nakafuku, M., Yamamori, B., Feng, J. H., Nakano, T., Okawa, K., Iwamatsu, A., and Kaibuchi, K. (1996). Regulation of myosin phosphatase by Rho and Rho-Associated kinase (Rho-kinase). *Science* **273,** 245–248.

Kinkema, M., and Schiefelbein, J. (1994). A myosin from a higher plant has structural similarities to class V myosins. *J. Mol. Biol.* **239,** 591–597.

Knight, A. E., and Kendrick-Jones, J. (1993). A myosin-like protein from a higher plant. *J. Mol. Biol.* **231,** 148–154.
Knight, M., Wright, J., and Wang, K. (1991). Nebulin as a length regulator of thin filaments of vertebrate skeletal muscles: Correlation of thin filament length, nebulin size, and epitope profile. *J. Cell Biol.* **115,** 97–107.
Labeit, S., and Kolmerer, B. (1995). Titins: Giant proteins in charge of muscle ultrastructure and elasticity. *Science* **270,** 293–296.
Labeit, S., Barlow, D. P., Gautel, M., Gibson, T., Holt, J., Hsieh, C. L., Francke, U. Leonard, K., Wardale, J., Whiting, A., and Trinick, J. (1990). A regular pattern of two types of 100-residue motif in the sequence of titin. *Nature (London)* **345,** 273–276.
Larson, R. E., Espindola, F. S., and Espreafico, E. M. (1990). Calmodulin-binding proteins and calcium/calmodulin-regulated enzyme activities associated with brain actomyosin. *J. Neurochem.* **54,** 1288–1294.
Lee, Y. C., and Wolff, J. (1982). Two opposing effects of calmodulin on microtubule assembly depend on the presence of microtubule-associated proteins. *J. Biol. Chem.* **257,** 6306–6310.
Lee, Y. C., and Wolff, J. (1984). Calmodulin binds to both microtubule-associated protein 2 and tau proteins. *J. Biol. Chem.* **259,** 1226–1230.
Lehman, W. (1991). Calponin and the composition of smooth muscle thin filaments. *J. Muscle Res. Cell Motil.* **12,** 221–224.
Lei, J., Tang, X., Chambers, T. C., Pohl, J., and Benian, G. M. (1994). Protein kinase domain of twitchin has protein kinase activity and an autoinhibitory region. *J. Biol. Chem.* **269,** 21078–21085.
Leto, T. L., Pleasic, S., Forget, B. G., Benz, E. J., Jr., and Marchesi, V. T. (1989). Characterization of the calmodulin-binding site of nonerythroid α-spectrin. Recombinant protein and model peptide studies. *J. Biol. Chem.* **264,** 5826–5830.
Li, X. L., and Bennett, V. (1996). Identification of the spectrin subunit and domains required for formation of spectrin/adducin/actin complexes. *J. Biol. Chem.* **271,** 15695–15702.
Linke, W. A., Ivemeyer, M., Olivieri, N., Kolmerer, B., Rüegg, J. C., and Labeit, S. (1996). Towards a molecular understanding of the elasticity of titin. *J. Mol. Biol.* **261,** 62–71.
Luby-Phelps, K., Hori, M., Phelps, J. M., and Won, D. (1995). Ca^{2+}-regulated dynamic compartmentalization of calmodulin in living smooth muscle cells. *J. Biol. Chem.* **270,** 21532–21538.
Lux, S. E., and Palek, J. (1995). Disorders of the red cell membrane. *In* "Blood: Principles and Practice of Hematology" (R. I. Handin, S. E. Lux, and T. P. Stossel, eds.), pp 1701–1818. Lippincott, Philadelphia.
Lynch, G., and Baudry, M. (1984). The biochemistry of memory: A new and specific hypothesis. *Science* **224,** 1057–1063.
Mabuchi, K., and Wang, C.-L. A. (1991). Electron microscopic studies of chicken gizzard caldesmon and its complex with calmodulin. *J. Muscle Res. Cell Motil.* **12,** 145–151.
Mabuchi, K., Li, Y. H., Tao, T., and Wang, C. L. A. (1996). Immunocytochemical localization of caldesmon and calponin in chicken gizzard smooth muscle. *J. Muscle Res. Cell Motil.* **17,** 243–260.
Maccioni, R. B., and Cambiazo, V. (1995). Role of microtubule-associated proteins in the control of microtubule assembly. *Physiol. Rev.* **75,** 835–864.
McIlroy, B. K., Walters, J. D., Blackshear, P. J., and Johnson, J. D. (1991). Phosphorylation-dependent binding of a synthetic MARCKS peptide to calmodulin. *J. Biol. Chem.* **266,** 4959–4964.

Madhavan, R., and Jarrett, H. W. (1994). Calmodulin-activated phosphorylation of dystrophin. *Biochemistry* **33,** 5797–5804.

Madhavan, R., Massom, L. R., and Jarrett, H. W. (1992). Calmodulin specifically binds three proteins of the dystrophin–glycoprotein complex. *Biochem. Biophys. Res. Commun.* **185,** 753–759.

Mani, R. S., McCubbin, W. D., and Kay, C. M. (1992). Calcium-dependent regulation of caldesmon by an 11-kDa smooth muscle calcium-binding protein, caltropin. *Biochemistry* **31,** 11896–11901.

Margolis, R. L., Rauch, C. T., and Job, D. (1986). Purification and assay of a 145-kDa protein (STOP145) with microtubule-stabilizing and motility behavior. *Proc. Natl. Acad. Sci. U.S.A.* **83,** 639–643.

Marston, S. B., and Lehman, W. (1985). Caldesmon is a Ca^{2+}-regulatory component of native smooth-muscle thin filaments. *Biochem. J.* **231,** 517–522.

Marston, S. B., and Redwood, C. S. (1991). The molecular anatomy of caldesmon. *Biochem. J.* **279,** 1–16.

Marston, S. B., Fraser, I. D. C., and Huber, P. A. J. (1994). Smooth muscle caldesmon controls the strong binding interaction between actin-tropomyosin and myosin. *J. Biol. Chem.* **269,** 32104–32109.

Matsudaira, P. (1991). Modular organization of actin crosslinking proteins. *Trends Biochem. Sci.* **16,** 87–92.

Matsudaira, P. T., and Burgess, D. R. (1979). Identification and organization of the components in the isolated microvillus cytoskeleton. *J. Cell Biol.* **83,** 667–673.

Matsuoka, Y., Hughes, C. A., and Bennett, V. (1996). Adducin regulation. Definition of the calmodulin-binding domain and sites of phosphorylation by protein kinases A and C. *J. Biol. Chem.* **271,** 25157–25166.

Medvedeva, M. V., Bushueva, T. L., Shirinsky, V. P., Lukas, T. J., Watterson, D. M., and Gusev, N. B. (1995). Interaction of smooth muscle caldesmon with calmodulin mutants. *FEBS Lett.* **360,** 89–92.

Mercer, J. A., Seperack, P. K., Strobel, M. C., Copeland, N. G., and Jenkins, N. A. (1991). Novel myosin heavy chain encoded by murine *dilute* coat colour locus. *Nature (London)* **349,** 709–713.

Mermall, V., McNally, J. G., and Miller, K. G. (1994). Transport of cytoplasmic particles catalysed by an unconventional myosin in living *Drosophila* embryos. *Nature (London)* **369,** 560–562.

Moerman, D. G., Plurad, S., Waterston, R. H., and Baillie, D. L. (1982). Mutations in the unc-54 myosin heavy chain gene of *Caenorhabditis elegans* that alter contractility, but not muscle structure. *Cell (Cambridge, Mass.)* **29,** 773–781.

Moncman, C. L., and Wang, K. (1995). Nebulette: A 107 kD nebulin-like protein in cardiac muscle. *Cell Motil. Cytoskeleton* **32,** 205–225.

Montell, C., and Rubin, G. M. (1988). The *Drosophila ninaC* locus encodes two photoreceptor cell specific proteins with domains homologous to protein kinasesand the myosin heavy chain head. *Cell (Cambridge, Mass.)* **52,** 757–772.

Mooseker, M. S., and Cheney, R. E. (1995). Unconventional myosins. *Annu. Rev. Cell Biol.* **11,** 633–675.

Mooseker, M. S., and Coleman, T. R. (1989). The 110-kD protein–calmodulin complex of the intestinal microvillus (brush border myosin I) is a mechanoenzyme. *J. Cell. Biol.* **108,** 2395–2400.

Nascimento, A. A. C., Cheney, R. E., Tauhata, S. B. F., Larson, R. E., and Mooseker, M. S. (1996). Enzymatic characterization and functional domain mapping of brain myosin-V. *J. Biol. Chem.* **271**, 17561–17569.

Nave, R., and Weber, K. (1990). A myofibrillar protein of insect muscle related to vertebrate titin connects Z band and A band: Purification and molecular characterization of invertebrate mini-titin. *J. Cell Sci.* **95**, 535–544.

Nave, R., Furst, D., Vinkemeier, U., and Weber, K. (1991). Purification and physical properties of nematode mini-titins and their relation to twitchin. *J. Cell Sci.* **98**, 491–496.

Ng, K. P., Kambara, T., Matsuura, M., Burke, M., and Ikebe, M. (1996). Identification of myosin III as a protein kinase. *Biochemistry* **35**, 9392–9399.

North, A. J., Gimona, M., Cross, R. A., and Small, J. V. (1994). Calponin is localised in both the contractile apparatus and the cytoskeleton of smooth muscle cells. *J. Cell Sci.* **107**, 437–444.

Novy, R. E., Liu, L.-F., Lin, C.-S., Helfman, D. M., and Lin, J. J.-C. (1993). Expression of smooth muscle and nonmuscle tropomyosins in *Escherichia coli* and characterization of bacterially produced tropomyosins. *Biochim. Biophys. Acta Protein Struct. Mol. Enzymol.* **1162**, 255–265.

Okabe, T., and Sobue, K. (1987). Identification of a new 84/82 kDa calmodulin-binding protein, which also interacts with actin filaments, tubulin and spectrin, as synapsin I. *FEBS Lett.* **213**, 184–188.

Patel, J., and Kligman, D. (1987). Purification and characterization of an Mr 87,000 protein kinase C substrate from rat brain. *J. Biol. Chem.* **262**, 16686–16691.

Pato, M. D., and Kerc, E. (1988). Purification of smooth muscle myosin phosphatase from turkey gizzard. *Methods Enzymol.* **159**, 446–453.

Perrin, D., Langley, O. K., and Aunis, D. (1987). Anti-α-fodrin inhibits secretion from permeabilized chromaffin cells. *Nature (London)* **326**, 498–501.

Petrucci, T. C., and Morrow, J. S. (1987). Synapsin I: An actin-bundling protein under phosphorylation control. *J. Cell Biol.* **105**, 1355–1363.

Pirollet, F., Derancourt, J., Haiech, J., Job, D., and Margolis, R. L. (1992). $Ca^{(2+)}$-calmodulin regulated effectors of microtubule stability in bovine brain. *Biochemistry* **31**, 8849–8855.

Pollard, T. D., and Cooper, J. A. (1986). Actin and actin-binding proteins. A critical evaluations of mechanisms and functions. *Annu. Rev. Biochem.* **55**, 987–1035.

Porter, J. A., and Montell, C. (1993). Distinct roles of the *Drosophila ninaC* kinase and myosin domains revealed by systematic mutagenesis. *J. Cell Biol.* **122**, 601–612.

Porter, J. A., Hicks, J. L., Williams, D. S., and Montell, C. (1992). Differential localizations of and requirements for the two *Drosophila ninaC* kinase/myosins in photoreceptor cells. *J. Cell Biol.* **116**, 683–693.

Porter, J. A., Yu, M., Doberstein, S. K., Pollard, T. D., and Montell, C. (1993). Dependence of calmodulin localization in the retina on the NINAC unconventional myosin. *Science* **262**, 1038–1042.

Porter, J. A., Minke, B., and Montell, C. (1995). Calmodulin binding to *Drosophila NinaC* required for termination of phototransduction. *EMBO J.* **14**, 4450–4459.

Pritchard, K., and Marston, S. B. (1991). Ca^{2+}-dependent regulation of vascular smooth-muscle caldesmon by S.100 and related smooth-muscle proteins. *Biochem. J.* **277**, 819–824.

Qin, Z., and Cafiso, D. S. (1996). Membrane structure of protein kinase C and calmodulin binding domain of myristoylated alanine rich C kinase substrate determined by site-directed spin labeling. *Biochemistry* **35**, 2917–2925.

7. Calmodulin-Binding Proteins of the Cytoskeleton 393

Qin, Z., Wertz, S. L., Jacob, J., Savino, Y., and Cafiso, D. S. (1996). Defining protein–protein interactions using site-directed spin-labeling: The binding of protein kinase C substrates to calmodulin. *Biochemistry* **35**, 13272–13276.

Quist, E. E., and Roufogalis, B. D. (1976). The relationship between changes in viscosity of human erythrocyte membrane suspensions and (Mg + Ca)-ATPase activity. *Biochem. Biophys. Res. Commun.* **72**, 673–679.

Rayment, I., Rypniewski, W. R., Schmidt-Bäse, K., Smith, R., Tomchick, D. R., Benning, M. M., Winkelmann, D. A., Wesenberg, G., and Holden, H. M. (1993). Three-dimensional structure of myosin subfragment-1: A molecular motor. *Science* **261**, 50–58.

Reddy, A. S. N., Safadi, F., Narasimhulu, S. B., Golovkin, M., and Hu, X. (1996). A novel plant calmodulin-binding protein with a kinesin heavy chain motor domain. *J. Biol. Chem.* **271**, 7052–7060.

Reinhard, J., Scheel, A. A., Diekmann, D., Hall, A., Ruppert, C., and Bähler, M. (1995). A novel type of myosin implicated in signalling by rho family GTPases. *EMBO J* **14**, 697–704.

Ridley, A. J., and Hall, A. (1992). The small GTP-binding protein rho regulates the assembly of focal adhesions and actin stress fibers in response to growth factors. *Cell (Cambridge, Mass.)* **70**, 389–399.

Ridley, A. J., Paterson, H. F., Johnston, C. L., Diekmann, D., and Hall, A. (1992). The small GTP-binding protein rac regulates growth factor-induced membrane ruffling. *Cell (Cambridge, Mass.)* **70**, 401–410.

Riseman, V. M., Lynch, W. P., Nefsky, B., and Bretscher, A. (1989). The calmodulin and F-actin binding sites of smooth muscle caldesmon lie in the carboxyl-terminal domain whereas the molecular weight heterogeneity lies in the middle of the molecule. *J. Biol. Chem.* **264**, 2869–2875.

Root, D. D., and Wang, K. (1994). Calmodulin-sensitive interaction of human nebulin fragments with actin and myosin. *Biochemistry* **33**, 12581–12591.

Rosahl, T. W., Spillane, D., Missler, M., Herz, J., Selig, D. K., Wolff, J. R., Hammer, R. E., Malenka, R. C., and Sudhof, T. C. (1995). Essential functions of synapsins I and II in synaptic vesicle regulation. *Nature (London)* **375**, 488–493.

Ruppert, C., Kroschewski, R., and Bähler, M. (1993). Identification, characterization and cloning of myr 1, a mammalian myosin-I. *J. Cell Biol.* **120**, 1393–1403.

Ruppert, C., Godel, J., Müller, R. T., Kroschewski, R., Reinhard, J., and Bähler, M. (1995). Localization of the rat myosin I molecules myr 1 and myr 2 and *in vivo* targeting of their tail domains. *J. Cell Sci.* **108**, 3775–3786.

Scaramuzzino, D. A., and Morrow, J. S. (1993). Calmodulin-binding domain of recombinant erythrocyte beta-adducin. *Proc. Natl. Acad. Sci. U.S.A.* **90**, 3398–3402.

Schneider, M. D., Sellers, J. R., Vahey, M., Preston, Y. A., and Adelstein, R. S. (1985). Localization and topography of antigenic domains wtihin the heavy chain of smooth muscle myosin. *J. Cell Biol.* **101**, 66–72.

Sellers, J. R. (1991). Regulation of cytoplasmic and smooth muscle myosin. *Curr. Opin. Cell Biol.* **3**, 98–104.

Sellers, J. R., and Goodson, H. V. (1995). Motor proteins 2: Myosin. *Protein Profile* **2**, 1323–1423.

Sellers, J. R., Cuda, G., Wang, F., and Homsher, E. (1993). Myosin-specific adaptations of motility assays. *In* "Motility Assays for Motor Proteins" (J. M. Scholey, ed.), pp. 23–49. Academic Press, San Diego.

Sellers, H. R., Goodson, H. V., and Wang, F. (1996). A myosin family reunion. *J. Muscle Res. Cell Motil.* **17,** 7–22.

Skripnikova, E. V., and Gusev, N. B. (1989). Interaction of smooth muscle caldesmon with S-100 protein. *FEBS Lett.* **257,** 380–382.

Smith, C. W., Pritchard, K., and Marston, S. B. (1987). The mechanism of Ca^{2+} regulation of vascular smooth muscle thin filaments by caldesmon and calmodulin. *J. Biol. Chem.* **262,** 116–122.

Sobue, K., and Sellers, J. R. (1991). Caldesmon, A novel regulatory protein in smooth muscle and nonmuscle actomyosin systems. *J. Biol. Chem.* **266,** 12115–12118.

Sobue, K., Takahashi, K., and Wakabayashi, I. (1985). Caldesmon150 regulates the tropomyosin-enhanced actin–myosin interaction in gizzard smooth muscle. *Biochem. Biophys. Res. Commun.* **132,** 645–651.

Song, H., Golovkin, M., Reddy, A. S. N., and Endow, S. A. (1997). In vitro motility of AtKCBP, a calmodulin-binding kinesin protein of *Arabidopsis*. *Proc. Natl. Acad. Sci. U.S.A.* **94,** 322–327.

Sri Widada, J., Asselin, J., Colote, S., Marti, J., Ferraz, C., Trave, G., Haiech, J., and Liautard, J. P. (1989). Cloning and deletion mutagenesis using direct protein–protein interaction on an expression vector. Identification of the calmodulin binding domain of α-fodrin. *J. Mol. Biol.* **205,** 455–458.

Stöffler, H. E., Ruppert, C., Reinhard, J., and Bahler, M. (1995). A novel mammalian myosin I from rat with an SH3 domain localizes to Con A-inducible, F-actin-rich structures at cell-cell contacts. *J. Cell Biol.* **129,** 819–830.

Strauss, J. D., and Murphy, R. A. (1996). Regulation of cross-bridge cycling in smooth muscle. In "Biochemistry of Smooth Muscle Contraction" (M. Barany, ed.), pp. 341–353. Academic Press, San Diego.

Swanljung-Collins, H., and Collins, J. H. (1991). Ca^{2+} stimulates the Mg^{2+}-ATPase activity of brush border myosin I with three or four calmodulin light chains but inhibits with less than two bound. *J. Biol. Chem.* **266,** 1312–1319.

Takahashi, K., Hiwada, K., and Kokubu, T. (1988). Vascular smooth muscle calponin. A novel troponin T-like protein. *Hypertension (Dallas)* **11,** 620–626.

Taniguchi, H., and Manenti, S. (1993). Interaction of myristoylated alanine-rich protein kinase C substrate (MARCKS) with membrane phospholipids. *J. Biol. Chem.* **268,** 9960–9963.

Thelen, M., Rosen, A., Nairn, A. C., and Aderem, A. (1991). Regulation by phosphorylation of reversible association of a myristoylated protein kinase C substrate with the plasma membrane. *Nature (London)* **351,** 320–322.

Theriot, J. A., Mitchison, T. J., Tilney, L. G., and Portnoy, D. A. (1992). The rate of actin-based motility of intracellular *Listeria monocytogenes* equals the rate of actin polymerization. *Nature (London)* **357,** 257–260.

Tilney, L. G., and Portnoy, D. A. (1989). Actin filaments and the growth, movement, and spread of the intracellular bacterial parasite, *Listeria monocytogenes*. *J. Cell Biol.* **109,** 1597–1608.

Tilney, L. G., Hatano, S., Ishikawa, H., and Mooseker, M. S. (1973). The polymerization of actin: Its role in the generation of the acrosomal process of certain echinoderm sperm. *J. Cell Biol.* **66,** 508–522.

Travé, G., Pastore, A., Hyvonen, M., and Saraste, M. (1995a). The C-terminal domain of alpha-spectrin is structurally related to calmodulin. *Eur. J. Biochem.* **227,** 35–42.

7. Calmodulin-Binding Proteins of the Cytoskeleton

Travé, G., Lacombe, P.-J., Pfuhl, M., Saraste, M., and Pastore, A. (1995b). Molecular mechanism of the calcium-induced conformational change in the spectrin EF-hands. *EMBO J.* **14,** 4922–4931.

Trinick, J. (1991). Elastic filaments and giant proteins in muscle. *Curr. Opin. Cell Biol.* **3,** 112–119.

Tsukita, S. (1985). Dsmocalmin: A calmodulin-binding high molecular weight protein isolated from desmosomes. *J. Cell Biol.* **101,** 2070–2080.

Turner, P. R., Fong, P. Y., Denetclaw, W. F., and Steinhardt, R. A. (1991). Increased calcium influx in dystrophic muscle. *J. Cell Biol.* **115,** 1701–1712.

Ursitti, J. A., Pumplin, D. W., Wade, J. B., and Bloch, R. J. (1991). Ultrastructure of the human erythrocyte cytoskeleton and its attachment to the membrane. *Cell Motil. Cytoskeleton* **19,** 227–243.

Uyeda, T. Q. P., Abramson, P. D., and Spudich, J. A. (1996). The neck region of the myosin motor domain acts as a lever arm to generate movement. *Proc. Natl. Acad. Sci. U.S.A.* **93,** 4459–4464.

Velaz, L., Hemric, M. E., Benson, C. E., and Chalovich, J. M. (1989). The binding of caldesmon to actin and its effects on the ATPase activity of soluble myosin subfragments in the presence and absence of tropomyosin. *J. Biol. Chem.* **264,** 9602–9610.

Vibert, P., Edelstein, S. M., Castellani, L., and Elliott B. W., Jr. (1993). Mini-titins in striated and smooth molluscan muscles: Structure, location and immunological cross-reactivity. *J. Muscle Res. Cell Motil.* **14,** 598–607.

Viel, A., and Branton, D. (1996). Spectrin: On the path from structure to function. *Curr. Opin. Cell Biol.* **8,** 49–55.

Wagner, M. C., Barylko, B., and Albanesi, J. P. (1992). Tissue distribution and subcellular localization of mammalian myosin I. *J. Cell Biol.* **119,** 163–170.

Walker, R. G., and Hudspeth, A. J. (1996). Calmodulin controls adaptation of mechanoelectrical transduction by hair cells of the bullfrog's sacculus. *Proc. Natl. Acad. Sci. U.S.A.* **93,** 2203–2207.

Wang, K., McCarter, R., Wright, J., Beverly, J., and Ramirez-Mitchell, R. (1993). Viscoelasticity of the sarcomere matrix of skeletal muscles. The titin–myosin composite filament is a dual-stage molecular spring. *Biophys. J.* **64,** 1161–1177.

Wang, P. Y., and Gusev, N. B. (1996). Interaction of smooth muscle calponin and desmin. *FEBS Lett.* **392,** 255–258.

Wang, Z. Y., Sakai, J., Matsudaira, P. T., Baines, I. C., Sellers, J. R., Hammer, III, J. A., and Korn, E. D. (1997). The amino acid sequence of the light chain of *Acanthamoeba* myosin IC/. *J. Muscle Res. Cell Motil.* **18,** 395–398.

Waterston, R. H., Thomson, J. N., and Brenner, S. (1980). Mutants with altered muscle structure of *Caenorhabditis elegans*. *Dev. Biol.* **77,** 271–302.

Wei, Q., Wu, X., and Hammer J. A. (1997). The predominant defect in dilute melanocytes is in melanosome distribution and not cell shape, supporting a role for myosin V in melanosome transport. *J. Muscle Res. Cell Motil.* **18,** 517–527.

Weil, D., Blanchard, S., Kaplan, J., Guilford, P., Gibson, F., Walsh, J., Mburu, P., Varela, A., Levilliers, J., Weston, M. D., Kelley, P. M., Kimberling, W. J., Wagenaar, M., Levi-Acobas, F., Larget-Piet, D., Munnich, A., Steel, K. P., Brown, S. D. M., and Petit, C. (1995). Defective myosin VIIA gene responsible for Usher syndrome type 1B. *Nature (London)* **374,** 60–61.

Weisenberg, R. C. (1972). Microtubule formation *in vitro* in solutions containing low calcium concentrations. *Science* **177,** 1104–1105.

Williams, B. D., and Waterston, R. H. (1994). Genes critical for muscle development and function in *Caenorbabditis elegans* identified through lethal mutations. *J. Cell Biol.* **124,** 475–490.

Winder, S. J., and Kendrick-Jones, J. (1995). Calcium/calmodulin-dependent regulation of the NH_2-terminal F-actin binding domain of utrophin. *FEBS Lett.* **357,** 125–128.

Winder, S. J., and Walsh, M. P. (1990). Smooth muscle calponin. Inhibition of actomyosin MgATPase and regulation by phosphorylation. *J. Biol. Chem.* **265,** 10148–10155.

Winder, S. J., Walsh, M. P., Vasulka, C., and Johnson, J. D. (1993). Calponin–calmodulin interaction: Properties and effects on smooth and skeletal muscle actin binding and actomyosin ATPases. *Biochemistry* **32,** 13327–13333.

Winder, S. J., Hemmings, L., Maciver, S. K., Bolton, S. J., Tinsley, J. M., Davies, K. E., Critchley, D. R., and Kendrick-Jones, J. (1995). Utrophin actin binding domain: Analysis of actin binding and cellular targeting. *J. Cell Sci.* **108,** 63–71.

Wirth, J. A., Jensen, K. A., Post, P. L., Bement, W. M., and Mooseker, M. S. (1996). Human myosin-IXb, an unconventional myosin with a chimerin-like rho/rac GTPase-activating protein domain in its tail. *J. Cell Sci.* **109,** 653–661.

Wolenski, J. S., Hayden, S. M., Forscher, P., and Mooseker, M. S. (1993). Calcium-calmodulin and regulation of brush border myosin-I MgATPase and mechanochemistry. *J. Cell Biol.* **122,** 613–621.

Wright, J., Huang, Q.-Q., and Wang, K. (1993). Nebulin is a full-length template of actin filaments in the skeletal muscle sarcomere: An immunoelectron microscopic study of its orientation and span with site-specific monoclonal antibodies. *J. Muscle Res. Cell Motil.* **14,** 476–483.

Xie, X., Harrison, D. H., Schlichting, I., Sweet, R. M., Kalabokis, V. N., Szent-Györgyi, A. G., and Cohen, C. (1994). Structure of the regulatory domain of scallop myosin at 2.8 Å resolution. *Nature (London)* **368,** 306–312.

Xu, P., Zot, A. S., and Zot, H. G. (1995). Identification of Acan125 as a myosin-I-binding protein present with myosin-I on cellular organelles of *Acanthamoeba. J. Biol. Chem.* **270,** 25316–25319.

Yamakita, Y., Yamashiro, S., and Matsumura, F. (1992). Characterization of mitotically phosphorylated caldesmon. *J. Biol. Chem.* **267,** 12022–12029.

Yamashiro, S., Yamakita, Y., Ishikawa, R., and Matsumura, F. (1990). Mitosis-specific phosphorylation causes 83K non-muscle caldesmon to dissociate from microfilaments. *Nature (London)* **344,** 675–678.

Yamashiro-Matsumura, S., and Matsumura, S. (1988). Characterization of 83-kilodalton nonmuscle caldesmon from cultured rat cells: Stimulation of actin binding of nonmuscle tropomyosin and periodic localization along microfilaments like tropomyosin. *J. Cell Biol.* **106,** 1973–1983.

Yan, Y., Winograd, E., Viel, A., Cronin, T., Harrison, S. C., and Branton, D. (1993). Crystal structure of the repetitive segments of spectrin. *Science* **262,** 2027–2030.

Zhang, J. Q., Luo, G., Herrera, A. H., Paterson, B., and Horowits, R. (1996). cDNA cloning of mouse nebulin—Evidence that the nebulin-coding sequence is highly conserved among vertebrates. *Eur. J. Biochem.* **239,** 835–841.

Zhu, Q., and Clarke, M. (1992). Association of calmodulin and an unconventional myosin with the contractile vacuole complex of *Dictyostelium discoideum. J. Cell Biol.* **118,** 347–358.

Zhu, T., Sata, M., and Ikebe, M. (1996). Functional expression of mammalian myosin Iβ: Analysis of its motor activity. *Biochemistry* **35,** 513–522.

Zhuang, S. B., Wang, E. Z., and Wang, C. L. A. (1995). Identification of the functionally relevant calmodulin binding site in smooth muscle caldesmon. *J. Biol. Chem.* **270,** 19964–19968.

8

Calmodulin and Ion Flux Regulation

PAUL C. BRANDT AND THOMAS C. VANAMAN

Department of Biochemistry
University of Kentucky Medical Center
Lexington, Kentucky 40536-0096

I. Introduction: The Nature and Physiological Importance of
   Ion Translocation.   397
II. Mechanisms of Cellular Calcium Regulation   398
   A. Different Pools Involved in Calcium Regulation.   398
   B. Different Types of Calcium Channels and Mechanisms for
      Their Regulation.   400
   C. Different Types of Pumps Found in Plasma Membrane vs
      Internal Membranes.   401
III. Plasma Membrane Ca^{2+}-ATPases (PMCAs): A Complex Family of Related
   Enzymes Whose Activities Are Differentially Regulated by Multiple Signaling
   Pathways in Animal Cells   403
   A. Structural Features of PMCA Family   403
   B. Multiple PMCA Isoforms Arising from Four Genes and
      Differential Splicing   413
   C. Catalytic and Regulatory Properties of PMCAs   427
   D. Regulation of PMCA Gene Expression   451
   E. Defining Physiological Function of PMCAs Using Genetic Approaches . . .   455
IV. Other Ion Channels Directly Regulated by Calmodulin   457
   References   458

I. INTRODUCTION: THE NATURE AND PHYSIOLOGICAL IMPORTANCE OF ION TRANSLOCATION

Calmodulin (CaM) was first identified due to its ability to regulate the concentrations of cAMP and cGMP through Ca^{2+}-dependent activation of the corresponding degradative and synthetic enzymes, as discussed elsewhere in this volume. Studies of the primary structure of calmodulin (Watterson *et al.*, 1976, 1980b) demonstrated its relatedness to troponin C, the dedicated "calcium receptive" protein responsible for the calcium-dependent regulation of skeletal muscle contraction. Initial demonstration of the highly conserved structure of calmodulin (Watterson *et al.*, 1980a) and its wide tissue and phylogenetic distribution (as reviewed in Klee and Vanaman, 1982) suggested the much wider range of regulatory functions for calmodulin which we now know that it possesses. Among these, none is more central to calcium signaling than the regulation of ion flux systems, particularly those involved in modulating the intracellular concentration of free calcium responsible for the calcium signal itself.

The maintenance of intracellular ion concentrations is accomplished through the actions of both passive and active transport systems present in cell membranes. This chapter focuses on those that appear to be regulated in a calcium-dependent fashion by direct and/or indirect actions of calmodulin. More general reviews of the major cellular ion flux systems have been published by a number of investigators (Carafoli, 1994; Strehler, 1996).

Very few ion translocases or channels are actually regulated by direct CaM binding. Many reports of calmodulin-mediated regulation of activity actually reflect the effects of calmodulin-dependent kinases and phosphatases. Of the translocases and channels that are truly regulated by direct interaction with calmodulin, the most well characterized by far are the plasma membrane Ca^{2+}-ATPases (PMCAs). For this reason the bulk of this chapter is dedicated to discussion of this complex family of enzymes.

II. MECHANISMS OF CELLULAR CALCIUM REGULATION

A. Different Pools Involved in Calcium Regulation

The number of different calcium pools that actually control cytosolic calcium levels in the cell seem to vary with cell type. Broadly speaking, there are, of course, two primary pools of calcium: calcium that enters the cytosol by crossing the plasma membrane from the extracellular space and calcium that enters the cytosol from internal stores. The internal calcium stores are generally considered to reside in the endoplasmic reticulum (ER), but evidence suggests that at least one, the nicotinic acid adenine dinucleotide phosphate (NAADP), responsive store is not in the ER (Lee and Aarhus, 1995). The nucleus also has substantial amounts of calcium associated with it, but whether these represent stores or transient changes in concentration is not clear. However, the components generally associated with regulation of a calcium store have been reported in nuclear membranes. A 1,4,5-inositol trisphosphate (IP$_3$ receptor (Malviya et al., 1990; Nicotera et al., 1990) and a PMCA-like activity (Nicotera et al., 1989; Yamaguchi and Oishi, 1993) have been found associated with nuclear membranes. The mitochondrion is also a major store of calcium, but it appears that this calcium is used primarily for the regulation of mitochondria-specific dehydrogenases (Carafoli, 1994).

The most well-characterized system for triggering the release of intracellular calcium is that mediated by IP$_3$. IP$_3$ is generated by the cleavage of phosphatidyl inositol-4,5-bisphosphate by phospholipase C β coupled to receptors through a G-protein system or PLCγ coupled to receptors through a phosphotyrosine on the receptor. IP$_3$ then diffuses through the cytosol where it binds to specific receptors in the ER membrane, stimulating opening of a calcium channel.

In addition to the IP$_3$ system, several other mechanisms are found in the cell that control the release of calcium from internal stores. Cyclic ADP ribose (cADPR) stimulates calcium release from the ER by binding to a specific receptor (Lee, 1991) that is very similar to the ryanodine receptor of the sarcoplasmic reticulum. This receptor is distinct from the IP$_3$ receptor, but its activity colocalizes with IP$_3$ receptor activity in fractionated microsomes (Lee and Aarhus, 1995). The cADPR receptor appears to be the target for calcium in calcium-induced calcium release (CICR), in which calcium influx at the plasma membrane stimulates release from internal stores (Lee et al., 1995) just as the ryanodine receptor does in muscle. Nicotinic acid adenine dinucleotide phosphate (NAADP) also releases calcium from internal stores (Lee and Aarhus, 1995). Very little is known about this mechanism of release, but the NAADP releasable calcium pool does not segregate with the IP$_3$ and cADPR receptors on density gradients of sea urchin egg microsomes, suggesting that it is not localized in the ER (Lee and Aarhus, 1995).

B. Different Types of Calcium Channels and Mechanisms for Their Regulation

Calcium channels can be divided into two fundamental types: voltage-operated calcium channels (VOCC) and receptor-operated calcium channels (ROCC). The VOCC are the traditional channels found in the plasma membrane primarily of excitable cells. These channels allow entry of calcium from the extracellular space in response to depolarization of the plasma membrane (Carafoli, 1994). A variety of these types of channels exist and are designated L, P, Q, N., and T. The different channels are distinguished by their responses to pharmacological agents and the magnitude and frequency of their opening on depolarization. The most well characterized of these channels is the L-type channel (Carafoli, 1994). L-type channels can be stimulated by phosphorylation with cAMP-dependent kinase and appear to be inactivated by elevated cytosolic calcium concentrations (de Leon et al., 1995). Numerous reports also have shown that CaM can enhance L-type channel opening by stimulating CaM kinase II to phosphorylate the channel (Anderson et al., 1994; Sihra and Pearson, 1995; Sperelakis et al., 1994; Xiong et al., 1994). The T-type calcium channel may also be activated by CaM kinase II-dependent phosphorylation (Lu et al., 1994), and some voltage-operated calcium channels may be inhibited by CaM kinase, such as in small lung carcinoma cells, but the specific type is not known (Williams et al., 1995). Thus, it appears that plasma membrane calcium channels are not generally regulated by direct calmodulin binding, but only through the action of CaM kinase II.

The most common types of plasma membrane receptor-operated Ca^{2+} channels are those triggered by neurotransmitters. The N-methyl-D-aspartate class of glutamate receptors, for example, allow passage of calcium through the plasma membrane when glutamate is bound. This type of receptor has an obligate need to bind glycine before it will open and the ion channel is effectively blocked by magnesium. Activation of this receptor is required for formation of the phenomenon known as long-term potentiation, which in turn appears to be necessary for the consolidation of short-term memory into long-term memory.

Internal membranes contain several calcium channels that are coupled to signals generated at the plasma membrane. The IP_3 receptor-coupled channel releases calcium from the endoplasmic reticulum into the cytosol on binding IP_3. Release from this channel is regulated by phosphorylation

8. CaM and Ion Flux Regulation

by cAMP-dependent kinase (Komalavilas and Lincoln, 1996; Quinton et al., 1996), cGMP (Komalavilas and Lincoln, 1996), tyrosine phosphorylation (Jayaraman et al., 1996), and calcineurin in conjunction with binding to FKBP12 (Cameron et al., 1995). Yamada et al. (1995) showed that the type 1 and 2 IP$_3$ receptors could bind calmodulin with a physiologically relevant K_d of 0.7 μM, but the type 3 receptor could not. Whether calmodulin binding regulates IP$_3$ receptor activity has not been demonstrated.

As noted, the cADPR receptor is homologous to the ryanodine receptor of muscle cells and is most likely the target of calcium in triggering CICR. The magnitude of the CICR response is modulated by both calmodulin and cADPR (Lee et al., 1995). Calcium-, cADPR-, and caffeine-mediated release through the cADPR-coupled calcium channel all have an obligate requirement for Ca^{2+}/calmodulin being associated with the receptor (Lee et al., 1995). Whether calmodulin itself is directly regulating calcium release through the cADPR-coupled channel is not clear, as the work of Takasawa et al. (1995) has shown that this activity is blocked by inhibitors of CaM kinase II.

C. Different Types of Pumps Found in Plasma Membrane vs Internal Membranes

As discussed in detail in the following section, three distinct calcium pumping ATPase systems are now known to exist (for review, see Carafoli and Guerini, 1993; Carafoli and Stauffer, 1994; MacLennan et al., 1992; Shull et al., 1992). The PMCAs transport calcium from the cell to the external environment and have thus been proposed to be responsible for overall calcium homeostasis. Plasma membrane Ca^{2+}-ATPases are 120- to 140-kDa proteins whose activities are regulated by interaction with calmodulin and, in some instances, by reversible phosphorylation as summarized in detail later. The somewhat smaller (~110 kDa) calcium pumps found in the sarcoplasmic reticulum in muscle and in ER of most animal cells, termed SERCAs, serve to resequester calcium into internal release compartments. The activity of these forms of the protein is also regulated by interaction with an ancillary protein, phospholamban, whose activity is controlled by phosphorylation. Although the amino acid sequences of the core enzyme portions of the PMCAs and SERCAs are quite highly conserved (vide infra), major differences are found in the amino- and carboxyl-terminal regions of these proteins involved in regulatory functions.

A third type of calcium-pumping ATPase has been identified on the basis of molecular cloning (Gunteski-Hamblin et al., 1992). This unique form of the pump, approximately the same molecular weight as the SERCA proteins, is present in all tissues and cell types thus far examined, where it appears to be localized in Golgi and related vesicular compartments (Antebi and Fink, 1992; Virk et al., 1985). It shows less than 30% identity to either SERCAs or PMCAs, but has substantial homology with the *Saccharomyces cerevisiae* secretory pathway ATPase PMR1/SSC1 (Rudolph et al., 1989), as well as with a similar protein cloned from cyanobacterium (Kanamaru et al., 1993). This has led to the proposal that it is a secretory pathway-specific membrane calcium ATPase (SPCA).

It should be noted that the presence of three distinct types of membrane calcium pumps distinguishes the calcium metabolism machinery from all other ion-translocating systems. Indeed, the level of diversity is greater still as at least two of the three major types, PMCAs and SERCAs (Wuytack et al., 1992), also exist as families of multiple isoforms derived by differential splicing of multiple gene products, as discussed later. The expression of different isoforms is tightly regulated in different cell types and during differentiation, presumably to meet differing requirements for the calcium regulatory cycle. This level of complexity parallels that observed for calcium channel species, in accord with the central role of calcium regulation in eukaryotes.

As first demonstrated by Schatzmann (1966), PMCAs are responsible for extruding calcium from the cell at the expense of ATP. Bulk calcium is removed at the plasma membrane by the action of the Na/Ca exchanger, which can reduce cytosolic calcium concentrations to about 1 μM (for review, see Carafoli et al., 1996). However, as this calcium concentration is not sufficient to stop most calcium-activated events, it is necessary to have an additional high-affinity calcium removal system. PMCAs provide this function by having a $K_{d,Ca^{2+}}$ of ~0.1 μM. The activity of PMCA is also stimulated by Ca^{2+}/calmodulin. Therefore, as the cytosolic calcium concentrations become sufficiently low to dissociate the Ca^{2+}/calmodulin complex, stimulated PMCA pumping of calcium is also stopped. This process serves as an autoregulatory mechanism for maintaining cytosolic calcium concentrations at levels below the $K_{d,Ca^{2+}}$ of calmodulin.

Studies of ATP-driven membrane ion translocases in general and Ca^{2+} pumps in particular have been in progress since the demonstration by Schatzmann (1966) of ATP-dependent Ca^{2+} extrusion from human red blood cell (RBC) ghosts. Numerous reviews of studies of the plasma membrane calcium pump have appeared and will not be recapitulated here. For a detailed historical perspective of this field, the reader is referred to Carafoli (1987, 1991a,b, 1994) and similar reviews.

The following sections will attempt to summarize current knowledge concerning the structural and functional properties of the derived set of plasma membrane calcium pump species found in animal cells. In particular, the differential regulation of their calcium extrusion activities and expression and the potential roles as components of the calcium signaling apparatus will be discussed. To place the plasma membrane Ca^{2+} pump in proper context, it is necessary to first consider its relationship to other calcium pumps found in internal membranes.

III. PLASMA MEMBRANE Ca^{2+}-ATPases (PMCAs): A COMPLEX FAMILY OF RELATED ENZYMES WHOSE ACTIVITIES ARE DIFFERENTIALLY REGULATED BY MULTIPLE SIGNALING PATHWAYS IN ANIMAL CELLS

A. Structural Features of the PMCA Family

1. Comparison of Membrane Calcium Pumping ATPases in Animal Cells

As noted earlier, the family of membrane calcium pumping ATPases is a large group of related enzymes that differ primarily in the way in which their activities are modulated and the membrane system to which they are localized by specific targeting signals in their linear sequences. The plasma membrane form of the enzyme is uniquely regulated by calmodulin. However, the many different isoforms of the PMCAs derived from the alternative processing of transcripts from four distinct genes are differentially responsive to different protein kinases and may differ in sensitivity to calmodulin as well. The SERCA forms of the pump appear to be regulated primarily by interaction with the protein phospholamban through a mechanism that may involve elements and mechanisms similar to those that operate in PMCAs, which are subject to control by phosphorylation. Little information is available, to date, on the regulation of the activity of SPCAs.

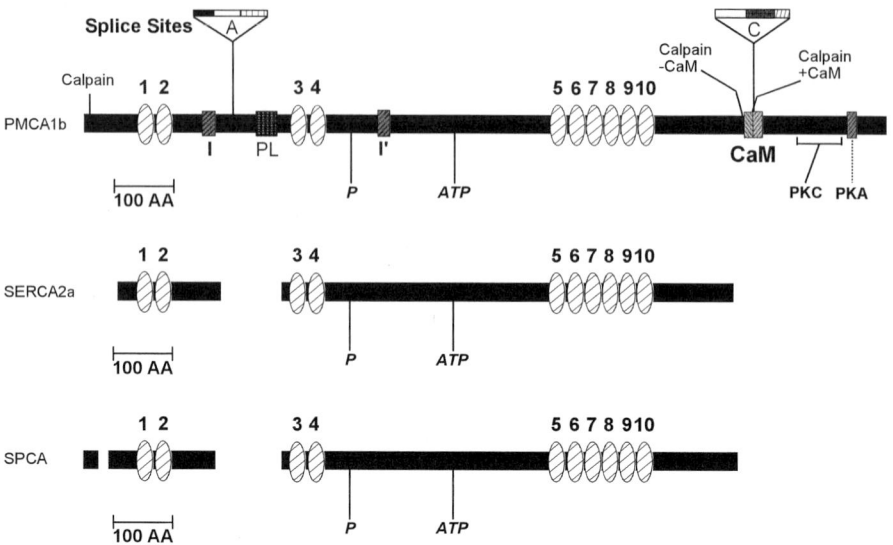

```
PMCA1b   mgdmannsvaysgvknslkeanhdgdfgitlaelralMELRSTDALrkIQESYGdvygic        60
serca2a  ------------------------------------MENAHTKTVEEVLGHFGV-----        18
spca     mkvarfqkipnvenetmipvlts---------------krASELAVSEVAGLLQAdlqng        45
                                                             TM1
PMCA1b   tklktspNEGLSGNPadLERREavFGKNfIPPKKPKTFLqLVWEALqDVTLIILeIAAIV       120
serca2a  -------NESTGLSLeqVKKLKerWGSNELPAEEgKTLLELVIEQFEDLLVrILLLAAcI        71
spca     lnksevshrrafhgw-----------NEFDIsEDEPLWKkYIsQFKnPLIMLLLASAVI         93
```

Fig. 1. (Top) Schematic representation of alignment of the major structural features of rat PMCA1b, SERCA2a, and SPCA based on primary sequence alignment shown in the bottom panel. Transmembrane domains (hatched ovals); calmodulin-binding inhibitory domains, I and I′; phospholipid-binding domain, PL; aspartyl phosphate, P; ATP-binding site, ATP; calmodulin-binding domain, CaM; cAMP-dependent kinase phosphorylation site, PKA; region containing the protein kinase C phosphorylation site(s), PKC; the A and C alternative splice sites and the sites of calpain cleavage with and without calmodulin. (Bottom) Primary sequence alignment of PMCA1b, SERCA2a, and SPCA using the MACAW program. Major structural features are boxed. Sequences in boldface type and with asterisks were referred to in the text. TM, transmembrane domain; dagger (†), amino acids in SERCA thought to be involved in Ca^{2+} transport.

8. CaM and Ion Flux Regulation

The common ancestry and catalytic functions of these enzymes are reflected in the high degree of homology found in the portions of these pumps that form the active unit required for ion translocation. Figure 1 shows an alignment of the primary structures of a representative isoform of each type of calcium pump from rat: PMCA1b (Shull and Greeb, 1988), SERCA2a (Gunteski-Hamblin *et al.*, 1988), and the putative secretory pathway calcium ATPase, SPCA (Shull *et al.*, 1992). The schematic presentation in the top panel of Fig. 1 shows the level of homology superimposed on a linear representation of the proposed functional domains of PMCAs. The actual residue alignments shown in the bottom panel of Fig. 1 were derived using the MACAW program. The alignments were optimized further by aligning shorter segments for maximum ho-

```
                                                                  TM2
PMCA1b    SLGLSFYqppegdnalcgevsvgEEEGEGeTGWIEGAAILLSVVCVVLVTAFNDWSKEKQ    180
SERCA2a   SFVLAWF----------------EEGEETiTAFVEPFVILLILVANAIVGVWQErNAENA    115
SPCA      SVLMRQFd-------------------DAVSITVAILIVVTVAFVQEYRSEKS          127

                                                               I-Domain
PMCA1b    FRGLqsrieqeqkftvirggqviqipvaDITVGDIAqVkYGDlLPADGiL--IqGNDLkI    238
SERCA2a   IEALkeyepemgkvyrqdrksvqrikaKDIVPGDIVeIaVGDKVPADIRLtsIKSTTLrV    175
SPCA      LEELsklvppechcvregklehtla--RDLVPGDTVcLsVGDRVPADLRL--FEAVDLsI    183


PMCA1b    DESSLTGESdhvkKSlDkdP-----------LLLSGThVmeGsGRMVVtAvGVNSqtGi    286
SERCA2a   DqSILTGESvsviKHTDPvPdpravnq-dKKNMLFSGTnIaaGKAmGVVVATGVNTEIGk    234
SPCA      DESSLTGETtpcsKVTaPqPaatngdlasRSNIAFMGTlVrcGKAKGIVIGTGeNSEFGe    243


PMCA1b    IFtllgAggeeeekkdekkkekknkkqdgaienrnkakaqdgaamemqplkseeggdgde    346
SERCA2a   IrdemvAt----------------------------------------------------    242
SPCA      vFkmmqAe----------------------------------------------------    251

                                                 TM3
PMCA1b    kdkkkanlpkkEKSvLQgKLTkLAvQIgKagllmsaitviilvlyfvidtfwvqkrpwla    406
SERCA2a   ---------EQERTPLQqKLDeFGeQLSKvisliciavwiinighfndpvhggswirg--    291
SPCA      ---------EApKTPLQksMDlLGkQLSfysfgiigiimlvgwllgkd-----------    290

                 TM4    †
PMCA1b    ectpiyiqyFVkFFiIGVTVLVVAVPEGLPLAVTISLAYsVKKMMKDNNLVRhLdacETM    466
SERCA2a   ---------AIyYFkIAVaLAVAAIPEGLPAVITTCLALGtRRMAKKNAIVRsLPSVETL    342
SPCA      ---------ILeMFtIsVSLAVAAIPEGLPIVVTVTLALGVmRMVKKrAIVKkLPIVETL    341

            Asp~P
              ↓
PMCA1b    GnaTAICSDKTGTLTmNrMTVvqayinekhykkvpepeaippnilsylVTG--------    517
SERCA2a   GCTSVICSDKTGTLTtNqMSVcrmfildkvegdtcslneft-------ITGstYApIGEV    395
SPCA      GCCNVICSDKTGTLTkNeMTVthiltsdglhae--------------VTGvgYNqFGEV    386
```

Fig. 1. Continued

```
                                                          I'
PMCA1b    ----------------IsvNCAytskilppekegglprh-VGnkTE CALlgfll dlkrd    559
SERCA2a   QkDDKpVkchqydglveLAtICALCNDsaldyneakgvyekVGeaTE tALtclve kmnvf   455
SPCA      IvDGDvVhgfynpavsrIVeAGCVCNDavirnntl------MGkpTE gALialam kmgld   440

                                    FITC
                                     ↓
PMCA1b    yqdvrneipeealykvyt-------------FnSvRKSMStvlknsdgsfri----FsKG     602
SERCA2a   dtelkglskieranacnsvikqlmkkeftleFSrDRKSMSvyctpnkpsrtsmskmFvKG      515
SPCA      glqqdyirkaeyp-----------------FSSEqKwMavkcvhrtqqdrpeic-FmKG       481

PMCA1b    AsEIILKKCfkilsangeakvfrprdrddivktviepmASegLRTIcLAfrdfpagepep      662
SERCA2a   ApEGVIDRCThirvgstkvpmtpgvkqkimsvirewgsGSdtLRCLALAThdnplrreem      575
SPCA      AyEQVIKYCTtynskgqtlaltqqqrdlyqqekaqm--GSagLRvLALASgpdlg-----      534
                           *   *                                *
PMCA1b    ewdnendvvt---GLTCIAVVGIeDPvRpEVpEAIKkCqrAGITVRMVTGDNINTArAIA      719
SERCA2a   hledsanfikyetNLTFVGCVGMLDPPRiEVassVKlCrqAGIrVIMITGDNKGTAVAIc      635
SPCA      -------------QLTLLGLVGIIDPPRtgVkEAVttliasGVSIKMITGDsqeTAIAIA      581

PMCA1b    TKCGILhpgEDflclegkdfnrrirnEkgEIEqeRIdkIWPKLRVLARsSPTdKhtLVKg      779
SERCA2a   rRIGIFgqdEDvTSKAFTG------rEFDELspsaqrdACLnARcFARVePSHKsKIVef      689
SPCA      SRLGLYsk----TSQSVSG------eEVDtMEvqHLsqIVPKVaVFYRASPrHKmKIIKs      631

                 FSBA-Site
                    *   *
PMCA1b    iidstvseqrqV VAvTGDGtNDGPALK KADVGFAMGiaGTDVAKEASDIILtDDNFTSIV     839
SERCA2a   -----LQSFDEI tAMTGDGVNDAPALK KsEIGIAMG-SGTaVAktASEMVLADDNFSTIV     743
SPCA      -----LQKNGSV VAMTGDGVNDAVALK aADIGVAMGqTGTDVCKEAaDMILVDDdFqTIM     686

                         † TM5                    †    † TM6
PMCA1b    kAVmwGRNVYdsIsk FLqFQLTvNVVAVIVAFtGAcItq dsPLkAVQM LWVNLIMDtLas    899
SERCA2a   aAVEEGRAIYNNmKq FIRYlISSNVGeVVcIFLtAALGF PeaLipVQL LWVNLVtDGLPA    803
SPCA      sAIEEGKGIYNNIKn FVRFQLSTsIAALtLIsLAtLMNF PnPLnAMQI LWINIIMDGpPA    746
```

Fig. 1. Continued

mology where necessary. This was particularly important for aligning the segments of PMCAs and SERCAs that have been proposed to be involved in forming the 10 putative transmembrane-spanning domains in the catalytic portion of the molecule. The sequences proposed to form these transmembrane elements in PMCAs and SERCAs are indicated above the sequences as TM1–10. Portions of the large cytoplasmic "transduction" domain of these enzymes involved in ATP binding and hydrolysis (FSBA and Asp ~ P sites) are also indicated. Regulatory domains that are specific to PMCAs—the autoinhibitory calmodulin-binding domain and related interacting sequences (I and I′) and sites for regulation by PKA and PKC—are also shown.

The most striking aspect of the comparison is the substantial diversity in the linear sequences of these three related proteins. The SERCAs

```
                                                         TM7
PMCA1b    lALAtEP PtesLLLRkPyGrnKPLISRtMM knILgh------------------------  935
SERCA2a   tALGFnP PDlDIMNKPPRNpKEPLISgWLF FRYLaigcyvgaatvgaaawwfiaadggpr  863
SPCA      qsLGVEP vDkDVIrKPPRNwKDsILTKNLI LKILvssiiivcgtlfvfwrelrdnvi---  803

                                          TM8
PMCA1b    -aFYQLvvvftllfagek FfdIDSGrnaplhappse hytivfntfvlmqlfneina rkih  994
SERCA2a   vsFYQLshflqckednpd FegVDCAifeSPYPMTM- ----alsvlvtiemcNALNS LSen  918
SPCA      ------------------ ----------TPRDTTM- ----tftcfvffdmfNALSS RSqt  830

                    TM9                                    TM10
PMCA1b    gernvfeGI fnNaIFCTIVLGTFVQIIIVQF gg--KpFscSeLSIe QWLWsIFLgmgtL  1052
SERCA2a   qsllrmppw eniwlvgs-ICLSMSLHFLILYV ePLPlIFQiTpLnLt QWLMVLkISlpVI   977
SPCA      ksvfei-GL csNkMFCYAVLGSIMGQLLVIYF pPLQKVFQtesLSIl DLLFLLGLTssVc   889

                                              CaM Binding Domain
PMCA1b    LWgqlisti ptSRLKflkeaghgtqkeeipeeelaedveeid haerelrrgqilwfrgln  1112
SERCA2a   LMdEtLKfV aRnyLEpaile---------------------- -------------------   997
SPCA      IVsEiIKkV eRSReKtqknttstpssflev------------ -------------------   919

                                          PKC Phosphorylation Region
PMCA1b    riqtq irvvnafrsslyeglekpesrssihnf mthpefriedsephipliddtdaeddap  1172
SERCA2a   ----- --------------------------- ------------------------------   997
SPCA      ----- --------------------------- ------------------------------   919

                PKA
                 ↓
PMCA1b    tkrnss ppps pnknnnavdsgihltiemnksatssspgsplhsletsl              1220
SERCA2a   --------------------------------------------------               997
SPCA      --------------------------------------------------               919
```

Fig. 1. Continued

and SPCAs have ~900–1000 amino acids whereas the PMCAs contain 200–300 additional residues. The latter are found primarily at the C terminus of the PMCAs where they form the bulk of the regulatory domain, providing sensitivity to both calmodulin binding and reversible phosphorylation. This area of PMCAs will be considered in detail later as it is also the region of greatest variability among different PMCA isoforms.

As might be expected, the regions that show the most homology among the three calcium pumps are those that appear to be involved in ATP binding and hydrolysis. For example, the aspartyl residue (D455 in PMCA1b), which forms the high-energy aspartyl phosphate intermediate during catalysis, is present in the center of a 10-residue sequence that is identical in all three proteins. Sequences upstream of this site, including the TM4 segment, also show a high degree of conservation. However, the three sequences diverge immediately C-terminal of the aspartyl phosphate site, with substantial gaps required to retain maximum downstream alignment. This pattern continues throughout this "active site" domain with short stretches of clear homology interspersed with regions of little or no homology among all three sequences.

As discussed in detail in Section III,C, the next notable region in the transduction domain is the short segment of amino acids labeled I' (557–563) identified (Falchetto *et al.*, 1991) as one of the two sites for internal interaction with the calmodulin-binding autoinhibitory domain of the C terminus of PMCAs. Including the two preceding residues, this short segment shows a higher level of similarity among the three proteins than surrounding sequences. Parallel homology in the other apparent site for autoinhibitory domain interaction, the I domain region (Falchetto *et al.*, 1991), is in accord with the observation (Chiesi *et al.*, 1991) that the CaM-binding domain of PMCAs can inhibit SERCAs. (The CaM-binding domain sequence shares homology with phospholamban, the regulator of SERCAs. The functional consequences of this similarity do not include calmodulin binding, as inhibition of SERCA by the PMCA CaM-binding domain is reversible by CaM whereas that by phospholamban is not.) If the structural similarities in the I and I' domain regions of PMCAs and SERCAs are responsible for this regulatory cross-reactivity, it is possible that SPCAs may also be regulated in a parallel fashion as these sequences are equally conserved in that calcium pump species. This proposal has yet to be tested experimentally.

The remainder of the C-terminal region of the large cytoplasmic domain is believed to form the ATP-binding site. Homologous structure prediction (Taylor and Green, 1989) indicates that this region of the molecule likely forms a region of β sheet, joined by α-helical segments and loop structures similar to the ATP-binding domains found in phosphofructokinase and adenylate kinase. As with the SR Ca^{2+}-ATPase (Mitchinson *et al.*, 1982; Pick and Karlish, 1982), the conserved lysyl residue at position 601 in PMCA1b is a site for labeling and inactivation by fluorescein isothiocyanate (FITC), a general ATP analog (Filoteo *et al.*, 1987). This conserved lysyl residue is at the beginning of the highly conserved 13-residue sequence Fs**K**GAsEIILKC. As with other ion transport ATPases, FITC labeling of this lysyl residue is blocked by ATP, indicating its location in the nucleotide-binding site involved in phosphoryl enzyme formation. However, additional FITC-reactive lysines may be present at this site as well. Adamo *et al.* (1996) have shown that replacement of the corresponding residue in hPMCA4b (K^{591}) with arginine yields an enzyme with only somewhat diminished activity (35%

of control), which is still subject to modification and inactivation by FITC. Similar conclusions have been made concerning additional lysyl residues in the ATP-binding site of the SR Ca^{2+}-ATPase based on labeling with pyridoxal ATP (Yamamoto et al., 1988; 1989) and 7-amino-4-methylcoumarin-3-acetylsuccinimidyl ester (Stefanova et al., 1993). In this regard, another reactive ATP analog, FSBA, has been shown to label the Na^+,K^+-ATPase at a lysyl residue (Ohta et al., 1986) proposed to be equivalent to that found in another conserved segment in the large cytoplasmic domain of the Ca^{2+}-ATPases adjacent to TM5 noted in Fig. 1. Indeed, the sequence of this site—AMTGDGVNDAPLK—is identical in all three calcium ATPases except for two conservative substitutions in PMCA1b.

Replacement of the aspartyl residue equivalent to D^{684} in rPMCA1b (asterisks in Fig. 1) by even such conservative substitutions as Asn or Glu led to a substantial loss of activity in both PMCA (Adamo et al., 1996) and SERCA (Clarke et al., 1990a). Replacement of the prolyl residue (also asterisks in Fig. 1) in SERCA two residues toward the C terminus with either Leu or Gly also largely inactivated the enzyme (Clarke et al., 1990a,b). This position is a Val in PMCA4b and its replacement with Pro also partially inactivated the enzyme (Adamo et al., 1996). Two aspartyl residues in the FSBA site (asterisks in Fig. 1) were also implicated by site-directed mutagenesis in ATP utilization by SERCAs but were not tested in PMCAs. Interestingly, all mutants tested with both enzymes showed an undiminished ability to form the aspartyl phosphate intermediate from ATP except for variants at the second aspartyl residue in the FSBA site (rPMCA1b D^{801}). Thus, only this aspartyl residue is directly implicated in nucleotide binding, whereas the other residues identified probably play a role in other aspects of activity (e.g., phosphoryl transfer).

The transmembrane segments of the ion transport ATPases must form the transmembrane pore structure through which ions are translocated. Numerous attempts have been made to identify the requisite amino acid side chains in the calcium pumping ATPases by a combination of homology search and site-directed mutagenesis. Although the general character of amino acids proposed to be involved in forming these segments is conserved in membrane calcium pumps, the level of actual homology differs substantially for different TM segments. TM1 and TM2 show a reasonable level of homology in all three pump subtypes shown in Fig. 1, but with no clusters of identical residues. The sequences proposed to form TM3 show little homology among any of the three sequences. However, the level of homology increases dramatically precisely at the beginning of the putative TM4 domain, which includes demonstrable regions of identity among the three isoforms. Substantial homology is also apparent for TM5 and TM6 extending up to the start of TM7 where the sequences appear to diverge substantially.

TM7 and TM8 show much less homology among all three proteins than TM4-7 whereas TM9 and TM10 again show considerable homology among all three proteins. The functional significance of this homology is unclear.

Site-directed mutagenesis has been performed to identify the residues in TM segments involved in forming the Ca^{2+}-binding sites used in Ca^{2+} translocation in both SERCAs (Clarke *et al.*, 1989) and PMCAs (Adebayo *et al.*, 1995; Guerini *et al.*, 1996). In general, these studies have focused on residues whose side chains possess potential calcium ligating oxygen (i.e., hydroxyl, carboxamide, and carboxyl) moieties positioned near the predicted center of the transmembrane segment. Muteins with very conservative replacements (e.g., Glu → Gln) as well as nonconservative ones (e.g., Ala replacement) have usually been examined in parallel. Measuring both calcium pumping activity and the formation of aspartyl phosphate intermediates for each mutein has provided detailed information concerning the precise step in the catalytic cycle altered by each substitution. Of particular use has been the fact that formation of the phosphoryl enzyme from ATP requires that calcium be bound to the translocation sites whereas bound calcium inhibits formation of E ~ P from free inorganic phosphate due to stabilization of the $E_2 \sim P_i$ conformation, which favors aspartyl phosphate hydrolysis. Therefore, measurement of the formation of phosphoryl enzyme intermediate in both directions provides a very precise estimate of the extent to which substitutions may affect calcium-binding sites involved in catalysis.

Applying these approaches to the SERCA pump, Clarke *et al.* (1989) identified individual glutamyl residues in TM4, TM5, and TM8 and adjacent asparaginyl, threonyl, and aspartyl residues in TM6 whose replacement with even conservative substitutions caused loss of Ca^{2+} binding as judged by the criteria set forth earlier. As shown in the SERCA and PMCA sequence alignments in Fig. 1, the glutamyl residue in SERCA TM4 is present in the pentapeptide sequence **PEGLP** (PMCA1b residues 432–436), which is conserved in PMCAs as well as SPCAs. Just as with SERCAs, replacement of this residue in PMCA4b with Gln completely abolished calcium translocation activity and aspartyl phosphate formation from ATP while conferring calcium independence to E ~ P formation from inorganic phosphate (Guerini *et al.*, 1996). The putative calcium-liganding glutamyl residue identified in SERCA TM5 aligns with alanyl residues in both PMCA (PMCA1b-866) and SPCAs, whereas that in TM8 appears to be Gln (PMCA1b 983) in PMCAs and Asp in SPCAs. Alanine replacement of the equivalent glutaminyl residue in PMCA4b led to a loss of properties associated with calcium binding (Guerini *et al.*, 1996). However, this mutation also appears to have affected proper insertion into the plasma membrane as the mutein was expressed exclusively in the ER, making conclusions on the role of this residue in forming calcium translocation sites equivocal.

The asparaginyl and aspartyl residues identified in SERCA TM6 are found in the highly conserved sequence $^{888}LW^V/_IN^L/_I^I/_V^M/_TD^{895}$. Replacement of either with a variety of substitutions led to inactivation of an apparent loss of calcium binding (Adebayo *et al.*, 1995; Guerini *et al.*, 1996). The intervening threonyl residue in this cluster in SERCAs is a Met in PMCAs (PMCA1b residue 894) as well as SPCAs that does not appear to be essential for calcium binding (Adebayo *et al.*, 1995).

Despite these analyses, knowledge of the nature of the calcium-binding site structures required for translocation formed by the TM segments and the residues involved in conferring high-affinity calcium transport activity is as yet only rudimentary. It should be noted that the C-terminal cytoplasmic domain of PMCA1b also possesses three high-affinity calcium-binding sites whose function remains obscure (Hofmann *et al.*, 1993).

A critical aspect of the structures of these different forms of calcium ATPases is the nature of sequences that dictate their membrane localization. As noted earlier, PMCAs have additional sequence elements compared to SERCAs and SPCAs both at the C terminus as well as at the N terminus where transmembrane domains important for targeting are usually found. This is also a region where products of the different PMCA genes that are differentially expressed and localized differ greatly in structure. Foletti *et al.* (1995) studied the membrane localization of a series of chimeras of SERCA1 and PMCA4 bearing either the opposing TM1 or TM1 and TM2 segments expressed in COS-7 cells to test whether this region is responsible for plasma membrane vs ER targeting. A strong ER retention signal was detected in the first 85 residues of SERCA1 by this approach, as the PMCA construct bearing this segment was retained in the ER. Interestingly, deletion of residues 18–75 of PMCA4b has been reported to lead to complete loss of activity without obvious impairment of membrane insertion or folding (Grimaldi *et al.*, 1996). However, analysis of localization of this mutein to plasma membrane was not examined.

Some evidence for plasma membrane targeting of PMCA by TM1 and TM2 was also obtained as a small portion of the SERCA chimeric construct did gain access to the plasma membrane. However, the bulk was retained in the ER even though it lacked the 85-residue amino-terminal region containing a known retention signal. This indicates that ER retention signals may reside elsewhere in the molecule. Indeed, studies with C-terminal truncation mutants of PMCA also show evidence of distal, possibly latent, ER retention signals (Zvaritch et al., 1995). However, such studies are difficult to interpret. For example, truncation of TM10 in PMCA by site-directed mutagenesis has been shown to lead to misinsertion of the remaining intact TM9 segment (Preiano et al., 1996). This observation suggests that the folding of this region of the molecule is critically dependent on the precise alignment of residues comprising TM9 and TM10 dictating conservation of structure. Also, even single point mutations in PMCA TM segments have been shown to cause improper processing of the PMCA polypeptide with ER retention (Guerini et al., 1996). Thus, analyses of this type must be interpreted with great caution.

Another very striking feature of the sequence comparisons shown in Fig. 1 is the fact that the SERCA and SPCA sequences end almost exactly at the beginning of the autoinhibitory CaM-binding domain of the PMCAs. This suggests strongly that this structure was acquired by an insertional event probably relatively late in the evolution of this family. This is in accord with the fact that calmodulin-regulated PMCAs are not found in plants or yeast, and their presence has not yet been confirmed in more primitive metazoa. Equally intriguing is the fact that the only large gap required internally to maintain alignment of the PMCA sequence with those of SERCA and SPCA occurs in the middle of the cytosolic domain between TM2 and TM3. This gap commences immediately following the PMCA I-domain segment. The portions of structure missing in the internal pump proteins include a site thought to be involved in the binding of acidic phospholipids leading to PMCA activation (Zvaritch et al., 1990). As discussed in detail in the following section, this region of PMCAs is also subject to insertion of additional coding segments at the "A" splice site involved in alternative splicing.

It should be noted that *direct* analysis of the topology and structure of the membrane calcium ATPase has thus far been performed by extensive partial proteolysis analyses (Benaim et al., 1984; Enyedi et al., 1993; James et al., 1989a,b; Niggli et al., 1981; Niggli and Carafoli, 1981; Papp et al., 1989; Zurini et al., 1984; Zvaritch et al., 1990) and more limited chemical labeling studies. This is due in large measure to the great difficulty in working with membrane proteins and to the limited quantities of pure material, particularly for PMCAs and SPCAs. Advances in producing purified PMCAs in quantity by recombinant methods holds promise for more rapid progress in such analyses in the immediate future.

B. Multiple PMCA Isoforms Arising from Four Genes and Differential Splicing

The PMCA family of enzymes is derived from four, and perhaps five, distinct genes. All of these genes have been localized in the human genome and thus far have not been associated with any known phenotype (Table I). The human PMCA1 gene was found on chromosome 12 at position 12q21–q23 (Olson *et al.*, 1991). PMCA2 was found on chromosome 3 (Brandt *et al.*, 1992a) and localized very close to the von Hippel–Lindau syndrome gene at 3p25–p26 (Latif *et al.*, 1993a), but was subsequently shown to be distinct from it (Latif *et al.*, 1993b). The rat PMCA2 gene has been localized to chromosomal band 4q41.3–q42.1 (Aldaz *et al.*, 1995). PMCA3 has been found on the X chromosome at position Xq28 (Wang *et al.*, 1994) whereas PMCA4 was found on chromosome 1 at 1q25–q32 (Olson *et al.*, 1991). Another putative PMCA gene, tentatively designated PMCA5, has been located on the X chromosome at Xq28, which puts it very close to PMCA3 but it appears to be a distinct PMCA gene (Heiss *et al.*, 1996). The authors speculate that this gene, or genes, surrounding it on the X chromosome may be a candidate for Barth syndrome or chondrodysplasia punctata. PMCA3 has been ruled

TABLE I

Chromosomal Localization of Human and Rat PMCA Genes

Gene	Chromosomal location	Reference
hPMCA1	12q21–q23	Olson *et al.* (1991)
hPMCA2	3p25–p26	Brandt *et al.* (1992a); Latif *et al.* (1993a,b)
rPMCA2	4q41.3–q42.1	Aldaz *et al.* (1995)
hPMCA3	Xq28	Wang *et al.* (1994)
hPMCA4	1q25–q32	Olson *et al.* (1991)
hPMCA5	Xq28	Heiss *et al.* (1996)

out as a candidate gene for X-linked myotubular myopathy (Smolenicka et al., 1996). In fact, no PMCA gene has yet been associated with a genetic disease.

The primary transcripts from all PMCA genes can be alternatively spliced at two sites designated A and C (Figs. 1 and 2). This alternative splicing can produce at least 25 different isoforms of the PMCA mRNAs. If splicing at these two sites is found to be undertaken independently, then the complexity of PMCA mRNAs will increase significantly. As might be expected with this extensive number of splicing options, a complex nomenclature for the mRNA products has evolved. Several different nomenclatures have been proposed, but the most commonly used one is that loosely based on the historical isolation of PMCA cDNAs. When Shull and Greeb (1988) isolated the first full-length PMCA1 cDNAs, they found two different spliced forms at the C splice site. The longer one was designated PMCA1a and the one lacking the alternatively spliced exon was designated PMCA1b (Fig. 2). From this point on it has been generally accepted for the C splice site that "b" indicates the isoform with no added exons and "a" refers to the site with the most added exons or the longest sequence from an internal splice site within an exon. Other splicing alternatives were then named starting with "c" as they were isolated. A similar nomenclature is followed at the A splice site. Isoforms with no internal inserted exons are designated "z" and those with the 42 nucleotide (nt) exon common to all PMCA genes inserted is designated "x". When additional exons are added to this site, a letter after "x" was chosen, as in the case of PMCA2. The generally accepted designations for different spliced forms are shown in Fig. 2.

A basic PMCA gene appears to consist of 22 or 23 exons based on the gene structures determined for human PMCA1 (Hilfiker et al., 1993) and rat PMCA3 (Burk and Shull, 1992). In the generic PMCA gene there are two sites that can undergo alternate mRNA splicing, designated A and C (Fig. 2). At the A site, exon 7 in human (or the homologous exon 8 in rat) can be alternatively spliced into the mRNAs of PMCA2, PMCA3, and PMCA4, depending on tissue and species (discussed later). However, in PMCA1 this exon appears to be constitutively spliced into the message. In PMCA2, the splicing at the A site is somewhat more complex in that two additional exons can be spliced with the exon 7 homolog to generate a third spliced isoform of this mRNA (Fig. 2) (Heim et al., 1992a). This region of alternative splicing is immediately adjacent to the sequences encoding the acidic lipid-binding site, but studies with alternatively spliced forms of PMCA2 mRNAs expressed in Sf9 (*Spodoptera frugiperda* fallarmy worm ovary) cells showed that the enzyme properties measured were invariant to stimulation by acidic lipids in all alternatively spliced forms (Hilfiker *et al.*, 1994). This does not preclude the possibility that the alternative splicing produces changes in activity or regulation that have not yet been tested. Examination of the intron upstream of exon 7 in the PMCA1 gene revealed sequences homologous to the exons alternatively spliced into the PMCA2 mRNAs, but they have never been observed to be included in PMCA1 mRNAs (Hilfiker *et al.*, 1993).

The PMCA1 gene apparently can use an alternative exon 1, designated exon 1* (Hilfiker *et al.*, 1993). However, use of exon 1* has thus far only been reported in porcine smooth muscle (De Jaegere *et al.*, 1990).

Near the 3' end of the PMCA genes is the C splice site. Splicing at the C site appears to be more complicated than at the A site in that a larger variety of alternatively spliced forms can be generated and more than one exon is involved for some PMCA isoforms. Possible alternative splicing patterns at this site are shown in Fig. 2. For all PMCA genes the C splice site is immediately adjacent to the sequence encoding the CaM-binding site. Because different spliced forms of the PMCAs are differentially susceptible to phosphorylation by different PKC isoforms (vida infra), the sequence encoded by alternatively spliced exons probably provides the specific PKC recognition sequences. Therefore, alternative splicing can provide a mechanism to control PMCA regulation by PKC. This would be similar to that found for PKA in the PMCA1 family where PMCA1b, PMCA1c, and PMCA1d can be phosphorylated and activated by PKA, but PMCA1a and PMCA1e are truncated proteins that do not contain the PKA phosphorylation site. As with the A splice site, splicing at the C site is developmentally controlled and tissue specific, as discussed in more detail later.

For the PMCA1 gene, a single exon (exon 21) can be alternatively spliced at site C using different internal donor sites to produce five different isoforms (Fig. 2). The human PMCA1 exon 21 is 154 nucleotides long, and internal splices within this exon yield insertion of exons of 87, 114, and 152 nucleotides (Fig. 2). Spliced variants also are formed that either lack or contain the entire 154 nucleotide exon. The addition of the full-length 154 nt exon or the 152 nt portion leads to a shift in the reading frame. This change in reading frame introduces in-frame stop codons that produce forms of the PMCA1a and PMCA1e proteins, respectively. These isoforms no longer contain the PKA phosphorylation

PMCA1 A Site

Exons		mRNA
Human		Designation
39 nt*	Exon 8	PMCA1x
Rat		
42 nt*	Exon 8	PMCA1x
*Exon is constitutively expressed		

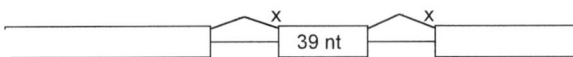

PMCA1 C Site

Exons		Exon	mRNA
Human		Compositon	Designation
154 nt*		154 nt	PMCA1a
87 nt	internal	87 nt	PMCA1c
114 nt	internal	114 nt	PMCA1d
152 nt	internal	152 nt	PMCA1e
		--	PMCA1b
Rat			
Same as human			
*Internal donor sites give rise to other inserts			

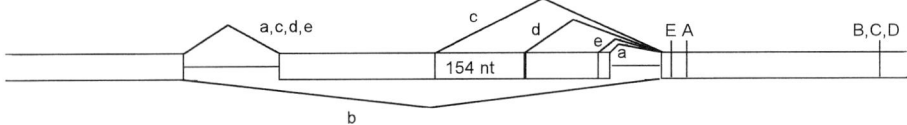

Fig. 2. Exon choices for splicing of the most prevalent PMCA mRNA isoforms. Lowercase letters indicate which isoforms use the adjoining exons. At the C site, uppercase letters indicate the position of a termination codon for a given isoform.

site because they terminate before the corresponding coding sequence. Conversely, the 87 and 114 nt inserts retain the reading frame, which leads to the insertion of an additional 29 and 38 amino acid residues, respectively, between the CaM-binding domain and the PKA phosphorylation site. As discussed in a later section, this insertion appears to alter

8. CaM and Ion Flux Regulation

PMCA2 A Site

Exons Human		Exon Composition	mRNA Designation
33 nt		42 nt	PMCA2x
60 nt		33, 60, 42 nt	PMCA2w
42 nt		--	PMCA2z
Rat			
		Same as human	

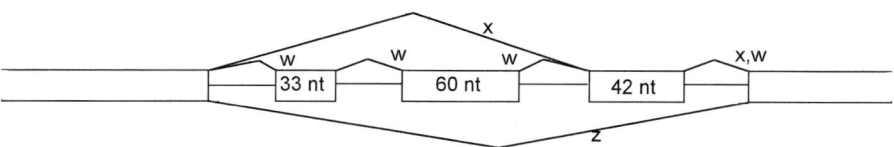

PMCA2 C Site

Exons Human		Exon Composition	mRNA Designation
172 nt		172, 55 nt	PMCA2a
55 nt		172 nt	PMCA2c
		--	PMCA2b
Rat			
		Same as human	

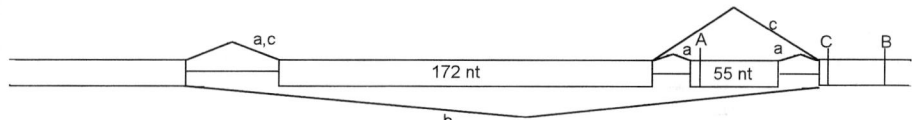

Fig. 2. Continued

CaM-binding affinity (Enyedi *et al.*, 1991) and could also change the effects of phosphorylation by PKA.

The splicing of the PMCA2 gene at the C site can involve two exons: a 172 nt exon that is structurally homologous to the 154 nt exon of PMCA1 and a 55 nt exon. The use of internal donor sites in the 172 or 55 nt exon analogous to that seen with PMCA1 has not been reported. PMCA2a incorporates both 172 and 55 nt exons, whereas PMCA2c incorporates just the 172 nt exon. PMCA2c has

PMCA3 A Site

Exons Human		Exon Composition	mRNA Designation
42 nt		42 nt	PMCA3x
		--	PMCA3z
Rat			
		Same as human	

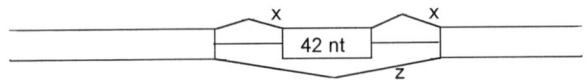

PMCA3 C Site

Exons Human		Exon Composition	mRNA Designation
		Same as rat	
Rat			
68 nt	Exon 22	154 nt	PMCA3a
154 nt*	Exon 23	87 nt	PMCA3c
114 nt	internal	114 nt	PMCA3d
87 nt	internal	--	PMCA3b
		154 nt and part of adjacent intron (hatched)	PMCA3e
		68 nt	PMCA3f
*Internal donor sites give rise to other inserts			

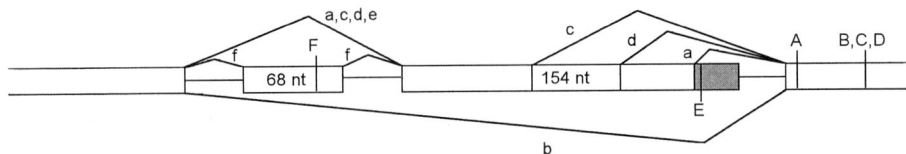

Fig. 2. Continued

PMCA4 A Site

Exons Human	Exon Composition	mRNA Designation
36 nt	36 nt	PMCA4x
	--	PMCA4z
Rat		
Same as human		

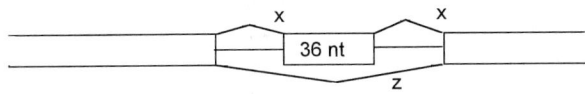

PMCA4 C Site

Exons Human	Exon Composition	mRNA Designation
178 nt	178 nt	PMCA4a
	--	PMCA4b
Rat		
175 nt	175 nt	PMCA4a
	--	PMCA4b

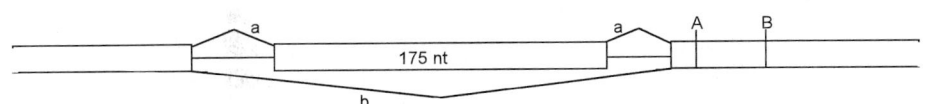

Fig. 2. Continued

only been observed in rat tissues so far. PMCA2b, as with all PMCA isoforms, uses neither exon.

Alternative splicing of the PMCA3 gene at the C site is the most complex of all the PMCA genes. As with PMCA2, there are two exons that can be alternatively spliced: a 68 nt exon and a 154 nt exon homologous to the 154 nt exon that is alternatively spliced in PMCA1. These are exons 22 and 23, respectively, in the rat gene (Keeton et al., 1993). Exon 23 has internal donor sites at the same positions as those of the PMCA1 exon; therefore, the nomenclature is analogous to that for the PMCA1 isoforms derived from the 154 nt exon. The 68 nt exon 22 is only spliced into isoform PMCA3f along with exon 23. Inclusion of exon 22 results in a truncated protein that terminates just after the calmodulin-binding domain (Fig. 2). Additionally, in one PMCA3 mRNA, a small piece of the intron following exon 23 is included in the spliced product to produce isoform PMCA3e (Fig. 2). PMCA3e is not homologous to PMCA1e, but results in a truncated protein that is shorter than PMCA3a, PMCA3b, PMCA3c, or PMCA3d.

Splicing at the PMCA4 C site is very simple compared to other PMCA genes. A single 175 nt exon that is most structurally homologous to the 172 nt exon of PMCA2 can be alternatively spliced into the PMCA4 mRNA. No internal donor sites have been reported. Addition of the 172 nt exon results in the shorter 4a protein, which lacks the tyrosine phosphorylation site at position 1176.

PMCA1 and PMCA4 have been reported to have an additional alternative splice site upstream of the C site, designated the B site (Howard *et al.,* 1993; 1994; Strehler, 1991). Alternative splicing at the B site has only been reported in human liver (Howard *et al.,* 1994) and intestine (Howard *et al.,* 1993). The putative 10th transmembrane domain is encoded by the 108 nt exon that can be alternatively spliced at this site.

1. Regulation of PMCA Isoform mRNA Splicing

With the multitude of alternatively spliced PMCA mRNAs, it might be expected that they would serve different roles in different cell types and under different conditions. Studies of the PMCA mRNA expression in tissues and clonally pure cell lines suggest that there is distinct cell-specific expression of PMCA mRNAs. Further, examination of PMCA expression in developing prenatal brain and excitable cells stimulated to differentiate *in vitro* has shown induction of alternative splicing of PMCA isoforms, supporting the hypothesis that new isoforms are required to perform unique functions associated with differentiated cells.

a. Tissue-Specific Splicing. The first detailed study of the tissue distribution of alternatively spliced PMCA mRNAs was reported by Brandt *et al.* (1992b). In this study, only splicing at the C site was examined, as alternate splicing at the A site had not yet been described. Using fetal human tissues, it was observed that PMCA1b and PMCA4b (then called 4a) were constitutively expressed in all tissues examined.

The general coexpression of PMCA1 and PMCA4 at the protein level has since been confirmed using isoform-specific antibodies (Stauffer *et al.*, 1995). However, some interesting and unexpected results were found using these antibodies. Immunoreactive PMCA1 was not found in heart and the choroid plexus, and PMCA4 was not found in the transformed embryonic kidney cell line 293, even though mRNAs for these isoforms had been found previously in these tissues and cells. The immunological determinations of PMCAs by Stauffer *et al.* (1995) used adult human tissues, whereas many of the studies examining PMCA mRNA isoform distribution used fetal material. Therefore, the differences in PMCA protein and mRNA expression may reflect developmental changes in PMCA expression. It may also reflect changes in the translational efficiency of different isoform mRNAs in different tissues. Adamo *et al.* (1992) have shown that the human PMCA4 mRNA has a start codon in a particularly bad context for efficient translation and, indeed, found that this isoform was poorly translated when transfected into COS-1 cells. It has been shown in adult heart that PMCA1b and PMCA1c mRNAs constituted 47% of the total PMCA mRNA and PMCA4a and PMCA4b constituted 51% (Stauffer *et al.*, 1993). However, immunological detection found only PMCA1 isoforms (Stauffer *et al.*, 1995), lending credence to the observation that PMCA4 mRNAs are poorly translated *in vivo*. Whether this poor translation is overcome by the addition of another translation factor, modification of an existing factor or increased mRNA content in cells where PMCA4 proteins are found has not been determined yet.

Another interesting aspect of PMCA mRNA isoform expression is that almost all isoforms are expressed in brain (Brandt and Neve, 1992; Stauffer *et al.*, 1993; Zacharias *et al.*, 1995). This most likely reflects the cell type diversity found there. Notably absent from brain was PMCA1d, which was found specifically in fetal skeletal and heart muscle (Brandt *et al.*, 1992b), but not in adult skeletal and heart muscle (Stauffer *et al.*, 1993).

As noted earlier, almost all tissues and cells examined contain mRNAs for PMCA1b or PMCA14b, and usually both. The other alternatively spliced forms of PMCA1 seem to be found only in excitable tissues (Brandt *et al.*, 1992b; Keeton *et al.*, 1993; Stauffer *et al.*, 1993). The form of PMCA4 alternatively spliced at the C site, PMCA4a (referred to as 4b in earlier papers), is found mainly in excitable tissues, but also in testes, stomach, colon, pancreas, and, to a lesser extent, lung (Brandt *et al.*, 1992b; Keeton *et al.*, 1993; Stauffer *et al.*, 1993).

PMCA2 and PMCA3 isoform mRNAs have a much more limited expression pattern. In addition to brain and spinal cord, PMCA2b was found only in fetal human liver at levels comparable to human fetal brain (Brandt et al., 1992b). However, in adult human liver it was less than 2% of the PMCA isoform complement as compared to 19% in the adult human cerebral cortex (Stauffer et al., 1993). In contrast, in adult rat liver it was a major isoform at levels comparable to adult rat brain (Keeton et al., 1993). In the cerebral cortex, PMCA2b is the only C site spliced form found and the A site is divided among PMCA2x (6% of the total PMCA mRNAs), PMCA2w (4%), and PMCA2z (9%) (Stauffer et al., 1993). Although PMCA2 in humans appears to have a tissue distribution limited to the central nervous system (CNS) and liver, PMCA2b in rats was found in uterus, brain, liver, kidney, spleen, and testes (Keeton et al., 1993), suggesting that it may be used as an alternative to an isoform used in human. Splicing at the A site of the PMCA2 mRNA has not been examined in rat.

PMCA3b was found only in human fetal brain, spinal cord, and adrenal gland (Brandt et al., 1992b). In adult human cerebral cortex, PMCA3b constitutes about 2% of the total PMCA message, with PMCA3e being an additional 3% and at the A site PMCA3x and PMCA3z constituting 4 and 2% of the total PMCA message population, respectively (Stauffer et al., 1993). PMCA3a is also found in human fetal spinal cord, at very low levels in human fetal brain, and is absent in human fetal adrenal cortex (Brandt et al., 1992b) and nearly absent in adult human cerebral cortex (Stauffer et al., 1993). In contrast, PMCA3a in adult rat is expressed in uterus, skeletal muscle, brain, lung, kidney, testes, stomach, small intestine, and colon, and PMCA3b is found primarily in skeletal muscle and brain with lesser amounts in kidney and testes. PMCA3d was found in brain, and PMCA3c was found in kidney and testes in low levels (Keeton et al., 1993). Interestingly, PMCA3f was found in adult skeletal muscle of rat (Keeton et al., 1993), but was only found in fetal human skeletal muscle by Stauffer et al. (1993).

The existence of PMCA2 and PMCA3 isoforms in rat tissues other than the CNS suggests a general trend wherein human tissues use a specific set of PMCA isoforms during development, which are turned off in adulthood. In contrast, the rat appears to use these isoforms in the mature animal as well. The reason for this switch of isoforms with maturation in humans is not known, but based on data from the rat, may not be essential.

It should be noted that some mRNAs are transported to the nerve terminal. It is possible that the trace amounts of PMCA mRNAs detected in some tissues by RT-PCR (reverse transcriptase-polymerase chain reaction) are due to mRNAs from innervation of these tissues.

8. CaM and Ion Flux Regulation

b. Tissue Subregion Distribution of PMCA mRNA Alternatively Spliced Isoforms. As almost all PMCA-spliced forms are found in brain, Zacharias *et al.* (1995) undertook a detailed study of isoform expression in various subregions of the adult human brain. They examined the frontal, parietal, occipital, and temporal cortices, caudate nucleus, cerebellum, globus pallidus, hippocampus, hypothalamus, inferior olivary nucleus, olfactory bulb, putamen, substantia nigra, and thalamus. Although definitive statements about amounts of each isoform in specific regions cannot be made because the PCR-based detection method is only semiquantitative under the conditions used, general trends can be seen. PMCA1a is the predominant isoform found in all regions of the brain. PMCA1b is found in most regions, but at levels below PMCA1a. PMCA1c and PMCA1e are found in various regions in a somewhat random distribution. For PMCA2 gene products, PMCA2b appears to be the predominate C-site spliced isoform in all brain regions whereas PMCA2z is the predominant A splice site isoform. Interestingly, only in the inferior olivary nucleus do all three of the PMCA2 A-site isoforms occur, although this region does not possess a particularly rich variety of neurons or glia. It would seem reasonable to expect that a certain cell type(s) in this region requires these alternate forms for functions specific to those cells. However, *in situ* hybridization is needed to determine which cell(s) uses these isoforms. Also, in the substantia nigra it appears that PMCA2x is used instead of PMCA2z. PMCA3x is the A-site splice form preferentially expressed in all brain regions, whereas the C-site spliced form used seems to be PMCA3a and PMCA3b in equal amounts. PMCA4x is the only A-site splice form and PMCA 4b is the only C-site isoform expressed in all brain regions. Interestingly, PMCA4a is expressed in a limited number of regions and its expression seems to loosely parallel that of PMCA1e, suggesting that they may exist as a heterodimeric pair.

Several studies have used *in situ* hybridization techniques to examine the distribution of PMCA mRNAs in brain sections and have localized some isoforms to some specific cell types. Unfortunately, these studies used probes that detect all spliced isoforms of the four PMCA families so that the cellular distributions of PMCA-spliced forms could not be determined. These studies have shown that all PMCAs mRNAs are most highly concentrated in neurons. The highest concentration of PMCA1 is found in pyramidal cells of the CA1 region of the hippocampus, but it is also distributed broadly throughout the brain, which probably reflects the putative function of PMCA1b as a "housekeeping" form of the pump (Stahl et al., 1992). PMCA2 isoforms were found most abundant in cerebellar Purkinje cells. PMCA3 mRNAs were found most prevalently in choroid plexus, the granule cell layer of the cerebellum and hippocampus and PMCA4 in the piriform cortex, neocortex, amygdala, and laminae 2 and 6 of the cerebral cortex (Stahl et al., 1994). Eakin et al. (1995) have used antibodies specific to PMCA3 to detect PMCA3 protein in brain sections and found that there was a strong correlation between PMCA3 protein and mRNA expression. PMCA3 protein appeared to translocate to nerve termini in the granule cells of the cerebellum and the pyramidal cells of the CA1 region in the hippocampus and was not prevalent in other brain regions. This observation is in contradiction with the work of Zacharias et al. (1995), which found PMCA3 mRNA in every brain region examined. Data of Eakin et al. (1995) suggest that the isoform(s) of PMCA3 produced in these cells is specific to the nerve terminals of a very limited population of neurons where it (they) may be involved in a highly specific function and lend support to the hypothesis that the large number of PMCA isoforms reflects their use in carrying out specific functions in specialized cells.

A fairly detailed analysis of PMCA distribution in kidney has been conducted by Magosci et al. (1992) and Caride et al. (1991). Magosci et al. (1992) showed that PMCA isoform mRNAs are localized to different regions of the kidney. The renal cortex was found to contain only PMCA1 and PMCA2 with trace amounts of PMCA3. Although the outer medulla had abundant PMCA3 in addition to PMCA1 and PMCA2, the inner medulla had only PMCA1. These studies were conducted before the extent of PMCA isoform diversity had been fully appreciated, so the experiments were conducted in such a way that individual spliced forms could not be distinguished. A more detailed analysis of PMCA3 and PMCA4 isoforms by Caride et al. (1995) found that PMCA3a and PMCA3c were the major PMCA3 mRNAs expressed in the outer medulla with a minor contribution of PMCA3b. PMCA4b was the major PMCA4 isoform in the outer medulla. Microdissection of the outer medulla found that PMCA3a was the exclusive PMCA3 isoform in the descending Henle's loop and PMCA4b was found in all microsections of the outer medulla.

Varadi et al. (1996) examined the distribution of PMCA mRNA isoforms in pancreatic β cells and the islets of Langerhans because changes in free intracellular calcium concentration are associated with regulation of insulin secretion. PMCA1b and PMCA2b were found in the islets of Langerhans as well as in an α-cell line and several different β-cell lines that were examined. A PMCA3 isoform was not found in any of the cells examined. Only PMCA4 was found to be different between the islet of Langerhans cells and the β-cell line RIN5mF where PMCA4a and PMCA4b and PMCA4b alone were expressed, respectively.

c. Developmental Regulation of PMCA RNA Splicing. i. Developing Brain. The first indication that control of spicing might be developmentally regulated was found in whole rat brain examined over a 38-day period (10 days prenatally and 28 days postnatally) (Brandt and Neve, 1992) as summarized in Table II. A significant conclusion drawn from this study was that the spliced mRNAs found in the mature brain were produced prior to birth. This showed that induction of new PMCA isoforms was not necessary for postnatal synaptic remodeling, where calcium has been shown to be required. It also implied that PMCAs might play a role in neuronal development and differentiation. As will be discussed later, this may be the case for PMCA1.

At the earliest prenatal time (E10) only the nonexon-containing forms of PMCA1 and PMCA3 were expressed (i.e., PMCA1b and 3b), suggesting that the factors carrying out this splicing may not have been transcribed at that point in development, i.e., they were functionally inactive or the specific cells types in which these spliced forms exist had not yet developed. Of these two isoforms, PMCA1b appeared to be more abundant than PMCA3b at day E10. This probably reflects its expression in all cells where it is thought to serve as a housekeeping enzyme. The PMCA2 gene was not transcribed until just before birth and at that point all PMCA2 C-site spliced forms were expressed. The late expression of PMCA2 may suggest that it is only required for postnatal events in the life of the animal. This fact, combined with the apparent localization of PMCA2 in dendritic spines, provides for the possibility that it may play a role in regulating calcium levels involved in long-term potentiation, the supposed mechanism underlying consolidation of short-term memory into long-term memory.

At embryonic day 10, PMCA1b is the most abundant isoform present, but by embryonic day 14 PMCA1a and PMCA1c are expressed. Concomitant with PMCA1a and PMCA1c expression, the PMCA1b pool decreases, suggesting that increased transcription does not accompany the induction of alternative splicing, at least for the PMCA1 gene. However, induction of PMCA3a splicing occurs at embryonic day 18, just 2 days before birth, and total PMCA3 mRNA production increases through birth. It is intriguing that induction of PMCA2a, PMCA2c, and PMCA3a splicing occurs at exactly the same time, suggesting that the signal respon-

TABLE II

Development Expression PMCA mRNAs in Whole Rat Brain[a]

Isoform	E10	E14	E16	E18	P2	P3	P4	P5	P8
PMCA1a	−	+	++	++	++	++	++	++	++
PMCA1b	+++	++	++	++	++	++	++	++	++
PMCA1c	−	++	++	++	++	++	++	++	++
PMCA1d	−	−	−	−	−	−	−	−	−
PMCA1e	nd	nd	nd	nd	nd	nd	nd	nd	nd
PMCA2a	−	−	−	+	+	++	++	++	++
PMCA2b	−	−	−	+	+	++	++	++	++
PMCA2c	−	−	−	+	+	++	++	++	++
PMCA3a	−	−	−	+	+	++	++	++	++
PMCA3b	+	+	++	+++	+++	+++	+++	+++	+++
PMCA3c	nd	nd	nd	nd	nd	nd	nd	nd	nd
PMCA3d	nd	nd	nd	nd	nd	nd	nd	nd	nd
PMCA3e	nd	nd	nd	nd	nd	nd	nd	nd	nd
PMCA3f	nd	nd	nd	nd	nd	nd	nd	nd	nd
PMCA4a	nd	nd	nd	nd	nd	nd	nd	nd	nd
PMCA4b	nd	nd	nd	nd	nd	nd	nd	nd	nd

[a] Summary of data presented by Brandt and Neve (1992). Relative amounts of each PMCA isoform detected by RT-PCR of rat brain at different stages of development are indicated by + symbols; nd, not determined; −, not detected; E, gestational day postappearance of vaginal plug; P, postnatal day.

sible for controlling their splicing may be the same. Based on the work of Hammes et al. (1994) and De Jaegere et al. (1993), the factor controling PMCA splicing is most likely myogenin.

At the time these studies were conducted, the existence of rodent PMCA4 had not been established. Cloning of the rat PMCA4 cDNA (Keeton and Shull, 1995) has shown that the primers and probes used by Brandt and Neve (1992) would not have allowed detection of PMCA4 due to dissimilarities of the sequences.

ii. Developmental Expression of PMCA mRNAs in Homogenous Excitable Cell Cultures. The study of induction of PMCA mRNA splicing in developing brain is complicated by the fact that the brain is a heterogeneous mixture of many different cell types and association of expression of a specific isoform mRNA to a particular cell type cannot easily be made. It was found that PMCA1b, PMCA1c, and PMCA1d were the

PMCA1-spliced forms expressed in muscle tissues (Brandt et al., 1992b; Strehler et al., 1989). Because PMCA1a and PMCA1c were found to be expressed later in brain development (Brandt and Neve, 1992), it was thought that PMCA1c and PMCA1d might be expressed as a function of differentiation in embryonic myocytes. When mouse C2 myocytes were differentiated by two different methods, serum deprivation or confluent contact, splicing of PMCA1c was induced when the cells fused and formed myotubes (Brandt and Vanaman, 1994). Surprisingly, PMCA1d was not induced when the myocytes differentiated. This study suggested that PMCA1d splicing may require another signal such as end plate formation with a motor neuron. Hammes et al. (1994), however, using two different myocyte cell lines, L6 and H9c2, found that PMCA1d was expressed when myotubes were formed. They also found that the splicing of PMCA4a was induced on differentiation. Hammes et al. (1994) further showed that transfection of rat skin fibroblasts with myogenin would also lead to induction of splicing of PMCA1c, PMCA1d, and PMCA4a, which are muscle specific and not normally found in fibroblasts. De Jaegere et al. (1993) had found splicing of PMCA1c, but not PMCA1d, on differentiation of the myocyte cell line BC_3H1 and this event was also myogenin dependent. De Jaegere et al. (1993) also found that the myogenin-dependent splicing was fairly unstable as treatment with cyclohexamide completely blocked splicing within 24 hr, presumably due to the degradation of myogenin.

The pheochromocytoma cell line, PC12, has also been examined for the differentiation-induced expression of PMCA mRNA splicing. PC12 cells maintain the phenotype of a chromaffin/sympathetic neuron precursor cell, but when treated with nerve growth factor (NGF) they will differentiate into a sympathetic-like neuron. Treatment with NGF resulted in splicing of PMCA1c, 2a and c, and 4b mRNAs (Hammes et al., 1994). Whether splicing associated with differentiation in PC12 cells is also under the control of myogenin has yet to be determined.

C. Catalytic and Regulatory Properties of PMCAs

1. General Mechanism and Kinetic Properties of Membrane Calcium Pumping ATPases

The calcium pumping ATPase of the plasma membrane belongs to the large family of "P"-type ion transport ATPases that share a common catalytic cycle and active site architecture. The essential feature of catalysis is the formation and hydrolysis of a high-energy acyl phosphate phosphoenzyme intermediate involving an invariant aspartyl residue. Calcium translocation is coupled to this reaction through two conformational states, most frequently termed E_1 and E_2, in which the calcium-binding sites face toward either the cytosol (E_1) or the extracellular (E_2) or luminal (in the case of internal membrane pumps) space. An interesting and useful aspect of this reaction is that it is readily reversible such that the aspartyl phosphate can be formed from inorganic phosphate in the absence of Ca^{2+} and this $E \sim P_i$ can be incorporated into ATP. Although this has no known physiological significance, it has been extremely useful in dissecting the pathway of catalysis in ion transport ATPases.

The diagram shown in Fig. 3 outlines the various proposed steps in the full catalytic cycle of ATP-driven on translocation as have been reviewed in numerous monographs and compendia. The discussion here is limited to aspects of the cycle that are important for understanding the regulation of calcium pumping activity by calmodulin, reversible phosphorylation, etc. For a detailed discussion of the mechanism of calcium pumping, the reader is referred to Jencks (1989) and Carafoli (1991a,b) and references therein.

Commencing with the unligated form of the protein in the initial E_1 conformation, calcium first binds to form the $E_1 \cdot Ca^{2+}$ complex, then ATP binds at a high-affinity site [$K_m \sim 1-3 \ \mu M$; (Mualem and Karlish, 1979; Rega and Garrahan, 1975)] and in the presence of Ca^{2+} (Katz and Blostein, 1975) forms, in a rapid reaction, the aspartyl phosphate intermediate first detected by Knauf et al. (1974). The K_m for Ca^{2+} for the activated form of the enzyme is in the range of 1 μM or below (Rega and Garrahan, 1975). As noted later, this value is substantially higher for the inhibited form of the enzyme (i.e., in the absence of calmodulin, acidic phospholipids, or regulatory phosphorylation).

Once formed, the bound calcium in the $E_1 \sim P \cdot Ca^{2+}$ complex is not free to dissociate and appears to be sequestered from solvent. The complex then isomerizes to the $E_2 \sim P \cdot Ca^{2+}$ conformation in which the K_d for calcium is 1–2 mM. Therefore, Ca^{2+} dissociates and $E_2 \sim P$ hydrolyzes. The E_2 conformation of the enzyme then isomerizes back to E_1, completing the cycle.

As noted in Fig. 3, a number of other species have the ability to modify or participate in various steps in the cycle. First and foremost is the divalent metal Mg^{2+}, which enhances the rate of phosphoryl transfer

8. CaM and Ion Flux Regulation

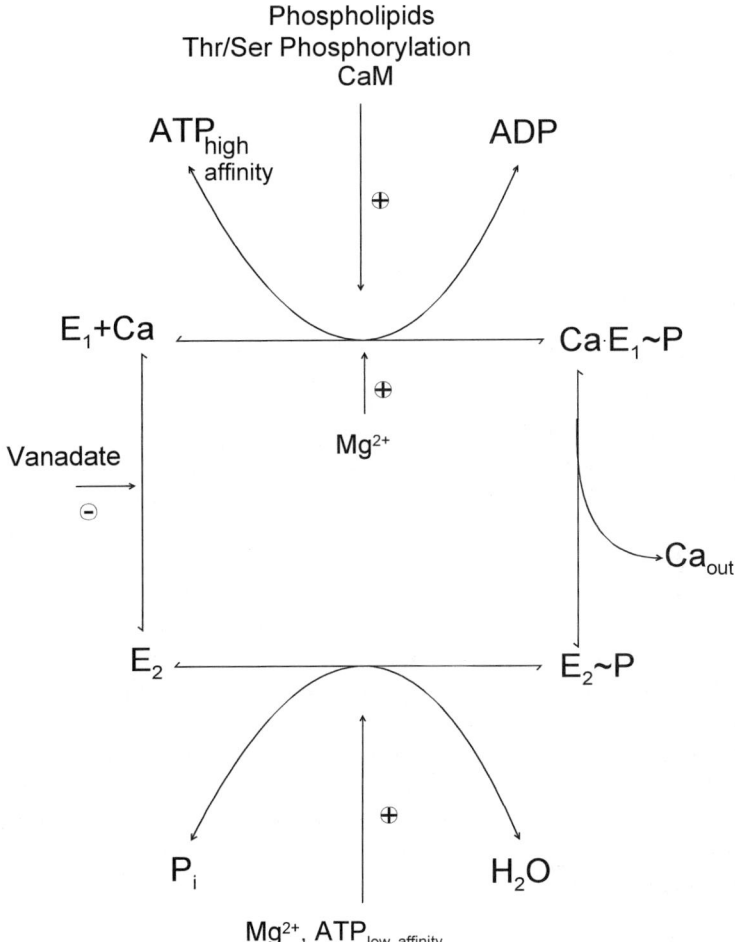

Fig. 3. The proposed reaction cycle for PMCAs and where various stimulatory and inhibitory agents interact with the enzyme.

from ATP to the active site aspartyl residue and from the resulting aspartyl phosphate to water (Garrahan and Rega, 1978; Rega and Garrahan, 1975). Mg^{2+} does not act as a simple cosubstrate (as is often the case with phosphotransferases) as it actually increases the K_m for ATP compared to that obtained with Ca^{2+} (Lacapere and Guillain, 1990). The phosphate analog, orthovanadate $[VO_3(OH)]^{2-}$, is a potent inhibitor of many ATPases (Bond and Hudgins, 1979), including not only those that transport ions but also ones involved in ATP-driven motility (i.e., myosin, dyneins). Micromolar concentrations of vanadate ion effectively inhibit the calcium pumping ATPases, including that of the plasma membrane (Bond and Hudgins, 1980). Binding appears to occur at the nucleotide-binding site of ion translocases (as reviewed by Jorgensen, 1982) but preferentially with the E_2 conformation, which vanadate stabilizes. Thus, it acts as a noncompetitive inhibitor of ATP as a substrate and also blocks the formation of the acyl enzyme from P_i (Barrabin et al., 1980).

There appear to be secondary sites for binding both of the substrates of the reaction, ATP and Ca^{2+}, which affect activity. As noted in Fig. 3, ATP binding at relatively low affinity sites (K_d ~ 1–2 mM) stimulates the hydrolysis of the E_2 ~ aspartyl phosphate intermediate (Rega and Garrahan, 1975). This may be a form of allosteric regulation by ATP. As noted in Section III,A, high-affinity calcium-binding sites are found in the C-terminal cytosolic domain of PMCAs, which have no known function.

Kosk-Kosicka et al. (1986) confirmed the various steps of the catalytic cycle set forth earlier using the purified human erythrocyte calcium pump. They obtained an initial $K_d \cong 1 \times 10^{-7}$ M for the binding of calcium to the calmodulin-free dephospho-E_1 form of the enzyme. The addition of ATP led to rapid formation of the phosphoenzyme intermediate with an apparent $K_d = 8.2$ μM for ATP and a rate constant for phosphoryl transfer from ATP to enzyme of ~10^2/sec, comparable to that found with the SR enzyme (Froehlich and Taylor, 1975). ADP then dissociates, leaving the E_1 ~ P·Ca^{2+} complex, which then rearranges to the E_2 ~ P complex, which was shown directly by Kosh-Kosicka et al. (1986) to possess a much lower affinity for Ca^{2+} (K_d 1–2 mM). Those investigators also noted that the free energy of hydrolysis of ATP (kcal/mol) is sufficient to support the difference in calcium affinities of the E_1 and E_2 forms of the enzyme (K_d values = 10^{-7} vs 10^{-3} → $\Delta G° \cong 6.1$ kcal/mol) necessary for the transfer of calcium up the concentration gradient from cytosol to external environment.

The mechanisms by which regulation of activity occurs are unique to each type of ion translocase. This is also true within the family of membrane calcium pumps where substantial differences in structure in regulatory domains exist. However, as will be noted later, the general mechanism by which PMCAs and SERCAs are regulated appears to be a common overall mechanism. As noted in the reaction cycle, activation of PMCAs either by direct interaction with calmodulin or by reversible Thr/Ser phosphorylation increases both the affinity for calcium binding as a substrate and V_{max} for pump turnover. It is assumed that this involves conformational changes related to the formation of the $Ca^{2+} \cdot E_1 \sim P$ complex through intramolecular actions discussed later. These same types of interactions appear to be involved in the activation of SERCAs by phospholamban.

2. Regulation of PMCA Activity

The plasma membrane calcium ATPase was the third enzyme recognized to be regulated by calmodulin. Bond and Clough (1973) first suggested the presence of a PMCA activator in RBCs. Simultaneous publications by Gopinath and Vincenzi (1977) and Jarrett and Penniston (1977) showed that calmodulin could activate the ATP-driven Ca^{2+} pumping activity of erythrocyte ghost preparations, and calmodulin was subsequently shown to be the erythrocyte activator (Jarrett and Penniston, 1978). Affinity chromatography based on reversible calcium-dependent interaction with calmodulin, first described by Watterson and Vanaman (1976), provided the basis for the isolation of essentially homogeneous PMCA solubilized with detergents from human erythrocyte membranes (Niggli et al., 1979a). (For a detailed discussion of purification, reconstitution, and assay of PMCA, the reader is referred to Gazzotti and Carafoli, 1994). This was the primary source of enzyme used in much of the work on the structure and function of PMCAs described later. It should be noted that the human erythrocyte proved to be an unusually good choice for this approach. With proper care removing bound calmodulin and other potential interfering materials, PMCA is the only high-affinity, calmodulin-binding protein present in detergent-solubilized extracts of stripped human RBC ghosts. Thus, it can be purified in a single step. The use of calmodulin–Sepharose affinity chromatography was a vast improvement over previous schemes based on classical biochemical approaches such as that originally used for the isolation of the porcine erythrocyte enzyme (Haaker and Racker, 1979).

Like most other tissues, the erythrocyte membrane contains a mixture of the two "housekeeping" PMCA isoforms 1b and 4b (Stauffer et al., 1995; Strehler et al., 1990). The inability to prepare pure individual isoforms hampered initial studies to examine isoform-specific regulatory and functional properties. Cloning of full-length cDNAs and the development of suitable systems for expressing them have been major advances in this work in recent years. Nonetheless, the bulk of the information concerning regulatory properties of PMCAs has come from analysis of the human erythrocyte enzymes.

As discussed in detail in the following paragraphs, PMCAs are subject to regulation by a number of factors in addition to calmodulin. These include regulatory phosphorylation and modulation by associated phospholipids that are reversible interactions. As with most calmodulin-regulated enzymes, irreversible activation of PMCA activity can also occur on partial proteolysis under conditions that specifically remove regulatory domains. The calcium-activated protease calpain appears particularly well suited for this conversion, raising the possibility that this is also a physiologically relevant pathway for regulating activity. Controlled proteolysis of purified PMCA using a variety of proteases has also been a valuable tool in mapping the functional regions of the molecule.

a. Activation by Calmodulin. As noted earlier, calmodulin activation of human erythrocyte PMCA was first demonstrated by Gopinath and Vincenzi (1977) and Jarrett and Penniston (1977) using stripped erythrocyte ghosts. Demonstration that the detergent-solubilized enzyme could still be activated (Lynch and Cheung, 1979; Niggli *et al.,* 1979b) indicated that the effect of calmodulin was direct and led the way to studies of the purified enzyme. Other studies showed that Ca^{2+}-replete calmodulin binds to the enzyme (either in intact membranes or as the purified enzyme reconstituted in phospholipid–detergent mixtures) with a K_d in the nanomolar range (Agre *et al.,* 1983; Graf *et al.,* 1980) forming a 1:1 molar complex (Hinds and Andreasen, 1981).

Binding of calmodulin leads to both increased Ca^{2+} affinity (decreased K_m) and an increased rate of transport. The K_m for calcium decreases from ~30 μM to as low as 0.25 μM on calmodulin binding, and the rate of ATP-dependent transport increases up to 10-fold (Jeffery *et al.,* 1981; Muallem and Karlish, 1981). It should be noted that the extent of calmodulin stimulation seen with purified, reconstituted PMCA depends on the type of phospholipids used in reconstitution due to the fact that acidic phospholipids also activate the enzyme fully (Niggli *et al.,* 1981; Niggli and Carafoli, 1981). As reviewed by Carafoli (1991a,b), PMCAs may also be activated to varying extents by acidic phospholipids in intact membrane preparations leading to different apparent levels of sensitivity to calmodulin binding as well as other types of regulatory interactions.

b. Regulation by Phosphorylation. Caroni and Carafoli (1981) first reported possible regulation of PMCA activity by reversible phosphorylation from studies with intact cardiac sarcolemmal membranes. After hydroxylamine treatment to specifically destroy the catalytic acylphosphate, the PMCA still retained substantial endogenous phosphate, which could be removed by phosphatase treatment resulting in reduced calcium pumping activity. Activity was returned by treatment with Mg^{2+} and ATP, resulting in the incorporation of hydroxylamine stable ^{32}P. This rephosphorylation was blocked by the PKA inhibitor peptide, PKI. Subsequent studies by Neyses *et al.* (1985) showed that treatment of both cardiac sarcolemmal and erythrocyte ghosts with the catalytic subunit of PKA led to increased activity primarily as a result of increased affinity for Ca^{2+}. Studies with the purified human erythrocyte enzyme (James *et al.*, 1989a,b) further refined this analysis, demonstrating that PKA treatment led to incorporation of ^{32}P into a single peptide at one site in the sequence APTKNRSS(P_i)PPPSPD. This site, indicated as PKA in Fig. 1, occurs near the C terminus of PMCA1b well downstream from the calmodulin-binding domain. This entire region of the molecule is missing in SERCAs and SPCAs and is not functional in PMCA2–4. The latter undoubtedly accounts for the fact that the maximal stoichiometry of phosphorylation observed in this study was 0.3 residues per mole, which approximates the ratio of PMCA1 to PMCA4 in the erythrocyte membrane (Strehler *et al.*, 1990). It also indicates that the effects of phosphorylation on activity may be greater than that observed. As discussed later, Dean *et al.* (1997) have shown that cAMP-dependent regulation of PMCA plays an important physiological role in platelet activation.

Far more evidence is available to suggest a role for PKC-dependent phosphorylation in regulating PMCA activity. A number of reports have appeared suggesting *in vivo* activation of calcium efflux in tissues treated with phorbol esters as has been reviewed by Monteith and Roufogalis (1995). Smallwood *et al.* (1988) first reported that treatment of the purified human erythrocyte PMCA as well as intact membranes with PKC and phorbol ester or diolein led up to a 5-fold stimulation of calcium ATPase activity and calcium transport. However, further activation was obtained with added calmodulin. In contrast to activation by calmodulin, however, the effect of PKC was to increase V_{max} rather than Ca^{2+}-binding affinity (Furukawa *et al.*, 1989; Smallwood *et al.*, 1988). Kuo *et al.* (1991) reported that PMA stimulated both enhanced PMCA expression and PMCA phosphorylation in rat aortic endothelial cells. A peak increase in activity of PMCA was 2.9-fold after 3 hr, whereas induction of PMCA expression peaked at 3.9-fold in 6 hr. Balasubramanyam and Gardner (1995) also have presented evidence for PKC-dependent activation of

PMCA-based calcium efflux in lymphocytes based on studies with human leukemic Jurkat T cells.

Studies of Wang *et al.* (1991) demonstrated that PKC phosphorylates purified human erythrocyte PMCA at two distinct sites located, on the basis on partial proteolysis studies, in the C-terminal calmodulin-binding domain. Phosphoamino acid analysis showed the presence of both PThr and PSer in the phosphorylated PMCA preparation. The site of threonine phosphorylation was inferred to be the QTQ sequence in the calmodulin-binding domain based on the fact that the synthetic peptide NRGLNRIQTQIKVVN was readily phosphorylated by treatment with PKC. Also, calmodulin inhibited phosphorylation of both the peptide and intact PMCA by PKC. It should be noted that relatively large amounts of PKC and long incubation times were required to achieve substantial phosphorylation. Nonetheless, this was assumed to be reasonable proof that the QTQ site was one of the two involved in PKC-dependent PMCA activation. Subsequent studies of Hofmann *et al.* (1994) further supported this notion by demonstrating that phosphorylation of the QTQ threonine inhibited the binding of CaM-binding domain peptides to calmodulin and reduced their effectiveness as inhibitors of calpain-activated PMCA preparations.

However, the work of Enyedi *et al.* (1996) strongly suggests that the QTQ site is not the relevant one for PKC phosphorylation and activation of PMCAs. Using membrane preparations containing various truncated constructs of hPMCA4b, expressed in COS-1 cells as substrates, they showed that phosphorylation by added PKC was lost completely on removal of the C terminus of PMCA commencing 11 residues downstream of the QTQ sequence. Even the presence of an additional 21 residues to the MTHP sequence (positions 1144–1147 in rPMCA1b; Fig. 1) did not restore phosphorylation fully to this construct. Functional studies showed that only the intact PMCA4b and a version truncated at approximately the position of the PKA site in PMCA1b (Fig. 1) showed substantial activation on PKC treatment. In addition, PKC activation was partial as an additional increase in Ca^{2+}-ATPase activity was obtained with added calmodulin in agreement with the original studies of Smallwood *et al.* (1988). This region of the sequence of PMCA4b contains a number of Thr and Ser residues that could be sites for PKC-mediated phosphorylation. However, the actual site(s) of phosphorylation was not identified by Enyedi *et al.* (1996). Nonetheless, their data, coupled with the fact that the native protein *in situ* was shown to be a much better PKC substrate than previously seen with the purified, reconstituted enzyme, strongly suggest that the actual site of PKC phosphorylation lies C-terminal to the QTQ sequence. As discussed in a following section, previous studies by Verma *et al.* (1994) had implicated this region C-terminal to the CaM-binding domain as an inhibitory domain. It is interesting to note that this region of PMCA was also proposed to be part of the inhibitory domain based on the observations made in initial PMCA cloning with expression libraries (Brandt *et al.*, 1988). This, coupled with the observation that the PKC site was as well as a site for tyrosine phosphorylation (vide infra) are probably located in a region very close to that of the PKA site in PMCA1b, suggests that this region of PMCAs forms a separate domain for regulation through reversible phosphorylation.

Studies in the authors' laboratory have demonstrated that tyrosine phosphorylation also occurs in the same region of PMCA as noted earlier. However, it has an unexpected and potentially important effect: inhibition of calcium pumping activity. Ca^{2+} signaling and tyrosine phosphorylation appear to be tightly coupled (Dash *et al.*, 1995) essential regulatory signals acting in concert to regulate the platelet activation cascade. Dean *et al.* (1997) have demonstrated that both cAMP-dependent and tyrosine kinase-mediated phosphorylation of PMCA occurs in human platelets in response to platelet inhibitors (PGE_1, forskolin) and activators (thrombin, aggregation), respectively. When human platelets, preloaded with $^{32}P_i$, were treated with forskolin or PGE_1 to elevate cAMP levels, significant increases in hydroxylamine stable ^{32}P labeling of PMCA was observed compared to controls. Treatment of isolated platelet membranes with the catalytic subunit of PKA also led to the incorporation of ^{32}P into PMCA, which was enhanced substantially by the pretreatment of membranes with type 2A protein phosphatase as observed in the initial studies with sarcolemmal membranes (Caroni and Carafoli, 1981). Protein phosphatase treatment also gave a 30% reduction in PMCA-specific Ca^{2+}-ATPase activity, and subsequent treatment with PKA caused a 1.4-fold increase.

Activation of platelets with thrombin or by spontaneous aggregation led to substantial phosphotyrosine incorporation into PMCA detected in antiphosphotyrosine immunoblots of total PMCA immunoprecipitates. PGE_1 treatment reduced PMCA PTyr content substantially whereas the tyrosine kinase inhibitor genistein eliminated it completely. Thrombin-stimulated PMCA tyrosine phosphorylation was transient, reaching a maximum within 5 min of thrombin addition and returning to basal levels by 10 min. Both the time course of PMCA tyrosine phosphorylation and the effects of PGE_1 and genistein were identical to those observed for focal adhesion kinase measured in parallel. Membranes isolated from thrombin-treated (5 min) platelets had a 40% lower PMCA-specific Ca^{2+}-ATPase activity than those from EGTA-treated control platelets. Incubation of purified platelet membranes with recombinant $pp60^{src}$ + ATP for 15 min led to a 75% reduction in initial activity compared to controls incubated in the absence of the kinase. Similar inhibition of PMCA activity of treated erythrocyte membranes was observed but occurred at a slower rate, requiring 1 hr of incubation. Inhibition of PMCA activity was not overcome by calmodulin despite the fact that the PTyr containing PMCA still bound calmodulin in a calcium-dependent manner at least as tightly as the dephosphoprotein. Reminiscent of observations with PKC discussed earlier, *purified* erythrocyte PMCA was phosphorylated at a much slower rate than seen with the intact membrane preparations. Whether this reflects the loss of essential regulatory factors or other tyrosine kinases during PMCA purification or differences in the conformation of PMCA in its natural lipid milieu is unclear.

The site of tyrosine phosphorylation was determined by analysis of the purified human erythrocyte PMCA treated *in vitro* with $pp60^{src}$ and [γ-^{32}P]ATP. A single ^{32}P-labeled peptide that contained only phosphotyrosine was isolated by HPLC (high-performance liquid chromatography) from Lys-C digests of this material. Its sequence was determined to be FGTRVLLLD(G)EVTP(Y)A, where parentheses indicate residues not identified in the sequence analysis. It is assumed that the penultimate tyrosine is the site of phosphorylation. This sequence is found in hPMCA4b in the same region as the PKA site in PMCA1b and the putative PKC site/inhibitory domain discussed earlier. This PTyr site is in a short segment specific to PMCA4 species (see Fig. 4). It should be noted that this is not a consensus $pp60^{src}$ phosphorylation site as it lacks the upstream basic residue. Interestingly, PMCA1 has a perfect $pp60^{src}$ consensus site indicted. However, there is no evidence that tyrosine phosphorylation occurs at this site.

Fig. 4. Comparison of the C-terminal regulatory region of all PMCAs downstream of the conserved CaM-binding domain. Termini are broken into classes, SLETSX and CIS, which are determined by the last amino acids of their sequence. Phosphorylation sites for PKA, PKC, and PMCA tyrosine kinase are boxed. The phosphorylated amino acid is indicated by an arrow, when known.

SLETSX Class of PMCA Termini

CaM-Binding Domain

```
hPMCA1b  HAERELRRGQILWFRGLNRIQTQ------------------------IRVVNAFRSSLYEGLEKPESRSSIHNF
rPMCA1b  HAERELRRGQILWFRGLNRIQTQ------------------------IRVVNAFRSSLYEGLEKPESRSSIHNF
rPMCA1c  HAERELRRGQILWFRGLNRIQTQMDVVNAFQSGGSIQGALRRQPSIASQHHD----IRVVNAFRSSLYEGLEKPESRSSIHNF
rPMCA1d  HAERELRRGQILWFRGLNRIQTQMDVVNAFQSGGSIQGALRRQPSIASQHHDVTNVSTPTHIRVVNAFRSSLYEGLEKPESRTSIHNF
rPMCA2b  HAERELRRGQILWFRGLNRIQTQ------------------------IRVVKAFRSSLYEGLEKPESRTSIHNF
rPMCA3b  HAERELRRGQILWFRGLNRIQTQ------------------------IRVVKAFRSSLYEGLEKPESKSCIHNF
rPMCA3c  HAERELRRGQILWFRGLNRIQTQMEVVSTFKRSGSFQGAVRRRSSVLSQLHD----IRVVKAFRSSLYEGLEKPESKSCIHNF
rPMCA3d  HAERELRRGQILWFRGLNRIQTQMEVVSTFKRSGSFQGAVRRRSSVLSQLHDVTNLSTPTHIRVVKAFRSSLYEGLEKPESKSCIHNF
rPMCA4b  HAEMELRRGQILWVRGLNRIQTQ------------------------IRVVKVFH-SFRDVIhksknqvSIHSF
hPMCA4b  HAEMELRRGQILWFRGLNRIQTQ------------------------IKVVKAFHSSLHESIqkpynqkSIHSF
```

```
                                                PTyr
                                                Kinase
         PKC Phosphorylation Region    PKA↓      Site↓                              SLETSX
hPMCA1b  MTHPEFRIEDSEPHIPLIDDTDAEDDAPTKRNSSPPPSP----NKNNNAVDSGIHLTIEMNKSATSSS------PGSPLHSLETSL
rPMCA1b  MTHPEFRIEDSEPHIPLIDDTDAEDDAPTKRNSSPPPSP----NKNNNAVDSGIHLTIEMNKSATSSS------PGSPLHSLETSL
rPMCA1c  MTHPEFRIEDSEPHIPLIDDTDAEDDAPTKRNSSPPPSP----NKNNNAVDSGIHLTIEMNKSATSSS------PGSPLHSLETSL
rPMCA1d  MTHPEFRIEDSEPHIPLIDDTDAEDDAPTKRNSSPPPSP----NKNNNAVDSGIHLTIEMNKSATSSS------PGSPLHSLETSL
rPMCA2b  MAHPEFRIEDSQPHIPLIDDTLEEDAALKQNSSPPSSI----NKNNSAIDSGINLJTDTSKSATSSS------PGSPIHSLETSL
rPMCA3b  MATPEFLINDYTHNIPLIDDTDV-DENEERLRAPPPPP----NQNNNAIDSGIYLTTHATKSATSSAFSSRPGSPLHSMETSL
rPMCA3c  MATPEFLINDYTHNIPLIDDTDV-DENEERLRAPPPPP----NQNNNAIDSGIYLTTHATKSATSSAFSSRPGSPLHSMETSL
rPMCA3d  MATPEFLINDYTHNIPLIDDTDV-DENEERLRAPPPPP----NQNNNAIDSGIYLTTHATKSATSSAFSSRPGSPLHSMETSL
rPMCA4b  MTQPEYAaddemsqsflnqeesssl---asksritkrlsdaetvsqnNTNNNAVDChqvqm-----------lashpnSPLQSQETPV
hPMCA4b  MTHPEFAieeelprtplldeeeenpdkaskfgtrvllldgevtpyaNTNNNAVDCnqv-----------qlpqsdSSLQSLETSV
```

Phosphorylation Regulation Domain

CIS Class of PMCA Termini

CaM-Binding Domain

```
rPMCA1a   HAERELRRGQILWFRGLNRIQTQMDVVNAFQSGGSIQGALRRQPSIASQHHDVTNVSTPT
rPMCA1e   HAERELRRGQILWFRGLNRIQTQMDVVNAFQSGGSIQGALRRQPSIASQHHDVTNVSTPT
rPMCA2a   HAERELRRGQILWFRGLNRIQTQIEVVNTFKSGASFQGALRRQSSVTSQSDVASLSSPS
rPMCA2c   HAERELRRGQILWFRGLNRIQTQIEVVNTFKSGASFQGALRRQSSVTSQSDVASLSSPS
rPMCA3a   HAERELRRGQILWFRGLNRIQTQMEVVSTFKRSGSFQGAVRRRSSVLSQLHDVTNLSTPT
rPMCA3f   HAERELRRGQILWFRGLNRIQTQvcwdgkkmlrttevg----------------------
rPMCA4a   HAEMELRRGQILWVRGLNRIQTQIEVINKFQTGASFKGVLRRQN--LSQQLDVklvpssy
```

```
                                              CIS
rPMCA1a   HVVFS---SSTASTPvgyp------SGECIS
rPMCA1e   HVVFS---SSTASTPvgsew-----------
rPMCA2a   RVSLSNALSSPTSLPpaAAGqg---------
rPMCA2c   RVSLSNALSSPTSLPpaAAGhprr---EGVP
rPMCA3a   HVTLSAA--KPTS----AAGnp---SGESIP
rPMCA3f   -------------------------------
rPMCA4a   seavasvrtspstssavtpppvgnqSGQSIS
```

Fig. 4. Continued

The physiological role of this apparent inhibition of PMCA activity is as yet unknown. However, it is tempting to speculate that this is an essential first step in mounting a calcium signal in platelets. PMCA can be activated by many factors, including the acidic phospholipids found on the inner leaflet of the plasma membrane. As noted by Carafoli (1991a,b), up to 50% of the PMCA in a given cell may be phospholipid activated at any given time, which may provide the mechanism for constantly blocking unwanted signaling events. Stimulus-activated inhibition of PMCA activity would thus be required to permit cytosolic calcium to rise during the initial phases of influx and release. This type of regulation is not likely to be restricted to platelets as PMCA tyrosine phosphorylation has been observed in excitable cells as well (P. C. Brandt, J. E. Sisken, and T. C. Vanaman, unpublished observations, 1997).

c. Other Potential Regulators of PMCA Activity. Although calmodulin binding and reversible phosphorylation are the two most widely accepted routes for regulating PMCA activity, several others deserve at least brief consideration here. These include activation by acidic phospholipids, partial proteolysis, other calmodulin-like regulatory proteins, and the PMCA oligomerization state.

Ronner *et al.* (1977) reported that lipids bearing a negative charge could activate the human erythrocyte membrane calcium pump, ablating the effects of calmodulin. However, it was studies of the purified human erythrocyte enzyme that fully demonstrated this effect. In the initial protocol for isolating the detergent-solubilized enzyme from erythrocyte ghosts by calmodulin–Sepharose affinity chromatography, Niggli *et al.* (1979a) chose to include phosphatidylserine in the detergent mixture in order to stabilize the enzyme. Unexpectedly, the isolated enzyme had lost calmodulin sensitivity despite the fact that it still bound calmodulin. Gietzen *et al.* (1980) reported that the same procedure but using phosphatidylcholine as the phospholipid yielded a fully calmodulin-activated enzyme. This apparent discrepancy was resolved in subsequent studies of the effects of phospholipids and fatty acids on the purified human erythrocyte enzyme. Niggli *et al.* (1981) confirmed that an enzyme reconstituted in phosphatidylcholine vesicles displayed the properties of the intact Ca^{2+} pump, including calmodulin activation. However, the addition of phosphatidylserine gave a pump that was fully activated and no longer calmodulin responsive. Other acidic (but not neutral or basic) lipids, including phosphatidylinositol, phosphatidic acid, cardiolipin, and even free fatty acids (oleic and linoleic acids), could activate the enzyme increasing V_{max} and decreasing the K_m for Ca^{2+}, giving values similar to those obtained with calmodulin.

Enyedi *et al.* (1987) demonstrated that phosphatidylinositol 4-phosphate shifted the purified erythrocyte PMCA K_m for calcium to 0.25 μM, below that obtained with calmodulin (0.5–0.7 μM) and equivalent to that of the enzyme fully activated by proteolysis. These results suggest that changes in the phospholipid composition of the plasma membrane could play a substantial role in regulating PMCA activity as summarized Carafoli (1991a,b), particularly the phosphatidylinositol phosphates known to turnover rapidly in membranes. Indeed, it has been calculated that as much as 50% of the total erythrocyte PMCA may be fully activated. The observations of Davis *et al.* (1991) that 1 μM inositol 1,4,5-trisphosphate and the corresponding 4,5-bisphosphate inhibited erythrocyte membrane Ca^{2+}-ATPase by 42 and 31%, respectively, are intriguing in this context, raising the possibility that generation of IP_3 through PIP_2 breakdown could both inhibit PMCA-mediated calcium extrusion and activate Ca^{2+} release in concert.

As noted in the following section, partial protcolysis has been shown to activate PMCAs *in vitro* completely under conditions that remove all regulatory properties. (This appears to be true for all calmodulin-regulated enzymes, reflecting the fact that the regulation of activity involves protease-sensitive autoinhibitory elements whose actions are alleviated by calmodulin binding.) However, PMCAs appears to be ideal for activation by the calcium-activated protease calpain, which cleaves the calmodulin-binding domain specifically (vide infra). Clearly, this would represent an irreversible activation of the pump. However, Salamino *et al.* (1994) has shown that PMCA is one of the preferred substrates for calpain in the intact erythrocyte, raising the possibility that such activation may be physiological under some condition (e.g., as a last line of defense under extreme conditions of oxidative or osmotic stress).

Periodic reports have shown that other members of the calmodulin family could interact with and possibly regulate PMCAs. For example, James *et al.* (1991) reported that the bovine 9-kDa vitamin D-dependent protein, calbindin$_{D9K}$, bound to the human erythrocyte PMCA in a calcium-dependent manner but with an affinity 5–10 times poorer than calmodulin. Cross linking with a photoactive calbindin$_{D9K}$ derivative labeled both the intact enzyme and a synthetic peptide representing the full-length calmodulin-binding domain. However, the poor affinity of binding plus the inability to demonstrate activation of PMCA activity by this binding suggested that this interaction is not physiologically relevant.

Reisner et al. (1992) reported that both purified erythrocyte membrane Ca^{2+}-ATPase and bovine brain 3′,5′-cyclic nucleotide phosphodiesterase (PDE) were stimulated in a dose-dependent, saturable manner by the vitamin D-dependent calbindin$_{D28K}$ from rat kidney. The concentration of calbindin$_{D28K}$ required for half-maximal activation of the Ca^{2+}-ATPase was 28 nM compared to 2.2 nM for calmodulin. However, maximal activation was equivalent to that obtained with excess CaM. Brain PDE also showed equivalent maximum saturable activation by calbindin ($K_{0.5}$ act. = 90 nM) or calmodulin ($K_{0.5}$ act. = 1.2 nM). Binding to both PDE and erythrocyte PMCA appeared to be calcium independent, however. Thus, the physiological significance of these results is also unclear.

No consideration of PMCA regulation or architecture would be complete without discussing its oligomeric structure and the effect of oligomerization on activity. Cavieres (1984) first used neutron inactivation analysis to study PMCA architecture in intact membranes that yielded a target size of ~250 kDa commensurate with the enzyme being a dimer. Kosk-Kosicka and Bzdega (1988) and Kosk-Kosicka et al. (1989) presented the first direct evidence for the existence of a dimer and demonstrated that PMCA oligomerization affects activity. Studies of the dependence of PMCA activity on enzyme concentration ± calmodulin as well as by fluorescence polarization using FITC-labeled PMCA preparations showed a definite transition from an inhibited, calmodulin-sensitive enzyme to one that was fully activated and calmodulin independent. The transformation occurred with half-maximal activation at 10–20 nM PMCA. Both forms of the enzyme retained the ability to bind calmodulin, but the binding of calmodulin was reported not to affect the monomer–oligomer equilibrium.

Coelho-Sampaio et al. (1991) demonstrated erythrocyte PMCA oligomerization using hydrostatic pressure dissociation and intrinsic fluorescence spectral measurements and fluorescence polarization of dansyl-labeled PMCA. The observed concentration dependence agreed with the initial studies of Kosk-Kosicka and Bzdega (1988) and Kosk-Kosicka et al. (1989), giving a dissociation constant for the monomer to oligomer transition of 6–9×10^{-8} M at atmospheric pressure. These data also strongly suggested that the oligomer was a dimer that was stabilized by Ca^{2+} but not calmodulin and was destabilized by vanadate. Sackett and Kosk-Kosicka (1996) have verified that PMCA exists in a monomer–dimer equilibrium by sedimentation analyses. Unlike previous work, their studies show that calmodulin, when present in molar excess, causes dissociation of the dimer to yield the monomeric PMCA–calmodulin complex.

Vorherr et al. (1991) also reported that calmodulin causes dissociation of PMCA into the monomeric state. Furthermore, they demonstrated that the calmodulin-binding domain is the site for PMCA dimerization on the basis of its inhibition by antibodies prepared to the synthetic CaM-binding domain peptide and the fact that the calpain-truncated form of the enzyme lacking the calmodulin-binding domain also failed to dimerize.

Overall, it is unclear what role this transition plays in the physiological regulation of PMCA activity. First, the concentration of the enzyme is exceedingly low in plasma membranes, probably well below that required for dimerization. Second, both the dimer and the calmodulin-bound monomer are activated forms of the enzyme. Therefore, a role for the calmodulin-dependent interconversion of the monomer and dimer in activation is not obvious.

d. PMCA Regulatory Domains. Studies of the molecular details of PMCA–calmodulin interaction have been extensive. Initial studies using partial proteolysis of intact erythrocyte membranes indicated that regulatory and catalytic functions could be uncoupled by limited trypsin or chymotrypsin treatment, indicating that they resided in different portions of the molecule (Enyedi et al. 1980). Subsequent studies using controlled proteolysis of the purified human erythrocyte enzyme (Benaim et al., 1984; Zurini et al., 1984) showed that the enzyme could be cleaved into various size fragments with varying functional properties. Short times of cleavage with dilute trypsin at 4° C yielded fragments primarily of 33.5 and 90 kDa with full retention of calcium pumping ATPase activity and sensitivity to calmodulin and phospholipids. The 33.5-kDa fragment, derived from the amino terminus of the pump (vide infra), was relatively stable to further digestion and, based on subsequent studies, must have remained associated with the larger fragment to yield a functional enzyme. However, further protease treatment caused removal of an additional 5-kDa segment from the 90-kDa fragment, producing an 85-kDa form that was still fully active and capable of binding calmodulin but no longer fully activated by it. Calmodulin binding was then lost with removal of an additional 4 kDa from this fragment, yielding what was assumed to be a catalytic core of ~81 kDa. Digestion of human erythrocyte PMCA with the calcium-activated protease calpain *in vitro* was

8. CaM and Ion Flux Regulation

shown to yield fragments of 124–125 kDa, which were catalytically active but fully independent of calmodulin (James *et al.*, 1989b).

Subsequent sequence analyses of both trypsin and calpain fragments in the context of cDNA-deduced PMCA structures (Zvaritch *et al.*, 1990) have provided a complete picture of the location of these fragments. This information and results of direct domain mapping studies performed using chemical labeling and site-directed mutagenesis have provided substantial understanding of the functional domains in the molecule. The 33.5-kDa trypsin fragment results from cleavage just amino-terminal to TM3 in the middle of the **PL** segment at the sequence ^{322}KAK $\downarrow$ AQDG^{328}*, extending back to the amino terminus of the molecule. The 90-kDa fragment extends from A^{325} in this sequence to almost the end of the C-terminal cytoplasmic domain, resulting from cleavage at a position equivalent to the PKA site in PMCA1b (^{1174}K $\downarrow$ RNS^{1177}). The 85- and 81-kDa fragments start at the same position as the 90-kDa fragment but end in the middle of (85 kDa- ^{1115}QTQIR $\downarrow$ VV^{1121}) or amino terminal to (81 kDa- ^{1075}TQK $\downarrow$ EEI^{1080}) the CaM-binding domain. The 124- and 125-kDa calpain fragments commence at the amino terminus of PMCAs and continue up to the beginning (^{1099}ELR $\downarrow$ RGQ^{1104}) or middle (^{1111}LN $\downarrow$ RIQTQ^{1117}) of the calmodulin-binding domain, depending on whether digestion was performed in the absence or presence of calmodulin, respectively. As expected, the former was fully activated and no longer could bind calmodulin whereas the slightly larger fragment was still partly inhibited and activated by CaM binding. It is clear from this work, as well as more recent studies summarized later, that sequences C-terminal to the ^{1115}QTQ^{1117} sequence shown here are involved in regulation by calmodulin and phosphorylation.

The ability to uncouple PMCA activity from regulatory properties by proteolytic cleavage at defined sites provided the necessary tools for direct detection of regulatory interaction domains in the proteins. The calpain-cleaved, 124-kDa protein lacking the C terminus encompassing the entire calmodulin reglatory domain was particulary useful in such studies. Such work was further enhanced by the development (Enyedi and Penniston, 1993) of a recombinant 120-kDa form of PMCA4b truncated by introduction of a stop codon immediately after that for the glutamyl residue in the sequence ^{1099}ELRR^{1102}. This is two amino acid residues amino-terminal to the site where calpain cleaves in the absence

---

* Numbering throughout the text is based on the deduced sequence of rat PMCA1b shown in Fig. 1.

of calmodulin. As with the calpain-treated human erythrocyte enzyme, the recombinant truncated 120-kDa hPMCA4b was calmodulin independent and fully activated. The ATP hydrolysis/translocation activity and calcium affinity of this construct are essentially identical to those of the fully calmodulin-stimulated, wild-type enzyme.

A combination of photochemical labeling approaches with both intact proteins and synthetic peptides and site-directed mutagenesis were used to identify the linear amino acid sequences in PMCA and calmodulin involved in regulation and the probable mechanism of regulation. In a series of studies with various derivatives of calmodulin, Vanaman and co-workers (Crocker et al., 1990; Dwyer et al., 1992; Imai et al., 1990; Mann and Vanaman, 1988, 1989) examined the interaction of calmodulin with the purified human erythrocyte PMCA as compared to cytosolic target proteins. Modifications with hydrophobic probes targeted to the hydrophobic pockets of calmodulin reduced the affinity of binding to both bovine brain PDE and human erythrocyte PMCA. However, the effects were much greater on binding to PDE than human erythrocyte PMCA. Photoaffinity labeling studies with a series of arylazide photoprobes of varying lengths and chemical reactivities, attached to different positions in the calmodulin molecule (Lys-21, -75, or -94), showed that the central α helix around Lys-75 must be in very close contact ($\sim$10 Å or less) with its binding site in PMCA, whereas Lys-21 and Lys-94, which are present in calcium loops on the surface of the molecule, are not.

James et al. (1988) used this same approach to isolate and characterize the calmodulin-binding domain of the human erythrocyte PMCA. They chose the Denny–Jaffe reagent (Jaffe et al., 1980), which combines a chemically reactive NHS ester (to permit covalent attachment at donor protein lysyl residues) attached through a cleavable linker arm to an iodinated photoactivatable azo compound for photochemical cross-linking. This permits transfer of the radiolabel from the donor to the target protein by cleavage after cross-linking. Photocross-linking of the purified human erythrocyte PMCA with a Denny–Jaffe-modified calmodulin derivative followed by cleavage of the linker and digestion of the PMCA with chymotrypsin yield one radioactive peptide on HPLC whose sequence was shown to be ELRRGQILWFRGLNRIQT-QIKVVNAFSSSLHEF. This sequence is identical to that of the hPMCA1b corresponding to residues 1099–1130 in the rat protein shown in Fig. 1. As noted by those authors, this sequence contains the now recognized calmodulin-binding domain motif consisting of an amphipathic α helix with essential aromatic residue clusters (usually Trp-Phe) imbedded in a hydrophobic face involved in binding into calmodulin hydrophobic pockets opposed by a highly positive face to interact with acidic residues in the central helical domain of calmodulin. Using synthetic peptides based on this sequence, Enyedi et al. (1991) demonstrated that the first 15 residues of this sequence were sufficient for high-affinity binding to calmodulin, forming a 1:1 molar complex, inhibiting its ability to activate the intact human erythrocyte PMCA. A 28-residue version of the CaM-binding domain peptide (^{1099}LR→SS^{1127} in rPMCA1b; Fig. 1) had the highest apparent affinity for CaM (IC$_{50}$ of 8 nM). One with the first 20 residues of this sequence (^{1099}LR→QTQIK^{1119}) still inhibited with an apparent $K_i = 40$ nM. Shorter versions were much less effective, as were derivatives in which the tryptophyl residue was replaced with a small aliphatic residue (alanine) demonstrating the importance of this residue in forming the binding structure. However, the most important aspect of this study was the demonstration that the 28-residue CaM-binding domain peptide inhibited the protease-activated form of the pump reconferring calmodulin sensitivity. This was shown to occur both with the calpain-activated 125-kDa form of the enzyme in intact erythrocyte ghosts and with the 33.5 + 81-kDa fragments derived by further digestion with trypsin or chymotrypsin. The CaM-binding domain peptide inhibited this activated enzyme, shifting both the K_m for Ca^{2+} and V_{max} toward the values for the intact form of the enzyme.

Subsequent studies using physical methods to assess the binding of synthetic peptides representing the PMCA CaM-binding domain (Vorherr et al., 1990) further demonstrated that the first 20 residues of this peptide are sufficient for binding to calmodulin with high affinity in a calcium-dependent manner. Evidence was also presented that the CaM-binding domain interacted with a sequence in PMCA just amino-terminal to itself, suggesting that this was the mechanism responsible for autoinhibition of PMCA activity. However, Falchetto et al. (1991) demonstrated that the primary inhibitory domain was located elsewhere by direct chemical labeling studies using photoactivatable peptide analogs. For this work, the Phe residue immediately after the single tryptophan in a 28-residue version of the CaM-binding domain peptide was replaced with a photoactivatable diazirine-Phe analog. The resulting peptide analog retained the ability to inhibit the calpain-activated (124 kDa) form of human erythrocyte PMCA. A radiolabeled version was used to photolabel the calpain-treated PMCA preparation, and the resulting labeled

PMCA fragment(s) was recovered, further digested, and labeled peptides isolated and sequenced. The primary site of photoincorporation was shown to be contained in the sequence CALLGF indicated as I' in Fig. 1. This sequence is situated in the middle of the proposed active site of the enzyme between the aspartyl phosphate and ATP-binding sites in the large cytoplasmic domain. Thus, interaction between this site and the C-terminally located calmodulin-binding domain was proposed to provide direct occlusion of the enzyme active site, analogous to the inhibitory mechanism proposed for calmodulin-regulated myosin light-chain kinase (Blumenthal *et al.*, 1988; Kemp *et al.*, 1987; Kennelly *et al.*, 1987; Pearson *et al.*, 1988). A subsequent study (Falchetto *et al.*, 1992) using the same approaches identified a second apparent binding site (Fig. 1, I) for the CaM-binding domain in the cytoplasmic domain between TM2 and TM3. The site is close to the site proposed for activation of activity by acidic phospholipids (Brodin *et al.*, 1992).

Vorherr *et al.* (1992) also used diazirine peptide probes to study interaction sites in calmodulin. They concluded that the N-terminal portion of the calmodulin-binding domain peptide commencing with ELRR bound the C-terminal hydrophobic pocket of CaM whereas the C-terminal region around the sequence -NAF-, contained only in the full-length 28-residue peptide, bound to the N-terminal CaM pocket.

As noted in Section III,A,1, SERCAs and SPCAs lack the C-terminal cytoplasmic domain identified in the studies described earlier as the calmodulin-binding domain and that involved in phosphorylation-based regulation. They also appear to lack the segment involved in activation by acidic phospholipids but retain both the I and I' domain sequences with a reasonable level of homology. Chiesi *et al.* (1991) have shown that the 28-residue CaM-binding domain peptide can also bind to and inhibit SR ATPase activity in intact SR membranes with an IC_{50} of 15 μM. This is presumably analogous to the inhibitory effects of phospholamban on the SR ATPase. Calmodulin reversed this inhibition and was shown to bind to the putative inhibitory sequence of phospholamban with a K_d of 0.7 μM. However, phospholamban could not replace the CaM-binding domain in inhibiting the activity of truncation-activated PMCA, and calmodulin does not reverse the phospholamban-mediated inhibition of SERCA activity. The latter requires direct phosphorylation of phospholamban by PKA and other protein kinases. Thus, phospholamban appears to replace the CaM-binding domain in the regulation of SERCAs, probably through binding to the same region of the transduction domain as in the case of PMCAs. The differences in structures

8. CaM and Ion Flux Regulation

of PMCA and SERCAs dictate the use of different regulatory molecules, but the overall mechanism of regulation is similar or identical.

Although the I and I' domains presumably interact with the CaM-binding domain as the primary mechanisms for inhibition of PMCA, the region C-terminal to the CaM-binding domain has also been implicated as an inhibitory domain. Indeed, such a role for this region of PMCAs was proposed in the initial report of PMCA cloning by Brandt *et al.* (1988) involving a partial cDNA from the C terminus of bovine PMCA4b isolated by screening expression libraries. A β-galactosidase-fusion protein prepared from this cDNA containing the C-terminal 71 residues of bovine PMCA4b was used to purify specific antibodies from a polyclonal rabbit antiserum prepared to the human erythrocyte protein. This antibody preparation gave partial activation of PMCA activity whereas those prepared to the 81-kDa trypsin fragment of the enzyme described earlier did not. Substantial similarities were noted between a portion of this PMCA sequence and the calcium-binding EF-hand regions of calmodulin. Thus, Brandt *et al.* (1988) hypothesized that this "inhibitory" domain might interact with a calmodulin-binding domain elsewhere in the molecule to provide inhibition of activity and that simple competitive binding of calmodulin with this structure would relieve the inhibition. Verma *et al.* (1994) provided further evidence for additional inhibitory activity in the C terminus of PMCA4b downstream from the calmodulin-binding domain. A PMCA4b construct truncated 10 amino acid residues C-terminal to the QTQ sequence possessed full calmodulin affinity but was still not completely inhibited compared to the full-length protein. Enyedi *et al.* (1996) have further extended this observation to demonstrate that full inhibition requires all but the last 48 residues of sequence, placing the second (partial) inhibitory domain immediately after the full calmodulin-binding domain and just amino-terminal to the PKA site in PMCA1b, including the calmodulin homolog sequence of Brandt *et al.* (1988). This region of PMCAs also appears to possess three high-affinity calcium-binding sites of unknown function (Hofmann *et al.*, 1993). As discussed earlier, it is likely that this second inhibitory region is linked to regulation by reversible phosphorylation.

e. Isoform-Specific Regulatory Properties. One of the most interesting and important discoveries concerning the plasma membrane calcium ATPases is the existence of multiple isoforms whose expression is tightly regulated at the level of both transcription and alternative splicing in a tissue and developmentally specific manner. These isoforms appear to be highly conserved in core catalytic properties commensurate with the fact that the structures of the catalytic domains are 95–99% identical in primary sequence. The bulk of the structural differences between PMCA isoforms occur in the primary sequences of their regulatory domains, which are also the regions of alternative splicing. Thus, the family of PMCA isoforms appears to have evolved as a set of enzymes whose regulatory properties are designed for specific physiological settings. As noted in the previous sections, it appears that PMCA isoforms differ in their sensitivity to regulation by different protein kinases as well as to calmodulin.

Figure 4 shows alignments of the C-terminal regulatory domains of the known PMCA isoforms, and Fig. 5 shows the corresponding I and I′ domains with which they are proposed to interact to inhibit activity. The latter are shown to simply make the point that they are highly conserved in structure. Indeed, there are only two nonconservative substitutions throughout the 71 residue I domain sequences from PMCA1–4. This is more conserved than the I′ domain sequence also shown. Note that alternative splicing does not effect either the I or the I′ domains.

In contrast, the sequences of C-terminal regulatory domains shown in Fig. 4 vary substantially both among different gene products and by virtue of alternative splicing. The first portion of the calmodulin-binding domain sequence is essentially invariant up to the QTQ, which represents the C-exon splice junction. Products that lack the C-exon-encoded peptide insert (the b isoforms) are the most commonly found. In all cases except for PMCA4, the C-exon has multiple internal splice junctions,

I Domain

```
hPMCA1a    QIPVADITVGDIAQVKYGDLLPADGILIQGNDLKIDESSLTGESDHVKKSLDKDPlLLSGTHVrEGSGRMV
hPMCA2a    QIPVAEIVVGDIAQVKYGDLLPADGLFIQGNDLKIDESSLTGESDQVRKSVDKDPMLLSGTHVMEGSGRML
rPMCA3a    QVPVAaLVVGDIAQVKYGDLLPADGVLIQGNDLKIDESSLTGESDHVRKSADKDPMLLSGTHVMEGSGRMV
hPMCA4a    QLPVAEIVVGDIAQVKYGDLLPADGILIQGNDLKIDESSLTGESDHVKKSLDKDPMLLSGTHVMEGSGRMV
```

I′ Domain

```
hPMCA1a    CALLGLLL
hPMCA2a    CgLLGFVL
rPMCA3a    CALLGFIL
hPMCA4a    CALLGFVt
```

Fig. 5. Comparison of the two putative CaM-binding, domain-interacting elements found in all PMCAs. Positions of these elements in the primary structure of PMCA1b are shown in Fig. 1 and are discussed in the text.

permitting the insertion of some or all of the segment. Incorporation of most or all of the C-exon causes a shift in the reading frame in every case, leading to truncation of the polypeptide chain due to in-frame stop codons. These truncated forms have only the alternative form of the variable end of the calmodulin-binding domain, which has been shown by Enyedi *et al.* (1991) to give substantially lower calmodulin-binding affinities than the b isoform structure. More importantly, they lack the entire phosphorylation domain found in full-length isoforms. Thus, unless other protein kinase sites are encoded in the unique frame-shifted segment just prior to the stop condons, it appears unlikely that these truncated isoforms will be subject to regulatory phosphorylation. It should also be noted that truncated PMCA forms have different C termini than those of full-length proteins. Although there is no known functional significance to this fact, the C termini encoded by all four primary gene products is conserved in the last eight to nine residues even though adjacent sequences can diverge as in the case of PMCA4. The resulting motif is termed SLETSX in Fig. 4. The corresponding C terminus of the truncated versions is termed the CIS motif in Fig. 5.

The best documented difference between different PMCA isoforms is the specific phosphorylation of PMCA1 by PKA. As shown in Fig. 4, PMCA1 has a consensus PKA phosphorylation site of $KRXS(P_i)$, but the essential serine is replaced by Ala or Arg/Gly in PMCA2, PMCA3, and PMCA4. In the different full-length spliced versions of PMCA1, the insertion of 29 (PMCA1c) or 38 (PMCA1d) amino acids moves the PKA site much farther away from the calmodulin-binding domain than in PMCA1b where PKA-mediated activation of activity has been demonstrated. However, the effect of these alterations in structure on regulation is unclear.

Studies of both intact PMCA isoforms and their calmodulin-binding domain regions have suggested some differences in their calmodulin-binding affinities. Hilfiker *et al.* (1994) reported that all hPMCA2 isoforms produced from full-length cDNAs in Sf9 cells had ~5-fold higher affinity for calmodulin (K_d = 8–10 nM) than hPMCA4 counterparts whose affinities (K_d = 40–50 nM) were identical to those reported for PMCA1. This is surprising in view of the fact that the CaM-binding domains of PMCA2 and PMCA4 are more homologous to each other than either is to PMCA1 or PMCA3. Enyedi *et al.* (1991) examined a series of calmodulin-binding domain peptides for both calmodulin-binding affinity and inhibitory activity with protease-activated PMCA. Peptides representing this same region of PMCA2 and PMCA4 up to the RSS sequence showed equal affinities for calmodulin as well as for inhibition of the proteolyzed enzyme. However, the PMCA1a analog containing acidic residues in place of Arg or Lys in the region immediately adjacent to the QTQ sequence had 10-fold lower calmodulin-binding affinities.

Attempts to examine the effect of different C-exon inserts on calmodulin binding and regulation are still in preliminary stages. The most direct analysis was that of Kessler *et al.* (1992) who examined the properties of the different versions of PMCA1 calmodulin-binding domains. Constructs producing the C termini of PMCA1a–d from approximately 30 residues amino-terminal to the start of the CaM-binding domain were expressed in *Escherichia coli,* purified by CaM affinity chromatography, and analyzed for calmodulin-binding properties. Unfortunately, aggregation (possibly dimerization) prevented accurate determination of K_d values for any of these constructs. However, decided differences were detected in the pH dependence of the dansyl–calmodulin binding. A synthetic peptide version of PMCA1b also showed this difference with K_d values for calmodulin binding being significantly different at pH 5.9 ($K_d = \sim 30$ nM) than at pH 7.2 ($K_d = 80$ nM). This difference is probably due to the presence of histidines in the PMCA1b sequence.

The C-exon insert sequences must have arisen by a tandem duplication mechanism as they are clearly repeats of the segment they displace (compare, for example, the sequence IRVVAFRSS immediately after the QTQ sequence in PMCA1b with that of the C-exon insert in PMCA1c and PMCA1d MDVVNAFQSG). Thus, the resulting C-exon containing isoforms have two homologs of this sequence. However, the C-insert domain lacks the basic residues found in the IRVVFRSS sequence. As noted earlier, the presence of this sequence alone, without the displaced original segment, leads to substantially reduced calmodulin affinity (Enyedi *et al.,* 1991), at least in the case of PMCA1 isoforms. Thus, the effect of the C-exon insertions is uncertain. A more detailed analysis of the properties of individual PMCA isoforms will be required to fully understand the functional significance of the structural differences observed for the different products and splicing variants.

3. Proposed Mechanism of Regulation

With all of this information now available, it is possible to propose a unified and relatively simple mechanism for the regulation of PMCA and SERCA activity that involves many of the same elements recognized in other second messenger-regulated enzymes. A relatively small autoinhibitory peptide, either part of the same polypeptide chain as in the case of the CaM-binding domain of PMCAs or a separate polypeptide in the case of SERCA inhibition by phospholamban, binds to the active site region of the pump, probably occluding it directly. Either phosphorylation of serine or threonine in the inhibitory domain or, alternatively, calmodulin binding to it in the case of PMCAs alone causes release of the interaction at the active site and relief of inhibition. (PMCAs also appear to have the additional alternative mechanism of reversal of inhibition by binding acidic phospholipid adjacent to the I inhibitor recognition element.) Relief of inhibition is the common mechanism through which all second messengers appear to act.

As noted in Section III,A,2,b, the specific role of tyrosine phosphorylation in inhibiting PMCA activity has just been identified and the precise mechanism is still under study. However, it would appear likely that this modification simply prevents dissociation of the inhibitory domain opposite to the activating effects of phosphorylation at Ser/Thr residues in adjacent positions in the molecule.

This model is a very attractive one, which appears to account for our existing knowledge about the structure and function of these enzymes. However, proof of this mechanism and the precise molecular details of these interactions await verification by the determination of three-dimensional structures of the respective protein complexes using physical methods. This will be a challenge given the large size and membrane localization of these enzymes.

D. Regulation of PMCA Gene Expression

Studies of PMCA mRNA production have concentrated largely on the processing of the primary transcript. However, regulation of transcription of PMCA genes has also been reported. The most obvious form of transcriptional control is the simple on/off regulation of various isoforms in different tissues. For example, the PMCA1 and PMCA4 genes are transcribed in almost all tissues examined, whereas PMCA2 and PMCA3 transcription is usually limited to a few select tissues and cells (vida supra). There are also examples of PMCA isoform transcription being regulated developmentally. For example, the PMCA2 gene in rat brain is not transcribed until just before birth, but PMCA1 and

PMCA3 are transcribed from at least embryonic day 10 (Brandt and Neve, 1992). Leaky transcription of PMCA genes has never been reported, even under the most sensitive PCR conditions. The mechanisms responsible for this tight regulation of PMCA transcription have not yet been elucidated.

Transcriptional regulation of the PMCA1 gene has been the most extensively studied. PMCA1 transcription, much like the enzyme activity itself, appears to be enhanced by a variety of stimuli, and no repressing stimulus has been reported. Activation of protein kinase C by the phorbol ester, PMA, was found to stimulate PMCA1 mRNA production 8- to 20-fold with maximal expression at 4 hr in rat aorta endothelial cells (Kuo et al., 1991). Concomitant PMCA protein production was seen with a peak at about 6 hr poststimulation, but the increase was only about 3.5-fold. As the antibody used in these studies would detect all PMCA isoforms, it is possible that PMCA1 levels were actually stimulated to a greater extent, but the overall response was lower due to the detection of other isoforms by the antibody. Kuo et al. (1991) also found that PMCA1 mRNA production was stimulated by cAMP, suggesting a role for CREB in transcriptional regulation, although Smith and Smith (1995) reported that elevated cAMP did not increase PMCA1 mRNA levels in arterial myocytes. This discrepancy may be due to different transcriptional signals used by the different cell types.

Angiotensin II was able to stimulate PMCA1 transcription in rat aorta endothelial cells to a similar extent as PMA (Kuo et al., 1991). Angiotensin II stimulation has been shown to act through PKC- and calcium-dependent mechanisms. Treatment of endothelial cells with ionomycin to mimic the calcium mobilization effect of angiotensin II resulted in no increase in PMCA1 mRNA accumulation, further confirming a role for PKC in stimulating transcription (Kuo et al., 1991). The PKC-dependent stimulation of PMCA1 transcription required *de novo* protein synthesis as the effect could be blocked by cyclohexamide. Kuo et al. (1991) speculated that the requirement for protein synthesis and stimulation by angiotensin II may indicate that the transcription factor, Jun, was being synthesized and used in PMCA1 transcription. Indeed, an AP-1-binding site was found in the 5'-flanking sequences of the mouse and human PMCA1 genes (Du et al., 1995; Hilfiker et al., 1993).

As noted earlier, there is variability of PMCA1 gene regulation among cell types, for example, in endothelial cells from two different sources, Kuo et al. (1993) found that while PMA and angiotensin II stimulated rat aorta endothelial cell production of PMCA1 gene transcription 8- to

20-fold, these same compounds resulted in only a 0- to 2-fold increase in rat brain-resistant vessel endothelial cells. In contrast, cAMP or thapsigargin stimulated PMCA1 transcription in rat brain-resistant vessel endothelial cells, but not in aortic endothelial cells.

As might be expected because of its role in regulating calcium levels, transcription of the PMCA1 gene is also controlled by vitamin D_3 or, more specifically, its metabolically active product 1,25α-dihydroxyvitamin D_3 [1,25-$(OH)_2D_3$]. 1,25-$(OH)_2D_3$ has been shown to increase PMCA1 mRNA levels in the intestine of rat (Armbrecht et al., 1994; Zelinski et al., 1991) and chicken (Cai et al., 1993). Pannabecker et al. (1995) have shown that the 1,25-$(OH)_2D_3$-dependent increase in PMCA1 mRNA is probably due to de novo nuclear transcription rather than message stablization. This observation is consistent with that of Kuo et al. (1991) for the PKC-dependent PMCA1 mRNA increase, which was blocked by actinomycin D. PMCA1 mRNA was also degraded at the same rate with or without PMCA treatment. Because the small intestine loses its ability to transport calcium with age, Armbrecht et al. (1994) examined PMCA1 mRNA expression in young and aged rats. The authors found that, indeed, in the duodenum, PMCA1 mRNA is about three times more abundant in young rats than adult rats. They found about a two-fold decrease in PMCA1 mRNA with age in the ileum. It appears that the age-associated decrease in serum 1,25-$(OH)_2D_3$ levels is responsible for decreased PMCA1 transcription in aged intestine, as administration of 1,25-$(OH)_2D_3$ to vitamin D-deficient young and aged rats yielded increased PMCA1 mRNA levels whose concentration correlated directly with the 1,25-$(OH)_2D_3$ dose to the same extent in both sets of animals.

Loss of the requirement for high concentrations of extracellular calcium has been a hallmark of cellular transformation known for many years (Boynton and Whitfield, 1978). Reisner et al. (1997) hypothesized that a possible loss of PMCA activity might be responsible for the lower extracellular calcium requirement by allowing basal calcium concentrations to be slightly elevated. Their examination of lung and skin fibroblasts transformed by simian virus 40 found that both mRNA and protein levels for PMCA1 and PMCA4, the only isoforms expressed in these cells, were decreased 1.5- to 10-fold compared to parental control cells. However, studies were not performed to determine whether the decreased PMCA mRNA levels were due to changes in gene transcription rather than increased degradation.

Examination of the 5'-flanking sequences of the PMCA1 gene has shown that it has consensus recognition sequences for the transcription factors AP-1, AP-2, SP-1, and CREB (Du *et al.*, 1995; Hilfiker *et al.*, 1993) and may contain a vitamin D_3 receptor recognition site (Kumar *et al.*, 1995). Du *et al.* (1995) showed that the control elements responsible for cAMP- and PMA-dependent induction of PMCA1 in the neuroblastoma cell line CLL131 were located in a 273-bp sequence upstream of the first transcriptional start site. Within this region are putative CREB- and AP1- binding sites, which are known to be involved in transcriptional activation of cAMP and PMA, respectively. Deletion of the promoter through the SP1-binding site at position −117 removed all basal expression of the promoter in CLL131 cells. The CREB site was located upstream of the deletion, but the authors did not confirm that cAMP could activate transcription from this site independently. Also, deletion of the sequence from −442 to −256 led to an increase in the basal promoter activity, suggesting that a negative regulatory element may be located in that region. Another general feature of the PMCA1 gene is that it exhibits the characteristics of a housekeeping gene, commensurate with the widespread distribution of PMCA1b mRNA. The regulatory region lacks a consensus TATA box and is CpG rich with multiple SP1 and GC box sequences (Du *et al.*, 1995; Hilfiker *et al.*, 1993), which are all hallmarks of housekeeping genes. The PMCA1 gene is also upregulated by treatment with thapsigarin, which means it may also be controlled by cytosolic-free calcium, inactivation of SERCAs, or the calcium influx factor.

The regulatory region of the PMCA3 gene is the only other isoform that has been characterized (Burk *et al.*, 1995). Like the PMCA1 gene, the PMCA3 promoter falls in a very GC-rich region, does not contain a TATA box, and has SP1 sites close to the putative transcription initiation site. However, PMCA3 is not constitutively expressed in most cells as is PMCA1. This change in constitutive expression may result from the fact that the putative PMCA3 promoter itself is not embedded in a CpG island (Burk *et al.*, 1995). PMCA3 gene products are found in muscle and brain, and an examination of the upstream flanking sequences found binding sites for the following muscle-specific transcription factors: 4 CarG sequences, 14 E-box sequences, 11 sites that match or nearly match the MCAT consensus sequences, and 2 MADS box sequences. An octamer and two POU consensus sites, which are used by brain-specific transcription factors, were also found in the PMCA3 gene (Burk *et al.*, 1995). The PMCA3 gene also contains a trinucleotide repeat of over 200 bp in the regulatory region. Because trinucleotide repeats have been associated with several genetic disorders and PMCA3 is expressed in muscle and maps to chromosomal position Xq28, it had been suggested that it might be a good candidate gene for X-linked skeletal muscle diseases such as myotubular myopathy (Burk et al., 1995).

E. Defining Physiological Function of PMCAs Using Genetic Approaches

Although much is known about the mRNA structure and expression of PMCA isoforms, little is known about the actual physiological role they play in the cell. This has become an increasingly perplexing question as PMCAs have generally been considered to provide a homeostatic function: mopping up calcium after a stimulatory event. The discovery of multiple PMCA genes, many alternatively spliced versions of the gene products, and unique tissue and developmental expression for these mRNAs have caused this original, simplistic view to be reevaluated. Attempts to elucidate the role of PMCA isoforms have been undertaken by overexpressing or blocking the expression of certain isoforms *in vivo*.

Several studies have overexpressed PMCA isoforms in mammalian eukaryotic cells to examine their properties (Adamo et al., 1992; Guerini et al., 1995; Heim et al., 1992b; Liu et al.,1996; Preiano et al., 1996). Most early studies were primarily interested in determining the biochemical properties of an overexpressed, purified single PMCA isoform and did not consider the physiological implications of this overexpression on the cell. Guerini et al. (1995) overexpressed PMCA4b in Chinese hamster ovary cells and found that there was an initial delay of entry into the cell cycle of about 24 hr, after serum deprivation. Once the cells had initiated division, their growth rate was not appreciably different from controls. Also, cells overexpressing PMCA4b showed downregulation in SERCA pump activity. Whether this occurred at the transcriptional, translational, or regulatory level was not determined, but as will be discussed later, overexpression of PMCA1a leads to a decrease in SERCA3 mRNA.

Liu et al. (1996) overexpressed PMCA1a in rat aortic endothelial cells and found some surprising results with respect to calcium signaling. As was found with the overexpression of PMCA4b by Guerini et al. (1995), SERCA pump activity was downregulated. Northern blot analysis showed that SERCA3 mRNA was strongly decreased in PMCA1a overexpressing cells, and antibodies that detected SERCA2 and SERCA3 isoforms confirmed that very little SERCA2 or SERCA3 was being produced in these cells. Accompanying the SERCA decrease was a substantial reduction in the IP_3 receptor, as well as a decrease in the amount of calcium in the internal stores. An increase in capacitative calcium influx was also found. In both cases where PMCA1a protein expression was increased ~20-fold (Liu *et al.*, 1996) or PMCA4b expression was increased ~4-fold (Guerini *et al.*, 1995) the actual increase of plasma membrane calcium removal only increased ~1.25-fold. This suggests that a mechanism controls the activity of the pump beyond the known regulators of activity (phosphorylation by PKC and PKA, calmodulin, acid phospholipids), which are all stimulatory. This control may be the inhibition of PMCA activity by tyrosine phosphorylation (Dean *et al.*, 1997). Along these same lines, Brandt *et al.* (1996) saw a small decrease in the rate of clearance of calcium after bradykinin stimulation in cells where all PMCA1 protein production had been blocked with antisense RNA, but saw no increase in total resting calcium levels. Release from bradykinin-stimulated IP_3-sensitive stores was unaltered, but release from ionomycin-sensitive stores was decreased. These observations serve to emphasize the dynamic ability of the cell to adjust to changes in calcium regulation through the use of a myriad of overlapping mechanisms.

Most studies that have examined the physiological implications of PMCA activity have been done by overexpressing a single isoform. Brandt *et al.* (1996) used the reverse strategy of treatment with antisense RNA to block production of all PMCA1 isoforms in the phenochromocytoma cell line PC6. As noted earlier, there was little change in the calcium-handling properties of the cell and, in particular, the resting calcium levels were the same in PMCA1 antisense and control cell lines. The gross morphological appearance of the undifferentiated cells not expressing PMCA1 isoforms was unchanged and their growth rates were unaltered, but time of entry into the cell cycle was not determined. However, when differentiated with NGF, cells lacking PMCA1 isoforms did not extend neurites. The molecular mechanism underlying this phenomenon is not understood, but it is interesting to note that antisense cell lines did not produce α_1-integrin, which is necessary for neurite extension in PC12 cells, the parental line of PC6 (P. C. Brandt, J. E. Sisken, and T. C. Vanaman, 1998). These results raise an interesting question apart from the effect on differentiation: If PMCA1b is the housekeeping isoform of the PMCA family, why is its complete elimination not lethal to the cell? These results again speak to the tremendous flexibility of the cell to adjust to changes in calcium regulatory mechanisms.

The only other study that has looked at the physiological consequences of blockage of production of a PMCA isoform examined the homologous system in yeast (Cunningham and Fink, 1994). In the yeast strain *S. cerevisiae,* the gene PMC1 is 41% homologous to PMCA1, but lacks a calmodulin-binding domain and sequences that would encode the lipid-binding domain. PMC1 is located in vacuole membranes and transports calcium into the vacuole. Deletion of the PMC1 gene led to the inability of the yeast to grow in medium containing 200 mM calcium, but allowed them to grow in medium containing $<1\mu M$ calcium. This is presumably due to their ability to regulate cytosolic calcium levels by pumping calcium into vacuoles. Calcium sensitivity could be compensated for by inhibiting calcineurin activity with cyclosporin A or FK506 or by disrupting the two calcineurin A genes or the calcineurin B gene. The target of calcineurin that is responsible for inhibiting yeast growth appears to be the *vcx1* gene product, which is a vacuolar calcium/proton exchanger (Cunningham and Fink, 1996).

IV. OTHER ION CHANNELS DIRECTLY REGULATED BY CALMODULIN

There are multitudinous reports of calmodulin regulating various ion channels. However, for the most part, in higher eukaryotes, calmodulin does not regulate channel activities by binding directly to the channel, rather it regulates the channel by activating CaM kinases and/or calcineurin.

In paramecium, calmodulin can regulate both Na^+ and K^+ channels by directly binding to them (Saimi and Kung, 1994). Binding of Ca^{2+}/calmodulin to the channel increases the probability of its opening on depolarization of the membrane. The mechanism of regulation of these channels appears to be very similar to that used by enzymes in controlling calmodulin-dependent stimulation. This is demonstrated, at least in part, by the fact that limited proteolysis of the paramecium potassium channel (Kubalski *et al.,* 1989) results in permanent, calmodulin-independent activation of the channel, similar to observations seen for PMCAs and calcineurin when C-terminal calmodulin-binding domains are cleaved off.

Interestingly, the Na$^+$ and K$^+$ channels of the paramecium each interact with different ends of the calmodulin molecule. Alteration in Na$^+$ or K$^+$ channel opening could be detected phenotypically by changes in the swimming behavior of the paramecium. Spontaneous mutations in calmodulin were found to be associated with each of these phenotypes (Kink et al., 1990; Ling et al., 1994). All mutations associated with altered Na$^+$ channel activity were clustered in the amino terminus of the molecule whereas all of those associated with K$^+$ channel activity were found at the carboxyl terminus. No mutations were found in the central helix.

Another set of ion channels, trp and trpl, are found in *Drosophila* retina as part of the photoreceptor system. In other cells, trp and trpl appear to be the capacitative calcium channel that opens when internal calcium stores have been depleted (Birnbaumer et al., 1996). Both trp and trpl have been reported to have putative calmodulin-binding domains (Phillips et al., 1992), but no functions have yet been associated with calmodulin binding.

ACKNOWLEDGMENTS

This work was funded by National Science Foundation grant IBN 9604729 (P.C.B.) and National Institutes of Health grant NS 21868 (T.C.V.).

REFERENCES

Adebayo, A. O., Enyedi, A. Verma, A. K., Filoteo, A. G., and Penniston, J. T. (1995). Two residues that may ligate Ca^{2+} in transmembrane domain six of the plasma membrane Ca^{2+}-ATPase. *J. Biol. Chem.* **270**, 27812–27816.

Adamo, H. P., Verma, A. K., Sanders, M. A., Heim, R., Salisbury, J. L., Wieben, E. D., and Penniston, J. T. (1992). Overexpression of the erythrocyte plasma membrane Ca^{2+} pump in COS-1 cells. *Biochem. J.* **285**, 791–797.

Adamo, H. P., Filoteo, A. G., and Penniston, J. T. (1996). The plasma membrane Ca^{2+} pump mutant lysine591 → arginine retains some activity, but is still inactivated by fluorescein isothiocyanate. *Biochem. J.* **317**, 41–44.

Agre, P., Gardner, K., and Bennett, V. (1983). Association between human erythrocyte calmodulin and the cytoplasmic surface of human erythrocyte membranes. *J. Biol. Chem.* **258**, 6258–6265.

8. CaM and Ion Flux Regulation

Aldaz, C. M., Yeung, R. S., Latif, F., Lerman, M. I., Xiao, G., Trono, D., and Walker, C. L. (1995). Colocalization of the rat homolog of the von Hippel Lindau (*VhI*) gene and the plasma membrane Ca^{++} transporting ATPase isoform 2 (*Atp2b2*) gene to rat chromosome bands 4q41.3 → 42.1. *Cytogenet. Cell. Genet.* **71,** 253–256.

Anderson, M. E., Braun, A. P., Schulman, H., and Premack, B. A. (1994). Multifunctional Ca^{2+}/calmodulin-dependent protein kinase mediates Ca^{2+}-induced enhancement of the L-type Ca^{2+} current in rabbit ventricular myocytes. *Circ. Res.* **75,** 854–861.

Antebi, A., and Fink, G. R. (1992). The yeast Ca^{2+}-ATPase homologue, PMR1, is required for normal Golgi function and localizes in a novel Golgi-like distribution. *Mol. Biol. Cell* **3,** 633–654.

Armbrecht, H. J., Boltz, M. A., and Wongsurawat, N. (1994). Expression of plasma membrane calcium pump mRNA in rat intestine: Effect of age and 1,25-dihydroxyvitamin D. *Biochim. Biophys. Acta* **1195,** 110–114.

Balasubramanyam, M., and Gardner, J. P. (1995). Protein kinase C modulates cytosolic free calcium by stimulating calcium pump activity in Jurkat T cells. *Cell Calcium* **18,** 526–541.

Barrabin, H., Garrahan, P. J., and Rega, A. F. (1980). Vanadate inhibition of the Ca^{2+}-ATPase from human red cell membranes. *Biochim. Biophys. Acta* **600,** 796–804.

Benaim, G., Zurini, M., and Carafoli, E. (1984). Different conformational states of the purified Ca^{2+}-ATPase of the erythrocyte plasma membrane revealed by controlled trypsin proteolysis. *J. Biol. Chem.* **259,** 8471–8477.

Birnbaumer, L., Zhu, X., Jiang, M., Boulay, G., Peyton, M., Vannier, B., Brown, D., Platano, D., Sadeghi, H., Stefani, E., and Birnbauer, M. (1996). On the molecular basis and regulation of cellular capacitative calcium entry: Roles for Trp proteins. *Proc. Natl. Acad. Sci. U.S.A.* **93,** 15195–15202.

Blumenthal, D. K., Charbonneau, H., Edelman, A. M., Hinds, T. R., Rosenberg, G. B., Storm, D. R., Vincenzi, F. F., Beavo, J. A., and Krebs, E. G. (1988). Synthetic peptides based on the calmodulin-binding domain of myosin light chain kinase inhibit activation of other calmodulin-dependent enzymes. *Biochem. Biophys. Res. Commun.* **156,** 860–865.

Bond, G. H., and Clough, D. L. (1973). A soluble protein activator of (Mg^{2+} plus Ca^{2+})-dependent ATPase in human red cell membranes. *Biochim. Biophys. Acta* **323,** 592–599.

Bond, G. H., and Hudgins, P. M. (1979). Kinetics of inhibition of NaK-ATPase by Mg^{2+}, K^+, and vanadate. *Biochemistry* **18,** 325–331.

Bond, G. H., and Hudgins, P. M. (1980). Inhibition of red cell Ca^{2+}-ATPase by vanadate. *Biochim. Biophys. Acta* **600,** 781–790.

Boynton, A. L., and Whitfield, J. F. (1978). Calcium requirements for the proliferation of cells infected with a temperature-sensitive mutant of Rous Sarcoma virus. *Cancer Res.* **38,** 1237–1240.

Brandt, P., and Neve, R. L. (1992). Expression of plasma membrane calcium-pumping ATPase mRNAs in developing rat brain and adult brain subregions: Evidence for stage-specific expression. *J. Neurochem.* **59,** 1566–1569.

Brandt, P., and Vanaman, T. C. (1994). Splicing of the muscle-specific plasma membrane Ca^{2+}-ATPase isoforms PMCA1c is associated with cell fusion in C2 myocytes. *J. Neurochem.* **62,** 799–802.

Brandt, P., Zurini, M., Neve, R. L., Rhoads, R. E., and Vanaman, T. C. (1988). A C-terminal, calmodulin-like regulatory domain from the plasma membrane Ca^{2+}-pumping ATPase. *Proc. Natl. Acad. Sci. U.S..A.* **85,** 2914–2918.

Brandt, P., Ibrahim, E., Bruns, G. A., and Neve, R. L. (1992a). Determination of the nucleotide sequence and chromosomal localization of the *ATP2B2* gene encoding human Ca^{2+}-pumping ATPase isoform PMCA2. *Genomics* **14**, 484–487.

Brandt, P., Neve, R. L., Kammesheidt, A., Rhoads, R. E., and Vanaman, T. C. (1992b). Analysis of the tissue-specific distribution of mRNAs encoding the plasma membrane calcium-pumping ATPases and characterization of an alternatively spliced form of PMCA4 at the cDNA and genomic levels. *J. Biol. Chem.* **267**, 4376–4385.

Brandt, P. C., Sisken, J. E., Neve, R. L., and Vanaman, T. C. (1996). Blockade of plasma membrane calcium pumping ATPase isoform I impairs nerve growth factor-induced neurite extension in pheochromocytoma cells. *Proc. Natl. Acad. Sci. U.S.A.* **93**, 13843–13848.

Brodin, P., Falchetto, R., Vorherr, T., and Carafoli, E. (1992). Identification of two domains which mediate the binding of activating phospholipids to the plasma-membrane Ca^{2+} pump. *Eur. J. Biochem.* **204**, 939–946.

Burk, S. E., and Shull, G. E. (1992). Structure of the rat plasma membrane Ca^{2+}-ATPase isoform 3 gene and characterization of alternative splicing and transcription products. Skeletal muscle-specific splicing results in a plasma membrane Ca^{2+}-ATPase with a novel calmodulin-binding domain. *J. Biol. Chem.* **267**, 19683–19690.

Burk, S. E., Menon, A. G., and Shull, G. E. (1995). Analysis of the 5′ end of the rat plasma membrane Ca^{2+}-ATPase isoform 3 gene and identification of extensive trinucleotide repeat sequences in the 5′-untranslated region. *Biochim. Biophys. Acta* **1240**, 119–124.

Cai, Q., Chandler, J. S., Wasserman, R. H., Kumar, R., and Penniston, J. T. (1993). Vitamin D and adaptation to dietary calcium and phosphate deficiencies increase intestinal plasma membrane calcium pump gene expression. *Proc. Natl. Acad. Sci. U.S.A.* **90**, 1345–1349.

Cameron, A. M., Steiner, J. P., Roskams, A. J., Ali, S. M., Ronnett, G. V., and Snyder, S. H. (1995). Calcineurin associated with the inositol 1,4,5-trisphosphate receptor–FKBP12 complex modulates Ca^{2+} flux. *Cell (Cambridge, Mass.)* **83**, 463–472.

Carafoli, E. (1987). Intracellular calcium homeostasis. *Ann. Rev. Biochem.* **56**, 395–433.

Carafoli, E. (1991a). Calcium pump of the plasma membrane. *Physiol. Rev.* **71**, 129–153.

Carafoli, E. (1991b). The calcium pumping ATPase of the plasma membrane. *Annu. Rev. Physiol.* **53**, 531–547.

Carafoli, E. (1994). The signaling function of calcium and its regulation. *J. Hypertens. Suppl.* **12**, S47–S56.

Carafoli, E., and Guerini, D. (1993). Molecular and cellular biology of plasma membrane calcium ATPase. *Trends Cardiovasc. Med.* **3**, 177–184.

Carafoli, E., and Stauffer, T. (1994). The plasma membrane calcium pump: Functional domains, regulation of the activity, and tissue specificity of isoform expression. *J. Neurobiol.* **25**, 312–324.

Carafoli, E., Garcia-Martin, E., and Guerini, D. (1996). The plasma membrane calcium pump: Recent developments and future perspectives. *Experientia* **52**, 1091–1100.

Caride, A. J., Penniston, J. T., and Rossi, J. P. (1991). The calmodulin-binding domain as an endogenous inhibitor of the *p*-nitrophenylphosphatase activity of the Ca^{2+} pump from human red cells. *Biochim. Biophys. Acta* **1069**, 94–98.

Caride, A. J., Chini, E. N., Yamaki, M., Dousa, T. P., and Penniston, J. T. (1995). Unique localization of mRNA encoding plasma membrane Ca^{2+} pump isoform 3 in thin descending loop of Henle. *Am. J. Physiol.* **269**, F681–F685.

Caroni, P., and Carafoli, E. (1981). Regulation of Ca^{2+}-pumping ATPase of heart sarcolemma by a phosphorylation-dephosphorylation process. *J. Biol. Chem.* **256,** 9371–9373.

Cavieres, J. D. (1984). Calmodulin and the target size of the (Ca^{2++} Mg^{2+})-ATPase of human red-cell ghosts. *Biochim. Biophys. Acta* **771,** 241–244.

Chiesi, M., Vorherr, T., Falchetto, R., Waelchli, C., and Carafoli, E. (1991). Phospholamban is related to the autoinhibitory domain of the plasma membrane Ca^{2+}-pumping ATPase. *Biochemistry* **30,** 7978–7983.

Clarke, D. M., Maruyama, K., Lor, T. W., Leberer, G., and MacLennan, D. H. (1989). Functional consequences of glutamate, aspartate, glutamine and asparagine mutations in the stalk sector of the Ca^{2+} ATPase of sarcoplasmic reticulum. *J. Biol. Chem.* **264,** 11246–11251.

Clarke, D. M., Loo, T. W., and MacLennan, D. H. (1990a). Functional consequences of alterations to amino acids located in the nucleotide binding domain of the Ca^{2+}-ATPase of sarcoplasmic reticulum. *J. Biol. Chem.* **265,** 22223–22227.

Clarke, D. M., Loo, T. W., and MacLennan, D. H. (1990b). Functional consequences of mutations of conserved amino acids in the beta-strand domain of the Ca^{2+}-ATPase of sarcoplasmic reticulum. *J. Biol. Chem.* **265,** 14088–14092.

Coelho-Sampaio, T., Ferreira, S. T., Benaim, G., and Vieyra, A. (1991). Dissociation of purified erythrocyte Ca^{2+}-ATPase by hydrostatic pressure. *J. Biol. Chem.* **266,** 22266–22272.

Crocker, P. J., Imai, N., Rajagopalan, K., Boggess, M. A., Kwiatkowski, S., Dwyer, L. D., Vanaman, T. C., and Watt, D. S. (1990). Heterobifunctional cross-linking agents incorporating perfluorinated aryl azides. *Bioconjugate Chem.* **1,** 419–424.

Cunningham, K. W., and Fink, G. R. (1994). Calcineurin-dependent growth control in *Saccharomyces cerevisiae* mutants lacking PMC1, a homology of plasma membrane Ca^{2+} ATPases. *J. Cell. Biol.* **124,** 351–363.

Cunningham, K. W., and Fink, G. R. (1996). Calcineurin inhibits VCX1-dependent H^+/Ca^{2+} exchange and induces Ca^{2+}-ATPases in *Saccharomyces cerevisiae*. *Mol. Cell. Biol.* **16,** 2226–2237.

Dash, D., Aepfelbacher, M., and Siess, W. (1995). The association of pp125^{FAK}, pp60^{Src}, CDC42Hs and Rap1B with the cytoskeleton of aggregated platelets is a reversible process regulated by calcium. *FEBS Lett.* **363,** 231–234.

Davis, F. B., Davis, P. J., Lawrence, W. D., and Blas, S. D. (1991). Specific inositol phosphates inhibit basal and calmodulin-stimulated Ca^{2+}-ATPase activity in human erythrocyte membranes *in vitro* and inhibit binding of calmodulin to membranes. *FASEB J.* **5,** 2992–2995.

Dean, W. L., Chen, D., Brandt, P. C., and Vanaman, T. C. (1997). Regulation of platelet plasma membrane Ca^{2+}-ATPase by cAMP-dependent and tyrosine phosphorylation. *J. Biol. Chem.* **24,** 15113–15119.

De Jaegere, S., Wuytack, F., Eggermont, J. A., Verboomen, H., and Casteels, R. (1990). Molecular cloning and sequencing of the plasma-membrane Ca^{2+} pump of pig smooth muscle. *Biochem. J.* **271,** 655–660.

De Jaegere, S., Wuytack, F., De Smedt, H., Van den Bosch, L., and Casteels, R. (1993). Alternative processing of the gene transcripts encoding a plasma-membrane and a sarco/endoplasmic reticulum Ca^{2+} pump during differentiation of BC3H1 muscle cells. *Biochim. Biophys. Acta* **1173,** 188–194.

de Leon, M., Wang, Y., Jones, L., Perez-Reyes, E., Wei, X., Soong, T. W., Snutch, T. P., and Yue, D. T. (1995). Essential Ca^{2+}-binding motif for Ca^{2+}-sensitive inactivation of L-type Ca^{2+} channels. *Science* **270,** 1502–1506.

Du, Y., Carlock, L., and Kuo, T. H. (1995). The mouse plasma membrane Ca^{2+} pump isoform 1 promoter: Cloning and characterization. *Arch. Biochem. Biophys.* **316,** 302–310.

Dwyer, L. D., Crocker, P. J., Watt, D. S., and Vanaman, T. C. (1992). The effects of calcium site occupancy and reagent length on reactivity of calmodulin lysyl residues with heterobifunctional aryl azides. Mapping interaction domains with specific calmodulin photoprobe derivatives. *J. Biol. Chem.* **267,** 22606–22615.

Eakin, T. J., Antonelli, M. C., Malchiodi, E. L., Baskin, D. G., and Stahl, W. L. (1995). Localization of the plasma membrane Ca^{2+}-ATPase isoform PMCA3 in rat cerebellum, choroid plexus and hippocampus. *Brain Res. Mol. Brain Res.* **29,** 71–80.

Enyedi, A., and Penniston, J. T. (1993). Autoinhibitory domains of various Ca^{2+} transporters cross-react. *J. Biol. Chem.* **268,** 17120–17125.

Enyedi, A., Sarkadi, B., Szasz, I., Bot, B., and Gardos, G. (1980). Molecular properties of red cell calcium pump. II. Effects of proteolysis, ptoteolytic digestion and drugs on the calcium-induced phosphorylation by ATP in inside/out red cell membrane vesicles. *Cell Calcium* **1,** 299–310.

Enyedi, A., Flura, M., Sarkadi, B., Gardos, G., and Carafoli, E. (1987). The maximal velocity and the calcium affinity of the red cell calcium pump may be regulated independently. *J. Biol. Chem.* **262,** 6425–6430.

Enyedi, A., Filoteo, A. G., Gardos, G., and Penniston, J. T. (1991). Calmodulin-binding domains from isozymes of the plasma membrane Ca^{2+} pump have different regulatory properties. *J. Biol. Chem.* **266,** 8952–8956.

Enyedi, A., Verma, A. K., Filoteo, A. G., and Penniston, J. T. (1993). A highly active 120-kDa truncated mutant of the plasma membrane Ca^{2+} pump. *J. Biol. Chem.* **268,** 10621–10626.

Enyedi, A., Verma, A. K., Filoteo, A. G., and Penniston, J. T. (1996). Protein kinase C activates the plasma membrane Ca^{2+} pump isoform 4b by phosphorylation of an inhibitory region downstream of the calmodulin-binding domain. *J. Biol. Chem.* **271,** 32461–32467.

Falchetto, R., Vorherr, T., Brunner, J., and Carafoli, E. (1991). The plasma membrane Ca^{2+} pump contains a site that interacts with its calmodulin-binding domain. *J. Biol. Chem.* **266,** 2930–2936.

Falchetto, R., Vorherr, T., and Carafoli, E. (1992). The calmodulin-binding site of the plasma membrane Ca^{2+} pump interacts with the transduction domain of the enzyme. *Protein Sci.* **1,** 1613–1621.

Filoteo, A. G., Gorski, J. P., and Penniston, J. T. (1987). The ATP-binding site of the erythrocyte membrane Ca^{2+} pump. Amino acid sequence of the fluorescein isothiocyanate-reactive region. *J. Biol. Chem.* **262,** 6526–6530.

Foletti, D., Guerini, D., and Carafoli, E. (1995). Subcellular targeting of the endoplasmic reticulum and plasma membrane Ca^{2+} pumps: A study using recombinant chimeras. *FASEB J.* **9,** 670–680.

Froehlich, J. P., and Taylor, E. W. (1975). Transient state kinetic studies of sarcoplasmic reticulum adenosine triphosphatase. *J. Biol. Chem.* **250,** 2013–2021.

Furukawa, K., Tawada, Y., and Shigekawa, M. (1989). Protein kinase C activation stimulates plasma membrane Ca^{2+} pump in cultured vascular smooth muscle cells. *J. Biol. Chem.* **264,** 4844–4849.

Garrahan, P. J., and Rega, A. F. (1978). Activation of partial reactions of the Ca^{2+}-ATPase from human red cells by Mg^{2+} and ATP. *Biochim. Biophys. Acta* **513,** 59–65.

Gazzotti, P., and Carafoli, E. (1994). Purification and reconstitution of the Ca^{2+}-pumping ATPase of red blood cells. *Methods (San Diego)* **6,** 3–10.

Gietzen, K., Tejcka, M., and Wolf, H. U. (1980). Calmodulin affinity chromatography yields a functional purified erythrocyte (Ca^{2+}, Mg^{2+})-dependent adenosine triphosphatase. *Biochem. J.* **189,** 81–88.

Gopinath, R. M., and Vincenzi, F. F. (1977). Phosphodiesterase protein activator mimics red blood cell cytoplasmic activator of $(Ca^{2+}\text{-}Mg^{2+})$-ATPase. *Biochem. Biophys. Res. Commun.* **77,** 1203–1209.

Graf, E., Filoteo, A. G., and Penniston, J. T. (1980). Preparation of ^{125}I-calmodulin with retention of full biological activity: Its binding to human erythrocyte ghosts. *Arch. Biochem. Biophys.* **203,** 719–726.

Grimaldi, M. E., Adamo, H. P., Rega, A. F., and Penniston, J. T. (1996). Deletion of amino acid residues 18–75 inactivates the plasma membrane Ca^{2+} pump. *J. Biol. Chem.* **271,** 26995–26997.

Guerini, D., Schroder, S., Foletti, D., and Carafoli, E. (1995). Isolation and characterization of a stable Chinese hamster ovary cell line overexpressing the plasma membrane Ca^{2+}-ATPase. *J. Biol. Chem.* **270,** 14643–14650.

Guerini, D., Foletti, D., Vellani, F., and Carafoli, E. (1996). Mutation of conserved residues in transmembrane domains 4, 6 and 8 causes loss of Ca^{2+} transport by the plasma membrane Ca^{2+} pump. *Biochemistry* **35,** 3290–3296.

Gunteski-Hamblin, A. M., Greeb, J., and Shull, G. E. (1988). A novel Ca^{2+} pump expressed in brain, kidney, and stomach is encoded by an alternative transcript of the slow-twitch muscle sarcoplasmic reticulum Ca-ATPase gene. Identification of cDNAs encoding Ca^{2+} and other cation-transporting ATPases using an oligonucleotide probe derived from the ATP-binding site. *J. Biol. Chem.* **263,** 15032–15040.

Gunteski-Hamblin, A. M., Clarke, D. M., and Shull, G. E. (1992). Molecular cloning and tissue distribution of alternatively spliced mRNAs encoding possible mammalian homologues of the yeast secretory pathway calcium pump. *Biochemistry* **31,** 7600–7608.

Haaker, H., and Racker, E. (1979). Purification and reconstitution of the Ca^{2+}-ATPase from plasma membrane of pig erythrocytes. *J. Biol. Chem.* **254,** 6598–6602.

Hammes, A., Oberdorf, S., Strehler, E. E., Stauffer, T., Carafoli, E., Vetter, H., and Neyses, L. (1994). Differentiation-specific isoform mRNA expression of the calmodulin-dependent plasma membrane Ca^{2+}-ATPase. *FASEB J.* **8,** 428–435.

Heim, R., Hug, M., Iwata, T., Strehler, E. E., and Carafoli, E. (1992a). Microdiversity of human-plasma-membrane calcium-pump isoform 2 generated by alternative RNA splicing in the N-terminal coding region. *Eur. J. Biochem.* **205,** 333–340.

Heim, R., Iwata, T., Zvaritch, E., Adamo, H. P., Rutishauser, B., Strehler, E. E., Guerini, D., and Carafoli, E. (1992b). Expression, purification, and properties of the plasma membrane Ca^{2+} pump and of its N-terminally truncated 105-kDa fragment. *J. Biol. Chem.* **267,** 24476–24484.

Heiss, N. S., Rogner, U. C., Kioschis, P., Korn, B., and Poustka, A. (1996). Transcription mapping in a 700-kb region around the DXS52 locus in Xq28: Isolation of six novel transcripts and a novel ATPase isoform (hPMCA5). *Genome Res.* **6,** 478–491.

Hilfiker, H., Strehler-Page, M. A., Stauffer, T. P., Carafoli, E., and Strehler, E. E. (1993). Structure of the gene encoding the human plasma membrane calcium pump isoform 1. *J. Biol. Chem.* **268,** 19717–19725.

Hilfiker, H., Guerini, D., and Carafoli, E. (1994). Cloning and expression of isoform 2 of the human plasma membrane Ca^{2+}-ATPase. Functional properties of the enzyme and its splicing products. *J. Biol. Chem.* **269,** 26178–26183.

Hinds, T. R., and Andreasen, T. J. (1981). Photochemical cross-linking of azidocalmodulin to the (Ca^{2+}, Mg^{2+})-ATPase of the erythrocyte membrane. *J. Biol. Chem.* **256,** 7877–7882.

Hofmann, F., James, P., Vorherr, T., and Carafoli, E. (1993). The C-terminal domain of the plasma membrane Ca^{2+} pump contains three high affinity Ca^{2+} binding sites. *J. Biol. Chem.* **268,** 10252–10259.

Hofmann, F., Anagli, J., Carafoli, E., and Vorherr, T. (1994). Phosphorylation of the calmodulin binding domain of the plasma membrane Ca^{2+} pump by protein kinase C reduces its interaction with calmodulin and with its pump receptor site. *J. Biol. Chem.* **269,** 24298–24303.

Howard, A., Legon, S., and Walters, J. R. (1993). Human and rat intestinal plasma membrane calcium pump isoforms. *Am. J. Physiol.* **265,** G917–G925.

Howard, A., Barley, N. F., Legon, S., and Walters, J. R. (1994). Plasma-membrane calcium-pump isoforms in human in rat liver. *Biochem. J.* **303,** 275–279.

Imai, N., Kometani, T., Crocker, P. J., Bowdan, J. B., Demir, A., Dwyer, L. D., Mann, D. M., Vanaman, T. C., and Watt, D. S. (1990). Photoaffinity heterobifunctional cross-linking reagents based on *N*-(azidobenzoyl)tyrosines. *Bioconjugate Chem.* **1,** 138–143.

Jaffe, C. L., Lis, H., and Sharon, N. (1980). New cleavable photoreactive heterobifunctional cross-linking reagents for studying membrane organization. *Biochemistry* **19,** 4423–4429.

James, P., Maeda, M., Fischer, R., Verma, A. K., Krebs, J., Penniston, J. T., and Carafoli, E. (1988). Identification and primary structures of a calmodulin binding domain of the Ca^{2+} pump of human erythrocytes. *J. Biol. Chem.* **263,** 2905–2910.

James, P., Vorherr, T., Krebs, J., Morelli, A., Castello, G., McCormick, D. J., Penniston, J. T., De Flora, A., and Carafoli, E. (1989a). Modulation of erythrocyte Ca^{2+}-ATPase by selective calpain cleavage of the calmodulin-binding domain. *J. Biol. Chem.* **264,** 8289–8296.

James, P. H., Pruschy, M., Vorherr, T. E., Penniston, J. T., and Carafoli, E. (1989b). Primary structure of the cAMP-dependent phosphorylation site of the plasma membrane calcium pump. *Biochemistry* **28,** 4253–4258.

James, P., Vorherr, T., Thulin, E., Forsen, S., and Carafoli, E. (1991). Identification and primary structure of a calbindin 9K binding domain in the plasma membrane Ca^{2+} pump. *FEBS Lett.* **278,** 155–159.

Jarrett, H. W., and Penniston, J. T. (1977). Partial purification of the Ca^{2+}-Mg^{2+}-ATPase activator from human erythrocytes: Its similarity to the activator of 3':5'-cyclic nucleotide phosphodiesterase. *Biochem. Biophys. Res. Commun.* **77,** 1210–1216.

Jarrett, H. W., and Penniston, J. T. (1978). Purification of the Ca^{2+}-stimulated ATPase activator from human erythrocytes. Its membership in the class of Ca^{2+}-binding modulator proteins. *J. Biol. Chem.* **253,** 4676–4682.

Jayaraman, T., Ondrias, K., Ondriasova, E., and Marks, A. R. (1996). Regulation of the inositol 1,4,5-trisphosphate receptor by tyrosine phosphorylation. *Science* **272,** 1492–1494.

Jeffery, D. A., Roufogalis, B. D., and Katz, S. (1981). The effect of calmodulin on the phosphoprotein intermediate of Mg^{2+}-dependent Ca^{2+}-stimulated adenosine triphosphatase in human erythrocyte membranes. *Biochem. J.* **194,** 481–486.

8. CaM and Ion Flux Regulation

Jencks, W. P. (1989). How does a calcium pump pump calcium? *J. Biol. Chem.* **264,** 18855–18858.
Jorgensen, P. L. (1982). Mechanisms of the Na$^+$, K$^+$ pump. Protein structure and conformations of the pure (Na$^+$ + K$^+$)-ATPase. *Biochim. Biophys. Acta* **694,** 27–68.
Kanamaru, K., Kashiwagi, S., and Mizuno, T. (1993). The cyanobacterium *Synechococcus* sp. PCC7942, possesses two distinct genes encoding cation-transporting P-type ATPases. *FEBS Lett.* **330,** 99–104.
Katz, S., and Blostein, R. (1975). Ca^{2+}-stimulated membrane phosphorylation and ATPase activity of the human erythrocyte. *Biochim. Biophys. Acta* **389,** 314–324.
Keeton, T. P., and Shull, G. E. (1995). Primary structure of rat plasma membrane Ca^{2+}-ATPase isoform 4 and analysis of alternative splicing patterns at splice site A. *Biochem. J.* **306,** 779–785.
Keeton, T. P., Burk, S. E., and Shull, G. E. (1993). Alternative splicing of exons encoding the calmodulin-binding domains and C termini of plasma membrane Ca^{2+}-ATPase isoforms 1, 2, 3 and 4. *J. Biol. Chem.* **268,** 2740–2748.
Kemp, B. E., Pearson, R. B., Guerriero, V., Jr., Bagchi, I. C., and Means, A. R. (1987). The calmodulin binding domain of chicken smooth muscle myosin light chain kinase contains a pseudosubstrate sequence. *J. Biol. Chem.* **262,** 2542–2548.
Kennelly, P. J., Edelman, A. M., Blumenthal, D. K., and Krebs, E. G. (1987). Rabbit skeletal muscle myosin light chain kinase. The calmodulin binding domain as a potential active site-directed inhibitory domain. *J. Biol. Chem.* **262,** 11958–11963.
Kessler, F., Falchetto, R., Heim, R., Meili, R., Vorherr, T., Strehler, E. E., and Carafoli, E. (1992). Study of calmodulin binding to the alternatively spliced C-terminal domain of the plasma membrane Ca^{2+} pump. *Biochemistry* **31,** 11785–11792.
Kink, J. A., Maley, M. E., Preston, R. R., Ling, K. Y., Wallen-Friedman, M. A., Saimi, Y., and Kung, C. (1990). Mutations in *Paramecium* calmodulin indicate functional differences between the C-terminal and N-terminal lobes *in vivo*. *Cell (Cambridge, Mass.)* **62,** 165–174.
Klee, C. B., and Vanaman, T. C. (1982). Calmodulin. *Adv. Protein Chem.* **35,** 213–321.
Knauf, P. A., Proverbio, F., and Hoffman, J. F. (1974). Electrophoretic separation of different phophoproteins associated with Ca-ATPase and Na, K-ATPase in human red cell ghosts. *J. Gen. Physiol.* **63,** 324–336.
Komalavilas, P., and Lincoln, T. M. (1996). Phosphorylation of the inositol 1,4,5-trisphosphate receptor. Cyclic GMP-dependent protein kinase mediates cAMP and cGMP dependent phosphorylation in the intact rat aorta. *J. Biol. Chem.* **271,** 21933–21938.
Kosk-Kosicka, D., and Bzdega, T. (1988). Activation of the erythrocyte Ca^{2+}-ATPase by either self-association or interaction with calmodulin. *J. Biol. Chem.* **263,** 18184–18189.
Kosk-Kosicka, D., Scaillet, S., and Inesi, G. (1986). The partial reactions in the catalytic cycle of the calcium-dependent adenosine triphosphatase purified from erythrocyte membranes. *J. Biol. Chem.* **261,** 3333–3338.
Kosk-Kosicka, D., Bzdega, T., and Wawrzynow, A. (1989). Fluorescence energy transfer studies of purified erythrocyte Ca^{2+}-ATPase. Ca^{2+}-regulated activation by oligomerization. *J. Biol. Chem.* **264,** 19495–19499.
Kubalski, A., Martinac, B., and Saimi, Y. (1989). Proteolytic activation of a hyperpolarization- and calcium-dependent potassium channel in *Paramecium*. *J. Membr. Biol.* **112,** 91–96.
Kumar, R., Strehler, E. E., and Cai, Q. (1995). *In* "Molecular Nephrology: Kidney Function in Health and Disease" (J. V. Boneventre and D. Schlondoroff, eds.), pp. 253–267. Dekker, New York.

Kuo, T. H., Wang, K. K., Carlock, L., Diglio, C., and Tsang, W. (1991). Phorbol ester induces both gene expression and phosphorylation of the plasma membrane Ca^{2+} pump. *J. Biol. Chem.* **266**, 2520–2525.

Kuo, T. H., Liu, B. F., Diglio, C., and Tsang, W. (1993). Regulation of the plasma membrane calcium pump gene expression by two signal transduction pathways. *Arch. Biochem. Biophys.* **305**, 428–433.

Lacapere, J. J., and Guillain, F. (1990). Reaction mechanism of Ca^{2+}-ATPase of sarcoplasmic reticulum. Equilibrium and transient study of phosphorylation with Ca. ATP as substrate. *J. Biol. Chem.* **265**, 8583–8589.

Latif, F., Duh, F. M., Gnarra, J., Tory, K., Kuzmin, I., Yao, M., Stackhouse, T., Modi, W., Geil, L., Schmidt, L., Li, H., Orcutt, M. L., Maher, E., Richards, F., Phipps, M., Ferguson-Smith, M., LePaslier, D., Lineham, W. M., Zbar, B., and Lerman, M. I. (1993a). von Hippel-Lindau syndrome: Cloning and identification of the plasma membrane Ca^{++}-transporting ATPase isoform 2 gene that resides in the von Hippel-Lindau gene region. *Cancer Res.* **53**, 861–867.

Latif, F., Tory, K., Gnarra, J., Yao, M., Duh, F. M., Orcutt, M. L., Stackhouse, T., Kuzmin, I., Modi, W., Geil, L., Schmidt, L., Zhou, F., Li, H., Wei, M. H., Chen, F., Glenn, G., Choyke, P., Walther, M. M., Weng, Y., Duan, D. R., Dean, M., Glavač, D., Richards, F. M., Crossey, P. A., Ferguson-Smith, M. A., Le Paslier, D., Chumakov, I., Cohen, D., Chinault, A. C., Maher, E. R., Linehan, W. M., Zbar, B., and Lerman, M. I. (1993b). Identification of the von Hippel-Lindau disease tumor suppressor gene. *Science* **260**, 1317–1320.

Lee, H. C. (1991). Specific binding of cyclic ADP-ribose to calcium-storing microsomes from sea urchin eggs. *J. Biol. Chem.* **266**, 2276–2281.

Lee, H. C., and Aarhus, R. (1995). A derivative of NADP mobilizes calcium stores insensitive to inositol trisphosphate and cyclic ADP-ribose. *J. Biol. Chem.* **270**, 2152–2157.

Lee, H. C., Aarhus, R., and Graeff, R. M. (1995). Sensitization of calcium-induced calcium release by cyclic ADP-ribose and calmodulin. *J. Biol. Chem.* **270**, 9060–9066.

Ling, K. Y., Maley, M. E., Preston, R. R., Saimi, Y., and Kung, C. (1994). New nonlethal calmodulin mutations in *Paramecium*. A structural and functional bipartition hypothesis. *Eur. J. Biochem.* **222**, 433–439.

Liu, B. F., Xu, X., Fridman, R., Muallem, S., and Kuo, T. H. (1996). Consequences of functional expression of the plasma membrane Ca^{2+} pump isoform 1a. *J. Biol. Chem.* **271**, 5536–5544.

Lu, H. K., Fern, R. J., Nee, J. J., and Barrett, P. Q. (1994). Ca^{2+}-dependent activation of T-type Ca^{2+} channels by calmodulin-dependent protein kinase II. *Am. J. Physiol.* **267**, F183–F189.

Lynch, T. J., and Cheung, W. Y. (1979). Human erythrocyte Ca^{2+}-Mg^{2+}-ATPase: Mechanism of stimulation by Ca^{2+}. *Arch. Biochem. Biophys.* **194**, 165–170.

MacLennan, D. H., Toyofuku, T., and Lytton, J. (1992). Structure–function relationships in sarcoplasmic or endoplasmic reticulum type Ca^{2+} pumps. *Ann. N.Y. Acad. Sci.* **671**, 1–10.

Magosci, M., Yamaki, M., Penniston, J. T., and Dousa, T. P. (1992). Localization of mRNAs coding for isozymes of plasma membrane Ca^{2+}-ATPase pump in rat kidney. *Am. J. Physiol.* **263**, F7–F14.

Malviya, A. N., Rogue, P., and Vincendon, G. (1990). Stereospecific inositol 1,4,5-[^{32}P] trisphosphate binding to isolated rat liver nuclei: Evidence for inositol trisphosphate

receptor-mediated calcium release from the nucleus. *Proc. Natl. Acad. Sci. U.S.A.* **87,** 9270–9274.
Mann, D. M., and Vanaman, T. C. (1988). Modification of calmodulin on Lys-75 by carbamoylating nitrosoureas. *J. Biol. Chem.* **263,** 11284–11290.
Mann, D. M., and Vanaman, T. C. (1989). Topographical mapping of calmodulin–target enzyme interaction domains. *J. Biol. Chem.* **264,** 2373–2378.
Mitchinson, C., Wilderspin, A. F., Trinnaman, B. J., and Green, N. M. (1982). Identification of a labeled peptide after stoicheiometric reaction of fluorescein isothiocyanate with the Ca^{2+}-dependent adenosine triphosphatase of sarcoplasmic reticulum. *FEBS Lett.* **146,** 87–92.
Monteith, G. R., and Roufogalis, B. D. (1995). The plasma membrane calcium pump—a physiological perspective on its regulation. *Cell Calcium* **18,** 459–470.
Muallem, S., and Karlish, S. J. (1979). Is the red cell calcium pump regulated by ATP? *Nature (London)* **277,** 238–240.
Muallem, S., and Karlish, S. J. (1981). Studies on the mechanism of regulation of the red-cell Ca^{2+} pump by calmodulin and ATP. *Biochim. Biophys. Acta* **647,** 73–86.
Neyses, L., Reinlib, L., and Carafoli, E. (1985). Phosphorylation of the Ca^{2+}-pumping ATPase of heart sarcolemma and erythrocyte plasma membrane by the cAMP-dependent protein kinase. *J. Biol. Chem.* **260,** 10283–10287.
Nicotera, P., McConkey, D. J., Jones, D. P., and Orrenius, S. (1989). ATP stimulates Ca^{2+} uptake and increases the free Ca^{2+} concentration in isolated rat liver nuclei. *Proc. Natl. Acad. Sci. U.S.A.* **86,** 453–457.
Nicotera, P., Orrenius, S., Nilsson, T., and Berggren, P. O. (1990). An inositol 1,4,5-trisphosphate-sensitive Ca^{2+} pool in liver nuclei. *Proc. Natl. Acad. Sci. U.S.A.* **87,** 6858–6862.
Niggli, V., and Carafoli, E. (1981). Interaction of the purified (Ca^{2+}, Mg^{2+})-ATPase from human erythrocytes with phospholipids and calmodulin. *Acta Biol. Med. Ger.* **40,** 437–442.
Niggli, V., Penniston, J. T., and Carafoli, E. (1979a). Purification of the (Ca^{2+}-Mg^{2+})-ATPase from human erythrocyte membranes using a calmodulin affinity column. *J. Biol. Chem.* **254,** 9955–9958.
Niggli, V., Ronner, P., Carafoli, E., and Penniston, J. T. (1979). Effects of calmodulin on the (Ca^{2+} Mg^{2+})-ATPase partially purified from erythrocyte membranes. *Arch. Biochem. Biophys.* **198,** 124–130.
Niggli, V., Adunyah, E. S., and Carafoli, E. (1981). Acidic phospholipids, unsaturated fatty acids, and limited proteolysis mimic the effect of calmodulin on the purified erythrocyte Ca^{2+}-ATPase. *J. Biol. Chem.* **256,** 8588–8592.
Ohta Y., Nishida, E., Sakal, H. (1996). Type II Ca^{2+}/calmodulin-dependent kinase binds to actin filaments in a calmodulin-sensitive manner. *FEBS Lett.* **208,** 423–426.
Olson, S., Wang, M. G., Carafoli, E., Strehler, E. E., and McBride, O. W. (1991). Localization of two genes encoding plasma membrane Ca^{2+}-transporting ATPases to human chromosomes 1q25-32 and 12q21-23. *Genomics* **9,** 629–641.
Pannabecker. T. L., Chandler, J. S., and Wasserman, R. H. (1995). Vitamin-D-dependent transcriptional regulation of the intestinal plasma membrane calcium pump. *Biochem. Biophys. Res. Commun.* **213,** 499–505.
Papp, B., Sarkadi, b., Enyedi, A., Caride, A. J., Penniston, J. T., and Gardos, G. (1989). Functional domains of the *in situ* red cell membrane calcium pump revealed by

proteolysis and monoclonal antibodies. Possible sites for regulation by calpain and acidic lipids. *J. Biol. Chem.* **264,** 4577–4582.

Pearson, R. B., Wettenhall, R. E., Means, A. R., Hartshorne, D. J., and Kemp, B. E. (1988). Autoregulation of enzymes by pseudosubstrate prototypes: Myosin light chain kinase. *Science* **241,** 970–973.

Phillips, A. M., Bull, A., and Kelly, L. E. (1992). Identification of a *Drosophila* gene encoding a calmodulin-binding protein with homology to the *trp* phototransduction gene. *Neuron* **8,** 631–642.

Pick, U., and Karlish, S. J. (1982). Regulation of the conformation transition in the Ca-ATPase from sarcoplasmic reticulum by pH, temperature, and calcium ions. *J. Biol. Chem.* **257,** 6120–6126.

Preiano, B. S., Guerini, D., and Carafoli, E. (1996). Expression and functional characterization of isoform 4 of the plasma membrane calcium pump. *Biochemistry* **35,** 7946–7953.

Quinton, T. M., Brown, K. D., and Dean, W. L. (1996). Inositol 1,4,5-trisphosphate-mediated Ca^{2+} release from platelet internal membranes is regulated by differential phosphorylation. *Biochemistry* **35,** 6865–6871.

Rega, A. F., and Garrahan, P. J. (1975). Calcium ion-dependent phosphorylation of human erythrocyte membranes. *J. Membr. Biol.* **22,** 313–327.

Reisner, P. D., Christakos, S., and Vanaman, T. C. (1992). *In vitro* enzyme activation with calbindin-D28k, the vitamin D-dependent 28 kDa calcium binding protein. *FEBS Lett.* **297,** 127–131.

Reisner, P. D., Brandt, P. C., and Vanaman, T. C. (1997). Analysis of plasma membrane Ca^{2+}-ATPase expression in control and SV40-transformed human fibroblasts. *Cell Calcium* **21,** 53–62.

Ronner, P., Gazzotti, P., and Carafoli, E. (1977). A lipid requirement for the (Ca^{2+} + Mg^{2+})-activated ATPase of erythrocyte membranes. *Arch. Biochem. Biophys.* **179,** 578–583.

Rudolph, H. K., Antebi, A., Fink, G. R., Buckley, C. M., Dorman, T. E., LeVitre, J., Davidow, L. S., Mao, J. I., and Moir, D. T. (1989). The yeast secretory pathway is perturbed by mutations in PMR1, a member of a Ca^{2+}-ATPase family. *Cell (Cambridge, Mass.)* **58,** 133–145.

Sackett, D. L., and Kosk-Kosicka, D. (1996). The active species of plasma membrane Ca^{2+}-ATPase are a dimer and a monomer–calmodulin complex. *J. Biol. Chem.* **271,** 9987–9991.

Saimi, Y., and Kung, C. (1994). Ion channel regulation by calmodulin binding. *FEBS Lett.* **350,** 155–158.

Salamino, F., Sparatore, B., Melloni, E., Michetti, M., Viotti, P. L., Pontremoli, S., and Carafoli, E. (1994). The plasma membrane calcium pump is the preferred calpain substrate within the erythrocyte. *Cell Calcium* **15,** 28–35.

Schatzmann, H. J. (1966). ATP-dependent Ca^{2+} extrusion from human red cells. *Experimentia.* **22,** 364–365.

Shull, G. E., and Greeb, J. (1988). Molecular cloning of two isoforms of the plasma membrane Ca^{2+}-transporting ATPase from rat brain. Structural and functional domains exhibit similarity to Na^+,K^+- and other cation transport ATPases. *J. Biol. Chem.* **263,** 8646–8657.

Shull, G. E., Clarke, D. M., and Gunteski-Hamblin, A. M. (1992). cDNA cloning of possible mammalian homologs of the yeast secondary pathway Ca^{2+}-transporting ATPase. *Ann. N.Y. Acad. Sci.* **671,** 70–80.

Sihra, T. S., and Pearson, H. A. (1995). Ca/calmodulin-dependent kinase II inhibitor KN62 attenuates glutamate release by inhibiting voltage-dependent Ca^{2+}-channels. *Neuropharmacology* **34,** 731–741.

Smallwood, J. I., Gugi, B., and Rasmussen, H. (1988). Regulation of erythrocyte Ca^{2+} pump activity by protein kinase C. *J. Biol. Chem.* **263,** 2195–2202.

Smith, L., and Smith, J. B. (1995). Activation of adenylyl cyclase downregulates sodium/calcium exchanger of arterial myocytes. *Am. J. Physiol.* **269,** C1379–C1384.

Smolenicka, Z., Guerini, D., Carafoli, E., Kress, W., and Liechti-Gallati, S. (1996). Detection of a new polymorphism in the plasma-membrane Ca^{2+}-ATPase isoform-3 gene and its exclusion as a candidate for X-linked myotubular myopathy (MTM1). *Hum. Genet.* **98,** 681–684.

Sperelakis, N., Xiong, Z., Haddad, G., and Masuda, H. (1994). Regulation of slow calcium channels of myocardial cells and vascular smooth muscle cells by cyclic nucleotides and phosphorylation. *Mol. Cell. Biochem.* **140,** 103–117.

Stahl, W. L., Eakin, T. J., Owens, J. W., Jr., Breininger, J. F., Filuk, P. E., and Anderson, W. R. (1992). Plasma membrane Ca^{2+}-ATPase isoforms: Distribution of mRNAs in rat brain by *in situ* hybridization. *Brain Res. Mol. Brain Res.* **16,** 223–231.

Stahl, W. L., Keeton, T. P., and Eakin, T. J. (1994). The plasma membrane Ca^{2+}-ATPase mRNA isoform PMCA 4 is expressed at high levels in neurons of rat piriform cortex and neocortex. *Neurosci. Lett.* **178,** 267–270.

Stauffer, T. P., Hilfiker, H., Carafoli, E., and Strehler, E. E. (1993). Quantitative analysis of alternative splicing options of human plasma membrane calcium pump genes. *J. Biol. Chem.* **268,** 25993–26003.

Stauffer, T. P., Guerini, D., and Carafoli, E. (1995). Tissue distribution of the four gene products of the plasma membrane Ca^{2+} pump. A study using specific antibodies. *J. Biol. Chem.* **270,** 12184–12190.

Stefanova, H. I., Mata, A. M., East, J. M., Gore, M. G., and Lee, A. G. (1993). Reactivity of lysyl residues on the $(Ca^{2+}-Mg^{2+})$-ATPase to 7-amino-4-methylcoumarin-3-acetic acid succinimidyl ester. *Biochemistry* **32,** 356–362.

Strehler, E. E. (1991). Recent advances in the molecular characterization of plasma membrane Ca^{2+} pumps. *J. Membr. Biol.* **120,** 1–15.

Strehler, E. E. (1996). Sodium–calcium exchangers and calcium pumps. *In* "Principles of Medical Biology," pp. 125–150. JAI Press.

Strehler, E. E., Strehler-Page, M. A., Vogel, G., and Carafoli, E. (1989). mRNAs for plasma membrane calcium pump isoforms differing in their regulatory domain are generated by alternative splicing that involves two internal donor sites in a single exon. *Proc. Natl. Acad. Sci. U.S.A.* **86,** 6908–6912.

Strehler, E. E., James, P., Fischer, R., Heim, R., Vorherr, T., Filoteo, A. G., Penniston, J. T., and Carafoli, E. (1990). Peptide sequences analysis and molecular cloning reveal two calcium pump isoforms in the human erythrocyte membrane. *J. Biol. Chem.* **265,** 2835–2842.

Takasawa, S., Ishida, A., Nata, K., Nakagawa, K., Noguchi, N., Tohgo, A., Kato, I., Yonekura, H., Fujisawa, H., and Okamoto, H. (1995). Requirement of calmodulin-dependent protein kinase II in cyclic ADP-ribose-mediated intracellular Ca^{2+} mobilization. *J. Biol. Chem.* **270,** 30257–30259.

Taylor, W. R., and Green, N. M. (1989). The predicted secondary structures of the nucleotide-binding sites of six cation-transporting ATPases lead to a probable tertiary fold. *Eur. J. Biochem.* **179,** 241–248.

Varadi, A., Molnar, E., and Ashcroft, S. J. (1996). A unique combination of plasma membrane Ca^{2+}-ATPase isoforms is expressed in islets of Langerhans and pancreatic beta-cell lines. *Biochem. J.* **314**, 663–669.

Verma, A. K., Enyedi, A., Filoteo, A. G., and Penniston, J. T. (1994). Regulatory region of plasma membrane Ca^{2+} pump. 28 Residues suffice to bind calmodulin but more are needed for full auto-inhibition of the activity. *J. Biol. Chem.* **269**, 1687–1691.

Virk, S. S., Kirk, C. J., and Shears, S. B. (1985). Ca^{2+} transport and Ca^{2+}-dependent ATP hydrolysis by Golgi vesicles from lactating rat mammary glands. *Biochem. J.* **226**, 741–748.

Vorherr, T., James, P., Krebs, J., Enyedi, A., McCormick, D. J., Penniston, J. T., and Carafoli, E. (1990). Interaction of calmodulin with the calmodulin binding domain of the plasma membrane Ca^{2+} pump. *Biochemistry* **29**, 355–365.

Vorherr, T., Kessler, T., Hofmann, F., and Carafoli, E. (1991). The calmodulin-binding domain mediates the self-association of the plasma membrane Ca^{2+} pump. *J. Biol. Chem.* **266**, 22–27.

Vorherr, T., Quadroni, M., Krebs, J., and Carafoli, E. (1992). Photoaffinity labeling study of the interaction of calmodulin with the plasma membrane Ca^{2+} pump. *Biochemistry* **31**, 8245–8251.

Wang, K. K., Wright, L. C., Machan, C. L., Allen, B. G., Conigrave, A. D., and Roufogalis, B. D. (1991). Protein kinase C phosphorylates the carboxyl terminus of the plasma membrane Ca^{2+}-ATPase from human erythrocytes. *J. Biol. Chem.* **266**, 9078–9085.

Wang, M. G., Yi, H., Hilfiker, H., Carafoli, E., Strehler, E. E., and McBride, O. W. (1994). Localization of two genes encoding plasma membrane Ca^{2+}-ATPase isoforms 2 (ATP2B2) and 3 (ATP2B3) to human chromosomes 3p26 → p25 and Xq28, respectively. *Cytogenet. Cell. Genet.* **67**, 41–45.

Watterson, D. M., and Vanaman, T. C. (1976). Affinity chromatography purification of a cyclic nucleotide phosphodiesterase using immobilized modulator protein, a troponin C-like protein from brain. *Biochem. Biophys. Res. Commun.* **73**, 40–46.

Watterson, D. M., Harrelson, W. G., Jr., Keller, P. M., Sharief, F., and Vanaman, T. C. (1976). Structural similarities between the Ca^{2+}-dependent regulatory proteins of 3′:5′-cyclic nucleotide phosphodiesterase and actomyosin ATPase. *J. Biol. Chem.* **251**, 4501–4513.

Watterson, D. M., Mendel, P. A., and Vanaman, T. C. (1980a). Comparison of calcium-modulated proteins from vertebrate brains. *Biochemistry* **19**, 2672–2676.

Watterson, D. M., Sharief, F., and Vanaman, T. C. (1980b). The complete amino acid sequence of the Ca^{2+}-dependent modulator protein (calmodulin) of bovine brain. *J. Biol. Chem.* **255**, 962–975.

Williams, C. L., Porter, R. A., and Phelps, S. H. (1995). Inhibition of voltage-gated Ca^{2+} channel activity in small cell lung carcinoma by the Ca^{2+}/calmodulin-dependent protein kinase inhibitor KN-62 (1-[*N*,*O*-bis(5-isoquinolinesulfonyl)-*N*-methyl-L-tyrosyl]-4-phenylpiperazine). *Biochem. Pharmacol.* **50**, 1979–1985.

Wuytack, F., Raeymaekers, L., De Smedt, H., Eggermont, J. A., Missiaen, L., Van Den Bosch, L., De Jaegere, S., Verboomen, H., Plessers, L., and Casteels, R. (1992). Ca^{2+}-transport ATPases and their regulation in muscles and brain. *Ann. N.Y. Acad. Sci.* **671**, 82–91.

Xiong, Z., Sperelakis, N., and Fenoglio-Preiser, C. (1994). Regulation of L-type calcium channels by cyclic nucleotides and phosphorylation in smooth muscle cells from rabbit portal vein. *J. Vasc. Res.* **31**, 271–279.

Yamada, M., Miyawaki, A., Saito, K., Nakajima, T., Yamamoto-Hino, M., Ryo, Y., Furuichi, T., and Mikoshiba, K. (1995). The calmodulin-binding domain in the mouse type 1 inositol 1,4,5-trisphosphate receptor. *Biochem. J.* **308,** 83–88.

Yamaguchi, M., and Oishi, K. (1993). Characterization of Ca^{2+}-stimulated adenosine 5′-triphosphatase and Ca^{2+} sequestering in rat liver nuclei. *Mol. Cell. Biochem.* **125,** 43–49.

Yamamoto, H., Tagaya, M., Fukui, T., and Kawakita, M. (1988). Affinity labeling of the ATP-binding site of Ca^{2+}-transporting ATPase of sarcoplasmic reticulum by adenosine triphosphopyridoxal: Identification of the reactive lysyl residue. *J. Biochem. (Tokyo)* **103,** 452–457.

Yamamoto, H., Imamura, Y., Tagaya, M., Fukui, T., and Kawakita, M. (1989). Ca^{2+}-dependent conformational change of the ATP-binding site of Ca^{2+}-transporting ATPase of sarcoplasmic reticulum as revealed by an alteration of the target-site specificity of adenosine triphosphopyridoxal. *J. Biochem. (Tokyo)* **106,** 1121–1125.

Zacharias, D. A., Dalrymple, S. J., and Strehler, E. E. (1995). Transcript distribution of plasma membrane Ca^{2+} pump isoforms and splice variants in the human brain. *Brain Res. Mol. Brain Res.* **28,** 263–272.

Zelinski, J. M., Sykes, D. E., and Weiser, M. M. (1991). The effect of vitamin D on rat intestinal plasma membrane CA-pump mRNA. *Biochem. Biophys. Res. Commun.* **179,** 749–755.

Zurini, M., Krebs, J., Penniston, J. T., and Carafoli, E. (1984). Controlled proteolysis of the purified Ca^{2+}-ATPase of the erythrocyte membrane. A correlation between the structure and the function of the enzyme. *J. Biol. Chem.* **259,** 618–627.

Zvaritch, E., James, P., Vorherr, T., Falchetto, R., Modyanov, N., and Carafoli, E. (1990). Mapping of functional domains in the plasma membrane Ca^{2+} pump using trypsin proteolysis. *Biochemistry* **29,** 8070–8076.

Zvaritch, E., Vellani, F., Guerini, D., and Carafoli, E. (1995). A signal for endoplasmic reticulum located at the carboxyl terminus of the plasma membrane Ca^{2+}-ATPase isoform 4CI. *J. Biol. Chem.* **270,** 2679–2688.

Index

A

A23187, 239, 240
Actin, 348, 375
Actin-binding proteins
  adducin, 375, 377–378
  of membrane cytoskeleton, 375
  spectrins, 375–377, 378
Actin filaments, 348, 379
Adducin, 375, 377–378
Adenylyl cyclases, 238–239
  Type 1, 239–248
  Type 3, 249–251
  Type 8, 248–249
ADP ribose (cADPR) receptor, 399, 401
AKAP79, 198
Alzheimer's disease, calcineurin and, 204
Ankyrin, 375
Apocalmodulin
  conformation, 21
  structure, 30, 31
AtKCBP, 366
Autoinhibition, 71, 86, 87, 109
Autoinhibitory domain, calcineurin and, 177–182
Autophosphorylation, 67, 323

B

bNOS, 290
Brain
  nitric oxide synthase, 314
  phosphodiesterase 1 isoforms expressed in, 267–271
  PMCA, 425

Brush-border myosin I, 43, 358
BTPDE1A1/2, 258–261, 267, 268

C

Caki, 144
Calbindin D9k, mutagenesis experiments, 24, 25
Calcineurin (CaN), 170, 350
  catalytic properties
    deactivation, 184
    inhibition, 186–190
    isoforms, 190–193
    mechanism, 183–184
    substrate specificity, 185–186
  functional domain organization
    autoinhibitory domain, 177–182
    B subunit, 175–177
    calmodulin and, 171–175
  physiological roles, 220–221
    antagonism of cAMP signal, 199–201
    cell death regulation, 218–220
    cytoskeleton regulation, 203–205
    gene expression regulation, 211–217
    ion channel and transporter regulation, 205–211
    isoforms, 192–193
    neurotransmitter release, 202–203
    postsynaptic potential, 201–202
    subcellular localization, 193–199
Calcium, 1
Calcium binding, 73–78
  affinity and, 21–22, 24–25
  cooperativity and, 22–23

473

Calcium binding (*continued*)
  enzyme recognition and, 8–9
  other ligands, 50–53
  peptides from calmodulin targets, 42–50
  selectivity and, 23–24
Calcium-binding proteins (CaBPs), 17–20
Calcium channels
  regulation, 398–403
  types, 400
Calcium-induced calcium release (CICR), 399
Calcium-loaded calmodulin
  peptide binding, 42
  structure, 27–30
    calcium trigger, 31–37
    energetics, 37–39
    HMJ model, 30–31
    methionines, 39–42
    three-dimensional structure, 27–30
Calcium-modulated signal transduction, 1
  cross-talk, 11–13
  disease and, 14
  energetics, 37–39
  molecular mechanisms, 4–14
Calcium regulation, pools involved in, 398–399
Calcium sensor
  calcium binding, 21–27
  calcium binding proteins, 17–20
  structural basis
    calcium-loaded calmodulin, three-dimensional structure, 27–30
    calcium trigger, 31–37
    energetics, 37–39
    HMJ model, 30–31
    methionines, 39–42, 49–50
  transduction of calcium signal, 42–57
Caldesmon, 367–369
Calmodulin, 1, 397–398
  adenylyl cyclases and, 238–251, 277
  amino acid sequence, 3–6
  calcineurin and, 169–220
  calcium binding, 21–27
  calcium-loaded *see* Calcium-loaded calmodulin
  as calcium sensor, 17–57
  caldesmon and, 369

  conformation, 20, 67–68
  cytoskeleton and, 203–205, 348–383
  EF-hand, 20
  history, 2, 7
  mechanism of action, 1–2
  name derivation, 3
  nebulin binding and, 373
  nitric oxide synthase and, 300–319
  phosphodiesterase (PDE) and, 2, 43, 251–278
  phylogenetic conservation, 3–4, 6
  plasma membrane Ca^{2+}-ATPases (PMCAs), 398, 401–402, 413–457
  signal transduction, 1, 4–14, 237–238
  structure, 43–45, 73, 74
    amino acid sequence, 73, 77, 81
    recognition sequences, 72–73
    regulatory segment, 71
  synapsin I and, 379
Calmodulin-binding microtubule-binding proteins, 380–382
Calmodulin-binding motor proteins
  kinesins, 366
  myosins, 350–366
Calmodulin-binding muscle proteins
  skeletal muscle proteins, 370
    connectin, 370–372
    dystrophin, 195, 349, 373–375
    nebulin, 349, 371, 372–373
    titin, 349, 370–372
    utrophin, 373–374
  smooth muscle proteins
    caldesmon, 367–369
    calponin, 204, 369–370
Calmodulin-binding proteins, 10, 11
  cytoskeleton, 348–350, 382–383
    adducin, 375, 377–378
    caldesmon, 367–369
    calponin, 204, 369–370
    connectin, 370
    desmocalmin, 380
    dystrophin, 195, 349, 373–375, 376
    kinesins, 349, 366
    MAPs, 380–381
    MARCKS protein, 378, 379–380
    membrane cytoskeleton, 348–349, 375–378
    Mip-90, 381–382

Index

muscle proteins, 349, 366–375
myosins, 43, 78, 79, 350–366
nebulin, 349, 371, 372–373
spectrins, 43, 375–377, 378
STOPs, 382
synapsin, 125, 378–379
titin, 349, 370–372
utrophin, 373–374
Calmodulin-dependent kinase kinases (CaMKK), 11
Calmodulin-dependent multiprotein kinase *see* Calmodulin-dependent protein kinase II (CaMPKII)
Calmodulin-dependent protein kinase I (CaMPKI), 11, 124–132
Calmodulin-dependent protein kinase II (CaMPKII), 10, 11, 45, 101
  amino acid sequence, 105–106
  calcium signal transduction and, 101–105
  as calmodulin-regulated enzyme, 105–114
  function, 101–105
  isoforms, 105, 114–115
  long-term potentiation, 101–103
  molecular genetics, 114–115
  nitric oxide synthase activity and, 319–320
  overexpression, 104
  phosphorylation, 13, 101, 103
  structure, 107–109
Calmodulin-dependent protein kinase III (CaMPKIII), 132–136
Calmodulin-dependent protein kinase IV (CaMPKIV), 11, 136
  as calmodulin-regulated enzyme, 138–139
  expression, 242
  function, 136–137
  molecular genetics, 139
  phosphorylation, 13
Calmodulin-dependent protein kinase kinases (CaMPKK), 140–143
Calmodulin-drug interactions, 50–53
Calmodulin gene, 4
Calmodulin kinase II *see* Calmodulin-dependent protein kinase II (CaMPKII)
Calmodulin-peptide complex, structure, 42–50, 73, 74
Calmodulin recognition peptides, 7–8, 67, 73
Calmodulin-regulated enzymes, 10, 69
Calmodulin-regulated protein kinases, 66–72, 145
caki, 144
calcium-dependent protein kinase II (CaMPKII), 10, 11, 13, 45, 101–115, 319–320
calmodulin-dependent protein kinase I (CaMPKI), 11, 124–132
calmodulin-dependent protein kinase III (CaMPKIII), 132–136
calmodulin-dependent protein kinase IV (CaMPKIV), 11, 13, 136–139, 242
calmodulin-dependent protein kinase kinases, 140–143
calmodulin recognition, 72–73
death associated protein (DAP) kinase, 141
enzyme recognition, 73–78
G-protein signaling, 144–145
MAP kinase cascade, 144–145
myosin light-chain kinase (MLCK), 10, 45, 78–100, 350
phosphorylase kinase, 116–124
segmental organization, 69–72, 83
Calpain I, 377
Calphostin, 240
Calponin, 204, 369–370
CaMKK *see* Calmodulin-dependent kinase kinases
CaMPK *see* Calmodulin-dependent protein kinase
CaMPKII gene, 114–115
CaMPKIV gene, 139
Carbachol, 239
Cardiovascular system
  nitric oxide and, 331–332
  phosphodiesterase 1 and, 274–277
Catalysis, enzyme based, 2
Caveolin, 295
Cell death, regulation of, 218–220
Central nervous system
  nitric oxide role in, 328
  neurodestruction, 329–330

Central nervous system (*continued*)
  neuroprotection, 330–331
  neurotransmitter release, 329
  signaling, 325, 326
  synaptic plasticity, 328–329
  nitric oxide synthase and, 314
  phosphodiesterase 1 isoforms expressed in, 267, 269–270, 271
CICR *see* Calcium-induced calcium release
Circadian rhythm, adenylyl cyclases and, 246–248
CnB subunit gene, 205
Colocalization, 197, 369
Connectin, 370
Contact map analysis, 33, 35, 37
CREB-327, 216
CREB-341, 216
Cross-talk, 11–13, 137, 324
CsA *see* Cyclosporin A
Cyclic nucleotide phosphodiesterase *see* Phosphodiesterase (PDE)
Cyclophilin A (CyP), calcineurin inhibition by, 186–190, 212
Cyclosporin A (CsA), calcineurin inhibition by, 186–190, 200–201, 211–213, 211–213, 215–216, 217, 218
CyP *see* Cyclophilin A
Cytoskeletal proteins, 2, 43, 348–350, 382–383
  adducin, 375, 377–378
  caldesmon, 367–369
  calponin, 204, 369–370
  connectin, 370
  desmocalmin, 380
  dystrophin, 195, 349, 373–375, 376
  kinesins, 349, 366
  MAPs, 380–381
  MARCKS protein, 378, 379–380
  membrane cytoskeleton, 348–349, 375–378
  Mip-90, 381–382
  muscle proteins, 349, 366–375
  myosins, 43, 78, 79, 350–366
  nebulin, 349, 371, 372–373
  spectrins, 43, 375–377, 378
  STOPs, 382
  synapsin, 125, 378–379
  titin, 349, 370–372
  utrophin, 373–374
Cytoskeleton, calcineurin regulation, 203–205

D

DARPP32, 199–201
DDM *see* Difference distance matrix
Death-associated protein (DAP) kinase, 141–144
Desmocalmin, 380
Difference distance (DD), defined, 33
Difference distance matrix (DDM), 33–35, 37
Dihydropyridines, interaction with calmodulin, 50
Disease, calcium-modulated signal transduction and, 14
DphKgamma gene, 117
Duchenne/Becker muscular dystrophies, 374
Duchenne muscular dystrophy, 349
Dynamin 1, calcineurin and, 195, 202–203
Dynamin 1 GTPase, 202
Dynein, 349
Dystrophin, 195, 349, 373–375, 376
Dystrophin gene, 374

E

ecNOS, 290
EF-2 kinase *see* Calmodulin-dependent protein kinase III (CaMPKIII)
EF-hand calcium binding proteins, 18, 20
8AC *see* Type 8 adenylyl cyclase
Electron transfer, in nitric oxide synthase, 315–318
Elongation factor-2 (EF-2), phosphorylation, 132
ENA1 gene, 206
Endothelial NOS (eNOS), 290, 291
Endotoxin shock, nitric oxide and, 331
eNOS, 290, 295, 301
  calcium binding and, 305–306
  inhibitors, 333
eNOS gene, 299

Index

Enzymes, 2
  calmodulin recognition sequences, 7
  peptide binding, 42–50
Erythrocytes
  peripheral proteins, 375
  PMCA, 431–432

F

F-actin, 377
Filament systems, 348
FK506, calcineurin inhibition by, 186–190, 200–201, 211–213, 215–216, 218
FKBP12, calcineurin inhibition by, 186–190, 212
Focal adhesion kinase (FAK), 78
Fodrin, 375, 376–377
FR-1AC, 241

G

Genetics. *see also* Molecular genetics
  calmodulin gene, 4
  mutagenesis experiments, 24–26, 87–95
  splicing of PMCA genes, 414–427
Genistein, 435, 436
Glutamic acid, mutation, 25, 26
Glycogen storage disease (GSD), 116–117
G-protein signaling, calmodulin-regulated protein kinase and, 144–145

H

H89, 240
HMJ structural model, 30–31

I

IL-2 gene, inhibition, 186–187, 211
Imidazoles, interaction with calmodulin, 50
Immune system, nitric oxide and, 332
Immunophilin, calcineurin inhibition by, 186–190, 211–213, 215–216
Immunosuppressant complexes, calcineurin inhibition by, 186–190, 211–213, 215–216
Inducible NOS (iNOS), 290, 291
Inflammatory system, nitric oxide and, 332
Inhibitor-1, 199–201
iNOS, 290, 301
  calcium binding and, 306–307
  effect on cardiovascular system, 331
  inhibitors, 333
iNOS gene, 298–299
Inositol triphosphate (IP3) receptor, 399, 400
Intermediate filaments, 348
Ion channels
  calcineurin and, 205–211
  regulation, 398
    calcium channels, 398–403
    sodium and potassium channels, 457–458
Ion translocation, 397–398
IQ motifs
  myosin I, 367
  myosin II, 354–356, 357
  myosin III, 362, 367
  myosin IV, 367
  myosin V, 363, 367
  myosin VI, 365, 367
  myosin VII, 367
  myosin VIII, 367
  myosin IX, 366, 367
  myosin X, 367
  myosin XI, 367
  myosin XII, 367
  neuromodulin, 54

J

J-8, interaction with calmodulin, 50

K

Kinesins, 349, 366
KN-62, 102–103
KN62, 240, 249
KRP, 78
KRP gene, 98, 99
KRP/telokin, 99
KT5926, 80

L

Learning
　1AC gene disruption and, 243–245
　calmodulin-dependent protein kinase II (CaMPKII), 101–103
Long-term depression (LTD), 201–202, 328–329
Long-term potentiation (LTP), 101–103
　1AC gene, 243–245
　calcineurin and, 201–202
　protein kinase C and, 241

M

macNOS, 290
MAP kinase cascade, calmodulin-regulated protein kinase and, 144
MAPs see Microtubule-associated proteins
MARCKS protein (myristoylated alanine-rich C kinase substrate), 378, 379–380
Melatonin, synthesis, 1AC and, 246–248
Membrane cytoskeleton, 348–349, 375–378
Memory, calmodulin-dependent protein kinase II, 101
Methionines, 39–42, 49–50
Microtubule-associated proteins (MAPs), 380–381
Microtubules, 348, 349, 379
Mip-90, 381–382
ML-9, 80
MLCK see Myosin light-chain kinase
Molecular genetics
　CaMPKII, 114–115
　CaMPKIV, 139
　MLCK, 97–100
　phosphorylase kinase, 117, 123–124
Multifunctional protein kinase see Calmodulin-dependent protein kinase II (CaMPKII)
Muscle proteins, 349, 366–375
Muscular dystrophy, 349, 374

Mutagenesis
　calcium ion affinity, 24–26
　CaMPKII, 109–114
　MLCK, 87–95
Myosin, 365
　phosphorylation, 79
　phylogenetic tree, 351
　scallop regulatory domain, 354
　structure, 350–357
Myosin I
　beta form, 361
　brush-border myosin I, 43, 358
　IQ motifs, 367
　isoforms, 360–361
　myosin II and, 358, 360
　myosin Ic, 360
　phylogenetic tree, 351, 357–359
　structure, 351
Myosin II, 78
　IQ motifs, 354–356, 357
　myosin I and, 360
　structure, 350–352, 351
Myosin III
　IQ motifs, 362, 367
　structure, 351, 361–362
Myosin IV
　IQ motifs, 367
　structure, 351
Myosin V
　enzymatic properties, 365
　function, 363–365
　IQ motifs, 363, 367
　structure, 351, 363
Myosin VI
　IQ motifs, 365, 367
　purification, 365
　structure, 351
Myosin VII
　IQ motifs, 367
　purification, 365
　structure, 351
Myosin VIII
　IQ motifs, 367
　structure, 351
Myosin IX
　IQ motifs, 366, 367
　purification, 365
　structure, 351

Index

Myosin X
  IQ motifs, 367
  structure, 351
Myosin XI
  IQ motifs, 367
  structure, 351
Myosin XII
  IQ motifs, 367
  structure, 351
Myosin light-chain kinase (MLCK), 45, 86
  amino acid sequence, 81–86
  calcium-modulated signal transduction, 78–81
  as calmodulin-regulated enzyme, 81–97
  function, 78–81, 350
  molecular genetics, 97–100
  subcellular localization, 10
Myosin light-chain kinase (MLCK) gene, 97–100
Myristoylated alanine-rich C kinase substrate see MARCKS protein
Myristoylation, 176, 192

N

Naphthalene sulfonamides, interaction with calmodulin, 50
ncNOS, 290
Nebulette, 373
Nebulin, 349, 371, 372–373
Neurodestruction, nitric oxide, 329–330
Neuromodulin, 53–54
Neuronal NOS (nNOS), 290, 291
Neurons
  nitric oxide role in, 328
    neurodestruction, 329–330
    neuroprotection, 330–331
    neurotransmitter release, 329
    signaling, 325, 326
    synaptic plasticity, 328–329
  phosphodiesterase 1 isoforms expressed in, 267, 269–270, 271
Neuroplasticity
  nitric oxide, 328–329
  PDE1A2 and, 270, 271
Neuroprotection, nitric oxide, 330–331
Neurotransmitter release, regulation of, 202–203, 329

NGF1 gene, expression, 215
NinaC, 361
Nitric oxide (NO), 288
  biochemical reactivity, 325–328, 333
  as messenger, 288–289
  physiology and pathophysiology
    cardiovascular system, 331–332
    central nervous system, 328–331
    immune system, 332
    inflammatory system, 332
    pharmacology and, 332–333
Nitric oxide synthase (NOS), 218, 288, 333–334
  biochemical properties, 295
  calmodulin and, 300
    binding significance, 312–319
    binding sites, 302–312
    history, 300–302
  catalytic mechanism, 290, 292–294
  dimerization, calmodulin regulation, 319
  domain organization, 290, 292
  genes, 296–299
  inhibitors, 332–333
  isoforms, 290–292
  localization, 295–296, 318–319
  purification, 294–295
  regulation by phosphorylation/dephosphorylation
    autophosphorylation, 323
    by calmodulin-dependent kinases and phosphatases, 319–320
    by cAMP/cGMP-dependent kinases, 321–322
    as control point, 323–324
    cross-talk, 324–325
    by protein kinase C, 320–321
    by tyrosine kinases, 322–323
  subcellular localization, calmodulin regulation, 318–319
Nitrovasodilators, 332
nNOS, 290, 301
  calcium binding and, 303–305
  inhibitors, 333
  phosphorylation, 323
nNOS gene, 298
Nonerythroid alpha-spectrin, 376–377
NOS genes, 296–299

NOS I, 290
NOS II, 290
NOS III, 290

O

Okadaic acid, 322
Olfactory signal transduction, 251, 270–272
1AC *see* Type 1 adenylyl cyclase

P

Paired helical filaments (PHFs), 204–205
Parvalbumin, 18
PDE *see* Phosphodiesterase
PDE1A gene, 256
PDE1B gene, 257, 277
PDE1C gene, 256
PDE1 gene, 253–254, 256
  expression, 274–275, 277
Peptides, from calmodulin targets, binding, 42–50
Phenothiazines, interaction with calmodulin, 50
PHFs *see* Paired helical filaments
PHKA gene, 117, 123
PHKB gene, 123
PHKG1 gene, 124
Phosphodiesterase (PDE), 2, 43, 251, 253, 277–278
  activation, 264–266
  expression, 267
    cardiovascular system, 274–277
    central nervous system, 267, 269–270, 271
    olfactory neurons, 270–273
    testis and sperm, 273–274
  inhibitors, 266, 276
  isoforms, 253, 256
    regulation by phosphorylation, 266–267, 268
  kinetic properties, 262–264
  nucleotide sequence, 255
  phylogenetic tree, 253, 254, 256
  structure and domain organizaiton, 257–262

Phosphoinositide cycle, 348
Phosphoprotein phosphatase 2B (PP2B), 170
Phosphorylase kinase (PhosK), 116–124
Phosphorylation
  autophosphorylation, 67, 323
  calcium-dependent, 11, 13
  myosin, 79
  PDE1A2 and, 270, 271
  PMCA regulation by, 432–433
  regulation of phosphodiesterase 1 isoforms by, 266–267, 268
  sites of, 71–72
  tyrosine, 435
Pineal gland, melatonin synthesis and, 246–248
PKC *see* Protein kinase C
Plasma membrane Ca^{2+}-ATPases (PMCAs), 398, 401–402
  calpain fragments, 443
  catalytic properties, 427
  gene expression, 451–455
  isoforms, 413, 447–450
  kinetics, 427–431
  molecular genetics, 413–427, 455–457
  physiological function, 455–457
  regulation, 431–432
    by acidic phospholipids, 439–440
    by calmodulin-dependent kinases and phosphatases, 432
    by calmodulin family members, 440–441
    isoform-specific regulatory properties, 447–450
    mechanism of, 450–451
    by oligomerization, 441–442
    by partial proteolysis, 440
    by phosphorylation, 432–439
    regulatory domains, 442–447
  structure, 403–413
  trypsin fragments, 443
Platelets, shape changing process, 80
PMCA1
  expression, 420–421
  structure, 449
PMCA2
  expression, 421–422
  structure, 449

Index

PMCA3
 expression, 421, 422
 structure, 449
PMCA4
 expression, 420–421
 function, 447
 structure, 449
PMCA1 gene, 413, 415–416
 expression, 451–455
 splicing, 420
PMCA2 gene, 413, 414–415
 expression, 451–455
 splicing, 417, 419
PMCA3 gene, 413–414
 expression, 451–455
 splicing, 419
PMCA4 gene, 413
 expression, 451–455
 splicing, 420
PMCA genes, 413–416
Postsynaptic potential, regulation, 201–202
Protein 4.1, 375
Protein 4.9, 375
Protein kinase C (PKC)
 nitric oxide synthase regulation and, 320–324
 stimulation of 1AC by, 241
Protein kinases, 68–69; see also Calmodulin-regulated protein kinases
Protein phosphatase I, 10
Pseudosubstrate hypothesis, 86–87, 109

Q

QTQ sequence, 434–435

R

Rac, 348
Receptor capping, 79
Receptor-operated calcium channels (ROCC), 400
Recoverin
 binding, 194
 structure, 31, 32
Relief-of-autoinhibition mechanism, 86
Retina, melatonin synthesis and, 246–248

Rho, 348
Rp-cAMP, 240

S

SCN see Suprachiasmatic nucleus
Semiopen conformation, calmodulin, 54–56
SERCA, 446–447
SERCA1, 411
SERCA2a, 405–409, 412
Serine/threonine kinases, 68, 69
Signal transduction see Calcium-modulated signal trandsuction
Site-directed mutagenesis, 24–26, 91
Skeletal muscle fiber, 349
Skeletal muscle myosin light-chain kinase (skMLCK), 68, 78
 calmodulin-binding domain, 43
 calmodulin recognition segment, 95–96
Smell, 251, 270–272
sm/nmMLCK gene, 97–99
Smooth muscle/nonmuscle myosin light-chain kinase (sm/nmMLCK), 68, 78
 amino acid sequence, 81–83
 mutagenesis experiments, 91–95
Spatial learning, 1AC gene disruption and, 243–245
Spectrins, 43, 375–377, 378
Sperm, phosphodiesterase 1 and, 273–274
Splicing, of PMCA genes, 414–427
Stable tubulin-only polypeptides (STOPs), 382
Staurosporines, 80
STE5, 197, 198
Suprachiasmatic nucleus (SCN), 246
Synapsin, 378–379
Synapsin I, 125, 379

T

Talin, 379
Targeting, 193–194
Testis, phosphodiesterase 1 and, 273–274
TFP see Trifluoperazine
3AC see Type 3 adenylyl cyclase
Titin, 349, 370–372
TKK1 gene, 206

Transduction, of calcium *see* Calcium-modulated signal transduction
Trifluoperazine (TFP), interaction with calmodulin, 50–51
Tropomodulin, 375
Tropomyosin, 369, 375
Troponin C, 2, 398
  HMJ model, 30–31
  mutagenesis experiments, 25
Tubulin dimers, 348, 349
TW240/260, 376
Twitchin, 371, 372
Type 1 adenylyl cyclase (1AC), 239
  developmental expression, 245–246
  disruption, 243–245
  inhibition, 241–242
  melatonin synthesis and, 246–248
  regulation, 239
  stimulation, 240–251
  tissue and cellular distribution, 242–243
Type 3 adenylyl cyclase (3AC), 249–251
Type 8 adenylyl cyclase (8AC), 248–249
Type II calmodulin kinase *see* Calmodulin-dependent protein kinase II (CaMPKII)
Tyrosine, phosphorylation, 435
Tyrosine kinases, 68, 322–323

U

Usher IB syndrome, 365
Utrophin, 373–374

V

Vanadate, 322
Vasodilators, 331–332, 333
Vinculin, 379
Voltage-operated calcium channels (VOCC), 400